Nuclear Structure Models

INTERNATIONAL WORKSHOP

Nuclear Structure Models

Oak Ridge, Tennessee   *March 16 – 25, 1992*

Editors

Ragnar Bengtsson

The University Isotope Separator at Oak Ridge (UNISOR), USA

Jerry Draayer

Louisiana State University, USA

Witold Nazarewicz

The Joint Institute for Heavy Ion Research, Oak Ridge, USA
and University of Warsaw, Poland

World Scientific

Singapore • New Jersey • London • Hong Kong

Published by

World Scientific Publishing Co. Pte. Ltd.

P O Box 128, Farrer Road, Singapore 9128

USA office: Suite 1B, 1060 Main Street, River Edge, NJ 07661

UK office: 73 Lynton Mead, Totteridge, London N20 8DH

INTERNATIONAL WORKSHOP ON NUCLEAR STRUCTURE MODELS

ISBN 981-02-1131-7

Printed in Singapore by Utopia Press.

PREFACE

The purpose of the Workshop was to bring together nuclear physicists from around the world representing different approaches to nuclear structure theory in an attempt bridge real and preceived communication gaps that exist between and among physicists advocating for different approaches. The Workshop lasted for ten days, from Monday, March 15, 1992 through Wednesday, March 25, 1992.

The Workshop was structured in a rather unusual way and this is reflected in the organization of the proceedings. It was divided into two parts. Part 1, which lasted six of the ten days, was devoted to presentations by spokespersons of ten different model types. Each spokesperson was allocated approximately two and one–half hours to present his or her model, inclusive of discussion periods. Background material provided by each spokesperson was distributed prior to the meeting. Complementary presentations by other participants were included as appropriated. Part 2 of the Workshop was devoted to evaluations of the different models presented in the first part. This was kept informal to encourage the maximum possible amount of open and frank interchange. The participants were assigned to teams, one team for each model type, and asked to evaluate that approach. Individuals were not assigned to teams based on their expertise, in fact, the organizers made an attempt to assign individuals to teams outside their specific area of expertise in an effort to encourage as much diversification and interaction as possible. Each of these teams elected a leader who during the evaluation period presented the team's view of the model they were asked to evaluate. The spokesperson for the model was then given an opportunity to respond to the evaluation. This format fostered strong but constructive discussions which are recorded in these proceedings.

The proceedings are organized under the ten sections corresponding to the different model types. There is an eleventh section which includes a discussion of other schematic models that did not fall naturally under one of the ten groups identified prior to the meeting by the organizers. Each section is organized in a way that reflects the structure of the Workshop. The contribution of the spokesperson is given first, that of the evaluation team second, and the spokesperson's rebuttal third, these being labeled a–c of their respective sections. Hence in Group 6 on the Kerman-Klein Method the contribution by Klein and Walet is labeled 6.1.a, that of the evaluation team headed by Heenen is labeled 6.1.b, and the rebuttal offered by Klein and Walet is labeled 6.1.c. If there were additional contributions within a group, these are labeled with 2 and higher numbers, for example, in Group 6 there is a contribution 6.2 by Walet and 6.3 by Zelevinsky. If a letter is missing, for example 1.1.c, it means the responsible person chose not to submit a response. In the case cited, for example, Heenen felt the evaluation was fair and required no further discussion.

The organizers felt strongly that the nuclear theory community needed a Workshop that would encouraged frank interchanges among different groups and set up the ground rules accordingly. Prior to the Workshop, it was referred to in a variety of ways, one being the "big nuclear theory fight" at Oak Ridge! In retrospect we wish to commend the participants on accepting their charge responsibly and promoting beyond our highest expectations the constructive exchange of scientific information and ideas. We also wish to encourage others to incorporate some of this philosophy into their meetings as we receive extremely favorable feedback from all participants.

We are indebted to the sponsors who made the Workshop possible. On a separate page we present a list of those organizations to which we are grateful for their support. We also wish to commend the Workshop secretaries, Jackie Smith and Carlene Stewart, for numerous helps that went well beyond the call of duty.

Ragnar Bengtsson
Jerry Draayer
Witek Nazarewicz

Oak Ridge, Tennessee
September, 1992

THE INTERNATIONAL WORKSHOP ON NUCLEAR STRUCTURE MODELS

is indebted to the following organizations
for their role as sponsors:

THE JOINT INSTITUTE FOR HEAVY ION RESEARCH
LOUISIANA STATE UNIVERSITY
THE UNIVERSITY ISOTOPE SEPARATOR AT OAK RIDGE (UNISOR)

OTHERS

OXFORD INSTRUMENTS NUCLEAR MEASUREMENT GROUP
DIGITAL EQUIPMENT CORPORATION
SOLON TECHNOLOGIES INCORPORATED
LECROY CORPORATION
BICRON CORPORATION
PULCIR INCORPORATED

CONTENTS

Preface v

1. SELF-CONSISTENT HARTREE-FOCK METHOD

1.1.a The Hartree-Fock+BCS and Generator Coordinate Methods 3
P.-H. Heenen, P. Bonche, J. Dobaczewski, H. Flocard, S. J. Krieger, J. Meyer, J. Skalski, N. Tajima and M. S. Weiss

1.1.b Evaluation of Self-consistent-Field Methods 23
C. R. Bingham, S. G. Rohoziński, A. Sustich and J. L. Wood

1.2 Nuclear Structure with the D1 Effective Interaction 26
J.-F. Berger, J. Decharge and D. Gogny

1.3 Microscopic Description of the Octupole Degree of Freedom with the Gogny Force 48
L. M. Robledo

2. PSEUDO-SPIN AND THE SYMPLECTIC MODEL

2.1.a Pseudo-Spin and the Symplectic Model 61
J. P. Draayer

2.1.b Evaluation of the Symplectic Model 93
K. Kumar, C. Baktash, D. H. Feng, M. W. Guidry and J.-Y. Zhang

2.1.c Comments on the Evaluation 95
J. P. Draayer

2.2 The Nuclear Symplectic Model 101
D. J. Rowe

3. COLLECTIVE MODELS

3.1.a Microscopic Theory of Nuclear Collective Dynamics 129
F. Sakata

3.1.b Evaluation of the Microscopic Theory of Nuclear Collective Dynamics 152
J. A. Maruhn, M. R. Strayer, P. Quentin and V. Zelevinsky

3.1.c Comments on the Evaluation 153
F. Sakata

3.2 The Geometric Collective Model 154
J. A. Maruhn

4. THE SHELL CORRECTION METHOD

4.1.a The Shell Correction Method 165
M. Brack

4.1.b Evaluation of the Shell Correction Method 188
S. A. Moszkowski, D. M. Cullen and J.-F. Berger

4.1.c Comments on the Evaluation 195
M. Brack

4.2 Applications to the Nilsson-Strutinsky Scheme 196
R. Wyss

5. FERMION DYNAMICAL SYMMETRY MODEL

5.1.a The Fermion Dynamical Symmetry Model 215
D. H. Feng, M. W. Guidry and C.-L. Wu

5.1.b Evaluation of the Fermion Dynamical Symmetry Model......... 216
M. W. Kirson

6. KERMAN-KLEIN SCHEME

6.1.a Three Decades of the Kerman-Klein Method in Quantum Mechanics, in Quantum Field Theory, and in Nuclear Physics ... 229
A. Klein and N. R. Walet

6.1.b Evaluation of the Kerman-Klein Method 253
P.-H. Heenen, W. Satuła and P. Semmes

6.1.c Comments on the Evaluation 256
A. Klein and N. R. Walet

6.2 Self-Consistent Determination of Collective Coordinates in the Theory of Large Amplitude Collective Motion 258
N. R. Walet

6.3 Generalized Density Matrix Revisited......................... 268
V. G. Zelevinsky

7. PARTICLE-CORE MODELS

7.1.a Particle-Core Coupling Models 287
P. B. Semmes

7.1.b Evaluation of the Particle-Core Models 306
R. Bengtsson and P. O. Lipas

7.2 Particle-Core Coupling 316
I. Hamamoto

8. SHELL MODEL APPROACHES

8.1.a *Monster* and *Vampir*: Variational Approaches to the Nuclear Shell Model 333
K. W. Schmid

8.1.b Evaluation of the Variational Approach to the Nuclear Shell Model 366
V. A. Madsen, Z. Szymański and E. C. Halbert

8.2 Shell-Model Calculations with Realistic Effective Interactions ... 370
E. Osnes, M. Hjorth-Jensen, T. Engeland and A. Holt

9. BOSON EXPANSION METHODS

9.1.a Reconstructing Broken Symmetries from Deformed Boson Expansions 387
E. R. Marshalek

9.1.b Evaluation of the Boson Expansion Theory 428
F. Sakata

9.1.c Comments on the Evaluation 431
E. R. Marshalek

10. INTERACTING BOSON MODEL

10.1.a The Interacting Boson Model 435
P. O. Lipas

10.1.b Evaluation of the Interacting Boson Model 454
D. J. Rowe, F. McGowan, S. Raman, R. Wyss and V. Zelevinsky

10.1.c Comments on the Evaluation 460
P. O. Lipas

10.2 Recent Issues of the Interacting Boson Model 462
T. Otsuka

11. SCHEMATIC MODELS

11.1 An Exact Solution to the Nuclear Cranking Model Hamiltonian 479
W. Satuła and Z. Szymański

11.2 Toy Models with Identical Bands 494
S. A. Moszkowski

11.3 Pairing and Identical Bands in a Toy Model 499
M. Mukerjee

List of Participants 505

GROUP 1

Selfconsistent Hartree–Fock Method

THE HARTREE-FOCK+BCS AND GENERATOR COORDINATE METHODS

P.-H. Heenen[(a)], P. Bonche[(b)], J. Dobaczewski[(c)], H. Flocard[(d)],
S.J. Krieger[(e)], J. Meyer[(f)], J. Skalski[(a)] N. Tajima[(b,d)] and M.S. Weiss[(e)]
(a) SPhNT U.L.B - C.P. 229 - 1050 Brussels, Belgium
(b) SPhT CEN Saclay - 91191 Gif sur Yvette Cedex, France
(c) ITPh Hoża 69, Warsaw University, PL 00681 Warsaw, Poland
(d) DPhT [7], IPN - 91406 Orsay Cedex, France
(e) LLNL, University of California, Livermore CA 94550, USA
(f) IPNL, IN2P3-CNRS / Université C. Bernard Lyon 1, 69622 Villeurbanne Cedex, France

ABSTRACT

We describe in this paper the method developped by our group to solve the Hartree-Fock +BCS equations with minimal symmetry requirements. We show that our method is very accurate with still reasonable computing time. We also show that our method can be generalised to study the nuclear dynamics within the generator coordinate method and we present some examples of applications.

1. The mean-field description of nuclear deformations

1.1. Historical background

First studies of deformed nuclei using the Hartree-Fock method date from the sixties [1]. They were mainly devoted to sd shell nuclei. The HF equations were solved by expansion of the individual wave functions on an oscillator basis, restricted to the sd shell. Nuclear interactions were specifically designed for such studies.

The seventies had witnessed a significant progress both in the determination of the effective interaction and in the elaboration of powerful schemes to solve efficiently and accurately the Hartree-Fock equations. Given an effective interaction, the Hartree-Fock scheme is tailored to provide the best possible mean-field. However, it cannot be used with interactions having a hard core like those of most bare nucleon-nucleon interactions. Since the works of Bruekner [2, 3], the solution to this problem is known at least in principle: in the Hartree-Fock equations one must replace the bare two-body force by a G-matrix. This not only removes the hard core but also includes within the mean-field potential the contribution (to all orders) of a class of dominant diagrams of the full many-body problem. Such Bruckner-Hartree-Fock calculations have been performed, starting for instance from the Reid soft core potential [4]. However, they led only to poor agreement with binding energies and root mean square (rms) radii, signaling that more complicated correlations had to be included. The local density approximation [5] was introduced to cure this deficiency. One may consider it as a phenomenological method mocking the effects of three-body correlations which are deemed responsible for the mediocre quality of

[7] Unité de recherches des Universités Paris XI et Paris VI associée au C.N.R.S..

G-matrix results. Such a phenomenology appeared necessary in view of the cost of many-body extensions of the G-matrix method.

Phenomenological effective two-body forces are an alternative answer to the same problem. They depend on a few parameters adjusted to reproduce binding energies and rms-radii of some selected nuclei. After more than a decade of maturation, it appears that two major types of effective interactions have been derived which can be said to fulfill the quality requirements one should expect for an effective interaction.

The first one, for which many different parametrizations exist, and which for technical reasons is the more commonly used, is that of Skyrme force [6]. The mathematical form of this force is extremely simple. It simulates the repulsive correlations, generated by the Pauli principle and the hard-core repulsion, by means of a local density dependance and yields a local mean-field which renders calculations easy and tractable.

In addition, it has been possible to establish the connection between this kind of interaction and the local density approximation [7]. Most of our calculations have been performed either with the SIII [8] or with the Skm* [9] parametrization of this force. These forces have been checked to reproduce with high accuracy the ground state and deformation properties of many nuclei, SIII being better suited for medium mass nuclei and Skm* for very deformed configurations.

The second family of interactions has been introduced to remove the difficulties encountered by zero-range forces in the calculation of pairing properties. Gogny [10] proposes a solution of this problem which keeps the density dependent part of the Skyrme force but replaces the zero-range rest of the interaction by a sum of two gaussians with spin-isospin exchange mixture. The resulting force, known as D1 [11, 12] proved to be as good as the best Skyrme parametrizations as far as mean-field properties were concerned and to provide, at the same time, a good description of pairing correlations. The major drawback of this interaction lies in the complexity it introduces in the mathematical structure of the HF (and HFB) equations, the associated HF mean-field being non-local.

In the same time, two different ways of solving the mean-field equations have been developed. The first one is the extension of the original methods of expanding the individual wave functions on oscillator bases, but now with a size large enough to be able to reach convergence. The symmetries of the oscillator bases have also been relaxed, in order to describe exotic nuclear shapes. The main difficulty encoutered in such calculations is that the HF energy has to be minimised with respect to all the oscillator frequencies, for each size of the basis, in order to achieve correct convergence.

The study of collisions using the time dependent Hartree Fock method (TDHF) has forced to find a way to describe nuclear shapes with minimal symmetry requirements. The method that has been set up for this purpose [13] [14] consists in a discretisation of the individual wave functions on a 3D rectangular mesh, with equidistant points along the three cartesian axes. The variationnal parameters are the values of the wave functions at the mesh points. TDHF studies were performed with simplified versions of the Skyrme force, without the spin orbit interaction. At the end of the seventies, we proposed to apply the same kind of technics to solve the static HF equations [15]. This method is very well suited to study the variation of the nuclear energy with respect to many different degrees of freedom using Skyrme like interactions. However, with presently available computational means, the same method cannot solve the HF equations obtained with a finite range interaction like the Gogny force. A seniority force must then be added to describe the correlations in the particle-particle channel.

1.2. Method of solution of the HF equations

The HF equations are a set of non-linear coupled integro-differential equations. The number of coupled equations can be large; a typical order of magnitude being the number of nucleons. This number can sometimes be significantly reduced by an appropriate use of the symmetries of the solution (rotational or time-reversal invariance for instance).

In this section we will briefly sketch the method developed by our group It relies on the specific properties of nuclear interactions of the Skyrme type.

The general form of a Skyrme interaction v is:

$$\begin{aligned} v \;=\; & t_0(1+x_0P_\sigma)\delta + \tfrac{1}{2}t_1(1+x_1P_\sigma)(\vec{k}^{\,2}\delta + \delta\vec{k}'^{\,2}) + \\ & t_2(1+x_2P_\sigma)\vec{k}.\delta\vec{k}' + \tfrac{1}{6}t_3(1+x_3P_\sigma)\rho^\alpha\delta + \\ & iW(\vec{\sigma}_1+\vec{\sigma}_2).\vec{k}\times\delta\vec{k}' \quad , \end{aligned} \tag{1}$$

where δ is the distribution δ depending on the relative coordinate between the nucleons, $\vec{k}$ is their relative momentum ($\vec{k}'$ acts on the left wave function), ρ is the one-body density and P_σ is the spin exchange operator. The Skyrme force depends on ten parameters t_0, x_0, t_1, x_1, t_2, x_2, t_3, x_3, α and W. They are fitted once for all to a few nuclear properties (binding energies of closed shell nuclei, saturation and compressibilty of nuclear matter...). The same set of parameters can be used for all the nuclear chart.

For such interactions, the HF energy E can be written as a space integral of a local functional:

$$E = \int \mathcal{H}(\vec{r})\, d\vec{r} \quad . \tag{2}$$

The explicit form of $\mathcal{H}(\vec{r})$ has been given in [16] and in [17] for time-reversal invariant and time-reversal non invariant nuclei respectively. It depends on several local densities which can be expressed in terms of the individual wave functions $\Phi_k(\vec{r},\sigma)$ and their occupation probabilities n_k as

$$\begin{aligned} \rho(\vec{r}) \;&=\; \textstyle\sum_{k\sigma} n_k \left|\Phi_k(\vec{r},\sigma)\right|^2 \\ \tau(\vec{r}) \;&=\; \textstyle\sum_{k\sigma} n_k \left|\vec{\nabla}\Phi_k(\vec{r},\sigma)\right|^2 \\ \vec{j}(\vec{r}) \;&=\; \tfrac{1}{2i}\textstyle\sum_{k\sigma} n_k \left(\Phi_k^*(\vec{r},\sigma)\,\vec{\nabla}\Phi_k(\vec{r},\sigma) - \Phi_k(\vec{r},\sigma)\,\vec{\nabla}\Phi_k^*(\vec{r},\sigma)\right) \\ \vec{\nabla}.\vec{J}(\vec{r}) \;&=\; -i\textstyle\sum_{k\sigma\sigma'} n_k \left(\vec{\nabla}\Phi_k^*(\vec{r},\sigma)\times\vec{\nabla}\Phi_k(\vec{r},\sigma')\right).\langle\sigma|\vec{\sigma}|\sigma'\rangle \\ \vec{\rho}(\vec{r}) \;&=\; \textstyle\sum_{k\sigma\sigma'} n_k\, \Phi_k^*(\vec{r},\sigma)\Phi_k(\vec{r},\sigma')\langle\sigma|\vec{\sigma}|\sigma'\rangle \end{aligned} \tag{3}$$

The Coulomb interaction between the protons must be added to the Skyrme force. The direct component of the Coulomb energy is calculated by solving the Poisson equation while the exchange contribution is evaluated by the Slater approximation.

The HF equations are obtained by varying E with respect to the individual wave functions:

$$(h\Phi_k)(\vec{r},\sigma) = \frac{\delta E}{\delta\Phi_k^*(\vec{r},\sigma)} \tag{4}$$

It takes the following form for a state Φ_k of isospin τ

$$\begin{aligned}(h\,\Phi_k)(\vec{r},\sigma) &= \sum_{\sigma'} \Big\{ -\vec{\nabla}\frac{\hbar^2}{2m_\tau^*(\vec{r})}\vec{\nabla}\,\delta_{\sigma\sigma'} + U_\tau(\vec{r})\,\delta_{\sigma\sigma'} + \vec{V}_\tau(\vec{r}).\langle\sigma|\vec{\sigma}|\sigma'\rangle \\ &+ i\,\vec{C}_\tau(\vec{r}).\vec{\nabla}\,\delta_{\sigma\sigma'} + i\,\vec{W}_\tau(\vec{r}).\left(\langle\sigma|\vec{\sigma}|\sigma'\rangle \times \vec{\nabla}\right)\Big\}\,\Phi_k(\vec{r},\sigma') \quad ,\end{aligned} \tag{5}$$

The effective mass m_τ^*, the spin scalar U_τ, spin vector $\vec{V}_\tau$, current $\vec{C}_\tau$ and spin-orbit $\vec{W}_\tau$ potentials are functions of the densities. Their explicit forms have been given in refs [16, 17].

In our method the HF equations are solved directly on a cartesian mesh. The number of variational parameters per orbital becomes of the order of the number of points used to discretize the problem. A typical dimension is between five and ten thousand and reaches in our largest calculations twenty thousand. The original point in our method is that we avoid the almost untractable diagonalization of the Hamiltonian h by noticing that nuclear properties are entirely determined by occupied orbitals. We therefore devised a technique which solves the problem for those orbitals only [18]. The reduction factor is about equal to the ratio of the number of space points to the number of single particle wave functions (the typical gain is a factor 10 to 40).

A realistic description of nuclei requires that one takes into account the residual pairing interaction. We have chosen to avoid the technical difficulties associated with HFB which is the natural extension of HF. Instead we introduce pairing effects by means of BCS either with a seniority or a zero range (δ force) interaction. The intensities G_n and G_p are parametrized with a simple formula adjusted to reproduce the quasi-particle energies of a selected set of nuclei. The corresponding HFBCS equations are solved self-consistently. A cut off in energy must be introduced to avoid divergence of the pairing energy when the single particle space is enlarged (see [16] for details). Finite range interactions like the Gogny force introduce a natural cut off in the two body matrix elements and do not require the introduction of such ad hoc cut off.

1.3. Principle of mesh calculations

A cartesian mesh is composed of a number of equally spaced points along the three directions Ox,Oy and Oz. The wave functions are described in terms of the values they take at the mesh points. Let us introduce Lagrange functions [19] by[1],

$$f_n(x) = \frac{1}{N}\frac{\sin \pi(x-n)}{\sin \frac{\pi}{N}(x-n)} \quad . \tag{6}$$

These functions have the following properties:

$$\begin{array}{rl} f_n(x_m) = & \delta_{nm} \\ \int_{-N}^{+N} f_n(x)\, f_m(x)\, dx & = \delta_{nm} \end{array} \quad . \tag{7}$$

If the spatial part of an individual wave functions is expanded on these Lagrange functions:

$$\Phi(\vec{r}) = \sum_{ijk} \Phi(ijk) f_i(x) f_j(y) f_k(z) \tag{8}$$

the coefficients $\Phi(ijk)$ of the expansion are its values at the mesh points. The overlap between two single particle wave functions takes then a very simple form:

$$\int \Phi_1(\vec{r})\Phi_2(\vec{r})\, d\vec{r} = \sum_{ijk} \Phi_1(ijk)\Phi_2(ijk) \tag{9}$$

This form suggests an approximated formula for the calculation of the integrals entering into the HF equations:

$$\int F(\vec{r})\, d\vec{r} = \sum_{ijk} F(ijk) \tag{10}$$

In fact this last formula is nothing but the Gauss quadrature formula associated with the cartesian mesh. The exact form of the derivatives of the wave functions can then be calculated using the form of the Lagrange functions. The kinetic energy operator along the x direction is given by:

$$\begin{array}{rll} T_{ij} & = \frac{\pi^2}{3}(1 - \frac{1}{N^2}) & i = j \\ & = (-1)^{i-j} \frac{2\pi^2}{N^2} \frac{\cos(\frac{\pi}{N}(i-j))}{\sin^2(\frac{\pi}{N}(i-j))} & i \neq j \end{array} \tag{11}$$

In this way, one can calculate the total energy as a function of the sole values of the individual wave functions at the mesh points. The HF equations are obtained by varying the total energy with respect to these values.

It has been shown in [19] that these differentiation formulae are equivalent to a calculation of the kinetic energy in momentum space using fast Fourier transformations. We shall refer to them as the Fourier formulae. The accuracy obtained in this way on all the terms of the Skyrme functionnal including first or second order derivatives is very high. However, such formulae have the disvantage that they make use of all the points along a line for each direction, while finite difference formulae only require a few of these points. We have therefore decided to use finite difference formulae for the iterative solution of the HF equations, reserving the more accurate

[1] We have taken a mesh size Δx equal to unity.

formulae to the GCM and the restorations of symmetries, where matrix elements are calculated between given single particle (or quasi particle) wave functions.

1.4. The constrained Hartree-Fock equations

Solving HF equations gives an approximation for the binding energy and for the wave function of the ground state. It gives also a description of the excited states built on few uncorrelated single-particle excitations. On the other hand, experimental evidence demonstrates that most of the nuclear dynamics is governed by slow (mostly quadrupole) collective motions. One of the attractive feature of the HF theory is its ability to incorporate the collective dynamics into an essentially independant particle model. This is achieved by the introduction of external constraints and leads to the constrained Hartree-Fock (CHF) equations.

The idea at the core of the CHF approach is that collective nuclear dynamics can be described in terms of a restricted number of degrees of freedom. In addition it is assumed that these degrees of freedom can be described by relatively simple (generally one-body) operators. The addition of these operators with appropriate Lagrange multipliers to the total binding energy, leads to a new variational principle in which the energy is minimized, subject to constraints which determine the values of the collective parameters. The knowledge of the variation of the binding energy in terms of the collective degrees of freedom provides then a crucial information on the characteristics of spectroscopy of the nucleus. Note that this way of treating the collective dynamics is typically adiabatic: all the degrees of freedom which are not constrained are fully relaxed and take their values leading to a minimum of the energy.

To be more specific, let us assume that the collective degree of freedom we want to study is associated with the measure of the operator $\hat{Q}$. We look then for a BCS wave function Ψ which minimizes the energy and such that the expectation value of the collective operator is equal to some chosen number q

$$\langle\Psi|\hat{Q}|\Psi\rangle = q \quad . \tag{12}$$

In the traditional method, this problem is solved by a minimization of the constrained energy $\mathcal{E}$

$$\mathcal{E} = \langle\Psi|\hat{H} - \mu\hat{Q}|\Psi\rangle \quad , \tag{13}$$

where μ is the Lagrange multiplier which must be adjusted to satisfy (12).

The method we just described works only when the constrained quantity $\mathcal{E}$ has a (local) minimum, i.e. when the curvature of the curve $E(q)$ is positive. Otherwise the nonlinear CHF equations are unstable. This problem can be avoided by replacing the linear constraint $\mu\hat{Q}$ by a quadratic one [20], so that the variational problem becomes

$$\mathcal{E} = \langle\hat{H}\rangle + \frac{C}{2}(\langle\hat{Q}\rangle - q_0)^2 \quad . \tag{14}$$

This amounts to the addition to $E(q)$ of a parabola whose axis is at q_0 and whose curvature is determined by the coefficient C. The resulting functional of $\mathcal{E}$ will always exhibit a minimum in the vicinity of q_0 provided C is large enough. In fact this quadratic constraint is somewhat equivalent to a linear one with a varying Lagrange multiplier $\mu = C(q_0 - q)$. In particular, the quadratic and linear constraints produce the same $E(q)$ when the latter is stable.

Very often an accurate description of the collective dynamics will require more than one degree of freedom. The problem will then be handled by a straightforward generalization of the method outlined above.

It should be mentionned that the relevant constraints are of a quite different nature depending on the physics one wants to describe. The above discussion implies that we are considering local operators acting on the shape of the nuclear density (quadrupole, octupole, ...). On the other hand to study the evolution of the nucleus with increasing spin, the natural constraining operator [20] is the projection of the total angular momentum on one of the inertial axes of the density; this is the so-called self-consistent cranking method. With the symmetries required in all our calculations, we can take these inertial axes along those of the intrinsic frame: Ox, Oy and Oz and select a cranking constraint proportional to $-\omega \hat{J}_z$. It may be noticed that while the multipole operators lead only to a breaking of the space symmetries (spherical symmetry or parity), the cranking constraint will additionally break the time-reversal invariance and will lead to structurally more complex CHF equations. For additional information we refer the reader to the references [17, 20].

1.5. Accuracy of mesh calculations

The accuracy of our method is illustrated on fig 1, where we have plotted the variation of the error on total energies with the mesh size. Accuracies obtained using either finite difference or Fourier formulae are shown. With a mesh size of 1.0 fm, the error is of the order of 0.4% (from 0.5MeV for ^{16}O to 10 MeV for ^{208}Pb) for finite difference formulae and of 0.1% using Fourier formulae. The relative error is nearly independant of the mass number, which means that the errors on total energies can be corrected by means of a function depending linearly on the number of nucleons. This leads to an error which is well below the uncertainities due to the nuclear interaction and the HFBCS method. More importantly, we checked [16] that the errors on relative energies are nearly two orders of magnitude lower. This is also illustrated on fig 2, where we compare the deformation energy curves of ^{194}Hg obtained with the SIII nuclear interaction from oblate to the prolate deformations corresponding to superdeformation. The curves obtained using either type of differentiation formulae are nearly indistiguishable.

Our method is moderatly time consuming. Of course, the number of iterations required to achieve convergence depends crucially on the starting point of the iterations. This starting point is most of the time obtained from a HFBCS calculation performed for a neighbouring nucleus. When such a wave function is not available, a starting wave function is constructed by diagonalisation of the Nilsson hamiltonian. To give some ideas on the computing time required for heavy nuclei, the calculation of the ground state of ^{208}Pb starting from Nilsson takes 200 seconds on a Cray2 with a mesh size of 1.0fm (corresponding to an error of 0.1% on the energy), assuming a nuclear density symmetric with repect to three planes (one cannot enforce spherical symmetry on a cartesian mesh). A factor 3 would be gained if the starting point would have been the wave function of a neighbouring nucleus. A recent paper [21] reports on a method of solution of the HF equations using the finite element method. Using this method, the time required to calculate the ground state of ^{208}Pb with an accuracy 0.03% was of 10 hours on Cray2. Our method leads to a similar accuracy (mesh size of 0.6fm) in the same conditions of calculation (no symmetry planes) in two hours of Cray2.

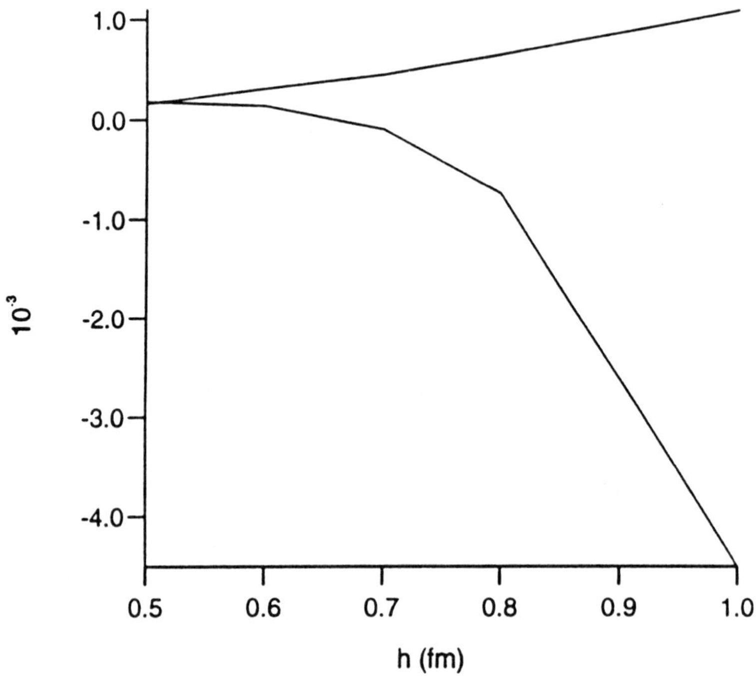

Figure 1

Relative errors (in units of 10^{-3}) on the HFBCS energies as a function of the mesh size. The upper curve is obtained with Fourier formulae for differentiation and the lower curve with finite difference formulae.

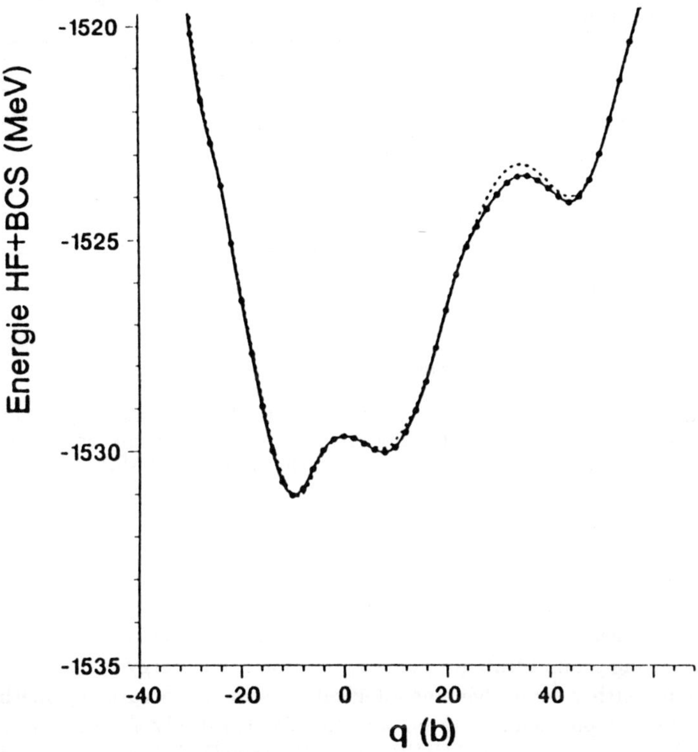

Figure 2

Variation of the energy of ^{194}Hg as a function of the axial quadrupole moment. The full line is obtained with the Fourier differentiation formulae and the dashed curve with the finite difference ones.

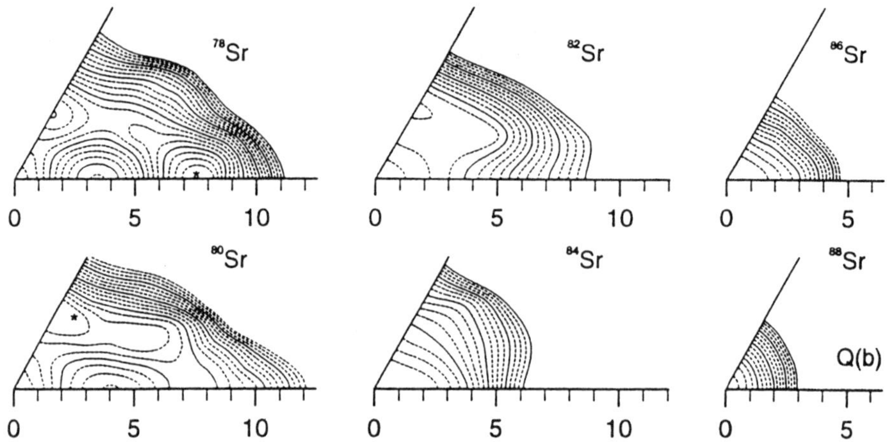

Figure 3

Variation of the energies of Sr isotopes as a function of the triaxial quadrupole moment. The energies are calculated using HFBCS wave functions projected on good particle number. The countour lines are separated by 0.250 MeV.

2. Beyond the mean field: the GCM

2.1. Introduction

The knowledge of the collective potential energy of nuclei as function of selected degrees of freedom allows one in general to classify spectra and to make predictions concerning ground state properties [22]. Potential energy surfaces (PES) can be constructed either semi-phenomenologically, as in the Strutinsky method, or in a more microscopic manner, from an effective two-body interaction with the Hartree-Fock, Hartree-Fock-BCS (HFBCS) or Hartree-Fock-Bogoliubov (HFB) methods. These methods have been tested in various regions of the nuclear chart with an overall good agreement with experimental data [23]. However, when the energy of a nucleus varies slowly with respect to some degrees of freedom, potential energy surfaces are clearly insufficient to make predictions on the ground state properties.

A good example of such a situation is provided by the neutron deficient strontium isotopes (fig. 3). Near the neutron shell closure ($N = 50$), a stiff minimum of the energy is obtained for spherical configurations. Similarly, for ^{78}Sr the ground state is very deformed with a landscape typical of a rotor. At intermediate neutron number, the energy surfaces exhibit two or three local minima (spherical, oblate and triaxial) nearly degenerated in energy. In such cases, one-body observables like mean square radii are largely undetermined.

To proceed further, a dynamical collective model is required. Kumar and Baranger proposed to calculate mass parameters and to construct a collective Bohr hamiltonian[24]. Such a method has been recently used with PES and mass parameters generated with the cranking method from wave functions obtained from HFB calculations[25]. This approach is somewhat indirect in that it requires both the choice of a method for evaluating mass parameters (cranking, adiabatic time dependent Hartree-Fock or gaussian overlap approximation) and a requantification to generate a spectrum. A more microscopic treatment is based on configuration mixing which for continuous collective variables can be reformulated as the generator coordinate method (GCM). Within the GCM, nuclear states correspond to the configuration mixings which minimize the expectation value of a microscopic hamiltonian within a collective space. In our work, the same hamiltonian is employed both in the GCM and in the constrained HFBCS equations used to construct the collective space.

A detailed study of the GCM for axial quadrupole deformations has been presented in [26]. It has been extended to study triaxal quadrupole deformations in [27]. This generalization brings in new problems. One of them is associated with the larger dimension of the variational space which now involves two collective variables: the deformation Q and the triaxiality angle γ. More important, one has to treat explicitly the degeneracies associated with the several labellings of the intrinsic axes of inertia.

2.2. The Generator Coordinate Method

Given a family of N-body wave functions $|\Phi(q)\rangle$ depending on a collective variable q, the GCM determines approximate eigenstates of the Hamiltonian $\hat{H}$ having the form:

$$|\Psi_k\rangle = \int f_k(q)|\Phi(q)\rangle \,\mathrm{d}q \quad , \tag{15}$$

These linear combinations are determined by a minimization of the expectation value of the two-body Hamiltonian $\hat{H}$:

$$E = \frac{\langle\Psi|\hat{H}|\Psi\rangle}{\langle\Psi|\Psi\rangle} \tag{16}$$

The weight function f_k (labelled according to the index k of the eigenstate) satisfies the Hill-Wheeler equation:

$$\int (H(q,q') - E_k\, I(q,q'))\, f_k(q')\,\mathrm{d}q' = 0 \quad . \tag{17}$$

in which the hamiltonian kernel H and the overlap kernel I are defined as:

$$I(q,q') = \langle\Phi(q)|\Phi(q')\rangle \tag{18}$$

$$H(q,q') = \langle \Phi(q)|\hat{H}|\Phi(q')\rangle \tag{19}$$

The Hill and Wheeler equation is solved in two steps. One first diagonalises the operator I :

$$\int I(q,q')\, u_k(q')\, \mathrm{d}q' = n_k u_k(q) \quad . \tag{20}$$

This permits to define the square root of I and to introduce a new collective wave function g_k:

$$g_k(q) = \int I^{1/2}(q,q') f_k(q')\, \mathrm{d}q \quad . \tag{21}$$

and also a transformation for any operator $\hat{O}$:

$$\tilde{O}(q,q') = \int I^{-1/2}(q,q_1) O(q_1,q_2)\, I^{-1/2}(q_2,q') \mathrm{d}q_1 \mathrm{d}q_2 \quad . \tag{22}$$

where $O(q,q')$ is calculated from the HFBCS wave functions:

$$O(q,q') = \langle \Phi(q)|\hat{O}|\Phi(q')\rangle \tag{23}$$

The Hill and Wheeler equation becomes:

$$\int \tilde{H}(q,q')\, g_k(q')\, \mathrm{d}q' = E_k\, g(q)_k \quad . \tag{24}$$

It is solved by discretization of the variables q [20]. Numerical problems arise because the overlap kernel has zero as an accumulation point of its eigenvalues. All the numerical problems posed by this problem have been addressed in an already published paper [26]. Note that the weight functions f_k are not orthornormal and cannot be interpreted as collective wave functions for the variable q. This role is assigned to the orthonormal set of functions g_k.

In the applications that we have performed up to now, the N-body wave functions have been constructed by means of the HFBCS method, with various constraints on the shape of the nuclear density. The same Skyrme plus pairing two-body Hamiltonian has been used in the HFBCS and the GCM calculations. Since the HFBCS wave functions are not eigenstates of the particle number operators, the GCM wave function does not have the right particle number. The correct mean particle number is obtained by the introduction of Lagrange multipliers in the Hill and Wheeler equation. Another way to correct the particle number consists in the projection of the BCS wave functions on good particle number by introducing gauge angles as additional generator coordinates.

It is useful to note that the GCM incorporates automatically the effects of ingredients of the Bohr collective Hamiltonian[22, 24] method such as the zero point motion energy and the effective collective masses. Moreover, it does not rely on assumptions such as the gaussian overlap approximation nor on the validity of

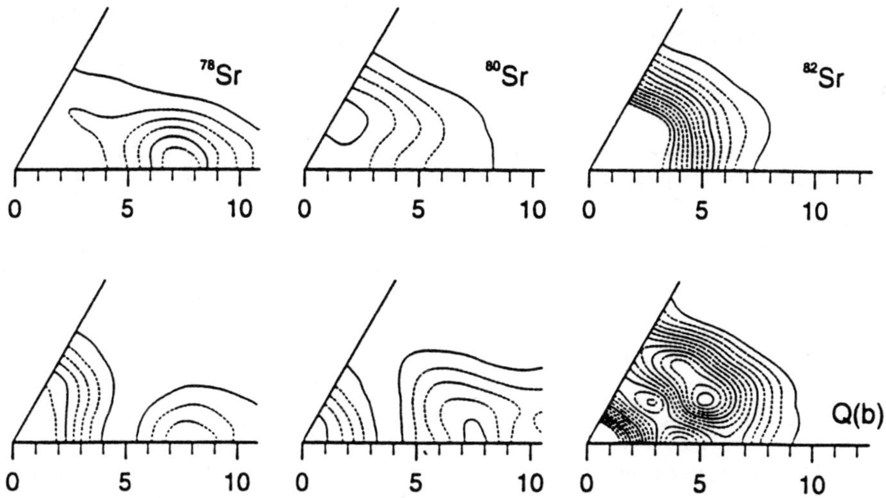

Figure 4

Square moduli of the ground and first excited GCM states of three Sr isotopes.

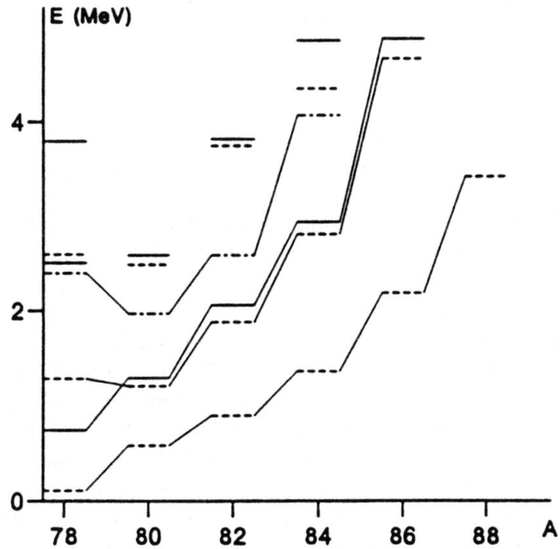

Figure 5

Spectra obtained using the GCM for the neutron deficient Sr isotopes. Full line: 0^+, dashed line: 2^+, dashed dotted line 3^+ (see text).

quadratic expansions of the kernels[20] which are often used to derive microscopically these ingredients[25].

3. Applications

We have developed several computer codes solving the coupled HFBCS equations with different symmetries of the nuclear density. Our first code was designed to solve the coupled HFBCS equations with symmetries of the density with respect to three planes [16], in which case the nuclear density contains only even multipoles. Time reversal invariance was inforced by considering that a wave function and its time reversed are simultaneously occupied. This code has been generalised either to consider nuclear states breaking the left right symmetry (two symmetry planes) [28] [29] or the time reversal invariance (rotating nuclei) [17] [30]. We have also developed codes which permit to calculate the non diagonal matrix elements required by the GCM, either without or with projection on particle number, for time reversal invariant wave functions.

Our first application [16] concerned the Kr, Sr, Zr and Mo isotopes, from neutron deficient to neutron rich ones. Fig 3 shows the variation of the energy of Sr isotopes as a function of the triaxial quadrupole deformation. These maps are not obtained in the same way as the one previously published: the HFBCS wave functions have been here projected on good particle number (to avoid a pairing collapse for some deformations which would have led to singularities in the energy maps, the Lipkin Nogami method has been used to construct the BCS states). An interesting result is that the topology of the surfaces is not strongly modified by projection [16, 27].

As discussed in the previous section, the topology of these surfaces indicates the necessity of going beyond a mean-field approximation for nuclei soft against deformations. This has been done in a recent publication [27] without particle number projection.

The GCM applied to the triaxial quadrupole dynamics requires to mix for each quadrupole moment the six possible orientations of the principal axes of the nucleus. We showed in [27] that such mixing is a first step in a restoration of rotationnal symmetry, with an accuracy decreasing with increasing deformations. On fig 4 are compared the square moduli of the ground and first excited states wave functions of ^{78}Sr, ^{80}Sr and ^{82}Sr. The ground state of ^{78}Sr is very deformed, its first excited state having its components with maximal weight at small deformations. The situation is reversed for ^{80}Sr, as one could expect from the energy maps. In the same way, ^{82}Sr is the softest of the three isotopes and the components of the first two states show a spreading over a large range of quadrupole moments.

The energies of the first GCM eigenstates for the six isotopes are plotted on fig. 5. As explained in ref [27], one can classify these eigenstates as approximations of 0^+, 2^+ and 3^+ states. The energies of the first (approximate) 0^+ and 2^+ states decrease steadily with decreasing neutron number. However, looking at fig. 4, one can see that the character of the first excited 0^+ state changes from ^{78}Sr and ^{80}Sr:

nearly spherical for the former and well deformed for the latter. Although mixing of the six possible orientations of the principal axes is a first approximation to a restoration of rotational invariance, the energies of these GCM eigenstates have still to be corrected for a spureous rotational energy, the correction being the largest for the most deformed configurations. This should lower by 1 to 2 MeV the energy of the most deformed state with respect to the nearly spherical one. One can thus predict that such correction will not modify the order of the ground and first excited states in ^{78}Sr, but will modify this order in ^{80}Sr, leading probably to a small energy difference between both states in this isotope. In the heavier isotopes, this rotationnal energy correction will lead to a compression of the spectra.

Most of our other applications of the GCM have been devoted to the study of superdeformation in isotopes in the mass region of ^{192}Hg. Some representative results are displayed on fig 6 and 7. In fig 6, we show the HFBCS energy curve obtained for ^{194}Hg with the Skyrme Skm* interaction (full curve) together with the GCM eigenstates obtained using the axial quadrupole moment as a generator coordinate (dots centered at the mean quadrupole moments of the GCM state). The excitation energy and the deformation of the superdeformed state are only weakly modified by the GCM. The square moduli of the collective wave functions of the two first and the sixth GCM eigenstates are plotted on fig 7. The oblate and prolate HFBCS minima are the main components of the ground and first excited GCM states respectively, the first excited state having although a non negligible oblate component. The sixth state is mainly located in the SD well. Similar results have been obtained for other isotopes [31]. In subsequent publications, we have studied the influence of octupole deformations [29] and of pairing vibrations [32] on SD states. The last example of GCM calculations that we give here concerns the effect of octupole correlations on SD states. On fig 8 are plotted the variation of energy of ^{194}Hg for a quadrupole moment corresponding to the SD minimum as a function of octupole deformations with μ=0,1 and 2 respectively. The HFBCS curve does not show a minimum in any of the three cases. The parity projected curves indicate the effect of dynamical octupole correlation, which is confirmed by the full GCM calculation. The octupole mode μ=0 is the most favourable one, leading to the lowest energy for the ground state and the lowest excitation energy for the negative parity states. Contrary to some other theoretical studies [33], the μ=1 mode does not seem to be competitive.

4. Conclusions

We have briefly described the actual status of mean-field and GCM calculations using Skyrme like effective interactions. As a matter of conclusion, let us quote a few remarks, indicating which directions of developments are underway or seem to be necessary in a near future.

As discussed in section 1, the collective path which is followed either in constrained HFBCS or in GCM is by construction an adiabatic path. However, the GCM formalism is very well adapted to the inclusion of non adiabatic effects. In

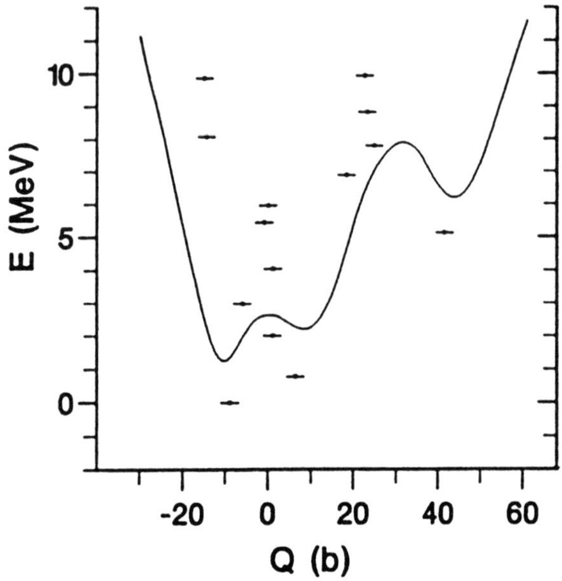

Figure 6

Variation of the HFBCS energy of ^{194}Hg as a function of the axial quadrupole moment. The bars indicate the energies of the GCM eigenstates. They are centered at the mean quadrupole moment of these states.

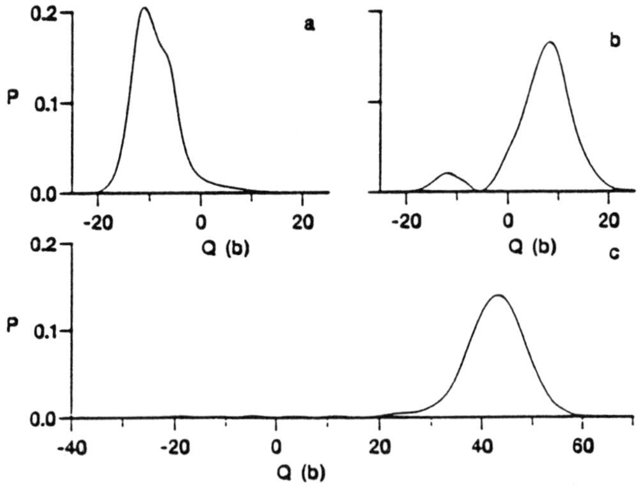

Figure 7

Square moduli of the collective ground, first excited and SD states obtained using the GCM.

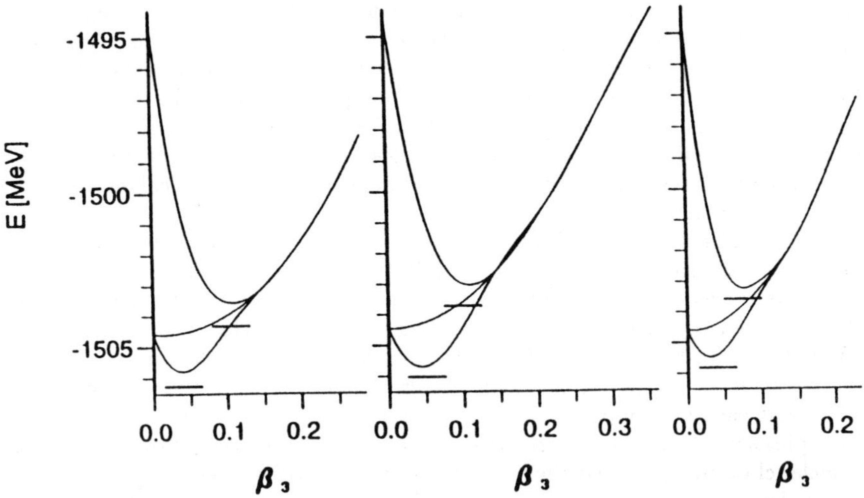

Figure 8

Variation of the HFBCS energy of ^{192}Hg as a function of the octupole moment for the μ=0, 1 and 2 modes (from the left to the right). The curves obtained by projecting the wave functions on good parity and of energies of the two first GCM states are also shown.

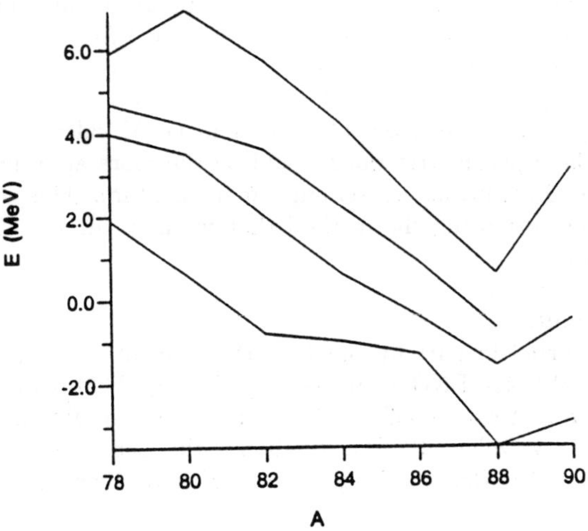

Figure 9

Differences between the experimental and the theoretical binding energies obtained using (from top to bottom), the HFBCS method, the GCM on HFBCS states, by projecting on particle number HF-Lipkin Nogami wave functions and by the GCM on these last wave functions.

particular, a generalisation of our method is under development and will enable us to introduce in the GCM configurations corresponding to quasi particle excitations [34]. Up to now, the collective path has been selected on physical arguments. A more satisfactory procedure would be to find the preferential modes of excitation of the nucleus, which could be done by methods like the adiabatic version of TDHF [20]. However, the interest of a selection of modes based on physical intuition is that it is possible to conserve a close connection with more phenomenoligical models, like the collective models.

The parametrizations of the Skyrme forces that we use have been adjusted to reproduce experimental data within the HF method. On fig. 9 are plotted the differences betzeen the experimental data and various theoretical calculations of the ground state energy of some Sr isotopes. Four different methods have been used: HFBCS, GCM on HFBCS wave functions, projection on particle number of BCS states constructed by the Lipkin Nogami method and GCM on these states. One can see that each new improvment of the method brings a gain of energy. In the most sophisticated calculation, the SIII interaction leads to a slight overbinding of the nuclei close to the closed shell, indicating that it will be necessary to slightly readjust the parameters of the interaction. However, the isospin dependance of the error is very similar in the four cases, execept that the larger error obtained for ^{80}Sr within the HFBCS method disappear in all the other cases. The main remaining error on these energies should be due to the breaking of rotationnal symmetry that remains in the triaxial quadrupole GCM. The partial symmetry restoration that has already been done should be sufficient for the isotopes with a predominently spherical configurations. One can thus expect that angular momentum projection will be more effective for the lightest isotopes, ^{78}Sr and ^{80}Sr, leading to a curve nearly flat as a function of neutron number. Two conclusions can be drawn from this plot. First, the parametrizations of the Skyrme force that we use do not show any pathology and can still be safely used in GCM calculations. Second, a readjustment of these parametrisations would be the more appropriate after to have included in the GCM projection on angular momentum. This could be done at a still reasonnable cost using the method that we have already developed for schematic interations [35].

5. Acknowledgements

This work was supported in part by the U.S. Department of Energy under Engineering contract NW 7405-ENG-48 in part by the Polish Ministry of National Education under Contract No. 20450 91 01 and in part by the PAI program of the Belgian Scientific Policy Office. H. Flocard thanks the Centre de calcul vectoriel pour la recherche and P-H Heenen the Université libre de Bruxelles for computational facilities.

6. References

1. G. Ripka *Adv. Nucl. Phys.* **1** (1968) 83
2. K. Brueckner, *Phys Rev* **97** (1955) 1356
3. B. Day, *Rev Mod Phys* **39** (1967) 719
4. K. Davies, R. McCarthy and P. Sauer, *Phys Rev* **C7** (1973) 943
5. J. Negele, *Phys Rev* **C1** (1970) 1260
6. T. Skyrme, *Phil Mag* **1** (1956) 1043
7. J. Negele and D. Vautherin, *Phys Rev* **C5** (1972) 1472
8. M. Beiner, H. Flocard, N. Van Giai and P. Quentin, *Nucl Phys* **A238** (1975) 29
9. J. Bartel, P. Quentin, M. Brack, C. Guet and H.-B. Håkansson, *Nucl. Phys.* **A386** (1982) 79
10. D. Gogny, in *Proc Int. Conf. on Nuclear Physics* eds J. De Boer and H. Mang, (North Holland, Amsterdam, 1973).
11. J. Decharge, M. Girod and D. Gogny *Phys Lett* **55B** (1975) 361
12. J. Decharge and D. Gogny *Phys Rev* **C21** (1980) 1568
13. H. Flocard, S. Koonin and M.S. Weiss, *Phys Rev* **C17** (1978) 1682
14. P. Bonche, B. Grammaticos and S. Koonin, *Phys Rev* **C17** (1978) 1700
15. P.-H. Heenen, H.Flocard and P. Bonche, in *Time dependent Hartree Fock method*, eds P. Bonche, B. Giraud and Ph Quentin (Editions de Physique, 1979) p. 175
16. P. Bonche, H. Flocard, P.-H. Heenen, S.J. Krieger and M.S. Weiss, *Nucl. Phys.* **A443** (1985) 39
17. P. Bonche, H. Flocard and P.-H. Heenen, *Nucl Phys* **A467** (1967) 115
18. K. Davies, H. Flocard, S. Krieger and M. Weiss, *Nucl Phys* **A342** (1982) 111
19. D. Baye and P.-H. Heenen, *J Phys A Mat Gen* **19** (1986) 2041
20. P. Ring and P. Schuck, *(The Nuclear Many-Body Problem)* (Springer Verlag, Berlin, 1980)
21. A.S. Umar, M.R. Straayer, J.S. Wu and M.C. Güçlü, *Phys Rev* **C44** (1991) 2512
22. A. Bohr and B. Mottelson, *(Nuclear Structure)* (vol2) (W. Benjamin, New York, 1975)
23. S. Aberg, H. Flocard and W. Nazarewicz, *Ann Rev Nucl Part Sci* **1990** 439
24. K. Kumar and M. Baranger, *Nucl. Phys.* **A92** (1967) 608
25. J.P. Delaroche, M. Girod, J. Libert and I. Deloncle, *Phys. Lett.* **B232** (1989) 145
26. P. Bonche, J. Dobaczewski, H. Flocard, P.-H. Heenen and J. Meyer, *Nucl. Phys.* **A510** 466 (1990)
27. P. Bonche, J. Dobaczewski, H. Flocard and P.-H. Heenen *Nucl Phys* **A530**

(1991) 149

28. P. Bonche, P.-H. Heenen, H. Flocard and D. Vautherin, *Phys Lett* **175** (1986) 387
29. P. Bonche, S.J. Krieger, M.S. Weiss, J. Dobaczewski, H. Flocard and P-H. Heenen *Phys Rev Lett* **66** (1991) 876
30. P. Bonche, H. Flocard and P-H. Heenen, *Nucl Phys* **A523** (1991) 300
31. P. Bonche, J. Dobaczewski, H. Flocard, P-H. Heenen, S.J. Krieger, J. Meyer and M.S. Weiss, *Nucl Phys* **A519** (1990) 509
32. J. Meyer, P. Bonche, J. Dobaczewski, H. Flocard and P.-H. Heenen *Nucl Phys* **A533** (1991) 307
33. R. Chasman *Phys Lett* **266B** (1991) 243
34. N.Tajima, P. Bonche, J. Dobaczewski, H. Flocard and P.-H. Heenen *Nucl Phys* in press (1992)
35. D. Baye and P.-H. Heenen, *Phys Rev* **C29** (1986) 1056

SELF-CONSISTENT-FIELD METHODS

SUMMARY OF DISCUSSION BY WORKING GROUP ONE

C. R. BINGHAM
Department of Physics, University of Tennessee
Knoxville, Tennessee 37996-1200, USA

S. G. ROHOZIŃSKI
Institute for Theoretical Physics, University of Warsaw, ul. Hoża 69
PL-00681 Warsaw, Poland

A. SUSTICH
Cyclotron Laboratory, Michigan State University
East Lansing, Michigan 48824, USA

and

J. L. WOOD
School of Physics, Georgia Institute of Technology
Atlanta, Georgia 30332, USA

ABSTRACT

Hartree-fock and Hartree-Fock-Bogolyubov (HFB) methods are like Motherhood: they cannot be criticized! One can only discuss details. Rhetorical questions (with the above in mind) were posed by Working Group One and received responses, details of which are briefly summarized below.

1. What Is New in HF(B)?

- Method of formulation - nothing
- Technology
  1. Computing power.
  2. Computation on a grid allows for any arbitrary deformation: $1\text{fm}^3 \equiv \hbar\omega$ in ^{40}Ca.
  3. Use of imaginary time to give faster convergence.
- Constraints
  1. Multidimensional constraints.
  2. $< Q_{22} >$ and $< Q_{30} >$ added to $< Q_{20} >$.
- Input - nothing:
  1. Input forces essentially the same (Skyrme or Gogny) for the last 15 years.

2. The initial wave functions are typically taken as harmonic oscillator wave functions.

- Generator coordinate methods (GCM)
  1. Widely explored
  2. Allows quantum mechanical calculation of large amplitude motion; hence, better than semi-classical result of TDHF.
  3. Hill-Wheeler equation solved but gausian overlap approximation (GOA) used as well.

2. What Problems Exist?

- Technology - Difficulties in handling off-diagonal matrix elements (non-local properties).
- Constraints
  1. The "fission mismatch" problem: must add a "matching coordinate".
  2. Choice of "the best" constraint: e.g., SCF + RPA at each iteration effectively scans landscape and suggests a different constraining operator to get there – feasible?
- Input (forces)
  1. Skyrme: must put in pairing "by hand".
  2. Extended Skyrme: too many parameters.
  3. Gogny and Skyrme need better isospin structure to describe nuclei far from stability: this will be taken care of as data from radioactive beam facilities become available.
  4. No energy dependence.
  5. Skyrme: very poor in describing magnetic properties.
- GCM
  1. Choice of collective coordinates: difficult for many dimensions.
  2. GOA depends on projection: fails for two local maxima, e.g. prolate and oblate deformation.
  3. GOA good in heavy nuclei, not so good in light nuclei.
  4. Unphysical states: must make correct projections.

3. Where Is HF(B) Going?

- What is driving it?

1. Better definition of constraints.
2. Description of extreme states: deformation (superdeformation), angular momentum, temperature (e.g., U+U collisions), N:Z ratio (nuclei far from β stability).

- What should/could drive it? (suggestions)

1. Input wave functions: choose symmetry based on successes of microscopic (algebraic) theories.
2. New constraints based on symmetry: would help to identify low-energy irreps.
3. Perform HF on a submanifold of Hilbert space by constraining to a symmetry given by an algebraic model; compare with HF on full Hilbert space: is there any evidence of the symmetry in the full calculation?
4. Microscopic theory of effective forces would give through HFB a microscopic foundation of the mean field.

NUCLEAR STRUCTURE
WITH THE D1 EFFECTIVE INTERACTION

J.-F.BERGER, J.DECHARGE
Commissariat à l'Energie Atomique
Service PTN, Centre d'Etudes de Bruyères-le-Châtel
B.P.12, 91680 Bruyères-le-Châtel, France

and

D.GOGNY
Commissariat à l'Energie Atomique
C.E.S.T.A.
33114 Le Barp, France

ABSTRACT

These notes present an overview of the microscopic approach of nuclear structure that one can achieve using the finite range effective force D1. The main characteristics of this interaction are first recalled. It is shown how the parametrization of the force has been chosen in order to describe not only the average nuclear field, but also pairing correlations as well as long range correlations associated with collective excitations. The techniques employed with the D1 force in mean field calculations (Hartree-Fock-Bogolyubov and constrained Hartree-Fock-Bogolyubov methods) and beyond the mean field approach (RPA and large amplitude collective motion theory) are presented. Finally a few of the numerous applications to specific nuclear systems performed during the last twenty years are listed.

1. Introduction

The search for a fundamental description of nuclei, that is one starting from the basic nuclear constituents and their mutual interactions, has been one of the main goals of nuclear physicists in the last 40 years. Such an approach clearly is appealing from a purely conceptual point of view. It may also be extremely useful in all instances where only scarce or indirect experimental information exists. For example, it is the ideal approach for describing exotic systems or transient nuclear states, and probably the only one capable of deriving the nuclear equation of state.

The approach presented in the following is an attempt to reach such a fundamental level of description in all nuclear situations where the mean field is a meaningful concept. It is based on the non-relativistic Density-Dependent Hartree-Fock (DDHF) theory employed with a phenomenological effective nucleon-nucleon (N-N) interaction. Its purpose is to derive in a self-consistent manner and from the sole input of the N-N interaction the average nuclear field and the different kinds of excitations this average field may be subject to. Its aim is to deduce the ground state properties of nuclei, the characteristics of low-lying excited states and the behaviour of nuclear systems during complex processes such as fission or low-energy nucleus-nucleus collisions.

Of course, this kind of approach is not as fundamental as those starting from a form of nuclear interaction reproducing N-N scattering cross-sections and/or including the influence of degrees of freedom other than nucleons. It is well known however that, due to the complexity of the many-body problem, calculations based on realistic models of N-N interaction are not able to give accurate results concerning the detailed structure of nuclear ground states and spectra. In this situation, the less ambitious point of view proposed by the DDHF theory seems to be the only possible way of achieving a quantitative microscopic description of nuclear structure and phenomena, at least for now. The most remarkable result of the extensive work performed along this line during the last 20 years has been the discovery that relatively simple parametrizations of the effective interaction inside the nuclear medium provide a surprisingly good account of a variety of nuclear properties. One of the purposes of the present talk is to illustrate this point.

Since the D1 force is the basic ingredient of the approach presented here, Section 2 is devoted to a brief description of this interaction and of its main properties. The fact that the parametrization of this interaction is suitable to describe correlations beyond the simple mean-field (Hartree-Fock) approach will be especially emphasized. This is important since the independent particle approximation is never a good one in nuclei. Pairing correlations are included by employing the full Hartree-Fock-Bogolyubov theory. Long range correlations associated with collective excitations are incorporated with the RPA approach in doubly magic nuclei or using a theory of large amplitude collective motion based on the Generator Coordinate (GC) method in other systems. A check of the validity of the effective interaction for describing long range correlations is obtained by examining the behaviour of the Landau parameters of the force. It appears that the finite range of the force and the density-dependence are crucial to obtain a correct description of both mean field properties and correlations beyond the mean field.

The Hartree-Fock-Bogolyubov (HFB) and constrained HFB procedures are the basic methods employed with the D1 interaction. They are used to derive the optimal nuclear average field and pairing field and to find how these fields change when nuclei are driven off equilibrium. The way these methods are implemented with D1 is described in Section 3. The techniques employed for including collective oscillations of the mean field of small (RPA) or large amplitude are also presented in this Section. In Section 4, a sample of results representative of the many studies performed with D1 is given. Finally Section 5 concludes this talk.

2.The D1 Interaction

The philosophy underlying the D1 force and the way it has been built is described in Ref.1. Only the main features of this interaction and the changes that have occurred since this initial publication are presented here.

2.1. Parametrization

The form of the parametrization of the D1 interaction is as follows:

$$\begin{aligned} v_{12} &= \sum_{j=1}^{2} e^{-(\vec{r}_1-\vec{r}_2)^2/\mu_j{}^2} \, (W_j + B_j P_\sigma - H_j P_\tau - M_j P_\sigma P_\tau) \\ &+ t_3(1+x_0 P_\sigma)\delta(\vec{r}_1-\vec{r}_2)\left[\rho(\frac{\vec{r}_1+\vec{r}_2}{2})\right]^{\alpha} \\ &+ iW_{LS} \overleftarrow{\nabla}_{12}\, \delta(\vec{r}_1-\vec{r}_2) \times \overrightarrow{\nabla}_{12} \cdot (\vec{\sigma}_1+\vec{\sigma}_2) \end{aligned} \tag{1}$$

The first term contains two finite ranges and the usual mixing of exchange operators P_σ, P_τ and $P_r = -P_\sigma P_\tau$ for spin, isospin and space variables respectively. The choice of a non zero-range central part is essential for being able to describe pairing correlations from the effective force itself. In fact, it is well-known that a N-N potential containing only zero-range and derivatives of zero-range components has a pathological behaviour in k-space and leads to a divergent expression for the pairing energy in superfluid systems. The additional complication introduced by the finite range, in particular the existence of a nontrivial exchange term in the mean field, is to a large extent compensated by the choice of gaussian spatial shapes.

The second term of D1 is a functional of the nuclear density ρ, with zero-range for simplicity. For $x_0 = 1$, a value usually adopted, this term acts only in the triplet even two-nucleon space ($S = 1$, $T = 0$), and therefore it does not contribute to the pairing field between identical nucleons.

The third term is a zero-range two-body spin-orbit interaction of the same form as in the Skyrme forces.

The parameters have been determined in the following way. The ranges μ_1 and μ_2, and the parameters x_0 and α are given initial values:

$$\mu_1 = 0.7\,fm, \quad \mu_2 = 1.2\,fm, \quad x_0 = 1, \quad \alpha = 1/3. \tag{2}$$

and t_3 is chosen to be around $1400\,MeV$. The eight parameters W_j, B_j, H_j and M_j are then calculated in order to reproduce:

i) the properties of nuclear matter at saturation: Fermi momentum $k_F \simeq 1.35\,fm^{-1}$, binding energy per particle $a_V \simeq 16\,MeV$, volume asymmetry energy $a_\tau \simeq 30\,MeV$ and incompressibility coefficient $K_\infty \simeq 200-220 MeV$.

ii) the binding energies and radii of ^{16}O and ^{90} Zr.

iii) and the values of pairing matrix elements of the form

$$\langle (nS)^2 0 \quad T=1 | v_{12} | (nS)^2 0 \quad T=1 \rangle, \quad nS = 1S \textit{ and } 2S. \tag{3}$$

These matrix elements are first given negative values which are subsequently adjusted to reproduce the experimental pairing gaps of a few Sn isotopes. Finally, the spin-orbit coefficient W_{LS} is chosen to reproduce the $p_{3/2} - p_{1/2}$ splitting in ^{16}O. Its value is found to be

$$W_{LS} = 130\,MeV. \tag{4}$$

In a second step the value of t_3 is fine-tuned in order to reproduce the binding energy of ^{208} Pb in a self-consistent DDHF calculation. The determination of the eight W_j, B_j, H_j and M_j is then repeated with this new value of t_3. The whole procedure is iterated until convergence is reached.

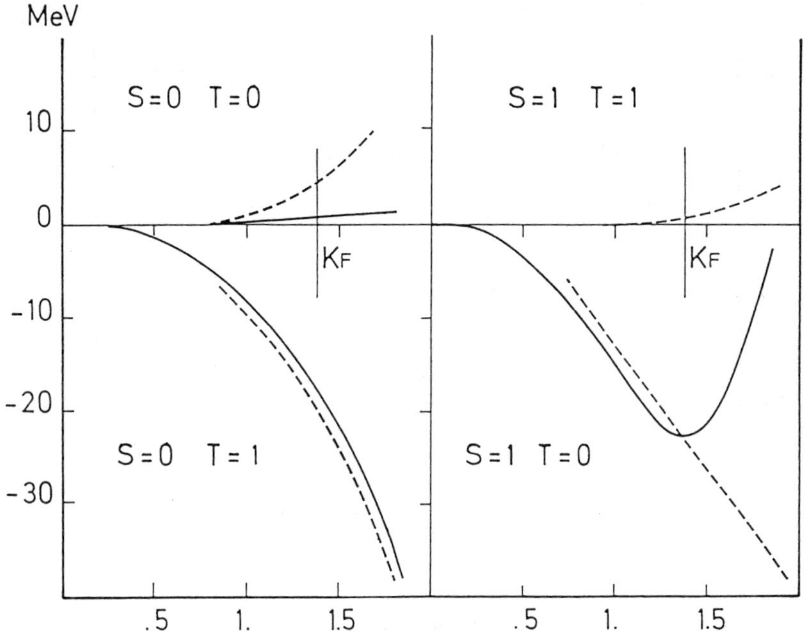

Figure 1: Contributions to the nuclear matter potential energy of the four spin-isospin components of the D1 force. The dashed lines represent the general trend obtained with realistic two-body forces and the Brueckner Hartree-Fock theory.

It turns out that the initial values adopted in (2) had not to be modified in order to satisfy all the conditions listed above. The parameters W_j, B_j, H_j, M_j and t_3 obtained with this fitting procedure are given in Ref.1.

More recently, a new constraint has been imposed on the parameters of D1. It concerns the surface coefficient a_S that can be defined, for instance, in semi-infinite nuclear matter. As a matter of fact, it appeared that the original D1 parameters significantly overestimated the empirical value ($\simeq 19\,MeV$) of this coefficient. As a consequence, they led to a much too high second fission barrier in ^{240}Pu (see Fig.2). The constraint on a_S has been taken into account by computing in an approximate way the binding energy of a spherical nucleus with a very large radius. The nine parameters W_j, B_j, H_j, M_j and t_3 involved in the fitting procedure have then been

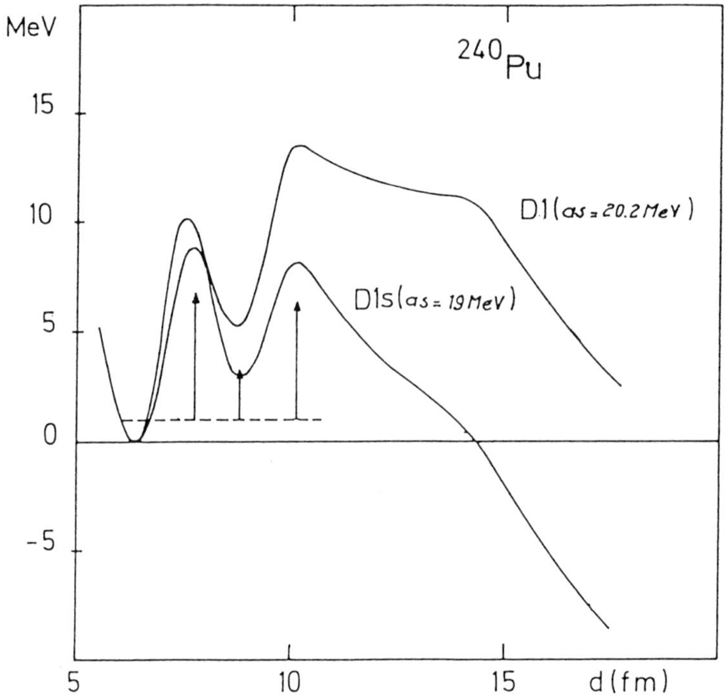

Figure 2: The fission barrier of ^{240}Pu with the parametrizations D1 and D1S. In the latter version of the interaction, the surface coefficient a_S has been imposed a lower value. The vertical arrows give an idea of the parabolic barrier heights usually deduced from experiment.

modified in order to decrease the value of a_S to the empirical value. This could be done without significantly changing the other properties of the force. Again, it turned out that the values (2) could be kept the same. This new version of the interaction D1 has been labeled D1S. The corresponding values of the parameters are, besides (2) and (4):

$$W_1 = -1720.30\, MeV,\ B_1 = 1300.00\, MeV,\ H_1 = -1813.53\, MeV,\ M_1 = 1397.60\, MeV,$$

$$W_2 = 103.64\, MeV,\ B_2 = -163.48\, MeV,\ H_2 = 162.81\, MeV,\ M_2 = -223.93\, MeV,$$

$$t_3 = 1390.60\, MeV.$$

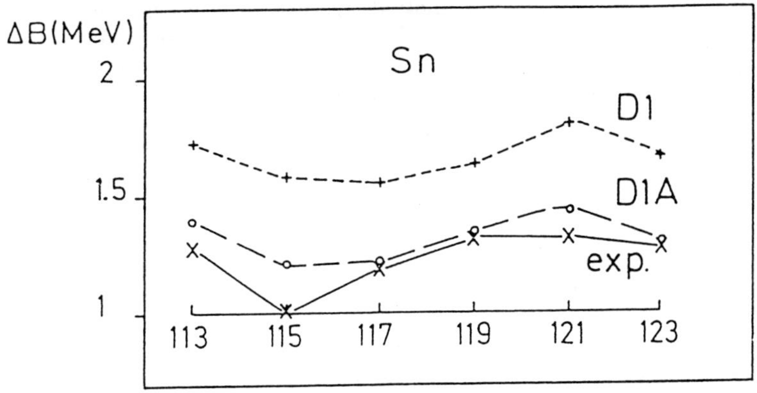

Figure 3: The odd-even mass differences for seven Sn isotopes calculated with D1. The curve labeled D1A has been obtained by increasing the pairing matrix elements used in the parameter fit by 3%.

2.2. Properties

2.2.1. Infinite Nuclear Matter

The main characteristics obtained in infinite nuclear matter with D1S are very close to those obtained with the first version of D1. They are:

$$a_V = -16.0\,MeV/nucleon, \quad k_F = 1.35\,fm^{-1}, \quad m^*/m = .67, \quad K_\infty = 209\,MeV,$$

where m^* is the nucleon effective mass. The saturation curves derived in infinite nuclear matter for the four spin-isospin spaces are indistinguishable from those of D1. They are shown in Fig.1.

The surface coefficient for D1S is $a_S = 19.0\,MeV$ instead of $20.2\,MeV$ for D1. The effect on the ^{240} Pu fission barrier (mass-asymmetry included) is illustrated in the Fig.2. One notes that reasonable barrier heights are derived with D1S. It can also be checked that the deformation properties given by this force (ground state and isomer deformation for instance) are in excellent agreement with experiment.

2.2.2. Pairing Properties

Fig.3 displays information about the pairing properties of D1. Those of D1S are similar. The Sn isotopes are chosen because they are spherical, a long series of them is known and their ground states contain strong neutron pairing correlations. In this figure, the odd-even mass differences

$$\Delta B = \frac{B_{A-1} + B_{A+1}}{2} - B_A, \qquad A \ \ odd \tag{5}$$

obtained in a full HFB calculation are plotted together with the experimental quantities for seven Sn isotopes. The HFB calculation for odd A are performed using the blocking approximation. The curve labeled D1 has been obtained with the D1 parameters. The curve labeled D1A has been derived with parameters fitted with values of the pairing matrix elements (3) increased by 3%. This latter curve gives an idea of the influence of these matrix elements in the fit of the parameters. One sees that the parameters actually taken for D1 and D1S lead to ΔB's approximately 300 *keV* above the experimental values. This deviation has been introduced in order to compensate for the well-known fact that the blocking approximation underestimates the binding energy of odd nuclei when quasi-particle-vibration coupling is present, which is the case in Sn isotopes. From the work of Kuo *et al.* [2] the correction should be a few hundred *keV*. The uncertainty in this value illustrates one of the difficulties encountered when determining a phenomenological effective force valid at a given level of approximation. In principle, quasi-particle-vibration coupling effects should be evaluated with D1 and included into the fit of the force.

2.2.3. Landau parameters

The Landau theory applied to infinite nuclear matter provides an extremely useful insight into the behaviour of phenomenological forces when employed in the description of collective excitations beyond the mean field approximation. Since such forces are determined from nuclear properties at equilibrium, there is no *a priori* guarantee that their particle-hole (p-h) matrix elements are suitable for computing collective spectra and correlations. A direct way of checking this is to compute nuclear collective excitations and to compare the result with experimental data. This will be discussed in Section 3. A numerically simpler test is to calculate the values of the Landau parameters of the N-N force and to check that they satisfy the stability conditions and the forward scattering amplitude sum rule that must be obeyed in infinite nuclear matter. The discussion that follows is based on the Ref.3. We refer the reader to this paper for details.

The Landau parameters express the p-h residual interaction of the N-N interaction in infinite nuclear matter for a vanishing absolute value of the momentum transfer $\vec{q} = \vec{k_p} - \vec{k_h}$. This part of the residual interaction governs the stability of the ground state and the propagation of small amplitude longitudinal collective excitations in the RPA approximation. It is important to note that the matrix elements of the p-h residual interaction include, besides the usual ones $\langle ph'|v_{12}|hp'\rangle$ and $\langle pp'|v_{12}|hh'\rangle$, also terms coming from the first and second derivatives of v_{12} with respect to the density (see (20) in Section 3.2.1). As will be seen, these contributions are essential in order to satisfy the forward scattering amplitude sum rule.

The Landau parameters can be written F_l^{ST}, where S and T are the spin and isospin of the p-h pairs, respectively, and l is the multipole associated with the dependence of the matrix elements on the angle between $\vec{k_p}$ and $\vec{k_h}$. One can show that F_1^{00}, F_0^{00} and F_0^{01} are related to the effective mass m^*, the compression modulus K_∞ and the asymmetry energy a_τ, respectively. The other F_l^{ST}'s describe properties of the nuclear medium not included in the determination of the effective interaction.

The first set of relations the F_l^{ST}'s must satisfy is deduced from the stability conditions of nuclear matter with respect to small fluctuations. It reads [4]:

$$1 + \frac{F_l^{ST}}{2l+1} > 0, \qquad any\ l,\ S,\ T \tag{6}$$

One readily checks that these relations are met by D1. Let us mention that this is not the case for some Skyrme forces which appear unstable in the spin channel [5].

The second relation is a consequence of the Pauli principle. It states that the scattering of a p-h pair must have zero amplitude when $\vec{k_p}$ and $\vec{k_h}$ are parallel. This leads to the forward amplitude scattering sum rule [4]:

$$\sum_{STl} \frac{F_l^{ST}}{1 + F_l^{ST}/(2l+1)} = 0 \tag{7}$$

The value obtained for this sum rule is -0.060 for D1. One may also compute this sum rule for other phenomenological effective forces. The values obtained for the Skyrme forces SIII and SkA are 1.39 and -0.284, respectively. It is tempting to relate these numbers with the density dependence of the three kinds of interaction: SkA and D1 whose density dependence is $\rho^{1/3}$ give a better result than SIII which depends linearly on the density. It may also be concluded that a finite range interaction leads to more realistic p-h matrix elements than a zero-range one.

Finally it is noted that the dependence of the force on the density is essential in order to meet the sum rule (7). In fact, recomputing the sum rule without including in the F_l^{ST}'s the rearrangement terms coming from the derivatives of D1 with respect to the density, gives 2.866.

3. Nuclear Structure with D1

Our study of nuclear ground states, excitations and processes in finite nuclei always begins with a self-consistent calculation of the nuclear mean-field. In doubly magic nuclei, where pairing correlations are absent, the Hartree-Fock (HF) method is employed in spherical symmetry to obtain a first approximation of the nucleus ground state (GS). In a second step, the small amplitude oscillations of the mean field around the HF minimum are analyzed using the RPA method. This gives the quantized energies of these oscillations, in particular the position of giant resonances, and the GS correlations coming from these fluctuations of the mean field. In all other nuclei, the Hartree-Fock-Bogolyubov (HFB) method is employed, and spherical, axial or triaxial symmetry may be imposed. Generally, constraints are added in order to explore the deformation properties of the mean field. The collective modes in these nuclei are found by solving a collective hamiltonian whose form is derived from a quasi-local approximation of the Generator Coordinate (GC) equations.

For the HF and GC formalism, we refer the reader to the contribution of P.-H.Heenen at this Workshop. We only recall here the form of the HFB equations, our method of solution and how we treat collective excitations in nuclei.

3.1. The Mean Field Approach

In the HFB theory the nuclear GS is approximated by a generalized Slater determinant $|\Psi\rangle$ which is the vacuum of quasi-particle (q-p) states

$$\eta_\mu = \sum_\alpha (U_{\mu\alpha} a_\alpha + V_{\mu\alpha} a_\alpha^+), \qquad \eta_\mu |\Psi\rangle = 0, \tag{8}$$

that can be interpreted as the individual excitations of a superfluid system. The a_α^+'s are any one-particle representation. By minimizing the total energy of the nucleus, the U and V are found as the solutions of the HFB equations [6]

$$\sum_\beta \begin{pmatrix} e & \Delta \\ -\Delta^* & -e^* \end{pmatrix}_{\alpha\beta} \begin{pmatrix} U \\ V \end{pmatrix}_{\mu\beta} = \begin{pmatrix} E_\mu & 0 \\ 0 & -E_\mu \end{pmatrix} \begin{pmatrix} U \\ V \end{pmatrix}_{\mu\alpha}. \tag{9}$$

Here e is the matrix of an HF-like hamiltonian:

$$e = t - \lambda_N - \lambda_Z + \Gamma^{D+E} + \Gamma^R \tag{10}$$

where t is the kinetic energy, Γ^{D+E} includes the direct and exchange component of the mean field:

$$\Gamma^{D+E}_{\alpha\beta} = \sum_{\gamma\delta} (\langle\alpha\gamma|v|\beta\delta\rangle - \langle\alpha\gamma|v|\delta\beta\rangle)\,\rho_{\delta\gamma}, \tag{11}$$

Γ^R is the rearrangement component of the mean field:

$$\Gamma^R_{\alpha\beta} = \frac{1}{2}\sum_{\gamma\delta\epsilon\eta} \langle\gamma\delta|\frac{\partial v}{\partial\rho_{\beta\alpha}}|\epsilon\eta\rangle\ (\rho_{\epsilon\gamma}\,\rho_{\eta\delta} - \rho_{\eta\gamma}\,\rho_{\epsilon\delta}), \tag{12}$$

and Δ is the matrix of the pairing field:

$$\Delta_{\alpha\beta} = \sum_{\gamma\delta} \langle\alpha\beta|v|\gamma\delta\rangle\ \kappa_{\delta\gamma}. \tag{13}$$

In (11)-(13) v is the sum of the effective nuclear interaction and of the Coulomb interaction between protons. The rearrangement field Γ^R arises because the effective interaction depends on the density. The density matrix ρ and the pairing tensor κ have their usual definition, and can be expressed as functions of the U's and V's:

$$\rho_{\delta\gamma} \equiv \langle\Psi|a_\gamma^+ a_\delta|\Psi\rangle = \sum_\mu V_{\mu\delta} V^*_{\mu\gamma} \qquad \kappa_{\delta\gamma} \equiv \langle\Psi|a_\delta a_\gamma|\Psi\rangle = \sum_\mu U^*_{\mu\delta} V_{\mu\gamma}. \tag{14}$$

The chemical parameters λ_N and λ_Z are chosen such that the numbers of protons and neutrons are those of the nucleus under study:

$$\langle\Psi|\hat{Z}|\Psi\rangle = Z, \qquad \langle\Psi|\hat{N}|\Psi\rangle = N. \tag{15}$$

Other types of constraints can be included, for instance constraints on specific types of nuclear deformation, in order to explore off-equilibrium collective properties of nuclei. This is done by adding to e, eq.(10), terms of the type: $-\ \lambda_j Q^{(j)}_{\alpha\beta}$, where the

$\hat{Q}^{(j)}$ are one-body operators, for example multipole operators (see below). In the same way as for particle numbers, the Lagrange parameters λ_j are determined from constraint conditions of the type

$$\langle\Psi|\hat{Q}^{(j)}|\Psi\rangle = q^{(j)}. \tag{16}$$

where $q^{(j)}$ is the value imposed on the collective parameter associated with $\hat{Q}^{(j)}$.

For solving the set of nonlinear equations (9), we expand the q-p states η_μ on finite sets of harmonic oscillator (HO) states a_α whose symmetry (spherical, axial, triaxial) and size is adapted to the nuclear problem under study. For very large deformation or for systems composed of two nuclei, as in fission or nucleus-nucleus collisions, a great deal of computational work is saved by expanding the q-p states on two sets of displaced HO states. In all cases the equations (9) are solved by iterative diagonalization.

Working with HO bases is much more efficient than solving the HFB equations in configuration space when a finite range interaction is employed. In fact, one can use in this case a finite separable expansion of the matrix elements of v. This expansion reads [7]:

$$\langle\alpha\gamma|v|\beta\delta\rangle \;=\; \sum_{\nu=|\alpha-\beta|}^{\alpha+\beta} T_{\alpha\beta\nu}\;\langle\gamma|f_\nu[v]|\delta\rangle \tag{17}$$

In this expression $T_{\alpha\beta\nu}$ is a universal coefficient which does not depend on v, and $f_\nu[v]$ is a functional of v having the structure of a one-body operator. The important point is that the sum in (17) is finite since $T_{\alpha\beta\nu}$ vanishes unless the HO indices α, β and ν satisfy a triangular inequality. (17) is therefore an exact expansion. Use of this relation greatly facilitates the storage of the matrix elements of v, and considerably speeds up the computation of the exchange and pairing fields. With this technique bases containing up to 15 shells in triaxial symmetry, 20 shells in axial symmetry and 24 shells in spherical symmetry can currently be used.

Let us mention that working with matrix techniques in a large size one-particle representation allows us to handle the constraints directly in their linear form. The method we use consists in taking advantage of the large size of the relevant one-quasi-particle space to invert the constraint conditions and solve them for the λ_N, λ_Z and λ_j, the values of N, Z and $q^{(j)}$ being imposed beforehand. In practice, the inversion is performed using a very fast perturbation technique. Namely, the Lagrange parameters are readjusted at each iteration up to first order in the deviation of the constraint values. As the iterative process converges, the mean values of the constraining operators get closer and closer to the imposed values and eventually reach them exactly. This method is far less delicate to manipulate than the technique of quadratic constraints that must be employed when the self-consistent equations are solved in configuration space (see P.-H.Heenen's contribution at this Workshop). We commonly use it to treat 5 simultaneous coupled constraints and can impose any values of the deformation parameters.

Besides the particle numbers, the constraints we usually include in HFB calculations are of the following type:

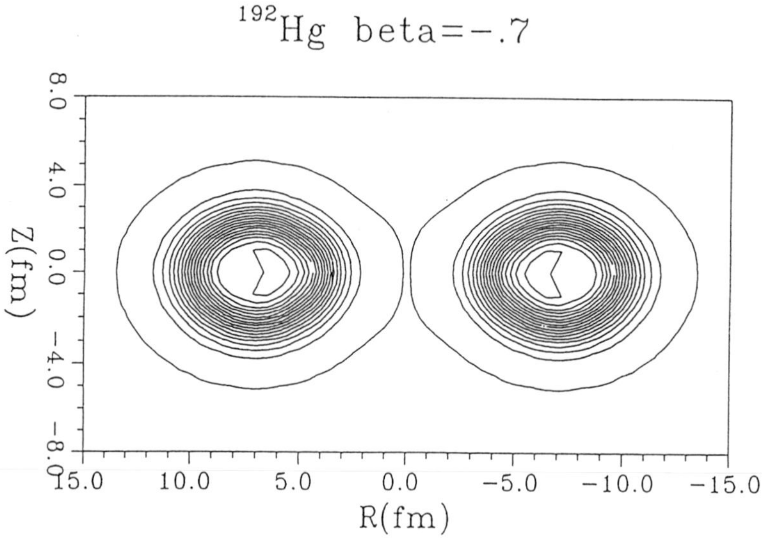

Figure 4: Nucleon spatial distribution of a ^{192}Hg nucleus contrained to an oblate deformation $\beta = -.7$ in axial symmetry. The nucleus is axial about the $R = 0$ axis.

i) axial quadrupole deformation with the constraining operator $\hat{Q}_{20} = r^2\ Y_2^0$;

ii) β - γ quadrupole deformations with the two operators $\hat{Q}_{20} = r^2\ Y_2^0$ and $\hat{Q}_{22} = r^2\ (Y_2^2 + Y_2^{-2})$;

iii) axial quadrupole-octupole deformations with the two operators $\hat{Q}_{20} = r^2\ Y_2^0$ and $\hat{Q}_{30} = r^3\ Y_3^0$;

iv) axial quadrupole-hexadecapole deformations with the two operators $\hat{Q}_{20} = r^2\ Y_2^0$ and $\hat{Q}_{40} = r^4\ Y_4^0$. The hexadecapole operator is used to control the necking-in of the shape of fissioning nuclei. A similar role can be played by an operator of the form e^{-z^2/a^2}, where a is of the order of 2 or 3 fm. The latter has been found particularly successful for studying the fusion of two ^{238}U nuclei [8];

v) nuclear volume in any symmetry, with the operator r^2.

Let us emphasize that including constraints in self-consistent calculations of the HF type is not equivalent to imposing a precise shape to the nucleus. Each constraint fixes only one relationship between the several hundreds of variational parameters that determine the nucleon orbitals. As an example Fig.4 gives the nucleon spatial distribution of a ^{192}Hg nucleus constrained to a large oblate deformation $\beta = -.7$ in axial symmetry. One observes that the nucleus does not adopt the shape of an oblate spheroid as would be expected. Instead, a toroidal shape is obtained. This is so because, among all the possible configurations having an axial quadrupole deformation $\beta = -.7$, the HFB procedure picks up the one having the maximum binding energy. One may say that such a toroidal shape represents the better compromise between volume, surface and Coulomb energy contributions for this nucleus at this deformation.

Let us end this Section on the mean field description of nuclei by one remark concerning the definition of the pairing energy. In the self-consistent HFB approach the pairing energy is defined as the contribution to the total binding energy of all correlated pairs, that is:

$$E_{pair}^{HFB} = \frac{1}{2}\sum_{\alpha\beta}\Delta_{\alpha\beta}\ \kappa_{\beta\alpha} = \frac{1}{2}\sum_{\alpha\beta\gamma\delta}\ \kappa_{\beta\alpha}^{*}\ \langle\alpha\beta|v|\gamma\delta\rangle\ \kappa_{\delta\gamma}. \qquad (18)$$

This pairing energy is not equal however to the gain in total energy E_{pair}^{gain} one obtains by switching on pairing correlations. The latter can be defined as the difference between the total energy obtained in a HF calculation and the total energy obtained in a full HFB one. Because of the self-consistent rearrangement of the average field due to the diffuseness of the Fermi surface, E_{pair}^{gain} is always smaller than E_{pair}^{HFB}. For instance in Sn isotopes $E_{pair}^{HFB} \simeq 16\ MeV$ while $E_{pair}^{gain} \simeq 3\text{–}4\ MeV$ [1]. This discussion shows that the pairing energy obtained in non self-consistent models as phenomenological shell models should not be compared with E_{pair}^{HFB}, but with E_{pair}^{gain}. Let us also note that pairing cannot be considered as a small effect in superfluid nuclei since its magnitude is actually measured by E_{pair}^{HFB}.

3.2. Beyond the Mean Field

As mentioned in the Introduction, mean-field solutions often are not good approximations of the nuclear GS. In doubly-magic nuclei, small amplitude oscillations of the average field induce correlations that may significantly affect GS observables. Open-shell nuclei often are soft against multipole deformations. The residual interaction between q-p can therefore mix states having strongly different deformations, and the GS cannot be approximated by a single generalized Slater determinant.

In this Section, we present the methods we use to deal with these two kinds of situations. Small amplitude collective vibrations are described with the RPA method. Large amplitude collective motion in soft nuclei are treated with an approximate GC configuration mixing technique that reduces to the diagonalization of a collective hamiltonian. In both cases, calculations yield, in addition to a correlated GS, a whole spectrum of collective oscillations. In the RPA case, some of the most collective of these excitations can be identified as the giant resonances observed experimentally. The comparison between observed and calculated transition energies and densities provide a test of the p-h matrix elements of the effective force in finite nuclei that supplements the one given by the Landau parameters. For instance, the energy of monopole resonances is sensitive to a compression modulus K_A that differs from the one defined in infinite nuclear matter [9]. Ground state correlations can also be computed and their effect on GS observables as, for example, on the charge density, can be compared with experimental data. The spectra obtained in soft nuclei are more difficult to relate to specific properties of the force because of the complexity of large amplitude oscillations and the approximations needed to describe them. One may say however that spectra and transition densities depend in an essential manner on the intensity of pairing correlations and on their variation with deformation, *via* the magnitude of the inertia parameters (moment of inertia, collective inertia tensor). Other properties such as the surface energy obviously can also play an important role. This remark emphasizes the importance of treating pairing correlations and deformations properties as accurately as possible.

After briefly recalling the RPA equations, we present here only a sample of the results that have been derived with D1. More detailed and extensive results can be found in the references [9–10]. The second part of this section is devoted to the theory of large amplitude collective motion we have employed.

3.2.1. RPA

In rigid nuclei, one looks at correlations generated by small variations of the HF density matrix. Up to second order the variation of the energy is

$$E[\rho_0 + \delta\rho] = E[\rho_0] + \frac{1}{2}\sum_{php'h'} (\delta\rho^*_{ph} \;\; \delta\rho_{ph}) \begin{pmatrix} A & B \\ B^* & A^* \end{pmatrix}_{ph,p'h'} \begin{pmatrix} \delta\rho_{p'h'} \\ \delta\rho^*_{p'h'} \end{pmatrix} \tag{19}$$

ρ_0 is the density matrix associated with the HF solution:

$$\frac{\partial E}{\partial \rho}[\rho_0] = 0$$

and $\delta\rho$ is a small variation. Therefore the matrix elements of $\delta\rho$ in the HF basis connect only particle and hole states.

The second term in (19) being a second derivative of the energy with respect to a density matrix, the A and B include matrix elements of the first and second derivatives of the effective interaction v_{12} with respect to the density. For instance, one has

$$A_{ph,p'h'} = (\epsilon_p - \epsilon_h)\delta_{pp'}\delta_{hh'} + \langle ph'|v_{12}|hp'\rangle + \sum_{\beta\delta}\left(\langle p\beta|\frac{\partial v_{12}}{\partial \rho_{p'h'}}|\widetilde{h\delta}\rangle + \langle h'\beta|\frac{\partial v_{12}}{\partial \rho_{hp}}|\widetilde{p'\delta}\rangle\right)\rho_{\delta\beta}$$
$$+ \frac{1}{2}\sum_{\alpha\beta\gamma\delta}\langle\alpha\beta|\frac{\partial^2 v_{12}}{\partial\rho_{hp}\partial\rho_{p'h'}}|\widetilde{\gamma\delta}\rangle\rho_{\delta\beta}\rho_{\gamma\alpha} \tag{20}$$

As mentioned in Section 2.2.3, these additional terms are essential in order to satisfy the forward scattering sum rule in nuclear matter. They are expected to play a similar important role in finite nuclei.

The normal modes of vibration of the nucleus are obtained after quantization of $\delta\rho_{ph}$. They can be described by quasi-boson operators θ_n^+ with amplitudes $X_{n,ph}$ and $Y_{n,ph}$ on p-h elementary excitations in the HF basis:

$$\theta_n^+ = \sum_{ph} X_{n,ph}\, c_p^+ c_h + Y_{n,ph}\, c_h^+ c_p . \tag{21}$$

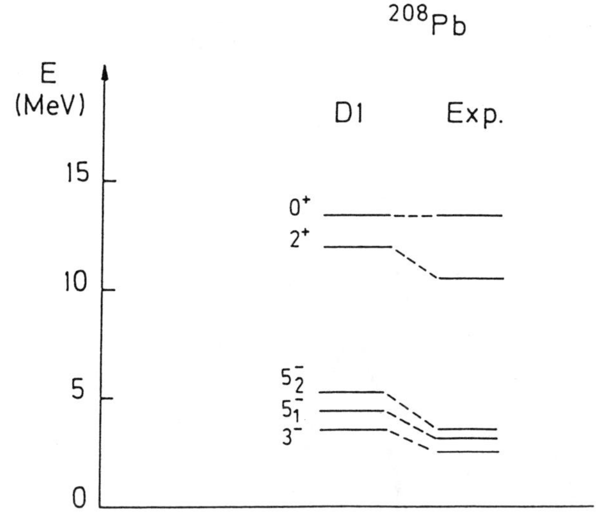

Figure 5: Low-energy collective states obtained with the RPA method in ^{208}Pb compared with experimental data.

The X, Y and the energies of the normal modes are solutions of the eigenvalue equation:

$$\sum_{p'h'} \begin{pmatrix} A & B \\ -B^* & -A^* \end{pmatrix}_{ph,p'h'} \begin{pmatrix} X \\ Y \end{pmatrix}_{n,p'h'} = E_n \begin{pmatrix} X \\ Y \end{pmatrix}_{n,ph} . \tag{22}$$

An example of spectrum of low-lying collective states obtained with D1 in ^{208}Pb is shown in Fig.5. One observes that the energy of the 0^+ breathing mode is very well reproduced. This agreement reflects the excellent properties of D1 regarding compressibility effects. The compression modulus obtained in ^{208}Pb is $K_A = 150\ MeV$ [9]. The other resonances shown in the figure are found at slightly too large energies. However, the transition densities computed for this levels agree quite well with experiment (see Fig.6), which shows that the collective strength distribution of these states is well reproduced by the calculation. Let us stress that these results have been derived from the D1 interaction in a completely microscopic way, without any phenomenological adjustment.

Let us now turn to GS correlations. In the RPA approximation, the nuclear GS is assumed to be the vacuum $|\tilde{0}\rangle$ of the operators (21) considered as exact bosons. A measure of the intensity of the GS correlations can be obtained by computing the occupation probabilities of the HF one-particle states in this vacuum. The expression $\langle\tilde{0}|c_\alpha^+ c_\alpha|\tilde{0}\rangle$ that seems the most natural for these quantities must be

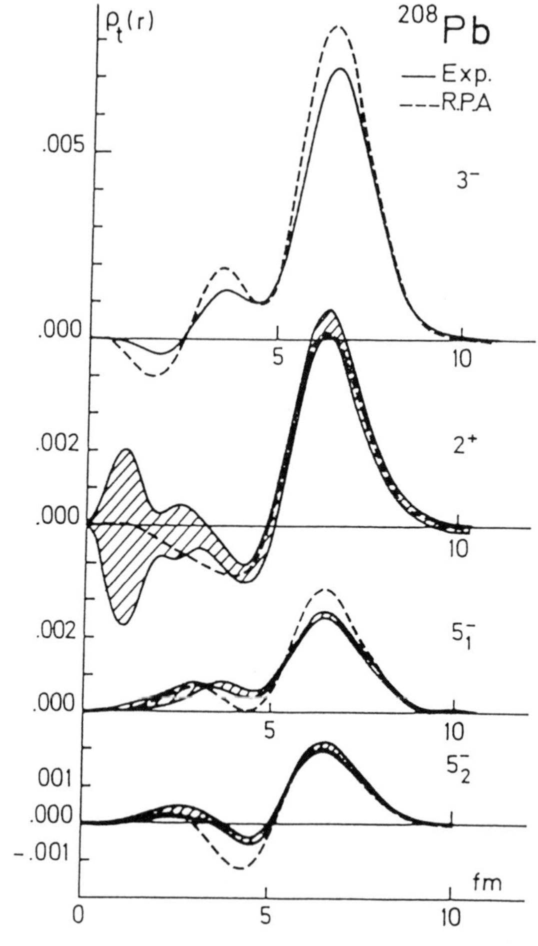

Figure 6: The transition densities obtained for four of the five states shown in Fig.5 in ^{208}Pb compared with experimental data.

taken with care since the quasi-boson approximation introduces a double counting in the evaluation of matrix elements of this kind. A corrective contribution must be applied in order to obtain the correct numbers [10]. Table I gives an example of the change in the occupation probabilities of the HF ^{208}Pb levels due to RPA correlations. One observes that 3 neutron levels are strongly depopulated. For protons, the $3s1/2$ is depopulated by more than 15%. The resulting effect on the charge density of ^{208}Pb is displayed in Fig.7. One sees that RPA correlations correct strongly the HF density profile.

Protons		Neutrons	
3p1/2	.035	4s1/2	.035
3p3/2	.037	2g7/2	.060
2f5/2	.053	3d5/2	.050
1i13/2	.038	1j15/2	.048
2f7/2	.058	1i11/2	.069
1h9/2	.071	2g9/2	.078
3s1/2	-.157	3p1/2	-.171
1h11/2	-.096	2f5/2	-.139
2d3/2	-.139	3p3/2	-.165
2d5/2	-.109	1i13/2	-.083
1g7/2	-.071	2f7/2	-.112
1g9/2	-.054	1h9/2	-.081

Table I

Change in the one-particle HF level occupation probabilities due to RPA GS correlations in ^{208}Pb.

Let us finally mention that GS correlations are essential to reproduce the sign of the difference between the charge radii of ^{48}Ca and ^{40}Ca. Experimentally, this difference is negative ($\Delta r = -.01\ fm$). HF calculations with D1 give $\Delta r = +.014\ fm$. Including RPA GS correlations yields $\Delta r = -.004\ fm$.

3.2.2. Large amplitude collective excitations

In soft nuclei, the HFB solution of smaller energy is not a good approximation to the GS since the residual interaction between q-p can couple states having very different deformations. A more appropriate ansatz for GS wave functions is to take in this case a superposition of these quasi-degenerate deformed states:

$$|\Psi>= \int f(q)\ |\Phi_q>\ dq\ . \tag{23}$$

q labels the set of deformations included in the superposition, the $|\Phi_q>$'s are deformed generalized Slater determinants obtained with the constrained HFB procedure, and $f(q)$ is a weight function to be determined. In the Generator Coordinate (GC) theory, the $f(q)$ are obtained from the well-known Griffin-Hill-Wheeler (GHW) equations [11] (see P.-H.Heenen contribution at this Workshop for details). Solving these equations yields in fact not only a GS wave function of the type (23) and a GS energy, but also a set of collective excited states whose structure may strongly differ from the one of the GS. For instance, one may obtain shape isomers or states containing both low and high deformation components (shape coexistence).

In general, the set of collective parameters q must include more than one type of deformation. For instance, nuclei soft against axial quadrupole deformation q_{20} (β deformation) are often soft with respect to non-axial deformation q_{22} (γ deformation). In addition, rotations, or equivalently the 3 other quadrupole degrees of

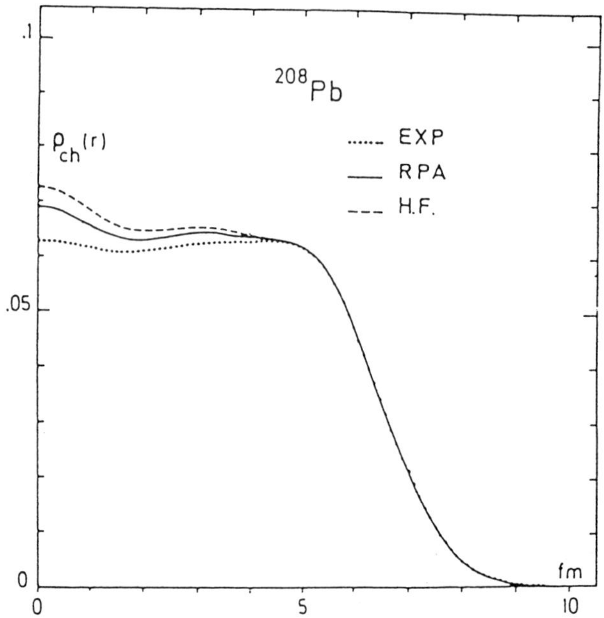

Figure 7: The charge density profiles obtained in ^{208}Pb with the Hartree-Fock method (dashed line) and after GS correlations have been included (solid line). The dotted line is the experimental charge density.

freedom q_{21}, q_{2-1} q_{2-2}, have to be included on the same footing when the expected collective states are not well-deformed. In this case (23) contains a five-fold integral that can be rewritten

$$|\Psi^{IM}> = \sum_{K=0}^{I} \int f_K^I(q) \ \ <\Omega\ |IMK> \ \ R(\Omega)\ |\Phi_q> \ \ dq\ d\Omega. \tag{24}$$

Here Ω is a set of 3 Euler angles, $<\Omega\ |IMK>$ is a rotor wave function (superposition of two D^I rotation matrices) and $q = (q_{20}, q_{22})$. The GHW equations are obviously extremely difficult to solve numerically in this case, especially with an interaction of finite range. For this reason, we have employed up to now a well-known technique (gaussian overlap approximation and expansion of the hamiltonian kernel up to second order in the non-locality) that permits one to evolve these equations into a partial differential equation [12]. The approximations involved in this step appear reasonable in most cases. However, they imply some form of adiabatic hypothesis. Hence, only low-lying collective states are expected to be realistically described with this formalism. With these approximations the GHW equation can be replaced by the eigen-equation:

$$\hat{\mathcal{H}}\,\bar{f}(q) = \;E\;\bar{f}(q).$$

where $\bar{f}(q)$ is a collective wave function related to $f(q)$ by means of a gaussian overlap kernel, and normalized to unity. $\hat{\mathcal{H}}$ is a collective Hamiltonian which appears directly in quantized form:

$$\hat{\mathcal{H}} = \sum_{\kappa} \frac{\hat{J}_{\kappa}^{2}}{2\mathcal{I}_{\kappa}(q)} - \frac{\hbar^{2}}{2} \sum_{j,k=1}^{2} \frac{\partial}{\partial q_{j}} (\mathcal{M}^{-1}(q))_{jk} \frac{\partial}{\partial q_{k}} + V(q) - \Delta E(q). \tag{25}$$

In this expression q_j and q_k are the β and γ quadrupole variables q_{20} and q_{22}, $\mathcal{I}_{\kappa}(q)$ are the 3 moments of inertia of the nucleus with respect to the principal axes of the nucleus, $\mathcal{M}^{-1}(q)$ is the inverse of the 2×2 inertia tensor relative to q_{20}-q_{22} motion and $V(q) = \langle \Phi_q | H | \Phi_q \rangle$ is the HFB deformation energy. $\Delta E(q)$ is a quantum correction, often denominated "zero-point energy correction", that emerges naturally from the GC equations. It arises because the deformed HFB states are not eigenstates of the constraining operators but fluctuate around the imposed mean values. Corrections for the translational and rotational degrees of freedom are also included in $\Delta E(q)$. They are computed as well as $\mathcal{I}_{\kappa}(q)$ and $\mathcal{M}(q)$ from the HFB self-consistent states. Expressions for these inertia parameters are obtained by expressing the response of the nucleus to small, given collective angular or deformation velocities. The resulting form, the so-called ATDHF inertia [13], has been shown to be more appropriate than the approximate form derived from the quasi-local expansion of the GC equations [14]. It involves moments of the generalized RPA matrix of the microscopic many-body Hamiltonian. In practical calculations, due to the huge size of this matrix, the moments are computed by neglecting the matrix elements of the residual interaction contained in the generalized RPA matrices (the so-called "cranking" approximation). While this approximation generally appears to be good for the vibrational masses $\mathcal{M}(q)$ (by comparing vibrational spectra with experiment), it seems to underestimate the values of the moments of inertia $\mathcal{I}_{\kappa}$. This discrepancy is roughly 25% in medium and heavy nuclei.

In specific cases, rotational degrees of freedom are expected to decouple from the vibrational ones. One example is the study of quadrupole-octupole vibrations around the isomeric minimum of actinides or light Hg isotopes (see L.M.Robledo contribution at this Workshop). In such cases, a simplified form of (24) can be adopted, based on the rotational model:

$$|\Psi^{IM}> = \;|IMK=0> \int f(q)\;|\Phi_q>\;dq, \tag{26}$$

where $q = (q_{20}, q_{30})$. Computations of collective spectra and wave functions then are much faster, since the hamiltonian is only a 2-D one.

In the 5-D case the collective Hamiltonian (25) is diagonalized using the numerical code built by K.Kumar [15] or the one developed by J.Libert [16]. The former one is based on an extremely efficient numerical technique working with triangular 2-D meshes. Libert's code expands collective wave functions on bases of

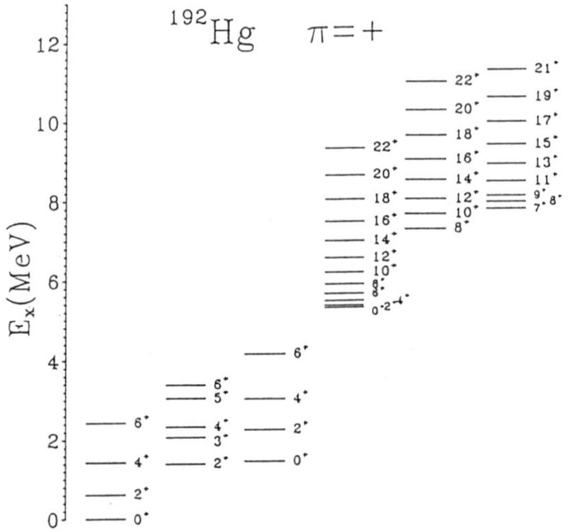

Figure 8: Collective states obtained in ^{192}Hg with the five-dimension quadrupole formalism described in the text. The three groups of states on the left are "normal" β and γ bands built on states having low deformation. The three bands on the right are super-deformed. They belong to an isomeric well in the potential energy surface at $\beta \simeq .55$.

appropriate functions that can be chosen of two-center type. This feature allows one to extend the calculation over a very broad range of deformation values. In particular, deformed and super-deformed states can be simultaneously obtained with a very good accuracy.

An example of this is given in Fig.8 in the case of ^{192}Hg. The six groups of states have been obtained from a single diagonalization of the collective hamiltonian. The three ones on the left have "usual", i.e. small β-deformation. The three ones on the right are three superdeformed (SD) bands. A similar result is obtained in ^{194}Hg. Let us note that three such SD bands have been experimentally identified [17] in this latter nucleus. It is interesting to remark that the calculation performed with D1 reproduces not only the number of observed bands, but also their precise characteristics [18]. It is not less interesting to remark also that this calculation predicts SD bands of collective nature, while usual interpretations describe them as rotational bands built upon simple quasi-particle excitations [19].

Finally, let us mention that transition probabilities between collective states of type (24) can in principle be computed in a straightforward manner. In reality, this is not an easy task since multiple integrals are involved. Generally this part of the calculation is much larger than the diagonalization of the collective hamiltonian itself. As a reward, the $B(E\lambda)$'s obtained with the present approach generally agree extremely well with the experimental data. This is true for transitions between excited states and the GS as well as for intra-band transitions or transitions between bands (see for instance Ref.18). In the same manner as in the RPA studies, this kind of agreement shows that the excited states derived with the present formalism have the correct collective structure. Let us emphasize that all the ingredients of the collective calculations are obtained from the HFB one quasi-particle states. There is therefore no other phenomenological parameters in this approach than those of the D1 interaction described at the beginning.

4. Main Applications

Since a detailed presentation of results is not the purpose of the present Workshop, we only list here the main topics the approach described above has been applied to. They are:

- The analysis of GS properties and of collective spectra and transitions in stable nuclei throughout the nuclear chart. As mentioned earlier, the RPA method or the large amplitude formalism described in Section 3.2.2 is employed.
- The study of actinide fission [20]. Two-dimensional quadrupole-octupole fission barriers in the $Z = 88-94$ region have been computed using the constrained HFB method. The satisfactory agreement obtained with the empirical data shows that D1 has the required properties for describing large deformations. Another study in this domain has been the analysis of the mechanism of scission by means of a time-dependent approach generalizing the large amplitude collective method described above.

In recent years, applications have concentrated essentially on nuclei or systems presenting exotic structure or features:

- Shape isomers and superdeformed bands (SD) in the Hg and rare earth regions (see Ref.18 and references therein). In addition to SD spectra and transitions, evaluations of isomer lifetimes are beginning to be made. As a first test of the method, the transition rate of the γ-branch of the ^{238}U isomer decay has been computed and compared with experiment [21].
- Exotic light nuclei such as the Ne, Mg, Si and S isotopes having $N = 20 \pm 2$ neutrons. First results of this study indicate, contrary to other works, that the $N = 20$ shell closure is preserved even at such large proton-neutron asymmetries.
- The ^{238}U-^{238}U quasi-molecule in connection with the anomalous e^+-e^- pairs observed in sub-threshold actinide-actinide collisions [8]. This study indicates that the energy source which generates these pairs may have a pure nuclear origin.

5. Conclusion

The D1 interaction proposed some 20 years ago seems to possess the essential characteristics necessary for extensive applications of self-consistent approaches to mean field structure problems and to extensions beyond the mean field. First, it is simple enough to render feasible self-consistent calculations where most spatial symmetries of the nucleon distributions are broken. At the same time, the results obtained for bulk nuclear properties, pairing effects, collective excitations and correlations are generally very satisfactory. In such an agreement, the finite range of the force plays a very important role. More precisely, while a finite range in the effective force does not seem to be mandatory for describing bulk properties at the simplest mean field level, it is obviously required with regard to pairing. Also, as evident in the discussion of Landau parameters,long range correlations are much better described when a finite range is used.

These characteristics are important in view of the many applications of microscopic approaches that can be envisaged with the present and future generation of computers. Among these applications, the most interesting ones concern nuclear systems having an unusual structure, such as super-deformed systems, shape and high spin isomers, nuclei close to the drip line and super-heavy nuclei. In these cases the experimental data may be very scarce or even nonexistent. It follows that approaches capable of describing in a realistic manner all important nuclear structure effects without any phenomenological input are required. In this perspective, self-consistent approaches employing effective forces such as D1 will certainly continue to be most useful and necessary.

We would like to thank C.R.Chinn for many helpful comments during the preparation of this text.

References

1. J.Dechargé and D.Gogny, *Phys. Rev. C* **21** (1980) 1568
2. T.T.S.Kuo, E.U.Baranger, and M.Baranger, *Nucl. Phys.* **79** (1966) 73
3. D.Gogny and R.Pagden, *Nucl. Phys. A* **293** (1977) 365
4. A.B.Migdal, *Theory of Finite Fermi Systems and Application to Atomic Nuclei*, (Interscience, NY, 1967)
5. S.O.Bäckman, A.D.Jackson, and J.Speth, *Phys. Lett.* **56 B** (1975) 209
6. J.G.Valatin, *Lectures in Theoretical Physics*, (University of Colorado Press, Boulder, 1961), Vol IV
7. D.Gogny, *Nucl. Phys. A* **237** (1975) 399
8. J.F.Berger, J.D.Anderson, P.Bonche, and M.S.Weiss, *Phys. Rev. C* **41** (1990) 2483
9. J.P.Blaizot, D.Gogny, and B.Grammaticos, *Nucl. Phys. A* **265** (1976) 315
10. J.Dechargé, M.Girod, D.Gogny, and B.Grammaticos, *Nucl. Phys. A* **358** (1981) 203; J.Dechargé, L.Sips and D.Gogny, *Phys. Lett. B* **98** (1981) 229

11. D.L.Hill and J.A.Wheeler, *Phys. Rev.* **89** (1953) 1102; J.J.Griffin and J.A.Wheeler, *Phys. Rev.* **108** (1957) 311

12. P.Ring and P.Schuck, *The Nuclear Many-Body Problem*, (Springer-Verlag, New York, 1980)

13. M.Baranger and M.Veneroni, *Ann. of Phys.* **114** (1978) 123

14. F.Villars, in *Nuclear Self-Consistent Fields*, G.Ripka and M.Porneuf editors, (North-Holland, 1975)

15. K.Kumar and M.Baranger, *Nucl. Phys. A* **92** (1967) 608

16. J.Libert and P.Quentin, *Z. Phys. A* **306** (1982) 315

17. M.A.Riley et al., *Nucl. Phys. A* **512** (1990) 178; E.A.Henry et al., *Z. Phys. A* **335** (1990) 361 ; F.S.Stephens et al., *Phys. Rev. Lett.* **64** (1990) 2623

18. M.Girod, J.P.Delaroche, J.Libert and I.Deloncle, *Phys. Rev. C* **45** (1992) 1420

19. C.F.Tsang and S.G.Nilsson, *Nucl.Phys. A* **140** (1970) 275; U.Götz et al., *Nucl. Phys. A* **192** (1972) 1; J.Dudek and W.Nazarewicz, *Phys. Rev. C* **31** (1985) 298 and references therein; J.Dudek et al., *Phys. Rev. Lett.* **59** (1987) 405; R.R.Chasman, *Phys. Lett. B* **219** (1989) 227; W.Nazarewicz, R.Wyss and A.Johnson, *Nucl. Phys. A* **503** (1989) 285; T.Bengtsson, *Nucl. Phys. A* **496**, (1989) 56; R.R.Chasman, *Phys.Lett. B* **242** (1990) 317

20. J.F.Berger, M.Girod, and D. Gogny, *Nucl. Phys. A* **428** (1984) 23c

21. C.R.Chinn, J.F.Berger, D.Gogny, and M.S.Weiss, *Phys. Rev. C* **45** (1992) 1700

MICROSCOPIC DESCRIPTION OF THE OCTUPOLE DEGREE OF FREEDOM WITH THE GOGNY FORCE.

L.M. Robledo
Departamento de Física Teórica C-XI, Universidad Autónoma de Madrid
E-28049 Madrid, Spain.

ABSTRACT

In this paper I describe how the standard nuclear physics microscopic techniques (Hartree- Fock- Bogoliubov mean fields, collective hamiltonians and projection methods) can be applied to the description of the octupole degree of freedom. These calculations have been carried out with the finite range, density dependent and parameter free Gogny force. I will mainly focus on the technical details of the calculations but a few representative examples from actual calculations will be given. Preliminary results showing the relevance of angular momentum projection in the calculation of transition probabilities will be discussed.

1. Introduction

In his talk[1], P.H. Heenen gave an overview of the many body techniques currently used in nuclear structure calculations of collective states. First, he discussed the mean field approach in its most general form (including pairing correlations and the rearrangement effect coming from the density dependence of the interaction). Then he went on to show how to introduce at the mean–field level the concept of collective variables (constrained mean fields) and how to introduce further correlations by mixing the different mean field wave functions (Generator Coordinate Method, GCM). He illustrated his presentation with specific examples obtained by using the Skyrme interaction. In the next talk[2] J.F. Berger gave a detailed account of the advantages and drawbacks of using the Gogny force in nuclear structure calculations. He then showed how to introduce additional correlations into the constrained mean field states by means of the so–called Collective Hamiltonian Method. This method can be justified as an approximate form of the GCM valid when the Gaussian Overlapp Approximation (GOA) is applicable and when the kernels are smooth functions of the collective variables. In the collective hamiltonian method (as in the GCM) one has first to introduce the relevant collective coordinates (usually shape parameters, like the quadrupole or octupole deformation) at the mean field level and then, by using the aforementioned approximations, one ends up with a Schrödinger equation for the collective variables. The collective parameters entering this equation (collective mass, zero point energy corrections, etc) can be derived from the quantities entering the GCM equation. It is also possible to extract them from the ATDHF theory but the two sets of parameters qualitatively differ. The preferred set of parameters is the so–called ATDHF+ZPE set which has been used for a long time by the Bruyeres–Le–Chatel group[3]. The reasons to use this set can be found in Berger's contribution[2].

In my talk I had the opportunity to demonstrate how the mean field theory and the collective hamiltonian method can be applied to the description of low–lying collective negative parity states by using the octupole degree of freedom. When the octupole degree of freedom is involved, the parity symmetry has to be broken. The parity symmetry can be restored in the collective hamiltonian method. However, it is very difficult to extract the contribution to the correlations of the parity projection and this is an important issue in the understanding of the physics of the octupole degree of freedom. We have carried out exact parity projected calculations after variation (the mean field is determined first and then projected) and the results show no qualitative agreement with the ones obtained with the collective hamiltonian method. The discrepancies can be understood if it is taken into account that the collective hamiltonian method includes –in an approximate way– many more correlations than the pure projection. The lesson we learn is that the correlations included with the parity projection are not enough to describe the physics of the octupole degree of freedom in the general case. Finally, we have questioned the suitability of the rotational formula in the calculation of transition probabilities for not very well quadrupole deformed nuclei. We have computed $B(E3)$ transition probabilities by projecting the intrinsic states onto the proper angular momentum, namely $I = 0$ for the positive parity intrinsic state and $I = 3$ for the negative parity one. Preliminary results for the ^{16}O nucleus have shown that the enhancement of the $B(E3)$ due to the angular momentum projection is of the order of four bringing the results of the projected calculation into an excellent agreement with the experimental result.

2. The mean field approximation

The mean field approximation is the starting point of any microscopic calculation in nuclear physics. It provides us with the ground state wave function and, additionally, the particle–hole excited states give an idea of the non collective low energy spectrum. When collective effects need to be considered it is first necessary to decide which are the collective degrees of freedom relevant to the problem (usually in an intuitive way). Once the collective operators are chosen (let us call them $\hat{Q}_\mu$) we can perform constrained mean field calculations in which the ground state wave function $|\Psi\rangle$ is determined with the additional constraints $\langle\Psi|\hat{Q}_\mu|\Psi\rangle = q_\mu$. In this way the behavior of the relevant mean field quantities –like the energy, dipole moment, etc– as a function of the collective coordinates q_μ can be obtained. By looking at this behavior it is possible to infer the relevance of the collective fluctuations around the mean field quantities. It also provides the starting point for more sophisticated methods like the GCM or the collective Schrödinger equation.

For the description of collective low–lying negative parity states we have chosen the axially symmetric octupole degree of freedom as the only relevant to the problem at hand. The constraining operator is then the multipole operator $\hat{Q}_{30} = \sqrt{(4\pi)/7}r^3 Y_{30}$. The octupole constraint breaks the parity symmetry of the wave functions giving rise to the possibility of admixtures of spurious states –associated to the center of mass motion– with the physical states. To prevent this mixing it is mandatory to constraint the position of the center of mass ($\langle \hat{z} \rangle$ in the axially symmetric case) at the origin of coordinates.

The constrained mean field equations have some particularities due to the use of the density dependent Gogny force, namely the appearance of the so–called rearrangement term in the mean field potential. When pairing correlations are taken into account, the constrained mean field equation (or Constrained Hartree- Fock-Bogoliubov, CHFB) is given by [4]

$$\begin{pmatrix} h & \Delta \\ -\Delta^* & -h^* \end{pmatrix} \begin{pmatrix} U \\ V \end{pmatrix} = E \begin{pmatrix} U \\ V \end{pmatrix}, \tag{1}$$

where the matrices h and Δ, including the rearrangement term, are given by

$$\begin{aligned} h_{ij} &= t_{ij} - \sum_{\mu} \lambda_\mu (Q_\mu)_{ij} + \sum_{kl} \langle ik|V|\tilde{jl}\rangle \rho_{lk} \qquad (2) \\ &+ \frac{1}{2} \sum_{klmn} \langle kl| \frac{dV(\rho)}{d\rho} \varphi_i^* \varphi_j |\tilde{mn}\rangle (\rho_{nl}\rho_{mk} + \frac{1}{2}\kappa_{nm}\kappa_{lk}^*) \\ \Delta_{ij} &= \frac{1}{2} \sum_{kl} \langle ij|V|\tilde{kl}\rangle \kappa_{kl}, \qquad (3) \end{aligned}$$

U and V are the CHFB wave functions and E are the quasiparticle energies. In the constraining operators Q_μ we have included, in addition to Q_3 and z, the particle number operators for protons and neutrons. The Lagrange multipliers λ_μ are determined by the corresponding constraints. The single particle wave functions are expanded in an axially symmetric harmonic oscillator basis. The parameters of the basis (the oscillator lengths along the z and perpendicular directions) are chosen as to minimize the energy of the system. (Usually such minimization is carried out for a set of selected constrained mean fields and for the others the interpolated values are used). The CHFB equation is solved iteratively, namely a convenient starting wave function is first used to compute h and Δ, then the CHFB matrix is diagonalized to obtain a new set of U and V which are used for the next iteration. The iterative process stops when, up to a given accuracy, the wave functions of a given iteration coincide with the ones of the previous iteration. In order to save computational time the CHFB equation is approximated by the so–called CHBF(BCS approximation) whereby in each iteration the h field is first diagonalized in order to obtain the single particle wave functions (HF basis) and then the nondiagonal matrix elements of the transformed Δ are neglected. The remaining equation (a BCS equation) is solved to obtain U and V. It has been tested[5] that the results of this procedure compare very well with the ones of the full HFB. In each iteration the particle number constraint is satisfied by means of the standard BCS procedure and the remaining constraints (in Q_3 and z) are treated perturbatively by using linear response theory.

As an example of the information that can be extracted from the CHFB calculations we represent in Fig 1 the mean field energies for several Ra, Th and Ba isotopes

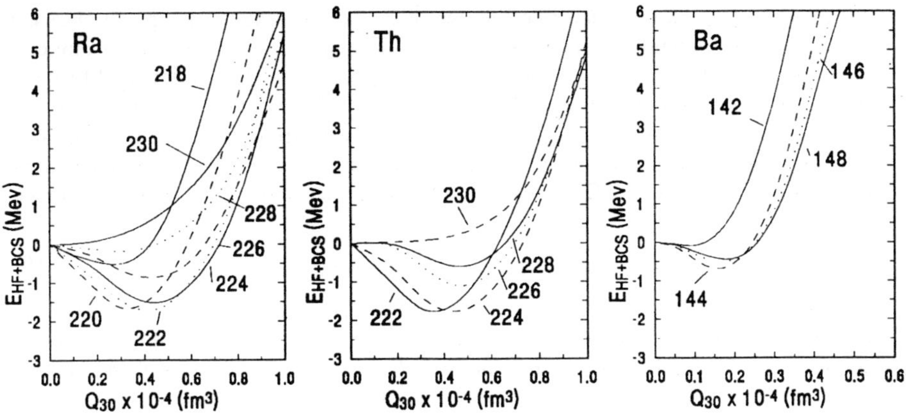

Figure 1: CHFB energy as a function of q_3 for several Ra, Th and Ba isotopes

as a function of the octupole deformation. The mean field solution is octupole deformed in many cases but the depth and width of the minima changes depending on the nucleus. For instance, the ^{228}Th nucleus has a octupole deformed mean field but it is so shallow and broad that it is conceivable that strong quantum mechanical correlations will influence that mean field result. Further results can be found in ref [6, 7].

3. Collective hamiltonian method

The information given by the CHFB is restricted to the energy and shape of the -generally- deformed ground state. To include further correlations we have used the q_3 collective degree of freedom to build up a collective hamiltonian in the spirit of the GCM. Assuming that the overlaps among different CHFB wave functions -corresponding to different q_3 values- are well described by gaussian shapes and that the energy kernels are well described in terms of a local expansion [10]; it is possible to reduce the GCM equation to a Schrödinger equation

$$\hat{\mathcal{H}}_{coll}\phi_\alpha(q_3) = \epsilon_\alpha\phi_\alpha(q_3), \tag{4}$$

where $\hat{\mathcal{H}}_{coll}$ is given by

$$\hat{\mathcal{H}}_{coll} = -\frac{1}{\sqrt{G(q_3)}}\frac{\partial}{\partial q_3}\sqrt{G(q_3)}\frac{1}{2B(q_3)}\frac{\partial}{\partial q_3} + V(q_3) - \epsilon_0(q_3). \tag{5}$$

The collective parameters $G(q_3)$, $B(q_3)$ and $\epsilon_0(q_3)$ are the metric, collective mass and zero point energy (ZPE) correction, respectively. The $V(q_3)$ potential is just the mean field energy. These quantities can be computed microscopically from the explicit expressions of the overlap and energy kernels of the GCM. It is also possible to obtain a collective hamiltonian like the one of Eq. (5) by requantizing the classical Adiabatic Time Dependent Hartree-Fock (ATDHF) equations of motion for one collective variable. However, the values of the collective parameters differ qualitatively from the ones computed in the GCM framework. For the collective parameters we will use the prescription used by the Bruyeres–le–Châtel group [3] which incorporates the advantages of both sets of parameters. This prescription (hereafter referred as ATDHF+ZPE) uses the ATDHF mass, the GCM metric and a ZPE correction given by

$$\epsilon_0(q_3) = \frac{1}{2} G_{GCM}(q_3) B^{-1}_{ATDHF}(q_3). \tag{6}$$

The exact calculation of the collective parameters is an impossible task nowadays because it implies the inversion of the HFB stability matrix. To compute them we have used the "Cranking approximation " [8, 9, 3] by which only the diagonal part of the stability matrix is taken into account. In the cranking approximation the ATDHF+ZPE collective parameters for one collective coordinate are given by

$$B(q_3) = \frac{M_{-3}(q_3)}{M^2_{-1}(q_3)}, \quad G(q_3) = \frac{M_{-2}(q_3)}{2M^2_{-1}(q_3)}, \tag{7}$$

where the quantities $M_{-n}(q_3)$ $(n = 1, 2, 3)$ are defined as

$$M_{-n} = \sum_{k,l} \frac{|Q_3^{20}|^2_{kl}}{(E_k + E_l)^n}; \tag{8}$$

The collective Schrödinger equation is invariant under the $q_3 \to -q_3$ operation which is the collective image of the parity operation. Therefore, we can label the collective wave functions ϕ_α according to their behavior under the $q_3 \to -q_3$ operation. The parity of ϕ_α under $q_3 \to -q_3$ corresponds with the parity of the corresponding wave function in real space $\Psi_\alpha = \int dq_3 f_\alpha(q_3)|q_3\rangle$. By this linear combination of $|q_3\rangle$ the parity symmetry of $|\Psi\rangle$ is then restored.

Expectation values and transition probabilities are easily given in the above formalism [11]. For example, the quadrupole moment q_2 of the nth excited state is given by

$$q_2 = \int dq_3 \sqrt{G(q_3)}\, |\phi_n(q_3)|^2\, Q_2(q_3) \tag{9}$$

with $Q_2(q_3) = \langle q_3|\hat{Q}_2|q_3\rangle$.

The reduced transition probabilities from the lowest 1^- and 3^- states to the ground state can be computed within the the Rotational Model approximation. For $K = 0$ bands, we obtain:

$$B(E1, 1^- \to 0^+) = \frac{e^2}{4\pi} \left| \int dq_3\, G^{1/2}\, \phi^*_{0^-}(q_3) D(q_3) \phi_{0^+}(q_3) \right|^2 \tag{10}$$

and

$$B(E3, 3^- \to 0^+) = \frac{e^2}{4\pi} \left| \int dq_3\, G^{1/2}\, \phi^*_{0^-}(q_3) (Q_{30})_{prot}(q_3) \phi_{0^+}(q_3) \right|^2, \tag{11}$$

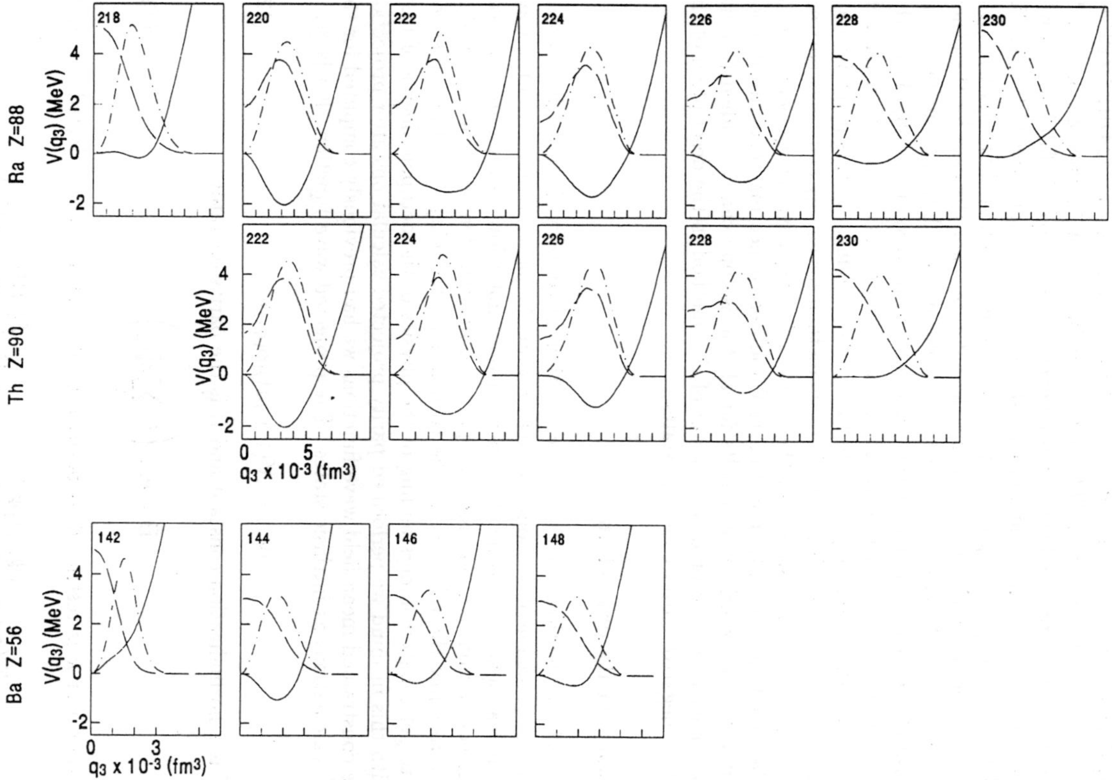

Figure 2: In this figure we depict the potential energy curves (full lines) as well as the positive (dashed) and negative (dash-dotted) wave function amplitudes corresponding to the lowest–lying collective states for three chains of isotopes.

where, as in Eq. (9), $D(q_3)$ and $Q_3(q_3)$ are the mean field expectation values of the dipole and octupole operators at the q_3 point.

To illustrate the method we present in Fig 2 the results of collective calculations[6, 7] for the q_3 degree of freedom carried out in the framework of the ATDHF+ZPE for several nuclei in the $Ra(Z = 88)$, $Th(Z = 90)$ and $Ba(Z = 56)$ chains of isotopes. In this figure we have represented the potential energy $V(q_3) - \epsilon_0(q_3)$ (full lines) as well as the positive (dashed) and negative (dash–dotted) parity amplitudes $G^{1/2}|\phi_\alpha|^2$. We observe how the ground state positive parity amplitude is centered around the mean field minimum for the nuclei with deeper wells but this is not the case for the others (for instance for ^{228}Th). The negative parity amplitudes are shifted to higher q_3 values because of symmetry considerations (they have to be zero at $q_3 = 0$). From this figure we learn that the collective correlations can distort considerably the mean field conclusions and, therefore, they must be taken into account. At this point we have to mention that the results obtained with this method for energy splittings between the positive parity ground state and the lowest–lying negative parity bandhead as well as the $B(E1)$ and $B(E3)$ transition probabilities connecting them are in good agreement with the experimental results for all the isotope chains studied (around 40 nuclei). This assess the predictive power of this method in the description of the properties of negative parity collective states.

4. Exact parity projection

As it has been mentioned in the previous section the collective hamiltonian method applied to the q_3 degree of freedom is able to restore the parity symmetry of the wave function in real space. However, this method does not allow to separate from the others the correlations coming from the parity projection. On the other hand, it is important to know the relevance of the correlations stemming from the projection in order to get a better understanding of the physics of the octupole degree of freedom. With this in mind we performed parity projected calculations after projection using the constrained mean field wave functions we had previously computed for the collective calculations as intrinsic states. The projected wave function $|\Psi_p\rangle$ is given by

$$|\Psi_p\rangle = \mathcal{N}(1 + p\hat{\Pi})|\phi(q_3)\rangle, \tag{12}$$

where $\mathcal{N}$ is a normalization constant and $\hat{\Pi}$ is the parity operator

$$\hat{\Pi} = \exp\left(i\pi\sum_k c_k^\dagger c_k\right) \tag{13}$$

and the sum over k is restricted to positive parity states. The projected energy is given by

$$E_p(q_3) = \frac{\langle\Psi_p|\hat{H}|\Psi_p\rangle}{\langle\Psi_p|\Psi_p\rangle} = \langle H\rangle + \Omega_p\left(\frac{\langle H\hat{\Pi}\rangle}{\langle\hat{\Pi}\rangle} - \langle H\rangle\right) \tag{14}$$

where

$$\Omega_p = p\frac{\langle\hat{\Pi}\rangle}{1+p\langle\hat{\Pi}\rangle}. \tag{15}$$

The intrinsic states associated to each parity are determined as the states corresponding to the minima of the corresponding projected energies.

The one body expectation values of positive parity operators $\hat{O}_+$ and transition matrix elements of negative parity operators $\hat{O}_-$ as well as the 0^+–0^- energy splitting* are given by

$$\langle\Psi_p|\hat{O}_+|\Psi_p\rangle = \langle\phi_p|\hat{O}_+|\phi_p\rangle + \Omega_p\left(\frac{\langle\phi_p|\hat{O}_+\hat{\Pi}|\phi_p\rangle}{\langle\phi_p|\hat{\Pi}|\phi_p\rangle} - \langle\phi_p|\hat{O}_+|\phi_p\rangle\right) \tag{16}$$

$$\langle\Psi_+|\hat{O}_-|\Psi_-\rangle = \frac{\langle\phi_+|\hat{O}_-|\phi_-\rangle - \langle\phi_+|\hat{O}_-\hat{\Pi}|\phi_-\rangle}{\sqrt{(1+\langle\hat{\Pi}\rangle_+)(1-\langle\hat{\Pi}\rangle_-)}} \tag{17}$$

$$\Delta\epsilon_{0^+-0^-} = E_+(q_3(+)) - E_-(q_3(-)) \tag{18}$$

where $q_3(+)$ and $q_3(-)$ are the q_3 values corresponding to the minima of the positive and negative parity projected energies. The reduced transition probabilities from the lowest 1^- and 3^- states to the 0^+ ground state can be computed within the rotational model approximation. For $K=0$ bands we obtain for the dipole and octupole transitions

$$B(E1, 1^- \to 0^+) = \frac{e^2}{4\pi}\left|\langle\Psi_+|\hat{D}|\Psi_-\rangle\right|^2 \tag{19}$$

$$B(E3, 3^- \to 0^+) = \frac{e^2}{4\pi}\left|\langle\Psi_+|\hat{Q}_{30}(\text{prot})|\Psi_-\rangle\right|^2 \tag{20}$$

here $\hat{D}$ and $\hat{Q}_{30}$ are the dipole and octupole operators, respectively.

In all the expressions above we have to compute mean values and transition matrix elements of multibody operators composed of the product of one–body ($\hat{O}$) or two–body ($\hat{H}$) operators times the parity operator $\hat{\Pi}$ between different CHFB wave functions. All these quantities can be computed with the help of the Generalized Wick Theorem and the Onishi formula [12, 13]. The explicit expressions in the parity projection case can be found in ref [14].

In these calculations there are two technical difficulties. The first one is related to the particle number of the projected wave functions. The intrinsic wave functions do not have good particle number (HFB wave functions) but they are built to have the right one on the average. However, the mean value of the particle number operators can change when computed with the parity projected wave functions. It has been shown in ref [14] that it is possible to take into account this fact perturbatively just

*The intrinsic wave functions are computed at $I=0$. In order to compute theoretically the 0^+-1^- splitting we add $I(I+1)/(2\mathcal{J})$ with $I=1$ to the 0^- energy to obtain the one for the $I^\Pi = 1^-$ state.

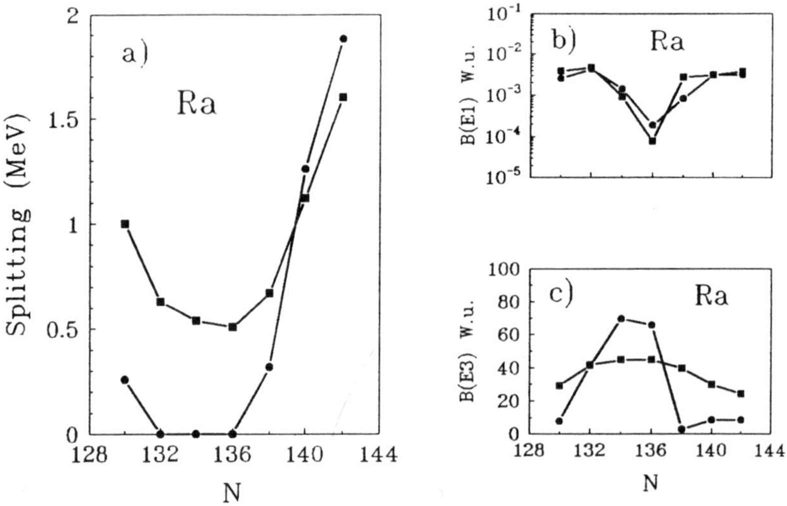

Figure 3: Comparison between the parity projected (full circles) and the collective hamiltonian method (full squares) results for the energy splitting, $B(E1)$ and $B(E3)$ transition probabilities.

by adding in the calculation of mean values and transition probabilities an extra term $-\lambda(\hat{N} - N_0)$ to all the operators used to compute them. The λ parameter depends on the operator considered and its explicit form can be found in [14]. It has been tested in our calculations that this correction is only important for the projected energy while for the multipole operators it is just of the order of one per cent or less. The second problem has to do with the density dependent term of the Gogny interaction. This term includes the mean field density $(\rho(\vec{r}))$ corresponding to the considered HFB state. However, it is not defined how to proceed in the calculation of $\langle \hat{H}\hat{\Pi} \rangle$. In our calculations we used two prescriptions, the first one is to keep using the mean field density in all the terms of the projected energy and the second one is to use $\rho_{\Pi} = \langle \hat{\rho}\hat{\Pi} \rangle / \langle \hat{\Pi} \rangle$ in the calculation of $\langle \hat{H}\hat{\Pi} \rangle$. The two prescriptions give the same qualitative results but they are quantitatively different.

The results obtained with the parity projection do not agree qualitatively with the ones obtained with the collective hamiltonian. In fig (3) the energy splittings, $B(E1)$ and $B(E3)$ transition probabilities computed with the parity projected method and with the collective hamiltonian for the Ra chain are depicted as a function of the neutron number N. It is seen that the parity projected energy splitting is zero for octupole deformed nuclei while it agrees quite well for non deformed ones. For the $B(E3)$ transition probabilities the situation is even worse. Surprisingly the $B(E1)$ transition probabilities show a quite good agreement. These discrepancies can be explained if it is argued that the parity projection does not incorporate as many octupole correlations as the collective hamiltonian method. This supposition has recently been corroborated [15] in a calculation where the Lipkin model was used as an octupole deformed system.

The conclusion is that the parity projection does not incorporate all the possible correlations present in the octupole degree of freedom.

5. Angular momentum projected transition probabilities

In this section I am going to make a few comments on some preliminary results we have obtained in comparing the transition probabilities computed with the rotational formula and the ones computed by projecting the intrinsic states onto good angular momentum. These calculations were triggered by the observation of very high $B(E3)$ transition probabilities in many nuclei that are not supposed to be octupole deformed (typical examples are the Lead isotopes around A=208, ^{148}Sm, ^{16}O, etc). When the collective hamiltonian method (also including in many cases the quadrupole degree of freedom) is used to compute the $B(E3)$ transition probabilities in those nuclei it is seen that the resulting figures are a factor of four or five smaller than the experimental ones. This is very surprising because our method works quite well for the neighboring octupole deformed systems and it is also supposed to work well for vibrational ones. However, after a closer examination we realized that in those nuclei where the discrepancies were bigger we were handling weakly quadrupole deformed or spherical nuclei. It is precisely in these kind of nuclei where the rotational formula is not supposed to work and where it is very important to consider wave functions with the proper symmetries.

To carry out the angular momentum projection we have used the standard procedure [10]. In our case this procedure is simplified by the use of axially symmetric wave functions which reduces the number of angular integrations to one. The final expression incorporating the collective wave functions is

$$\langle \Psi_-^{J=\lambda}||Q_3||\Psi_+^{J=0}\rangle = \frac{\int dq_3\sqrt{G}\phi_-(q_3)\phi_+(q_3)\langle q_3\hat{P}^{J=\lambda}||\hat{Q}_\lambda||\hat{P}^{J=0}q_3\rangle}{\left(\int dq_3\sqrt{G}|\phi_+|^2\langle\hat{P}^{J=0}\rangle\int dq_3\sqrt{G}|\phi_-|^2\langle\hat{P}^{J=\lambda}\rangle\right)^{1/2}} \tag{21}$$

where

$$\begin{aligned}\langle \Psi_1^{I_1}||Q_\lambda||\Psi_2^{I_2}\rangle &= (-)^{I_1}\sum_\mu (2I_1+1)(2I_2+1)\begin{pmatrix} I_1 & \lambda & I_2 \\ 0 & \mu & -\mu \end{pmatrix} \\ &\times \int_0^\pi d\beta \sin\beta d^{I_2}_{-\mu 0}(\beta)\langle\phi_1|Q_{\lambda\mu}e^{-i\beta\hat{J}_y}|\phi_2\rangle. \end{aligned} \tag{22}$$

The most difficult part of the calculation is to compute the effect of the $e^{-i\beta\hat{J}_y}$ operator on the CHFB wave functions. We had to express the original wave functions in a spherical harmonic oscillator basis starting from the usual expansion in the axial basis. After this expansion is achieved the calculation can be carried out along standard ways.

As the results are preliminary I will not extend very much on details. Just to mention that for the ^{16}O nucleus we obtain a projected $B(E3)$ of 17.3 W.u. while the collective hamiltonian method gives 3.5 W.u. and the experimental result is

14.1 W.u.. The enhancement due to the projection is in this case of a factor of five and the effect of the angular momentum projection is seen to be crucial to bring the theoretical result closer to the experimental one. At present we are working to extend these calculations to heavier nuclei and we plan to report the results in the near future.

Acknowledgements

Most of the work presented in this contribution has been done in collaboration with Luis Egido and the angular momentum projection part in collaboration with Yang Sun. I would like to express my gratitude to J.F. Berger, M. Girod J.P. Delaroche and D. Gogny of the Centre d'Etudes Nucleaires de Bruyeres le Châtel for many helpful discussions and suggestions. This work was supported in part by DGIC y T, Spain, under project No. PB88 - 0177.

References

1. P.H. Heenen, Contribution to this conference.

2. J.F. Berger, Contribution to this conference.

3. J.F. Berger, M. Girod and D. Gogny. Nucl. Phys. **A428** (1984) 23

4. D. Gogny. *Nuclear Selfconsistent fields.* Eds. G. Ripka and M. Porneuf (North Holland 1975).

5. M. Girod and B. Grammaticos. Phys. Rev. **C27** (1983) 2317

6. J.L. Egido and L.M. Robledo. Nucl. Phys. **A494** (1989) 85

7. J.L. Egido and L.M. Robledo. Nucl. Phys. **A518** (1989) 475

8. P.G. Reinhard and K. Goeke. J. of Phys. **G4** (1978) 245

9. M. Girod and B. Grammaticos. Nucl. Phys. **A330** (1979) 40

10. P. Ring and P. Shuck. *The Nuclear Many Body Problem* . Springer–Verlag Edt. New York 1980.

11. B. Nerlo–Pomorska, K. Pomorski, M. Brack and E. Werner. Nucl. Phys. **A462** (1987) 252

12. N. Onishi and S. Yoshida. Nucl. Phys **80** (1966) 367

13. R. Balian and E. Brézin. Nuovo Cim. **64B** (1969) 37

14. J.L. Egido and L.M. Robledo, Nucl. Phys. **A524** (1991) 65

15. L.M. Robledo, to appear in Phys Rev C.

GROUP 2

Pseudo–spin and the Symplectic Model

PSEUDO-SPIN AND THE SYMPLECTIC MODEL

J. P. DRAAYER
Department of Physics and Astronomy
Louisiana State University, Baton Rouge, LA 70803

ABSTRACT

The origin and consequences of pseudo-spin symmetry in nuclear physics, which is exact for an oscillator potential with one-body orbit-orbit (v_{ll}) and spin-orbit (v_{ls}) interaction strengths in the ratio $\mu \equiv 2v_{ll}/v_{ls} = 0.5$, are considered. Specifically, the $v_{ls} \approx 4v_{ll}$ condition is shown to be consistent with relativistic mean-field results and a pseudo–*LS* shell-model coupling scheme. The complementarity of good pseudo-spin symmetry and a residual quadrupole-quadrupole interaction is then explored; in this case, the pseudo-spin picture extends to pseudo–SU(3), which embodies features found in an extended triaxial quantum rotor description of a many-particle dynamics. The symplectic extension of the pseudo-SU(3) scheme means quadrupole phenomena can be described without interdicting effective charges. The addition of intruder levels to this scheme, yields a complete shell-model picture for studying a variety of nuclear phenomena, including superdeformation.

1. Pseudo-spin and Related Symmetries

The pseudo-spin concept follows naturally from the single-particle shell-model scheme, so we will begin with it, the simplest of all microscopic theories which accounts for the major shell closures and correctly predicts the ground state spin and magnetic moments of most odd-A nuclei. A convenient choice for the central field is the three-dimensional isotropic harmonic oscillator (H_0). This is usually augmented with one-body spin-orbit ($l\cdot s$) and orbit-orbit (l^2) interactions,

$$H = H_0 + Cl\cdot s + Dl^2. \tag{1-1}$$

The l^2 term ($v_{ll} \equiv D < 0$) pushes high angular momentum states down relative to those with lower l values, a feature that occurs automatically when a more realistic form like a Woods-Saxon well is used as the central potential. The $l\cdot s$ term ($v_{ls} \equiv C < 0$) — coupling space and spin degrees of freedom — is required to achieve shell closures at the magic nucleon numbers 2, 8, 20, 50, 82, 126 and 184. Unfortunately, the required value for C is so large that the spin-orbit interaction actually destroys the underlying SU(3) symmetry of the isotropic oscillator for all but light ($A \lesssim 28$) nuclei, and renders it of little apparent value in attempts at unraveling the structure of heavy ($A \gtrsim 100$) systems. Specifically, for heavy nuclei the $j = n+1/2$ orbital of the n-th oscillator shell (which includes levels with $j = l\pm1/2$ and $l = n, n-2, ..., 1$ or 0) is pushed down among the orbitals of the next lower shell. This yields new shells with normal parity $j = 1/2, 3/2, ..., n-1/2$ orbitals; plus a $j = n+3/2$ unique parity intruder from the shell above.

In this section we will show that this seemingly unfavorable scenario gives way to a much more favorable one for heavy nuclei, because $C \approx 4D$ or the Nilsson parameter $\mu = 2D/C \approx 0.5$. First of all, we will see that this leads to good pseudo-spin symmetry, because when $C \approx 4D$ (an empirical result deducible from single-particle energy systematics) the splitting generated by the $l \cdot s$ and l^2 interactions can be duplicated by a pseudo-oscillator Hamiltonian plus a pseudo l^2 interaction with (at most) a very small symmetry-breaking pseudo $l \cdot s$ term. We will then go beyond the empirical evidence and show that the $C \approx 4D$ condition is actually consistent with relativistic mean-field predictions. In the case of the many-particle extension of the single-particle picture, we will discover that when the residual deformation-inducing quadrupole-quadrupole interaction dominates over the pseudo l^2 term (which must be the case when well-developed rotational bands are present) each pseudo-spin symmetry will have associated with it an yrast band that is dominated by its leading pseudo-SU(3) irrep, the symmetry group of the pseudo-oscillator. We will also briefly consider the intruder's role in building up quadrupole collectivity, a subject only now being given careful consideration by algebraic structure theorists.

1.1 Spin-Orbit Doublets

The quickest way to gain an appreciation for the importance and simplicity of the pseudo-spin concept is by plotting the eigenvalues of the spherical (β=0) single-particle Hamiltonian of (1-1) as a function of the parameter $\mu = 2D/C$ which measures the relative strength of the l^2 and $l \cdot s$ terms. For the special value $\mu = 0.5$, the orbital pairs with $j = l+1/2$ and $j = (l+2)-1/2$ are degenerate for all l values. Furthermore, the splitting of these degenerate pairs follows an $\tilde{l}(\tilde{l}+1)$ rule where $\tilde{l}$ is the average l value of the pair: $\tilde{l} = [l+(l+2)]/2 = l+1$. (Throughout, we will use a tilde (~) over a quantity to denote its pseudo form.) This association is fixed as a special *normal* ↔ *pseudo* unitary transformation that makes this the degeneracy of pseudo spin-orbit partners, $\tilde{j} = \tilde{l} + \tilde{s}$ where $j = \tilde{j}$, $\tilde{l} = l + 1$ and $\tilde{s} = 1/2$. For the single-particle basis states of the three-dimensional isotropic harmonic oscillator this transformation has the form,

$$|\tilde{n}(\tilde{l},\tilde{s})\tilde{j}\tilde{m}\rangle = U_{njm,\tilde{n}\tilde{j}\tilde{m}}(l,\tilde{l})\,|n(l,s)jm\rangle,$$

$$U_{njm,\tilde{n}\tilde{j}\tilde{m}}(l,\tilde{l}) = \delta_{n-1,\tilde{n}}\,\delta_{j,\tilde{j}}\,\delta_{m,\tilde{m}}\,\delta_{l\pm1/2,\tilde{l}\mp1/2}\,. \tag{1-2}$$

From the structure of $U_{njm,\tilde{n}\tilde{j}\tilde{m}}(l,\tilde{l})$, it should be clear that this unitary transformation amounts to a simple relabeling of the basis states with levels of the n-th shell (except $j = n+1/2$) associated with levels of the $\tilde{n}$-th shell of another oscillator, $\tilde{n} = n-1$. A label-independent operator form which effects this special *normal* ↔ *pseudo* unitary transformation is known: $U_{njm,\tilde{n}\tilde{j}\tilde{m}}(l,\tilde{l}) = 2(\eta\cdot\xi-2l\cdot s+3)^{-1/2}(\xi\cdot s)$, where η and ξ create and annihilate oscillator quanta, and l and s are the single-particle orbital angular momentum and spin operators, respectively. This new structure was dubbed the *pseudo* oscillator by its originators, even though its algebraic properties are identical to those of the *normal* oscillator.

Under this mapping, the single-particle Hamiltonian transforms as follows:

$$H_0 + Cl\cdot s + Dl^2 \;\underset{\text{— pseudo} \rightarrow}{\overset{\leftarrow \text{normal —}}{\longrightarrow}}\; \tilde{H}_0 + (4D-C)\tilde{l}\cdot\tilde{s} + D\tilde{l}^2 + (\hbar\omega+2D-C). \quad (1\text{-}3)$$

Since the $(\hbar\omega+2D-C)$ term is a constant, the pseudo form, $\tilde{H} = \tilde{H}_0 + \tilde{C}\tilde{l}\cdot\tilde{s} + \tilde{D}\tilde{l}^2$, has the *same* excitation spectrum as the normal one ($H = H_0 + Cl\cdot s + Dl^2$) when $\hbar\tilde{\omega} = \hbar\omega$, $\tilde{C} = (4D-C)$ and $\tilde{D} = D$. This transformation is important, because for heavy ($A \gtrsim 100$) systems, $C \approx 4D$, so $\tilde{C} \approx 0$. Specifically, $\mu_\nu \approx 0.4$ and $\mu_\pi \approx 0.6$ (ν for neutrons and π for protons); this places medium and heavy mass nuclei very close to the exact pseudo-spin limit ($\mu = 0.5$) of the theory. Indeed, the average μ value is almost exactly 0.5. The familiar single-particle shell-model Hamiltonian for medium and heavy nuclei can therefore be replaced by a less familiar, but equivalent, pseudo form which is inherently simpler due to its much smaller spin-orbit interaction strength.

Actually, an appreciation for the importance of the pseudo-spin picture was first gained in a different way, through consideration of diagrams which show eigenvalues of the Nilsson Hamiltonian,

$$H = H_0 + Cl\cdot s + Dl^2 - \tfrac{1}{2}\chi(\kappa\beta)r^2Y_{20}(\theta,\varphi), \quad (1\text{-}4)$$

plotted as a function of the deformation β. This form is the single-particle Hamiltonian of (1-1) augmented with a deformed (prolate) field. This type of single-particle Hamiltonian may be obtained by averaging over one of the Q^c operators in the $Q^c\cdot Q^c$ product: $Q^c\cdot Q^c \rightarrow Q^c\cdot\langle Q^c\rangle \rightarrow (\kappa\beta)r^2Y_{20}(\theta,\varphi)$, where $\kappa = 3k(16\pi/5)^{1/2}$, and $k^2 = (5/9\pi)(A\overline{r^2})^2$. Clearly, (1-4) reduces to the simpler (1-1) expression in the special limit $\chi \rightarrow 0$. The Nilsson levels are normally labelled with their asymptotic (cylindrical geometry) quantum numbers $[Nn_z\Lambda]\Omega$:

- N is the principal quantum number and is therefore equal to the total number of oscillator quanta;
- $n_z = N, N\text{-}1, \ldots$, is the number of oscillator quanta along the body-fixed symmetry axis (chosen to be the z-th direction);
- $\Lambda = \pm n_\perp, \pm(n_\perp\text{-}2), \ldots, 1$ or 0 is the projection of the orbital angular momentum along the symmetry axis with $n_\perp = (N\text{-}n_z)$; and
- $\Omega = \Lambda + \Sigma$ where $\Sigma = \pm 1/2$ is the projection of the spin on the body-fixed symmetry axis.

In the pseudo-spin picture, these levels are relabeled by $[\tilde{N}\tilde{n}_z\tilde{\Lambda}]\tilde{\Omega}$ where $\tilde{N} = N-1$, and the indexing of the levels is the same as for the normal scheme: $\tilde{n}_z = \tilde{N}, \tilde{N}\text{-}1, \ldots$; $\tilde{\Lambda} = \pm\tilde{n}_\perp, \pm(\tilde{n}_\perp\text{-}2), \ldots, 1$ or 0; and $\tilde{\Omega} = \tilde{\Lambda} + \tilde{\Sigma} = \Omega$ where $\tilde{\Sigma} = \pm 1/2$ is the projection of the pseudo-spin on the body-fixed symmetry axis and $\tilde{\Omega} = \Omega$, because the total angular momentum and its projection are conserved under the *normal* $\leftrightarrow$ *pseudo* transformation. This relabeling of the orbitals yields pseudo-spin partners, $\tilde{\Omega}=\tilde{\Lambda}\pm 1/2$ which are nearly degenerate in energy. The near degeneracy of the pseudo-spin partners extends over the full range of β values, actually improving as a function of increasing deformation.

1.2 Mean-field Results

The pseudo-spin concept may be better understood by comparing an intuitive result for D with relativistic nuclear mean-field predictions for C. The origin of the l^2 term in H is in the flatness of the mean field in the interior region, as compared with the quadratic oscillator form ($V(r) = (1/2)M\omega^2r^2$). In the large mass limit ($A \to \infty$) the potential approaches that of a spherical well of finite depth. If this spherical well is replaced by one with an infinite depth, the single-particle energies are given by

$$E_{nl} = \frac{\hbar^2}{2MR^2} x_{nl}^2 \qquad (1\text{-}5)$$

where M is the nucleon mass, R is the radius of the well, and the x_{nl} are zeroes of spherical Bessel functions. These zeroes are approximately given by the result $x_{nl}^2 \approx [(n/2+1)\pi]^2 - l(l+1)$. Therefore, we have

$$D = \frac{-\hbar^2}{2MR^2}. \qquad 1\text{-}6)$$

A more complete theory for D using the Klein-Gordon equation leads to the same conclusion when the kinetic energy is a small fraction of the nucleon mass.

Next, consider the strength of the spin-orbit coupling. Starting with the Dirac equation (with only the time component of the scalar and vector potentials taken into account) and using a non-relativistic reduction of the relativistic mean field theory, the spin-orbit interaction is given by

$$V_{ls} = \frac{\hbar^2}{2M} \frac{2}{r} \frac{d}{dr}\left(\frac{1}{1 - B\rho/\rho_0}\right) l \cdot s\,. \qquad (1\text{-}7)$$

In this expression, ρ and ρ_0 are respectively the nucleon density at radius r and the nuclear matter density. The dimensionless quantity B in (5) is related to the strength of the scalar and vector coupling constants. The spin-orbit strength C can be obtained from the average of C over the region inside radius R,

$$C = \frac{-\hbar^2}{2MR^2} \frac{6B}{1 - B}. \qquad (1\text{-}8)$$

The fact that $d\rho/dr$ vanishes everywhere, except near the surface of the nucleus, was used in determining this result.

It follows from equations (1-6) and (1-8) that the ratio

$$\mu = \frac{2D}{C} = \frac{1 - B}{3B} \qquad (1\text{-}9)$$

is independent of mass number. Furthermore, to obtain $\mu = 0.5$ requires $B = 0.4$. In the simplest version of the theory, $B = (1/2)(B_s + B_v)$ with its scalar (i=s) and vector (i=v) components given by $B_i = g_i^2\rho_0/\mu_i^2 Mc^2$ where μ_i and g_i respectively denote meson masses and coupling constants. Using this expression for B, the Nambu–Jona-Lasinio (NJL) model — which in its modern form starts with massless quarks and generates hadron masses out of the vacuum by spontaneous symmetry-breaking, and which has also been used to predict the coupling constants and masses appearing in a relativistic nuclear field theory — gives the result $\mu = 0.686$. Results for the original Walecka model and a derivative coupling model due to Zimanyi and Moszkowski — which gives a more realistic equation of state for nuclear matter; which includes the effect of nucleon recoil; and when extended to include exchange correlations — also yield reasonable results for μ.

1.3 Pseudo-LS Coupling

As indicated above, the pseudo-spin scheme organizes the normal parity $j = 1/2, 3/2, ..., n\text{-}1/2$ levels of the n-th oscillator shell into a pseudo shell with $\tilde{n} = n\text{-}1$. For example, the normal parity levels of the n = 4 shell are mapped onto the ($3\tilde{p}_{1/2}$, $3\tilde{p}_{3/2}$, $1\tilde{f}_{5/2}$, $1\tilde{f}_{7/2}$) orbitals of an $\tilde{n}$ = 3 shell. In fact as we have already seen, this mapping of single-particle orbitals defines the pseudo-spin coupling scheme. To grasp its full significance, recall that the symmetry group for the usual many-particle generalization of the single-particle theory with particles distributed among the lowest available single-particle levels, is the unitary group $U(k\Omega)$ where $\Omega=(n+1)(n+2)/2$ is the spatial degeneracy of the n-th oscillator shell and k=2 or 4 for a spin or spin-isospin formulation of the theory. The $U(\Omega) \otimes U(k)$ direct product subgroup of this $U(k\Omega)$ group separates the full (k by Ω) dimensional space into its spatial (Ω) and spin (k=2) or spin-isospin (k=4) parts. Irreps of $U(\Omega)$, which are labeled by a Young pattern $[f] = [f_1,f_2,...,f_\Omega]$, specify the space symmetry; irreps $[f^c] = [f_1^c,f_2^c,...,f_k^c]$ of U(k), which must be related to the [f] of $U(\Omega)$ by row-column interchange in order to insure the overall antisymmetry in $U(k\Omega)$ required by the Pauli Exclusion Principle, label the complementary spin or spin-isospin symmetry. The breakup of the full model space into irreps of $U(\Omega) \otimes U(k)$ and its subgroups is shown schematically in Figure 1.

An important difference between the two one-body interactions is $l \cdot s$ couples different spatial symmetries, while l^2 does not. The [f] and therefore $[f^c]$ are good quantum numbers when the strength of all terms like $l \cdot s$ that couple different spatial symmetries is small relative to the strength of others like l^2 which do not mix these symmetries. While this is not the case for the many-particle extension of the normal single-particle Hamiltonian ($H = H_0 + C l \cdot s + D l^2$), it is for its pseudo counterpart ($\tilde{H} = \tilde{H}_0 + \tilde{C}\tilde{l} \cdot \tilde{s} + \tilde{D}\tilde{l}^2$). This feature is important, because it means we can partition the full model space into disjoint (pseudo) subspaces {$[\tilde{f}]$ and $[\tilde{f}^c]$ of $\tilde{U}(\tilde{\Omega}) \otimes \tilde{U}(\tilde{k})$, respectively} that have much smaller dimensions than can be realized with the normal scheme. In addition (as is known to be the case for the surface-delta interaction, and as is approximately so for the quadrupole-quadrupole interaction) if the residual two-body interaction is a pseudo-space scalar operator, it actually reinforces the goodness of the pseudo-space and pseudo-spin or pseudo-spin-isospin symmetries labelled by $[\tilde{f}]$ and $[\tilde{f}^c]$, respectively. The end result of partitioning the space via $\tilde{U}(\tilde{\Omega}) \otimes \tilde{U}(\tilde{k})$, rather than by $U(\Omega) \otimes U(k)$, is that we can deal with a subspace of the full space comprised of a collection of only a few pseudo-spatial symmetries rather than the full space which is much too large for most practical applications.

For heavy nuclei, the valence protons and neutrons occupy different shells, so we must apply an identical particle ($\tilde{k}$=2) formulation to each. Specifying the $[\tilde{f}]$ and $[\tilde{f}^c] = [\tilde{f}_1^c, \tilde{f}_2^c]$ labels is then equivalent to giving the total number of normal parity particles and their pseudo spin: $\tilde{m} = \tilde{f}_1^c + \tilde{f}_2^c$ and $\tilde{S} = (\tilde{f}_1^c - \tilde{f}_2^c)/2$. We partition the normal parity $\tilde{m}$-particle space, with $\tilde{m} = \tilde{m}_\pi$ for protons and $\tilde{m} = \tilde{m}_\nu$ for neutrons, into subspaces with $\tilde{S} = 0, 1, 2, 3, ..., \tilde{S}_{max}$ for $\tilde{m}$ even and $\tilde{S} = 1/2, 3/2, 5/2, ..., \tilde{S}_{max}$ for $\tilde{m}$ odd, where $\tilde{S}_{max}$ is the minimum of $\tilde{m}/2$ and $\tilde{N}\text{-}\tilde{m}/2$. This means that $\Delta\tilde{S}$ (proton *and*

Major Oscillator Shell

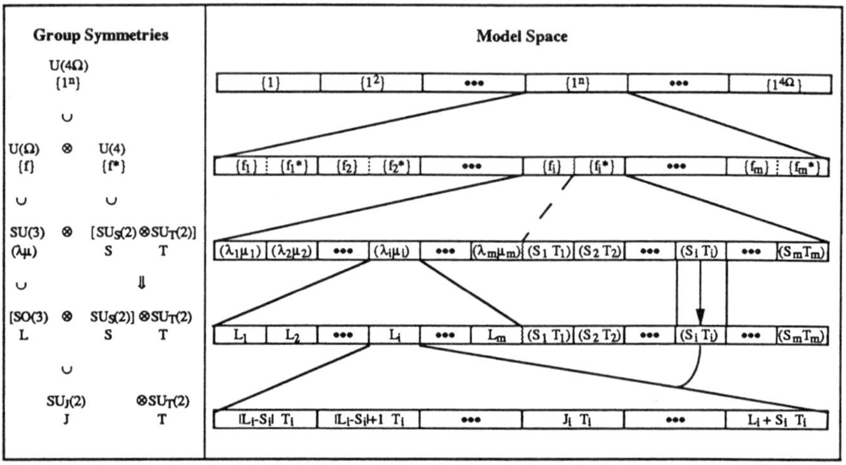

Figure 1. Group symmetries of a typical LST coupled shell-model scheme for a fermion description of the dynamics. U(4Ω) is the symmetry group of the N-th shell: $\Omega = \Sigma(2l+1)=(N+1)(N+2)/2$ is the spatial degeneracy and 4 comes from the spin-isospin degrees of freedom. SU(3) is an intermediate (non-exact) symmetry of the spatial $[U(\Omega) \supset SU(3) \supset SO(3)]$ chain. The total angular momentum J of $SU_J(2)$, which is an exact symmetry because of rotational invariance, results from the coupling of the total angular momentum L [SO(3)] to the total spin S [$SU_S(2)$]. Irreducible representations of the various groups are indicated on the left.

neutron) is always an integer; also, there is a complementary set of spatial configurations for each $\tilde{S}$. Therefore, to the extent pseudo-spin symmetry is good, we expect to observe sets of states, such as rotational sequences, that differ in *total* angular momenta ($\mathbf{J} = \tilde{\mathbf{L}} + \tilde{\mathbf{S}}$) by integer (even-A compared with even-A) or half-integer (odd-A with even-A) amounts. De-excitation gamma-ray spectra differing from one another by unit spin alignment — $\Delta E_i(J) = \Delta E_j(J+1)$ for the i-th and j-th rotational sequences, where $\Delta E_\alpha(J) = E_\alpha(J+1) - E_\alpha(J-1)$ — are therefore consistent with good pseudo-spin symmetry.

It is important to understand that this alignment feature can be either proton or neutron in origin, or a combination of the two. This follows because the many-particle basis states are π–ν coupled configurations: $|\Psi^J\rangle = |[(\tilde{\alpha}_\pi \tilde{L}_\pi \tilde{\alpha}_\nu \tilde{L}_\nu)^{\tilde{L}} \times (\tilde{S}_\pi \tilde{S}_\nu)^{\tilde{S}}]^J\rangle$, where $\tilde{\alpha}_\kappa$ labels multiple occurrences of the $\tilde{L}_\kappa$ values, $\kappa = (\pi,\nu)$. The full symmetry group of this combined π–ν system is $[\tilde{U}_\pi(\tilde{\Omega}_\pi) \otimes \tilde{U}_\pi(2)] \otimes [\tilde{U}_\nu(\tilde{\Omega}_\nu) \otimes \tilde{U}_\nu(2)]$. This direct product structure can be reordered as for $|\Psi^J\rangle$ so the pseudo-space and pseudo-spin associations are made first: $[\tilde{U}_\pi(\tilde{\Omega}_\pi) \otimes \tilde{U}_\nu(\tilde{\Omega}_\nu)] \otimes [\tilde{U}_\pi(2) \otimes \tilde{U}_\nu(2)]$. In this expression, $\tilde{\Omega}_\kappa = (\tilde{n}_\kappa+1)(\tilde{n}_\kappa+2)/2$ is the pseudo-space degeneracy of the $\kappa = (\pi,\nu)$ sub-shell. Simple alignment is consistent with good total pseudo-spin symmetry, $\tilde{S} = \tilde{S}_\pi \times \tilde{S}_\nu$, provided the π–ν interaction — like the π–π and ν–ν terms — conserves $\tilde{S}$. The real quadrupole-quadrupole, $Q^\pi \cdot Q^\nu$, which only weakly (compared to the symmetry-preserving $\tilde{Q}^\pi \cdot \tilde{Q}^\nu$ interaction) couples configurations with different $\tilde{S}$ symmetry, is such an interaction.

1.4 Pseudo-SU(3) Picture

The importance of the SU(3) model for light nuclei follows from the dominance of the collective quadrupole-quadrupole interaction ($Q^c \cdot Q^c$) over the one-body $l \cdot s$ and l^2 terms, as well as over all other two-body forms. Even though the spin-orbit interaction is relatively strong, yrast states of nuclei such as ^{20}Ne and ^{24}Mg are typically 60-80% pure leading representations of SU(3). This comes about primarily because $Q^c \cdot Q^c$ conserves spatial symmetry; secondly, because SU(3) is a subgroup of U(Ω) with second order invariant $C_2 = (1/4)(Q^a \cdot Q^a + 3L^2)$ where $Q^a Q^a$ is the single-shell $0\hbar\omega$ approximation for $Q^c \cdot Q^c$; and finally, because the expectation value of $Q^a \cdot Q^a$ is proportional to the square of the deformation, so each spatial symmetry [f] is subdivided into (λ,μ) irreps of SU(3) with the most deformed of these lying lowest and the least deformed highest. We now want to show that these same arguments apply for heavy deformed nuclei if the *normal* $\leftrightarrow$ *pseudo* transformation is applied.

The amount and sharpness of the separation — first into irreps of U(Ω) and then by irreps of SU(3) — depends on the relative strength of the symmetry-preserving and symmetry-breaking interactions; and this, in turn, depends upon whether or not the available space supports strongly deformed configurations. Systems such as ^{20}Ne and ^{24}Mg have leading (λ,μ) irreps with relatively large deformation, so $Q^a \cdot Q^a$ overpowers the other interactions and yrast states have good [f] and (λ,μ) quantum labels. We now argue that the pseudo-SU(3) scheme, in which $\widetilde{SU}(3)$ stands in the same relationship to $\tilde{U}(\tilde{\Omega})$ as SU(3) does to U(Ω), provides a similar explanation for observed quadrupole collectivity in heavy deformed nuclei. There are, of course, some very important differences that must be considered:

- the valence neutrons and protons for heavy nuclei of the rare earth and actinide regions occupy different major shells;
- the $Q^a \cdot Q^a$ interaction cannot be expressed solely in terms of the quadratic invariants of $\widetilde{SU}(3)$ and $\widetilde{SO}(3)$;
- whereas the coefficient D of l^2 is positive for the ds shell, it is negative for rare earth and actinide nuclei.

We can address the first of these three differences by noting that a very simple calculation can be carried out demonstrating that the collectivity of the system is not decreased, even when the protons and neutrons are in different shells, so long as the two species interact (albeit even weakly) through their quadrupole fields. Even when the valence protons and neutrons occupy different major shells and the separate two-body proton-proton ($V^{\pi\pi}$) and neutron-neutron ($V^{\nu\nu}$) interactions are dominated by pairing, a small $V^{\pi\nu} = Q^a_\pi \cdot Q^a_\pi$ interaction between the two suffices to drive the whole system towards the strong-coupled pseudo-SU(3) limit of the theory.

To consider the second matter, we simply note that the pseudo-spin scheme is an excellent starting point for a many-particle description of heavy nuclei, whether or not they are deformed. In particular, the $\tilde{l} \cdot \tilde{s}$ interaction term is weak relative to $\tilde{l}^2$ [$\tilde{C}/\tilde{D} = (4D-C)/D \approx (C/D)/5$] so the pseudo-spin symmetry is approximately good. Also, since the surface-delta interaction — which is known to be a good effective interaction for many applications — is a pseudo-spin scalar operator, the residual two-body interaction is not expected to change this picture very much, though it will induce some mixing among the different pseudo-space symmetries. As an example, although $Q^a \cdot Q^a$ is not an invariant of $\widetilde{SU}(3)$, it is transformed under the *normal* $\leftrightarrow$ *pseudo* mapping into its pseudo counterpart plus small corrections,

$$Q^a \cdot Q^a = \kappa \tilde{Q}^a \cdot \tilde{Q}^a + \cdots . \tag{1-10}$$

Indeed, within the leading pseudo-space symmetry, the sum of all correction terms has been shown to induce less than a 1% change in calculated excitation energies and electromagnetic transition strengths.

To explore the third and final question concerning the relevance of the pseudo-SU(3) scheme for strongly deformed nuclei with $D < 0$ (a condition responsible for destroying SU(3) in the fp-shell, for example), we considered the many-body problem with the Hamiltonian,

$$\tilde{H} = \tilde{H}_0 + \tilde{C}\sum_i \tilde{l}\cdot\tilde{s} + \tilde{D}\sum_i \tilde{l}_i^2 - \frac{1}{2}\chi\tilde{Q}^a\cdot\tilde{Q}^a = \tilde{H}_0 + \tilde{C}\sum_i \tilde{l}\cdot\tilde{s} + \tilde{D}\sum_i \tilde{l}_i^2 - \frac{1}{2}\tilde{\chi}(4\tilde{C}_2 - 3\tilde{L}^2). \tag{1-11}$$

The last form for $\tilde{H}$ follows because $\tilde{Q}^a \cdot \tilde{Q}^a = 4\tilde{C}_2 - 3\tilde{L}^2$ within a major shell of the oscillator. Since the $\tilde{C}_2$ and $\tilde{L}^2$ interactions are diagonal in an $\widetilde{SU}(3)$ basis, they split but do not break the pseudo-oscillator symmetry; because $\tilde{\chi}$ is always positive, the $\widetilde{SU}(3)$ representations with the largest eigenvalues for $\tilde{C}_2$ — and hence the greatest intrinsic deformations ($\beta^2 \sim \tilde{\beta}^2 \sim \langle\tilde{Q}^a\cdot\tilde{Q}^a\rangle = \langle 4\tilde{C}_2\rangle$ in $J = \tilde{L} = 0$ states) — lie lowest. Energies of calculated 0^+ states of this Hamiltonian, (1-11), with $\tilde{C} = 0$ and $\tilde{D} < 0$ show that under conditions very similar to those for pseudo-SU(3) applications in rare earth and actinide nuclei, the symmetry-breaking *decreases* sharply as the strength of the deformation-inducing quadrupole-quadrupole interaction *increases*.

The pseudo-spin scheme considers the highest $j = n+1/2$ state associated with the n-th major shell of the oscillator to have defected out of the valence space — pushed down among occupied levels of the (n−1)-th shell below by the strong spin-orbit interaction. Furthermore, the $j = n+3/2$ level from the (n+1)-th shell immediately above is considered to intrude into the valence space — playing only a passive role in the dynamics of low-lying excitations, because it has the opposite parity to its new-found partners. It is usually assumed that particles distributed in the unique parity intruder levels enter only as J=0 coupled pairs for configurations below the backbending region, contributing binding energy to the system but nothing to the dynamics. Through the backbending region and beyond, however, alignment of the angular momentum of these pairs (J≠0) sets in, and the intruder level can no longer be considered passive. We have some preliminary results which challenge this simple picture. In particular, the results suggest that the coupling of the intruder to its natural partners, even though these may be as much as a full major shell ($1\hbar\omega$) away, is strong and leads to a sizable contribution (roughly in proportion to the fraction of particles in the intruder levels) to the quadrupole moment and therefore E2 strengths of the many-particle system.

Reproducing experimental results such as backbending and electromagnetic transition strengths is a necessary, but not sufficient condition for a successful theory; it is important that the Hamiltonian and basis space choices are reasonable and complementary. This can be insured, for example, by checking to make sure calculated results are stable under enlargement of the space, and by not allowing the Hamiltonian to change significantly as one moves through a series of neighboring nuclei. This is particularly important for algebraic theories, as we claim they are microscopic, and it is well-known that agreement can be forced in isolated cases by the *right* choice for the interaction parameters. Theories producing eigenstates that have small overlap with those determined in larger and more sophisticated calculations should be discounted. Regarding this matter, we need to emphasize that the pseudo-SU(3) scheme yields stable solutions in heavy deformed nuclei because:

- validity of the pseudo-spin symmetry concept;
- complementary nature of the rotor and SU(3) algebras;
- dominance of the quadrupole-quadrupole interaction.

These three features are foundation elements undergirding the theory. The first has already been given careful consideration. The second will be considered at length in the next section. The dominance of the quadrupole-quadrupole interaction follows from a consideration of M1, E2, and M3 transitions in rare earth and actinide nuclei cast in the framework of the pseudo-SU(3) picture.

2. Quantum Rotor and SU(3) Symmetry

The *macroscopic* and *microscopic* options for interpreting nuclear phenomena have existed since the earliest period (1935 – 1955) in the history of nuclear physics, and we continue to use these labels in a general way for characterizing models in use or under development today. Of course, there is no longer a macroscopic-microscopic dichotomy, as most modern theories fit solidly into the intermediate region between the extremes. For example, the simplest Bohr-Mottelson rotation-vibration model for even-A systems, and its Nilsson extension for strongly deformed odd-A nuclei have evolved into a variety of self-consistent theories, both simple and sophisticated. We will refer to these as *geometric* models, because they use (β,γ)-shape variables for describing quadrupole deformation in an intrinsic principal-axis frame, and a quantum top dynamics for a description of the accompanying rotational motion. From the shell-model perspective, there are now a variety of *algebraic* models exploiting group symmetries in order to focus more sharply on nuclear collective degrees of freedom. As applications to superdeformed phenomena are explored, both approaches are coming under very sharp and critical testing, because this regime of strong deformation focuses on the unique situation of collective rotational phenomena in isolation from other normally important effects. The challenge facing nuclear structure physicists is the need to adequately incorporate *microscopic* features in a *geometric* theory or *macroscopic* features in an *algebraic* one.

Significant progress towards bridging this gap between the macroscopic and microscopic pictures of nuclear structure has recently been made. In this section, we will review this progress for rotational motion; extensions along similar lines for vibrations and the coupling of rotations and vibrations are still under development. Specifically, a simple one-to-one mapping between the triaxial quantum rotor problem and an SU(3) shell-model theory will be established. In order to accomplish this, the quantum rotor problem will be reformulated to identify the corresponding many-particle, shell-model theory. An explicit connection between the (β,γ)-shape variables of a macroscopic approach and the (λ,μ)-irrep labels of the microscopic theory follows from this development. This $(\beta,\gamma)\leftrightarrow(\lambda,\mu)$ mapping establishes an unambiguous link between the geometric and algebraic approaches to the structure of nuclei. Numerous examples can be found in the nuclear physics literature (covering excitation energies and transition strengths) demonstrating the quality of agreement between the two theories under the mapping. Because this matter is so important, we will show how the equivalence of these approaches arises at the level of the symmetry algebras of the two theories, even though our intent is to avoid using advanced group theory arguments when possible. We will also show how the theory can be used to identify a specific linear combination of shell-model operators expected to play a significant role in nuclear K-band splitting.

2.1 Triaxial Quantum Rotor

The triaxial rotor was one of the first problems considered after the introduction of the quantum theory. Interest in it peaked during the late 20's and early 30's as it was used as a model for studying molecular rotations. The Hamiltonian is given by

$$H_{ROT} = A_1 I_1^2 + A_2 I_2^2 + A_3 I_3^2 \tag{2-1}$$

where I_α (α=1, 2, and 3) is the projection of the total angular momentum of the rotor on the α–th body-fixed symmetry axis and A_α is the corresponding inertia parameter: $A_\alpha = 1/2\mathcal{I}_\alpha$ where $\mathcal{I}_\alpha$ is the moment of inertia of the system about the α–th principal axis. Here we will deviate from the practice of using **J** for the angular momentum and use **I** and **L** instead, to respectively denote body-fixed and lab-frame angular momentum operators. Later, when spin is included in the formulation, we will return to the use of $\mathbf{J} = \mathbf{I} + \mathbf{S} = \mathbf{L} + \mathbf{S}$. Note that **I** (vector) = **L** (vector); the use of different symbols (rather than primed and unprimed operators, for example) for the body-fixed and lab-frames is for clarity only.

The actual values for the A_α in (2-1) depend on details of the model. Two extremes can be identified: rigid rotation and irrotational flow. A system is rigid if all of its constituent particles are locked in place and participated equally in the rotational motion — as for a simple top or a molecule. It is irrotational if the rotation is simply a manifestation of a rearrangement of the constituent particles — as happens when a water balloon is tossed in the air. Nuclear rotations are somewhere between these two limits. The more deformed the system, the more rigid-rotor-like it seems to become. As an example, recall that superdeformed bands suggest the underlying configurations have rigid-rotor characteristics.

The familiar principal-axis form for the rotor Hamiltonian (2-1) can be rewritten in a less unfamiliar but nonetheless important and very revealing frame-independent representation by introducing three special scalar operators:

$$L^2 = \sum_\alpha L_\alpha L_\alpha = \sum_\alpha I_\alpha^2; \tag{2-2a}$$

$$X_3^c = \sum_{\alpha,\beta} L_\alpha Q^c_{\alpha,\beta} L_\beta = \sum_\alpha \lambda_\alpha I_\alpha^2; \tag{2-2b}$$

$$X_4^c = \sum_{\alpha,\beta,\gamma} L_\alpha Q^c_{\alpha,\beta} Q^c_{\beta,\gamma} L_\gamma = \sum_\alpha \lambda_\alpha^2 I_\alpha^2. \tag{2-2c}$$

The L_α and $Q^c_{\alpha,\beta}$ in (2-2) are Cartesian forms of the total angular momentum and collective quadrupole operators, respectively. (*Remember the superscript* c *is appended to* Q *to denote the collective quadrupole operator which has non-vanishing matrix elements between major shells* (n' = n, n±2), *in contrast with the algebraic quadrupole operator* (Q^a) *which has non-vanishing matrix elements only within a major shell,* n' = n.) The last expression given for each scalar in (2-2) is the form these operators take in the body-fixed, principal-axis system: $\langle Q^c_{\alpha,\beta}\rangle^{BF/PA} = \lambda_\alpha \delta_{\alpha,\beta}$. This result presumes the eigenvalues of the quadrupole operator are sharp. Equations (2-2) can be inverted to yield an expression for the I^2_α in terms of L^2 and the X^c operators:

$$I^2_\alpha = [(\lambda_1\lambda_2\lambda_3)L^2 + (\lambda^2_\alpha)X^c_3 + (\lambda_\alpha)X^c_4]/D_\alpha$$

$$\text{where} \quad D_\alpha \equiv 2\lambda^3_\alpha + \lambda_1\lambda_2\lambda_3. \tag{2-3}$$

Substituting this result for the I^2_α into (2-1) yields

$$H_{ROT} = aL^2 + bX^c_3 + cX^c_4, \tag{2-4}$$

where a, b and c depend on the inertia parameters and the eigenvalues of Q^c,

$$a = \sum_\alpha a_\alpha A_\alpha, \quad a_\alpha = \lambda_1\lambda_2\lambda_3/D_\alpha; \tag{2-5a}$$

$$b = \sum_\alpha b_\alpha A_\alpha, \quad b_\alpha = \lambda^2_\alpha/D_\alpha; \tag{2-5b}$$

$$c = \sum_\alpha c_\alpha A_\alpha, \quad c_\alpha = \lambda_\alpha/D_\alpha. \tag{2-5c}$$

The eigenvalues and eigenstates of Hamiltonians (2-1) and (2-4) with parameters related through (2-5) are the same.

2.2 *Rotor Dynamics and SU(3) Symmetry*

The reason for casting the rotor Hamiltonian into a frame-independent form — when its principal-axis representation is clearly simpler — is so its shell-model image can be readily identified. To achieve this, note that the many-body form for the L_α and $Q^c_{\alpha,\beta}$ operators are simply their single-particle forms summed over all particles,

$$L_\alpha = \sum_i l_\alpha(i) \quad \text{and} \quad Q^c_{\alpha,\beta} = \sum_i q^c_{\alpha,\beta}(i). \tag{2-6}$$

It would seem from this that the many-body, shell-model image of the rotor Hamiltonian is (2-4), with the L_α and $Q^c_{\alpha,\beta}$ operators interpreted as in (2-6). However, this interpretation ignores the underlying shell structure and the fermion character of the many-body system. In particular, it is important to remember that while the L_α have non-vanishing matrix elements only within a major oscillator shell, the $Q^c_{\alpha,\beta}$ couple shells differing by two quanta (n' = n, n±2). Indeed, the off-diagonal (n' = n±2) matrix elements of the $Q^c_{\alpha,\beta}$ are about equal in size to the diagonal (n' = n) ones. It follows from this that operators like $Q^c \cdot Q^c$ and the X^c operators (even if used only as residual interactions) can actually destroy the shell structure. One way around this dilemma is to set the matrix elements of Q^c coupling to different major shells to zero, which has the effect of transforming the collective model quadrupole operators ($Q^c_{\alpha,\beta}$) into their algebraic counterparts ($Q^a_{\alpha,\beta}$). As noted above, these $Q^a_{\alpha,\beta}$ operators, along with the L_α operators, generate SU(3), the symmetry algebra of the isotropic harmonic oscillator Hamiltonian. In this case, the shell-model Hamiltonian is given by,

$$H_{SU3} = H_0 + aL^2 + bX^a_3 + cX^a_4. \qquad (2\text{-}7)$$

Before determining shell-model values for the λ_α, which will also establish a connection between the (β,γ)-shape variables of the collective model and the (λ,μ)-irrep labels of the shell-model approach and thereby complete the $H_{ROT} \leftrightarrow H_{SU3}$ correspondence, it is instructive to point out the relationship of these two theories at a more fundamental level. This can be accomplished most simply by comparing the algebras of their symmetry groups. The symmetry group of the quantum rotor is the semi-direct product $T_5 \wedge SO(3)$ where T_5 is the group generated by the five components of the collective quadrupole operator (Q^c_μ); as usual, SO(3) is generated by the angular momentum operators (L_μ). (*The semi-direct product symbol* $\wedge$ *denotes the fact that the generators of this special group structure divide into two sets, each separately closed under commutation, but the commutator of an element of one with and element of the other yielding an element of only one (not both) of the sets, symbolically:* $[T_5, T_5] \rightarrow T_5$ *and* $[SO(3), SO(3)] \rightarrow SO(3)$ *but* $[T_5, SO(3)] \rightarrow T_5$, *see below.*) The generators of SU(3), on the other hand, are the Q^a_μ and L_μ operators. If Q^x denotes a generic quadrupole operator, then the commutation relations of the L and the Q^x operators are given by

$$[L_\mu, L_\nu] = -\sqrt{2}\langle 1\mu, 1\nu | 1, \mu+\nu\rangle L_{\mu+\nu}, \qquad (2\text{-}8a)$$

$$[L_\mu, Q^x_\nu] = -\sqrt{6}\langle 1\mu, 2\nu | 2, \mu+\nu\rangle Q^x_{\mu+\nu}, \qquad (2\text{-}8b)$$

$$[Q^x_\mu, Q^x_\nu] = c\langle 2\mu, 2\nu | 1, \mu+\nu\rangle L_{\mu+\nu}, \qquad (2\text{-}8c)$$

where $c = +3\sqrt{10}$ for SU(3) $[Q^x=Q^a]$ and $c = 0$ for $T_5 \wedge SO(3)$ $[Q^x=Q^c]$. In (2-8) the <-,-|-> symbol denotes an ordinary [SO(3)] Clebsch-Gordan (Wigner) coefficient. Actually, there is one additional case, $c = -3\sqrt{10}$ for Sl(3,R) $[Q^x=Q^b \sim (x_i p_j + p_j x_i)]$ which identifies the symmetry associated with shear degrees of freedom. All three groups, SU(3), $T_5 \wedge SO(3)$, and Sl(3,R), are subgroups of the symplectic group Sp(3,R), which we will consider in the next section.

One can see from these commutation relations how the SU(3) algebra of the isotropic oscillator goes over into $T_5 \wedge SO(3)$ of the rotor. Specifically, if Q^a is renormalized by dividing it by the square root of the second order invariant of SU(3) $[Q^a \leftarrow Q^a/\sqrt{C_2}$ where by definition the invariant $C_2 = (Q^a \cdot Q^a - 3L^2)/4$ commutes with both the Q^a and L operators] the first (a) and second (b) commutators given in (2-8) remain unchanged, while the $L_{\mu+\nu}$ on the right-hand-side of the third (c) goes over into $L_{\mu+\nu}/C_2$. For low L values in large SU(3) irreps, $L_{\mu+\nu}/C_2 \rightarrow 0$ and the SU(3) algebra reduces to $T_5 \wedge SO(3)$. This renormalization of the Q^a is called a group contraction process; the arguments presented show the SU(3) algebra reduces to the algebra of $T_5 \wedge SO(3)$ in the contraction limit, and consequently, the SU(3) theory reduces to that of the rotor.

2.3 *Shape Variables (β,γ) ↔ (λ,μ) Irrep Labels*

The parameters a, b and c of H_{SU3} (2-7) are related to the A_α of H_{ROT} through (2-5). However it is necessary to have shell-model values for the λ_α to complete the mapping. The simplest and most direct way this can be achieved is by requiring a correspondence between invariants of the rotor and shell-model theories. This is a very natural requirement, since two theories describing the same physical phenomena should have the same constants of the motion. Because SU(3) is a rank two group it has two invariants: C_2, with eigenvalue $\lambda^2+\lambda\mu+\mu^2+3(\lambda+\mu)$; and C_3, with eigenvalue $(\lambda-\mu)(\lambda+2\mu+3)(2\lambda+\mu+3)$, where λ and μ are the usual SU(3) irrep labels — λ+μ and μ specify respectively the number of boxes in the first and second rows of a Young tableau labelling of the SU(3) irreps. Note that C_2 is of degree two in the generators of SU(3) while C_3 is of degree three. The symmetry group of the rotor $[T_5 \wedge SO(3)]$ also has two invariants: traces of the square {Trace[$(Q^c)^2$]} and the cube {Trace[$(Q^c)^3$]} of the collective quadrupole matrix. The eigenvalues of these invariant operators are $\lambda_1^2+\lambda_2^2+\lambda_3^2 \rightarrow (k\beta)^2$ and $\lambda_1\lambda_2\lambda_3 \rightarrow (k\beta)^3\cos(3\gamma)$, respectively. The requirement of a linear correspondence between these two sets of invariants, one for SU(3) and the other for $T_5 \wedge SO(3)$, leads to the following linear relation between the λ_α of $T_5 \wedge SO(3)$ and the (λ,μ) of SU(3):

$$\lambda_1 = -(\lambda-\mu)/3, \tag{2-9a}$$

$$\lambda_2 = -(\lambda+2\mu+3)/3, \tag{2-9b}$$

$$\lambda_3 = (2\lambda+\mu+3)/3. \tag{2-9c}$$

This correspondence, in turn, sets up a direct relationship between the (β,γ) shape variables of the collective model and the (λ,μ) irrep labels of SU(3),

$$\beta^2 = (4\pi/5)(A\overline{r^2})^{-2}[\lambda^2 + \lambda\mu + \mu^2 + 3(\lambda + \mu) + 3], \tag{2-10a}$$

$$\gamma = \tan^{-1}\left(\frac{\sqrt{3}(\mu + 1)}{2\lambda + \mu + 3}\right). \tag{2-10b}$$

The (β,γ) and (λ,μ) relationship is shown schematically in Figure 2 on a traditional (β,γ) or polar plot with β the radius vector and γ the azimuthal angle. The Cartesian components of β are given by

$$k\beta_x = k\beta\cos(\gamma) = (2\lambda+\mu+3)/3, \tag{2-11a}$$

$$k\beta_y = k\beta\sin(\gamma) = (\mu+1)/\sqrt{3}, \tag{2-11b}$$

where $k^2 = (5/9\pi)(A\overline{r^2})^2$. These relations shows that each (λ,μ)-irrep corresponds to a unique value for the pair (β,γ). Because the λ and μ are representation labels of SU(3) they can only be nonnegative integers. Furthermore, for any particular nucleus the group structure, $U(\Omega) \supset SU(3)$ dictates a set of allowed (λ,μ) irreps for each [f] value. The $[f] \to (\lambda,\mu)$ branching rule is called a group plethysm; tables as well as a computer code for determining the set of (λ,μ) irreps in [f] are available. For example, for the ds-shell nucleus ^{24}Mg, the most important $U(\Omega{=}6)$ symmetry is [f] = [44] and this contains the following SU(3) irreps: (λ,μ) = (8,4); (7,3); (8,1); (4,6); (5,4);$(6,2)^2$; (3,5); (4,3); $(5,1)^2$; (0,8); $(2,4)^2$; (3,2); $(4,0)^2$; (1,3); and (0,2) where the superscript denotes the number of times the (λ,μ) irrep occurs (its multiplicity). The $[f] \to (\lambda,\mu)$ reduction means the microscopic fermion structure as determined through group plethysms places constraints on the allowed values for (β,γ) — in sharp contrast with the macroscopic liquid drop picture which takes β and γ to be continuous variables. This is then a very clear and significant difference in the macroscopic and microscopic approaches — the microscopic structure, together with the branching rules this implies, dictates the (λ,μ) values that can occur; and this, in turn, implies constraints should to be placed on the allowed (β,γ) shapes of a macroscopic theory.

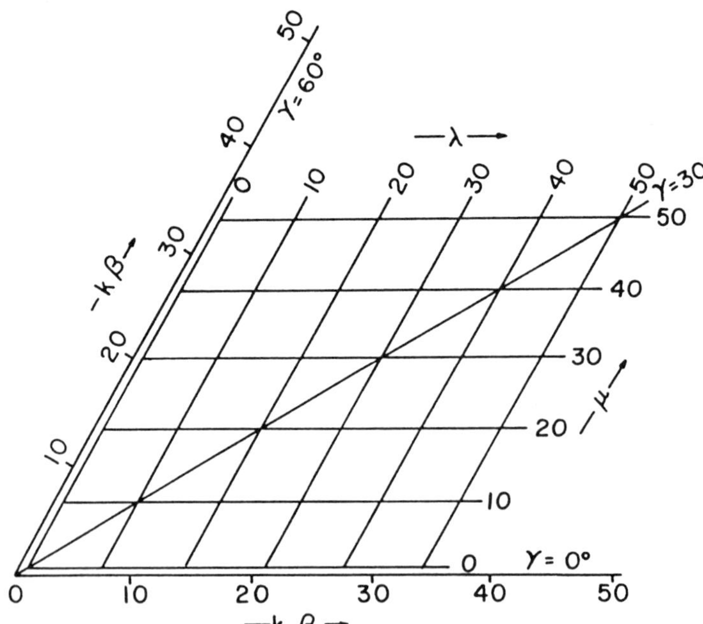

Figure 2. A traditional plot with β the radius vector and γ the azimuthal angle which shows that relationship between the (β,γ) collective model shape variables and the SU(3) irreducible representations labels (λ,μ). Under the rotor-SU(3) mapping, a linear relationship connects the two, see (2-4), (2-5), (2-9) and (2-11).

This correspondence between (λ,μ) and (β,γ) means that the potential energy surface concept, so much a part of the macroscopic picture, can be given a shell-model interpretation. The usual procedure used for determining the potential energy surface is to assume the Hamiltonian is a five-dimensional oscillator based on the quadrupole shape excitations, plus an anharmonic residual interaction that is a polynomial function of maximum phonon order k in the scalars $[\beta^2$ and $\beta^3\cos(3\gamma)]$ of the theory,

$$H = H_0 + \sum_{p,q}^{2p+3q\leq k} C_{p,q}\beta^{2p+3q}[\cos(3\gamma)]^q. \tag{2-12}$$

This Hamiltonian is diagonalized in a basis built by considering all possible phonon states up through some maximum number (n) of quanta. The coefficients $C_{p,q}$ of the residual interaction are determined by fitting to known experimental data on excitation spectra and transition strengths. This approach is insensitivity to k and n, provided both are sufficiently large; results for rare earth and actinide nuclei are given in the literature for k and n values up to 6 and 32, respectively. The k=6 terms are need to get a quadratic dependence on $\cos(3\gamma)$, and hence the possibility of a nonzero triaxial equilibrium shape. The energy surface is the expectation value of H in the calculated eigenstates plotted as a function of β and γ. If this procedure is revisited from the shell-model perspective suggested above, the allowed β and γ values must (at a minimum) be restricted by the $[f] \to (\lambda,\mu)$ constraint to a sub-plane of the full (β,γ)-plane. Some (β,γ) values are ruled out by the Pauli Principle. Indeed, since within a shell there is always a leading (λ,μ) for which C_2 has a maximum, the shell structure places a constraint on the maximum β deformation. The only way around this dilemma is to allow for the excitation of nucleons into higher shells, thereby including particle-hole configurations.

3. Symplectic Shell-Model Picture

Understanding collective phenomena from a shell-model perspective has been a long-standing goal of nuclear structure physics. Although all algebraic theories (boson and fermion) share this objective, neither deals directly with the fact that reproducing observed E2 transition strengths in deformed nuclei (which even in light systems can be more than two orders of magnitude greater than simple single-particle estimates) requires extending the picture of single-shell valence spaces ($0\hbar\omega$ theory) to include configuration admixtures from higher-lying configurations ($n\hbar\omega$; n = 2, 4, etc.). A $0\hbar\omega$ theory cannot (without a large renormalization of the quadrupole operator) account for the observed E2 transition strengths. A simple argument can be given which demonstrates the inadequacy of a $0\hbar\omega$ shell-model theory for describing deformed phenomena. The SU(3) coupling scheme organizes a model space according to the deformation of its many-particle configurations. The maximum eigenvalue which Q_0^a can have is $\varepsilon_{max} = 2\lambda+\mu$ where (λ,μ) is the leading m-particle SU(3) representation. If all the valence nucleons could be put into the most deformed single-particle level [a scenario ruled out by the Exclusion Principle for m > 4 (spin-isospin) or m > 2 (spin)] one obtains $(\lambda,\mu) = (\eta m,0)$ for that leading configuration where η is the oscillator shell number. The corresponding maximum eigenvalue for Q_0^a would then be $\varepsilon_{max} = 2\eta m$; while this is usually larger than the allowed $2\lambda+\mu$ values, it places an upper limit on the deformation and shows that the E2 transition strengths ($\sim \langle Q^c\cdot Q^c\rangle$) can grow no stronger than quadratically with m in a $0\hbar\omega$ shell-model theory. The same conclusion follows directly from the fact that Q^a is a one-body operator. Observed E2 strengths show a growth with m as one moves away from a closed shell that is much stronger than this. This rapid buildup in strength can only come about through coherent admixtures from configurations with larger values for η; this means higher-lying, multi-$\hbar\omega$ configurations must be included from the outset if calculated eigenstates are to be considered realistic wavefunctions of the physical system.

The symplectic model, sometimes also called the microscopic collective model, is a shell-model scheme that extends the Elliott SU(3) model to include multiple $2\hbar\omega$ excitations of the monopole (l=0) and quadrupole (l=2) type. It can be used to describe both low-lying *and* giant resonance states of light ($A \lesssim 28$) nuclei that are strongly deformed. The multi-$\hbar\omega$ feature of the model means that the action of the real collective quadrupole operator, $Q^c = (16\pi/5)^{1/2}\, r^2 Y_2(\hat{r})$, and not only its symmetrized intra-shell $0\hbar\omega$ preserving part, $Q^a = (16\pi/5)^{1/2}\, [r^2Y_2(\hat{r}) + p^2Y_2(\hat{p})]/2$, can be fully accommodated. The symmetry algebra of the scheme generates a realization of the non-compact (infinite dimensional irreps) symplectic group Sp(3,R) which has the Elliott SU(3) as its maximal compact subgroup. (More will be said about this structure below.) One can appreciate the fact that Q^c (in contrast with Q^a) involves couplings to shells that differ by $\pm 2\hbar\omega$ by looking at the simpler and more familiar one-dimensional case. The matrix elements $\langle n'|x^2|n\rangle$ of the x^2 operator are nonvanishing for $n' = n\pm 2$ as well as for $n' = n$, in contrast with those of its symmetrized form $(x^2+p^2)/2$ which has non-zero matrix elements only for $n' = n$ because this combination is simply the oscillator Hamiltonian. (Couplings to states with $n' = n \pm 1$ are ruled out by parity considerations.) As we will see, this coupling to higher shells means that each allowed $0\hbar\omega$ SU(3) shell-model irrep (parent configuration) has associated with it a tower of states (infinite in number) that span its Sp(3,R) extension. The point is that these symplectic spaces are complete with respect to the action of the real quadrupole operator. This means intra-band and inter-band E2 transition strengths between low-lying, as well as giant resonance, states can be reproduced within the framework of the model without introducing proton and neutron effective charges. And as we will see below, it is also possible to define a symplectic extension for the pseudo SU(3) scheme — called the pseudo symplectic model — which can be applied to the study of giant resonance and related phenomena in heavy ($A \gtrsim 100$) nuclei.

3.1 Tenets of the Theory

As indicated above, the symplectic model extends the SU(3) picture by allowing for inter-shell $2\hbar\omega$ excitations of the monopole and quadrupole type. This feature is essential for modeling a system's quadrupole coherence. Specifically, if Q^c and Q^a denote the collective and algebraic forms for the quadrupole operator as defined above, then $Q^c = Q^a + (3/2)^{1/2}(B_2^+ + B_2^-)$ where B_2^+ and B_2^- are pure $2\hbar\omega$ quadrupole raising and lowering operators, respectively. There are two types of raising (B_l^+) and lowering (B_l^-) operators in the symplectic scheme: the monopole (l=0) and quadrupole (l=2) parts of the cartesian tensor operator form, $Q^c_{\alpha,\beta} = (x_\alpha x_\beta - r^2/3\ \delta_{\alpha,\beta})$. The 6 raising ($B^+_{lm}$) and 6 lowering ($B^-_{lm}$) operators ($l = 0$ and 2 with projections $m_l = -l, -l+1, ..., l-1, +l$), together with the number operator N (differing only by an additive constant from the oscillator Hamiltonian) as well as the 3 angular momentum operators L_α and the 5 quadrupole operators Q^a_μ, form a 21 dimensional set which

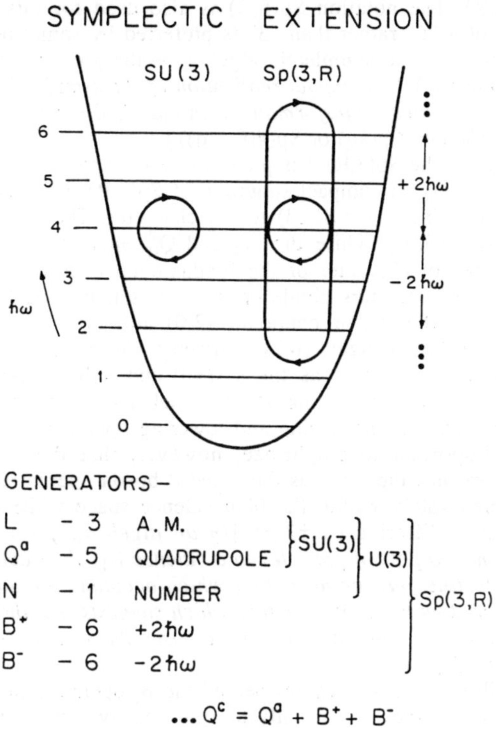

Figure 3. Schematic plot illustrating the symplectic [Sp(3,R)] extension of the Elliott [SU(3)] shell-model scheme. Sp(3,R) is a non-compact group with 21 generators while SU(3) is its largest compact subgroup with 8 generators. The additional operators generate particle-symmetric couplings to shells that are multiples of 2ħω removed (up and down) from the 0ħω valence space.

closes under commutation of its members. This set of 21 operators generates a realization of the non-compact Sp(3,R) (symplectic) group. Since this 21-dimensional Lie algebra contains the 8-dimensional SU(3) sub-algebra generated by the 3 angular momentum operators L_α and the 5 quadrupole operators Q^a_μ, SU(3) is a subgroup of Sp(3,R). Other subgroups, such as the symmetry group of the quantum rotor ($T_5 \wedge$ SO(3) which is generated by the 3 angular momentum operators L_α and the 5 quadrupole operators Q^c_μ, are also contained in the rich Sp(3,R) group structure. However, SU(3) is the largest (maximal) compact (finite dimensional irreps) subgroup of Sp(3,R). The notation Sp(6,R) is also used for this symplectic group structure. The use of a "6" rather than "3" is preferred by some, because there are 6 fundamental operators in the symplectic algebra — the 3 coordinates (x_α) and the 3 (conjugate) momenta (p_α). [*A compact realization of the symplectic group also arises in a description of canonical transformations in classical mechanics and that context presents a similar choice*: Sp(#q) or Sp(#(p + q)).]

The structure of the Sp(3,R) group and its relation to SU(3) is shown in Figure 3. Since SU(3) is the largest compact subgroup of Sp(3,R), it is useful to decompose the Sp(3,R) generators into their SU(3) tensor character. The number operator is an SU(3) scalar [$(\lambda,\mu) = (0,0)$], while the L_α and Q^a_μ operators generate SU(3) and therefore have the tensor character of the fundamental 8 dimensional $(\lambda,\mu) = (1,1)$ irrep. The remaining 12 operators divide up into two sets of 6 each: the raising (B^+_l) operators which have the tensor character (2,0) and the adjoint lowering (B^-_l) operators which have the conjugate (0,2) symmetry. The B_l operators therefore have the same SU(3) tensor character as the s (l=0) and d (l=2) boson creation and annihilation operators of the interacting boson model (IBM). This feature comes about because in both applications the raising and lowering operators generate two-phonon excitations. It is important to emphasize, however, that the B_l operators of the symplectic model are *not* the same as the s and d boson operators of the IBM since the latter are defined solely within the $0\hbar\omega$ valence space while the former couple major oscillator shells differing by $\pm 2\hbar\omega$. [*In an algebraic fermion model* (AFM) *picture, the addition of a pair of particles to a valence* (η) *shell adds* $2\eta\hbar\omega$ *of energy to the system* (*neglecting zero point motion which accounts for an additional* $3\hbar\omega$). *This is to be compared with the* IBM *picture which suggests that the addition of a pair is effected by the* s^+ *or a* d^+ *operator which increases the system's energy by* $2\hbar\omega_{IBM}$. *Clearly,* $\hbar\omega_{IBM} \neq \hbar\omega_{AFM} \equiv \hbar\omega$.]

We now will explore the physics behind the B_l operators in greater depth, and ultimately discover why they can usually be replaced by simpler boson forms. First note that the B_l operators, as for Q^a and Q^c, are standard one-body objects which act in the same way on all particles. As a consequence, these operators do not change the particle permutation symmetry when they are applied to a many-particle wavefunction — the labelling of basis states by the irreducible representation label $[1^m]$ of $U(k\Omega)$ survives, even though the application of a B^+_l operator to a $0\hbar\omega$ configuration implies the promotion of a particle into a higher-lying shell $2\hbar\omega$ up. The system has no way of knowing which of the particles gets promoted; all particles participate equally to the excitation process. In this regard it is important to emphasize that the B^+_l operators generate very special (coherent) excited state configurations, and not generate generic $2\hbar\omega$ shell-model states. For example, a $2\hbar\omega$ excitation generated by promoting two particles across a single major shell (rather than a single particle across two major shells) cannot be represented in terms of a B^+_l operator acting on a $0\hbar\omega$ state; particle-hole (p-h) states of this type represent independent (excited but non-symplectic) shell-model configurations. Another example is a $4\hbar\omega$ configuration generated via a hexadecapole excitation operator — such a configuration cannot be represented as a double symplectic $(B^+_2 \times B^+_2)^4$ excitation $\{r^4Y_4(\hat{r}) \neq [r^2Y_2(\hat{r}) \times r^2Y_2(\hat{r})]^4\}$. Negative parity configurations must also to added to the symplectic picture, just as they are in the usual shell-model scheme. In general, any (non-spurious) excited (1p-1h, 2p-2h, 3p-3h, etc.) configuration that the B^-_l operator destroys is an independent shell-model configuration, however symplectic excitations can be superimposed on all such independent shell-model configurations in exactly the same way as they are added to $0\hbar\omega$ configurations.

Because the B_l operators act only on spatial coordinates, they are spin or spin-isospin scalar objects that preserve the $[\tilde{f}]$, as well as the $[1^m]$ symmetry label of a many-particle basis state. Since the spatial symmetry group $U(\Omega)$ and the spin or spin-isospin symmetry group $U(k)$ are complementary within antisymmetric irreps of $U(k\Omega) \supset U(\Omega) \otimes U(k)$, the $U(\Omega)$ and $U(k)$ irreps which enter into the construction of many-particle basis states are conjugates of one another. This means that the [f] symmetry label is also good; the B_l operators are block-diagonal within irreducible representations of $U(\Omega)$ as well as $U(k)$. This property of the symplectic generators is extremely important — they do not involve spin or isospin degrees of freedom, and therefore do not couple different space-spin/space-spin-isospin symmetries of a standard LS/LST-coupled shell model. This feature means the symplectic scheme is actually a very simple and straightforward extension of the ordinary shell-model picture; everything one has learned about shell-model symmetries survives, and is expanded to include vertical (multi-$\hbar\omega$) excitations as required to accommodate large deformation and enhanced quadrupole transition strengths. Looking ahead, one can anticipate that this feature also insures the success of the pseudo symplectic model (which involves symplectic excitations superimposed on the pseudo-spin and pseudo SU(3) pictures) — quadrupole coherence built in via pseudo symplectic operators will not destroy the underlying pseudo-spin symmetry.

We will now consider the structure of the many-particle basis states in greater detail. Each application of a $B_l^{\dagger}(B_l^{-})$ adds(subtracts) $2\hbar\omega$ of energy to(from) the system. It follows from this that the application of a B_l^{-} operator to *any* $0\hbar\omega$ basis state destroys it because these $0\hbar\omega$ states cannot give up $2\hbar\omega$ of energy without shifting a valence particle into an already occupied core configuration as this would violate the Exclusion Principle. The usual $0\hbar\omega$ shell-model states are therefore lowest-weight (LW) states with respect to the B_l^{-} operators of the symplectic algebra. As for the familiar angular momentum scenario, all the states belonging to a particular $0\hbar\omega$ SU(3) irrep (labelled (λ_s,μ_s) in what follows) can all be generated from the unique state which is lowest weight with respect to the action of the generators of the SU(3) subgroup of Sp(3.R): $C_{ij}|(\lambda_s,\mu_s)LW\rangle = 0$, for all $i<j$. Applications of the SU(3) raising operators, $C_{i,j}$ with $i>j$, to the lowest-weight state suffice to generate all states of the (λ_s,μ_s) irrep (equal in number to the dimensionality, $\dim(\lambda_s,\mu_s) = (\lambda_s+\mu_s+2)(\lambda_s+1)(\mu_s+1)/2$, of the representation) just as in the SU(2) case. For example, the C_{21} operator shifts a phonon from the 1 direction to the 2 direction. These shifts conserve the number of oscillator quanta and can be considered horizontal displacements in the $0\hbar\omega$ plane. Each application of a $B_l^{\dagger}$ operator adds $2\hbar\omega$ of energy to the system, but in this case [in contrast with the raising operators of SU(2) and SU(3)] there is no cap on the number of allowed applications — the entire set of states (infinite in number because Sp(3,R) is non-compact) belonging to a symplectic irrep can be generated from $|(\lambda_s,\mu_s)LW\rangle$, a single unique $0\hbar\omega$ configuration that is lowest weight — first with respect to the symplectic lowering operators, and second with respect to SU(3). (This argument extends to excited non-symplectic shell-model configurations.) It follows that Sp(3,R) irreps can be identified by the (λ_s,μ_s) label of this lowest-weight state. A note of caution is in order, because an arbitrary $2\hbar\omega$ excitation could be spurious; fortunately, the $B_l^{\dagger}$ operators of the symplectic model act only on internal coordinates of a system, hence this is not a problem.

A boson realization of the symplectic $B^{\pm}_{lm}$ generators can be introduced as well. This simplification leads to a more tractable version of the theory. In this case, a formal group theoretical expansion-contraction procedure such as the one presented in Section 2 for SU(3) can be invoked to accomplish the transformation. Specifically, it can be shown that the generators of the Sp(3,R) group transform as follows:

$$
\begin{aligned}
B^{\pm}_{lm} &\rightarrow \sqrt{(4/3)\,N_s}\, b^{\pm}_{lm}\,,\\
N &\rightarrow N_s + 2N_b\,,\\
L_m &\rightarrow L^s_m + L^b_m\,,\\
Q^a_m &\rightarrow Q^s_m + Q^b_m\,,
\end{aligned}
\qquad (3\text{-}1)
$$

where l=(0,2) and the $b^{\pm}_{lm}$ operators satisfy the usual boson commutation relations. (These AFM boson operators should not to be confused with the similarly labeled IBM boson operators.) The s and b indices (subscripts and superscripts) are used to distinguish the $0\hbar\omega$ shell-model and boson parts, respectively, of the N, L and Q^a operators. The N_s operator counts the total number of oscillator bosons in the $0\hbar\omega$ system, including the zero point energy, and is therefore a large number, while N_b counts the number of boson excitations, n for a $2n\hbar\omega$ theory. The 2 multiplying N_b in the expression for N enters, because each boson carries $2\hbar\omega$ units of energy. The L and Q operators have, in addition to their usual $0\hbar\omega$ shell-model parts, an added boson contribution which derives from the symplectic extension. It is the boson extension of the quadrupole operator that leads to the large enhancements observed in E2 transitions strengths. The contraction procedure replaces the symplectic algebra by the direct sum of a 6-dimensional Heisenberg-Weyl algebra and su(3) [sp(3,R) $\rightarrow$ w(6) + su(3)] while the group itself reduces to a semi-direct product structure, Sp(3,R) $\rightarrow$ W(6) $\wedge$ SU(3).

This structure leads to a simple yet elegant scheme for labelling basis states of the contracted symplectic model,

$$
\begin{array}{ccccccccc}
U_b(6) \wedge U_s(3) & \rightarrow & SU_b(3) \wedge SU_s(3) & \rightarrow & SU(3) & \rightarrow & SO(3) & \rightarrow & SO(2)\\
N_b \quad N_s & & (\lambda_b\mu_b) \quad (\lambda_s\mu_s) & \rho & (\lambda\mu) & \kappa & L & & M
\end{array}
\qquad (3\text{-}2)
$$

which explicitly demonstrates the manner in which the usual $0\hbar\omega$ shell-model result is augmented with the boson extension. The $U_b(6) \rightarrow SU_b(3)$ reduction is the same as one encounters in the U(6) $\rightarrow$ SU(3) limit of the IBM. For example, the (λ_b,μ_b) irreps that can occur in a space with N_b bosons follows the rule for the decomposition of a symmetric U(6) irrep [n], built out of n products of (2,0) SU(3) irreps, into a direct sum of SU(3) representations:

$$
\prod^{n} [\otimes\ (2,0)] = \sum_{n_1\geq n_2\geq n_3\geq 0,\ \text{all } n_i \text{ even}}^{n_1+n_2+n_3=2n} [\oplus\ (n_1-n_2, n_2-n_3)]. \qquad (3\text{-}3)
$$

Accordingly, the basis states for identical nucleon configurations of the symplectic model can be labelled as

$$|\Psi^{m[f]}_{\gamma JM_J}\rangle \equiv |m[f]\alpha[N_b(\lambda_b\mu_b)N_s(\lambda_s\mu_s)]\rho(\lambda\mu)\kappa LSJM_J\rangle \tag{3.4}$$

where we have replaced the K label with a running index κ to denote an orthonormal set of bases states. The κ=1 state corresponds to the normalized K=0 or 1 configuration obtained when the minimum of (λ,μ) is even or odd, respectively; the κ=2 state refers to the normalized K=2 or 3 configuration orthogonal to the κ=1 solution; etc. If there are no boson (N_b=0) excitations, this multi-ħω scenario reduces to the usual 0ħω shell-model result.

To summarize, the structure of the set of states (infinite in number and quadrupole coherent) associated with every allowed (0ħω or excited, but then independent and non-symplectic) SU(3) shell-model configuration should now be clear. We refer to the collection of all 0ħω configurations as the 0ħω *horizontal* shell-model space and the towers built on each of the SU(3) (λ_s,μ_s) irreps as the *vertical* extension of that irrep. Each vertical extension can be partitioned into horizontal slices with the states within each slice representable as a homogeneous polynomial function in the B^+_{lm} [symplectic → Sp(3,R)] or b^+_{lm} [boson approximation → W(6)] raising operators acting on a parent 0ħω (λ_s,μ_s) shell-model configuration. The degree of these homogeneous polynomials in the raising operators identifies the horizontal slice to which the basis state belongs; for example, it belongs to the 2nħω slice of the (λ_s,μ_s) irrep if the polynomial acting on the (λ_s,μ_s) 0ħω configuration is of order n in the raising operators. Remember as well that in each (λ_s,μ_s) irrep there is one state which is also lowest-weight with respect to SU(3). It follows that the full irrep space, its horizontal as well as its vertical extensions, can be generated by applying the raising operators of SU(3) to span its horizontal extent, and the B^+_{lm} [Sp(3,R)] or b^+_{lm} [W(6)] raising operators to generate its vertical extension. Because of this structure, when dealing with interactions (see below) it is useful to speak in terms of those that effect only horizontal mixing, such as pairing, and those like the quadrupole-quadrupole interaction which effect vertical mixing but do not break SU(3) within the horizontal slices. Some interactions generate both horizontal *and* vertical mixing and these present the greatest challenge to the structure theorist as they are the most difficult to handle.

3.2 Symplectic Model Hamiltonian

An appropriate Hamiltonian for describing rotational phenomena within the symplectic model consists of the harmonic oscillator, which provides the background shell structure; the quadrupole-quadrupole interaction, $Q^c \cdot Q^c$; a residual interaction that should include (for example) the single-particle orbit-orbit and spin-orbit terms, as well as pairing and other interactions involving currents and a hexadecapole-hexadecapole form as needed to get M1 and E4 transition strengths correct, respectively; etc. However, most applications of the theory are much less ambitious than this, restricting the interaction to terms that can be expressed solely in terms of generators of the Sp(3,R) algebra. The problem with proceeding to more realistic Hamiltonian forms is that codes for determining matrix elements of operators like pairing between basis states of the SU(3) scheme are not generally available — even for leading irreps of the $0\hbar\omega$ horizontal space. Despite this limitation, however, much can and has been learned about the buildup of quadrupole collectivity using the simpler operator forms.

There is another problem, because although observed E2 transition strengths in light nuclei (^{24}Mg, for example) can be reproduced with only the harmonic oscillator H_0 and $Q^c \cdot Q^c$ terms taken into account ($H = H_0 - (1/2)\chi Q^c \cdot Q^c$), the required value for χ is so large that the $Q^c \cdot Q^c$ interaction actually destroys the mean field of the oscillator established by H_0. To render $Q^c \cdot Q^c$ an appropriate residual interaction, it is necessary to first remove its contribution to the mean field. Once this is done, it becomes an acceptable residual interaction compatible with the structure established by H_0. An appropriate form is therefore

$$H = H_0 - \frac{1}{2}\chi[Q^c \cdot Q^c - (Q^c \cdot Q^c)_{\text{shell}}] + H_r, \tag{3-5}$$

where $Q^c \cdot Q^c$ has been replaced by $Q^c \cdot Q^c - (Q^c \cdot Q^c)_{\text{shell}}$ and $(Q^c \cdot Q^c)_{\text{shell}}$ is an operator that reproduces traces of $Q^c \cdot Q^c$ in all major shells of the oscillator. Simple analytic expressions can be given for the trace-equivalent forms of most operators. In particular, for $Q^c \cdot Q^c$ one finds that

$$(Q^c \cdot Q^c)_{\text{shell}} = \frac{65}{7}N^2 + \frac{250}{7}N + \frac{10}{3}NN_s + \frac{15}{4}C_2^s + 10N_s, \tag{3-6}$$

where N counts the total number of oscillator quanta [$N=N_s+2N_b$, with N_s counting quanta up through the $0\hbar\omega$ level and N_b counting those associated with the $2\hbar\omega$ symplectic excitations, see (3-1)], and C_2^s is the second order Casimir invariant of SU(3) at the $0\hbar\omega$ level with eigenvalue $[(\lambda_s+\mu_s+3)(\lambda_s+\mu_s)-\lambda_s\mu_s]$ in the (λ_s,μ_s) irrep.

The procedure one uses to obtain the boson form for Hamiltonian (3-5) is straightforward. First of all, each operator is expressed in normal ordered form: after making the substitution $Q^c = Q^a + \sqrt{6/2}(B^- + B_2^-)$, the $2\hbar\omega$ raising generators (B_l^+) are arranged to be in leftmost positions, and the lowering generators (B_l^-) are shifted into rightmost positions. After this normal ordering of the raising and lowering operators, substitutions (3-1) are made. This procedure yields the following boson form for the Hamiltonian:

$$\begin{aligned} H &= H_0 - \frac{1}{2}\chi[Q^c \cdot Q^c - (Q^c \cdot Q^c)_{\text{shell}}] + H_r, \\ &\xrightarrow{\text{boson approx}} 2\hbar\omega N_b + \chi\Big[(\frac{5}{3}N_s + \frac{75}{7})N_b + \frac{25}{14}N_b^2 \\ &\qquad - \frac{1}{2}Q^a \cdot Q^a - \sqrt{2N_s}(b^+ \cdot Q^a + Q^a \cdot b) \\ &\qquad + 10\sqrt{N_s}(b_{00}^+ + b_{00}) - N_s(b^+ \cdot b^+ + b \cdot b) \\ &\qquad - 2N_sN_d + \frac{5}{4}C_2(\lambda_s,\mu_s)\Big] \\ &\qquad + aL^2 + bX_3^a + cX_4^a, \end{aligned} \tag{3-7}$$

where C_2 is the second order Casimir operator of the SU(3) algebra, N_d counts the number of $l = 2$ bosons, the expansion of $(Q^c \cdot Q^c)_{shell}$ is included, and all additive constant terms have been dropped. A specific choice for H_r is included in this expression, namely, the interaction that we determined in Section 2 to be the image of the rotor — see (2-7). In this form, matrix elements of the Hamiltonian [in the boson basis given by (3-2) and (3-4)] can be easily evaluated and the matrix diagonalized.

An estimate for the strength χ of the quadrupole-quadrupole interaction can be determined by equating the difference in the expectation value of H in the 0^+ state of the $0\hbar\omega$ (λ_s,μ_s) irrep, and the so-called stretched $2\hbar\omega$ $(\lambda_s+2,\ \mu_s)$ irrep to the excitation energy, $80A^{-1/3}$, of the giant monopole resonance:

$$\chi = \frac{A^{-1/3}}{4\lambda_s + 2\mu_s + \frac{15}{4} + N_s(\langle N_d\rangle - 5/6)} . \tag{3-8}$$

This result follows because in $L^\pi = 0^+$ states $Q^a \cdot Q^a = 4C_2$ and matrix elements of the X_3 and X_4 operators, like those of L^2, vanish. The only unknown in (3-8) is the expectation value of N_d in the $2\hbar\omega$ space, and an analytic expression for it in terms of the (λ_s,μ_s) can be given:

$$\langle N_d\rangle = \frac{2\lambda_s(\lambda_s+2) + \mu_s(2\lambda_s+1)}{3(\lambda_s+1)(\lambda_s+\mu_s+2)} . \tag{3-9}$$

Though this estimate for χ has been shown to be quite good, it can be improved by using it to calculate eigenfunctions of the Hamiltonian in (3-7) with a=b=c=0 in the 0^+ and 2^+ spaces and calculating the $2_1^+ \rightarrow 0_1^+$ reduced E2 transition probability. If the E2 strength is smaller (bigger) than the experimental result, a bigger (smaller) value of χ should be chosen. This step can be repeated until agreement with the experimental result is achieved. The assumption of a linear relation between E2 values and χ suffices in interpolating between results. The residual interaction can be ignored in this iterative process — even though it changes the excitation energy of the 2_1^+ state, it has virtually no effect on the calculated $2_1^+ \rightarrow 0_1^+$ transition strength.

Values for the parameters b and c of the residual interaction in (3-7) can be determined from the splitting of the lowest (2_1^+) and first excited (2_2^+) J=2 states. Specifically, in establishing a direct mapping between rotor and SU(3) Hamiltonians, an operator expression was found for the square of the projection of the angular momentum on an intrinsic body-fixed symmetry axis, (2-3). In that expression, the λ_α denote principal moments of inertia, and were related to the SU(3) irrep labels λ and μ by requiring invariants of the rotor and SU(3) to map onto one another, (2-9). To the extent that the 2_2^+ state is a K=2 band-head configuration, it follows that

$$b = \frac{\alpha\lambda_3}{2\lambda_3^2+\lambda_1\lambda_2} \quad \text{and} \quad c = \frac{\alpha}{2\lambda_3^2+\lambda_1\lambda_2} \quad \text{where} \quad \alpha = (E_{2_2^+} - E_{2_1^+})/4. \tag{3-10}$$

Note, in particular, that the b/c ratio is a constant given by $\lambda_3 = (2\lambda+\mu+3)/3$.

In a similar fashion, the parameter a that multiplies L^2 in (3-7) can be given in terms of the experimental value for the inertia parameter $(\hbar^2/2I)$,

$$a = \hbar^2/2I - \frac{3}{2}\chi + \frac{\alpha\lambda_1\lambda_2}{2\lambda_3^2+\lambda_1\lambda_2}. \tag{3-11}$$

The second term in this expression reduces the parameter a from its $(\hbar^2/2I)$ value because a L^2 contribution also enters through $Q^a \cdot Q^a = 4C_2 - 3L^2$. The third term is required for the L^2 dependence in the K^2 operator. If the leading irrep has $\mu=0$, the theory predicts an absence of a K=2 band associated with the K=0 ground band. When this happens — for example, for ^{20}Ne of the ds-shell with $(\lambda,\mu) = (8,0)$ and for the ^{238}U case with $(\lambda,\mu) = (54,0)$ — it is simplest to set $\alpha = 0$ in both (3-10) and (3-11). The parameter, a, is then just the moment of inertia reduced by $(3/2)\chi$, the quadrupole-quadrupole contribution. In general, the parameter, a, is the least well-determined by these simple arguments. This is both understandable and fortunate: understandable as it involves a sum of three different terms and makes use of a replacement of $Q^c \cdot Q^c$ by $Q^a \cdot Q^a$; and fortunate, because the structure of the calculated eigenstates is independent of the value or the parameter, a.

3.3 Pseudo Symplectic Scheme

As discussed in Section 2, for heavy ($A \gtrsim 100$) nuclei there are two valence shells, one for protons (π) and one for neutrons (ν). Furthermore, each of these two spaces is comprised of a set of normal parity orbitals which may be identified as members of a pseudo shell ($\tilde{n}_\alpha = n_\alpha - 1$) and a unique parity intruder level with ($j_\alpha = n_\alpha + (3/2)$ from the shell above, where $\alpha=(\pi,\nu)$. If the normal (n) parity part of each space is partitioned into irreps of pseudo SU(3), and a seniority scheme is used for the single j shell, unique (u) parity configurations, then the many-particle basis states can be given as angular-momentum-coupled products of four distinct and separate (proton and neutron) subspaces: $|\psi_J\rangle = |\{[(J_{n_\pi} \times J_{n_\nu})J_n] \times [(J_{u_\pi} \times J_{u_\nu})J_u]\}J\rangle$. When the residual proton-neutron interaction is of the quadrupole-quadrupole type (regardless of the specific nature of the interaction in the separate subspaces) yrast states of strongly deformed even-even nuclei *below* the backbending region are well-represented by strong coupled pseudo SU(3) basis states $[(\lambda_\pi,\mu_\pi) \times (\lambda_\pi,\mu_\pi) \rightarrow (\lambda,\mu)]$ and seniority zero-coupled ($J_{u_\pi} = 0$ and $J_{u_\nu} = 0$, so $J_u = J_{u_\pi} \times J_{u_\nu} = 0$) unique parity configurations. Of course, the structure of states *through and beyond* the backbending region will be more complicated. In particular, since they lie at an excitation energy that is comparable to the pairing gap, seniority two ($J_u = 2, 4, \ldots$) and higher unique parity configurations must be involved.

The pseudo symplectic scheme extends the pseudo SU(3) picture by allowing for inter-shell $2\hbar\omega$ excitations of the monopole and quadrupole type among the normal parity orbitals, just as the symplectic scheme does for light ($A \lesssim 25$) nuclei. In this case, however, the underlying symmetry group is pseudo-Sp(3,R) which contains pseudo SU(3) as a subgroup. The generators of the pseudo symplectic symmetry groups (one for the proton space and another for the neutrons) include the 3 angular momentum operators ($\tilde{L}_\alpha$) and 5 quadrupole operators ($\tilde{Q}^a_\alpha$) — which together generate an $\widetilde{SU}_\alpha(3)$ subgroup — plus 6 raising $\tilde{B}^+_\alpha$ and 6 lowering $\tilde{B}^-_\alpha$ operators. In total, these 21 operators generate the non-compact $\widetilde{Sp}_\alpha(3,R)$ extension of the compact $\widetilde{SU}_\alpha(3)$ group. Since multiple $2\hbar\omega$ excitations in the normal parity spaces are included, it would seem that unique parity excitations of this type should also be considered, as well as mixing between the two. But it can be argued as before that this is not necessary for states in even-even nuclei which lie below the backbending region, since the unique parity parts of these wavefunctions will be dominated by pairing correlations which differ from quadrupole correlations, because they require the involvement of higher shells for strength enhancement. Only for states immediately below and above the pairing gap (which coincides with the onset of backbending) or for odd-A nuclei are unique parity configurations expected to play an important role in the dynamics. However, we must point out that as in the $0\hbar\omega$ case, this assumption has not be thoroughly tested.

The pseudo symplectic model Hamiltonian is taken to be the sum of proton and neutron pseudo harmonic oscillator Hamiltonians, plus a real quadrupole-quadrupole interaction acting between the particles, including π-ν terms as well as the identical particle π-π and ν-ν interactions. And as indicated above, contributions to the dynamics from protons and neutrons in the unique parity levels have been suppressed thus far. Also, to insure that the mean field of the oscillator is conserved under the quadrupole-quadrupole perturbation, as in the case of light nuclei, the $\tilde{Q}^c_\alpha \cdot \tilde{Q}^c_\alpha$ interactions are replaced by $\tilde{Q}^c_\alpha \cdot \tilde{Q}^c_\alpha - (\tilde{Q}^c_\alpha \cdot \tilde{Q}^c_\alpha)_{shell}$ where $(\tilde{Q}^c_\alpha \cdot \tilde{Q}^c_\alpha)_{shell}$ is an operator that reproduces single shell traces of the $Q^c_\alpha \cdot Q^c_\alpha$ operator in the α-th subspace. Thus we have

$$H = \tilde{H}_0 - \frac{1}{2}\chi[Q^c \cdot Q^c - (Q^c_\pi \cdot Q^c_\pi)_{shell} - (Q^c_\nu \cdot Q^c_\nu)_{shell}] + H_r, \tag{3-12}$$

where

$$\tilde{H}_0 = \tilde{H}_{0\pi} + \tilde{H}_{0\nu} \text{ and } Q^c = Q^c_\pi + Q^c_\nu . \tag{3-13}$$

As before, the superscript c in these equations denotes the collective quadrupole operators as opposed to the symmetrized algebraic ones of a $0\hbar\omega$ pseudo-S(3) scheme. An appropriate residual interaction, H_r, is also included and in applications has been chosen to be a rotor-like form (see (3-7) and (3-15), below).

We can express the mass quadrupole operators Q^c_π and Q^c_ν as of generators of the real (not pseudo) symplectic algebras: $Q^c_\alpha = Q^a_\alpha + (3/2)^{1/2}(B^+_{2\alpha} + B^-_{2\alpha})$. And as for the pseudo SU(3) case, these operators ($O = Q^a, B^+, B^-$) can all be expanded in terms of pseudo SU(3) tensors as

$$O = \kappa\tilde{O} + \cdots , \tag{3-14}$$

where $\tilde{O}$ has the same tensorial character as O and the ellipses represents other tensors which arise because the normal $\leftrightarrow$ pseudo transformation is not symmetry preserving. Once again, the constant κ in this expansion is always greater than unity, ranging from ~1.4 for $O = Q^a(\tilde{n} = 0)$ to ~1.1 for $O = B^+_{2m}(\tilde{n} = 6)$. An average κ value for actinide nuclei is 1.14. The expansion coefficients of the non-generator parts are usually less than ten percent of the leading (generator) term. In pseudo SU(3) applications, these terms have been found to have no noticeable effect on the spectrum and yield less than a one percent change in calculated electromagnetic transition rates. In applications that employ the symplectic extension, it is therefore customary to drop the additional tensors.

Neglecting higher order terms, the Hamiltonian (3-12) can be rewritten as

$$H = \hbar\omega\tilde{N} - \frac{1}{2}\chi\left[\tilde{Q}^c\cdot\tilde{Q}^c - (\tilde{Q}^c_\pi\cdot\tilde{Q}^c_\pi)_{shell} - (\tilde{Q}^c_\nu\cdot\tilde{Q}^c_\nu)_{shell}\right] + a\tilde{L}^2 + b\tilde{X}^a_3 + c\tilde{X}^a_4. \quad (3\text{-}15)$$

In this expression, the average value for the κ introduced in (3-14) is absorbed into the χ, and the last three terms are an explicit form for H_r (as in (3-7), these can be used to fix the K-band splitting and moment of inertia of the system). Note that $\tilde{L}$ is the total pseudo angular momentum operator while the $\tilde{X}^a_3$ and $\tilde{X}^a_4$ terms are third and fourth order rotational scalars built out of the algebraic quadrupole (Q^a) operators: $\tilde{X}^a_3 = [\tilde{L} x \tilde{Q}^a x \tilde{L}]$ and $\tilde{X}^a_4 = [(\tilde{L} x \tilde{Q}^a)^1 x (\tilde{Q}^a x \tilde{L})^1]$ — see (2-2). As in the pseudo SU(3) case, this special form for the residual interaction means the moment of inertia and band splitting of the low-lying states can be adjusted without otherwise affecting the dynamics. In particular, note that matrix elements of this H_r vanish in 0^+ states. In analogy with (3-7) for the normal symplectic case, one can write down a boson expansion form for (3-15). In this case the eigenvalue of the $\tilde{N}_s$ operator is much larger, and accordingly the boson approximation much better, because one is closer to the contraction limit of the theory. And as for light nuclei, an estimate for the strength χ of the quadrupole-quadrupole interaction can be made by equating the difference in the expectation value of H in the 0^+ state of the $0\hbar\omega$ $(\tilde{\lambda},\tilde{\mu})$ irrep and the stretched $(\tilde{\lambda}+2,\tilde{\mu})$ pseudo SU(3) irrep at $2\hbar\omega$ to the excitation energy, $80A^{-1/3}$, of the giant monopole resonance [see (3-9)]. And as before, this estimate for χ can be improved upon by using it to calculate eigenfunctions of the Hamiltonian, (3-15), with $a=b=c=0$ in the 0^+ and 2^+ spaces and calculating the $2^+_1 \to 0^+_1$ reduced E2 transition probability. In fact, all the expressions and procedures given for determining the parameters χ, a, b, and c in the normal symplectic case apply without change to the pseudo symplectic situation. For example, if the E2 strength is smaller (larger) than the experimental result a larger (smaller) value of χ should be chosen, etc.

3.4 Miscellaneous Applications

Space does not permit an extended discussion of applications of the model. The most successful of these applications involves theoretical prediction of E2 transition strengths. For a discussion of this we refer the reader to the nuclear physics literature on this subject. Here we will consider a related feature of the theory — the fact that a direct connection between collective and shell-model theories of nuclear structure can be established via a consideration of energy surfaces by combining the shell-model rotor formulation presented in Section 2 with the symplectic scheme given above. To see this, consider Figure 4(A) which gives a schematic representation of the expectation value of H as a function of deformation in $J^\pi=0^+$ states of a system. The curve shows a sharp rise beyond the minimum in the potential at deformation $\beta = \beta_0$, which is obtained for the leading $0\hbar\omega$ representation of SU(3). This rise comes about because configurations with $\beta > \beta_0$ exist only in the higher $2n\hbar\omega$ $(n \neq 0)$ spaces. In fact, the shape of the curve in the $\beta > \beta_0$ domain can be understood as resulting from

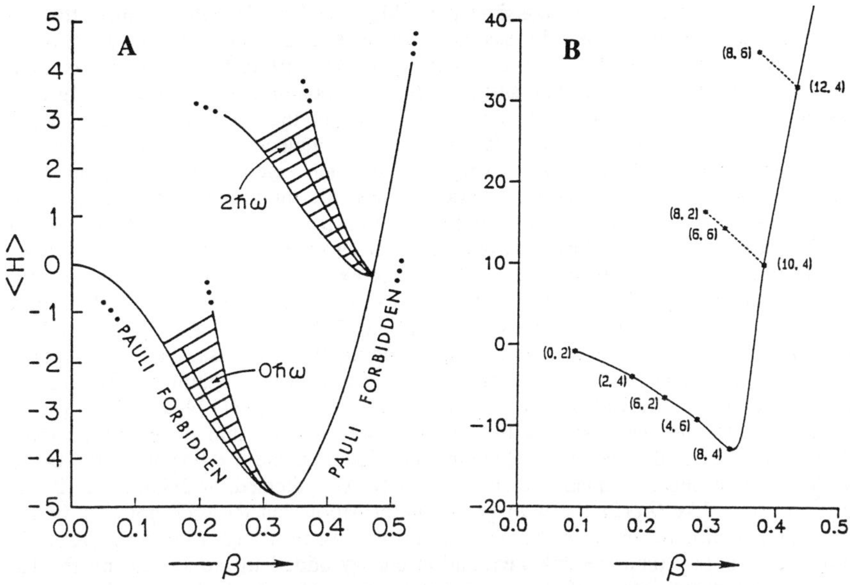

Figure 4. **(A)** Schematic diagram illustrating a shell-model interpretation of the collective-model potential energy surface. The curve to the left of the minimum is defined by 0ħω irreps of SU(3) while the curve to the right is determined by the 2nħω irreps of the symplectic model. **(B)** Expectation value of H in L=0 basis states for the indicated irreps of ^{24}Mg. The curve cuts through the potential energy surface with γ=20.6° for the (λ,μ)=(8,4) irrep. The results for the expectation of H in calculated eigenstates is similar: the binding is greater so the minimum is deeper and it is shifted to the right (larger deformation) due to the mixing of SU(3) irreps in the symplectic band.

competition between a strong and near-linear rise with deformation due to the oscillator term in the Hamiltonian [$\beta \sim \sqrt{C_2} \sim (\lambda_s+2n)$ so $\Delta\langle H_0\rangle \sim 2n \sim (\beta-\beta_0)$], and a weaker (but nonetheless important) negative quadratic dependence on deformation coming from the $-(1/2)\chi Q^c\cdot Q^c$ term in H [$Q^c\cdot Q^c \sim \beta^2 \sim C_2$ so $\Delta\langle(1/2)Q^c\cdot Q^c\rangle \sim (2n)^2 \sim (\beta-\beta_0)^2$]. In contrast with this behavior for $\beta > \beta_0$, the shape of the curve in the $\beta < \beta_0$ domain is defined solely by 0ħω (n = 0) irreps less deformed than the leading irrep; the curvature comes solely from the $Q^c\cdot Q^c$ term, which as for $\beta > \beta_0$, is very nearly an negative quadratic function of β. Of particular importance is the fact that the curve's rise to the right ($\beta > \beta_0$, 2nħω irreps) as well as the left ($\beta < \beta_0$, 0ħω irreps) of the $\beta = \beta_0$ minimum, is due to the Pauli Principle, and not to higher order terms in the potential as the collective model suggests. {*Recall that in order to realize a non-zero equilibrium deformation* (β_0,γ_0) *in the collective model picture requires* $\beta^4=(\beta^2)^2$, $\beta^6=(\beta^2)^3$, $\beta^3\cos3\gamma$, *and even* $(\beta^3\cos3\gamma)^2$ *terms in the potential, in addition to the* β^2 *dependence that comes from the oscillator structure. Stable triaxial* $(\gamma_0\neq0)$ *shapes require a sixth order theory since the dependence on* γ *enters first through the third order invariant* [$(Q^c\times Q^c)\cdot Q^c \sim \beta^3\cos3\gamma$]}.

As an explicit example, consider the ^{24}Mg case. To adequately display the full energy surface requires a three-dimensional plot since, in general, <H> is a function of both β and γ. The curve shown in the figure must therefore to be interpreted as a cut ($\gamma \approx 0$) through a three-dimensional surface. The corresponding curve for the ^{24}Mg case was generated by calculating the expectation value of the Hamiltonian ($H = H_0 - (1/2)\chi[Q^c \cdot Q^c - (Q^c \cdot Q^c)_{shell}]$) in basis states of the symplectic scheme. It is shown in Figure 4(B). The resultant curve is very similar when eigenstates rather than basis states are used to calculate the expectation values; the minimum is then deeper and shifted towards larger deformations due to SU(3) irrep mixing. The $0\hbar\omega$ SU(3) irreps for ^{24}Mg are $(0s)^4(1p)^{12}(2s0d)^8$ configurations, and follow from the decomposition of representations of U(6) into irreps of SU(3). For example, the most symmetric U(6) spatial symmetry ([f]=[44]) contains the following SU(3) irreps with $J^\pi=0^+$ states: $\{(\lambda_s,\mu_s)\}=\{(8,4);(4,6);(0,8);(6,2)^2;(2,4)^2;(4,0)^2;(0,2)\}$, where the superscript labels multiplicity. Of all the allowed $0\hbar\omega$ irreps, including those in less symmetric spatial symmetries, the (8,4) yields the maximum β value and for this leading SU(3) irrep $\gamma = 20.6°$. The other allowed spatial symmetries, {[f]} = {[431],[422],[4211],[332], [3221],[2222]}, are associated with $\{(\lambda_s,\mu_s)\}$ sets which yield enveloping curves lying above that of the leading [44] symmetry. Note that the $\beta < \beta_0$ part of the curve approximates a negative quadratic function of β. As indicated, at $2n\hbar\omega$ (n=1, 2, ...) there are irreps like (λ_s+2n,μ_s) with greater deformation, but these do not drop below the (λ_s,μ_s) minimum, because the increase in binding resulting from the increase in deformation is less than the $2n\hbar\omega$ excitation energy added to the system by the H_0 term. As noted previously, if this were not the case, the residual interaction would destroy the shell structure and invalidate the theory. A stable equilibrium deformation results from competition among the various allowed $0\hbar\omega$ and $2n\hbar\omega$ configurations.

This rather simple picture also suggests a method for gaining a microscopic interpretation of superdeformation; however, merely adding symplectic excitations to $0\hbar\omega$ SU(3) shell-model configurations is not sufficient to achieve axis ratios in excess of ~1.3 without destroying the underlying shell-structure. To gain larger deformations, for instance axis ratios of ~1.5 to ~2.0, requires symplectic bands built on excited particle-hole configurations, since the latter are typically more strongly deformed than those that are available within the lowest $0\hbar\omega$ space. A well-known example is the first excited 0_2^+ state in ^{16}O at 6.049 MeV, which features a strong 4p-4h configuration dominated by the leading SU(3) irrep in the product configuration $[(1p^{-4})(0,4) \times (2s0d)^4(8,0)]$, namely the (8,4) representation which is coincidentally the same irrep as the leading $0\hbar\omega$ SU(3) symmetry for ^{24}Mg. As another example, consider ^{24}Mg: the leading 10-particle, 2-hole configuration has $(\lambda,\mu) = (14,2)$ as compared to $(\lambda,\mu) = (8,4)$ for the ground state band. The band head for this configuration (without symplectic enhancement) has an axis ratio nearly 40% larger than that of the ground state band: $\sqrt{C_2(14,2)/C_2(8,4)} = 1.36$. If adding symplectic excitations generated by the B_l^+ operators to this particle-hole configuration generates an additional 30–40% increase, one ends up well into the superdeformed region with an axis ratios lying in the 1.7 ~ 1.8 range. In heavy nuclei, there is a second method to enhance the deformation which involves folding in contributions from the intruder levels. For example, should the nucleons in the intruder levels contribute maximally to the deformation (that is, as pure SU(3) representations) the $[n_\pi \times n_\nu] \rightarrow [(18,0) \times (36,0)] \rightarrow (54,0)$ irrep of ^{238}U would swell to $\{[n_\pi \times n_\nu] \times [u_\pi \times u_\nu]\} \rightarrow \{[(18,0) \times (36,0)] \times [(16,2) \times (34,4)]\} \rightarrow \{(54,0) \times (50,6)\} \rightarrow (104,6)$ which would easily account for observed E2 rates without any renormalization whatsoever.

4. Acknowledgments

The research work reported on in this paper was support by the U. S. National Science Foundation. Discussions with the following individuals are gratefully acknowledged: J. P. Elliott (Sussex, England); M. Moshinsky (UNAM, Mexico); J. Dudek (Strasbourg, France); J. Dobaczewski (Warsaw, Poland); W. Nazarecwicz (Oak Ridge, USA and Warsaw, Poland); and G. Rosensteel (Tulane, USA). The results include original contributions spearheaded by collaborators S. A. Moszkowski (UCLA, USA); O. Castaños and P. Hess (UNAM, Mexico); and students C. Bahri, J. Escher and H. Naqvi of LSU.

5. References (Global by Section)

5.1 Section 1: Pseudo-spin and Related Symmetry

— Shell-model background

M. G. Mayer, Phys. Rev. **75** (1949) 1969; **78** (1950) 16.
O. Haxel, J. H. D. Jensen and H. E. Suess, Phys. Rev. **75** (1949) 1766; Z. Physik **128** (1950) 295.
S. G. Nilsson, Dan. Vid. Selsk. Mat. Fys. Medd. **29** (1955).

— Spin-orbit doublet picture

A. Arima, M. Harvey and K. Shimizu, Phys. Lett. **B30** (1969) 517.
K. T. Hecht and A. Adler, Nucl. Phys. **A137** (1969) 129.

— Theoretical underpinnings

A. Bohr, I. Hamamoto and B. R. Mottelson, Phys. Scr. **26** (1982) 267.
J. P. Draayer, in *Proceedings of Symmetries in Physics, Cocoyoc, Morelos, México, June 3-7, 1991*, ed. A. Frank and K. B. Wolf (Sringer-Verlag, Berlin, tbp).
O. Castaños, M. Moshinsky and C. Quesne, in *Proceedings of the International Symposium on Group Theory and Special Symmetries in Nuclear Physics, Ann Arbor, Michigan, USA, September 19-21, 1991*, ed. J. P. Draayer and J. Jänecke (World Scientific, Singapore, tbp).
C. Bahri, J. P. Draayer and S. A. Moszkowski, Phys. Rev. Lett. **68** (1992) 2133.

— Mean-field references

Y. Nambu and G. Jona-Lasinio, Phys. Rev. **122** (1961) 345; **124** (1961) 246.
H. Reinhardt, Phys. Lett. **B188** (1987) 263.
B. D. Serot and J. D. Walecka, in *Advances in Nuclear Physics*, ed. J. W. Negele and E. Vogt (Plenum Press, New York, 1986), Vol. 16.
J. Zimanyi and S. A. Moszkowski, Phys. Rev. C **42** (1990) 1416.

— Pseudo-SU(3) extension

R. D. Ratna Raju, J. P. Draayer and K. T. Hecht, Nucl. Phys. **A202** (1973) 433.
J. P. Draayer and K. J. Weeks, Ann. Phys. **156** (1984) 41.
O. Castaños, J. P. Draayer and Y. Leschber, Ann. Phys. **180** (1987) 290.

— Miscellaneous applications

A. Arima and D. Strottman, *A Compilation of SU(3) Wave Functions Obtained from Realistic Matrix Elements*, Oxford University Report No. **46**, 1973.
D. Braunschweig, K. T. Hecht and J. P. Draayer, Phys. Lett. **B76** (1978) 538.
R. D. Ratna Raju, K. T. Hecht, B. D. Chang and J. P. Draayer, Phys. Rev. C **20** (1979) 2397.
J. P. Draayer, C. S. Han, K. J. Weeks and K. T. Hecht, Nucl. Phys. **A365** (1981) 127.
K. J. Weeks, C. S. Han and J. P. Draayer, Nucl. Phys. **A371** (1981) 19.
J. P. Draayer, K. J. Weeks and K. T. Hecht, Nucl. Phys. **A381** (1982) 1.
K. J. Weeks and J. P. Draayer, Nucl. Phys. **A393** (1983) 69.
J. P. Draayer and K. J. Weeks, Phys. Rev. Lett. **51** (1983) 1422.
K. J. Weeks, J. P. Draayer and K. T. Hecht, Nucl. Phys. **A414** (1984) 42.
Y. Leschber and J. P. Draayer, Phys. Lett. **B190** (1987) 1.
O. Castaños, Y. Leschber and J. P. Draayer, Nucl. Phys. **A473** (1987) 494.
O. Castaños and J. P. Draayer, Nucl. Phys. **A491**(1989) 349.
H. G. Börner, J. Jolie, S. J. Robinson, B. Krusche, R. Piepenbring, R. F. Casten, A. Aprahamian and J. P. Draayer, Phys. Rev. Lett. **66** (1991) 691.
P. Rochford and J. P. Draayer, Ann. Phys. **214** (1992) 341.

— Corroborating references

W. Nazarewicz, P. J. Twin, P. Fallon and J. D. Garrett, Phys. Rev. Lett. **64** (1990) 1654.
F. S. Stephens *et al.*, Phys. Rev. Lett. **64** (1990) 2623; **65** (1990) 301.
S. Raman *et al.*, Atom. Data and Nucl. Data Tables, **42** (1989) 1.
J. Dudek, B. Herskind, W. Nazarewicz, Z. Szymanski and T. R. Werner, Phys. Rev. C **38** (1988) 940.

5.2 Section 2: Quantum Rotor and SU(3) Symmetry

— Background references

G. Racah, in *Group Theoretical Concepts and Methods in Elementary Particle Physics*, ed. F. Gürsey (Gordon and Breach, New York, 1964) 31.
R. N. Sen, *Construction of Irreducible Representations of SU(3)*, thesis (Hebrew University, Jerusalem, 1963).
J. P. Elliott, Proc. Roy. Soc. London **A245** (1958) 128, 562.
V. Bargmann and M. Moshinsky, Nucl. Phys. **18** (1960) 697; **23** (1961) 177.
M. Moshinsky, Rev. Mod. Phys. **34** (1962) 813.
J. P. Elliott and M. Harvey, Proc. Roy. Soc. **A272** (1963) 557.
J. P. Elliott and C. E. Wilsdon, Proc. Roy. Soc. **A302** (1968) 509.
M. Harvey, in *Advances in Nuclear Physics*, ed. M. Baranger and E. Vogt (Plenum Press, New York, 1968), Vol. 1, pp. 67-182.

K. T. Hecht, in *Annual Review of Nuclear Science* (Academic Press, New York, 1973) Vol. 23, pp. 123-161.

— Infrastructure developments

J. D. Vergados, Nucl. Phys. **A111** (1968) 681.
J. P. Draayer and Y. Akiyama, J. Math. Phys. **14** (1973) 1904.
Y. Akiyama and J. P. Draayer, Comp. Phys. Commun. **5** (1973) 405.
R. Gilmore and J. P. Draayer, J. Math. Phys. **26** (1985) 3053.
J. P. Draayer and G. Rosensteel, Nucl. Phys. **A439** (1985) 61.
H. De Meyer, G. Vanden Berghe and J. Van der Jeugt, J. Math. Phys. **26** (1985) 3109.

— Rotor–SU(3) mapping

Y. Leschber and J. P. Draayer, Phys. Rev. C **33** (1986) 749.
J. P. Draayer and Y. Leschber, in *Symmetries in Science II,* eds. B Grüber and R. Lenczewski (Plenum, New York, 1986) 127.
Y. Leschber, *Algebraic Image of the Asymmetric Top*, thesis (Louisiana State University, Baton Rouge, LA, 1986).
Y. Leschber, J. P. Draayer and G. Rosensteel, J. Phys. G: Nucl. Phys. **12** (1986) L179.
Y. Leschber and J. P. Draayer, Phys. Lett. **B190** (1987) 1.
J. P. Draayer, Y. Leschber and O. Castaños, Ann. UMCA (Lublin, Poland), Sec. AAA, Vol. XLI/XLII (1987) 97.
Y. Leschber, Hadronic J. Supp. **3** (1987) 95.
O. Castaños, J. P. Draayer and Y. Leschber, Computer Phys. Commun. **52** (1988) 71.
O. Castaños, J. P. Draayer and Y. Leschber, Z. Phys. **A329** (1988) 33.

— K–band operator

H. A. Naqvi and J. P. Draayer, Nucl. Phys. **A516** (1990) 351.
H. A. Naqvi and J. P. Draayer, Nucl. Phys. **A536** (1992) 297.

5.3 Section 3: Symplectic Shell-Model Picture

— Theoretical underpinnings

G. Rosensteel and D. J. Rowe, Phys. Rev. Lett. **38** (1977) 10.
G. Rosensteel and D. J. Rowe, Ann. Phys. **96** (1976) 1; **123** (1979) 36; **126** (1980) 198, 343.
D. J. Rowe, Rep. Prog. Phys. **48** (1985) 1419.

— Pseudo-symplectic scheme

O. Castaños and J. P. Draayer, Nucl. Phys. **A491** (1989) 349.
C. Bahri, J. P. Draayer, O. Castaños, and G. Rosensteel, Phys. Lett. **B234** (1990) 430.

O. Castaños, P. O. Hess, J. P. Draayer, and P. Rochford, Nucl. Phys. **A524** (1991) 469.

— Miscellaneous applications

J. P. Draayer and G. Rosensteel, Nucl. Phys. **A386** (1982) 189.
G. Rosensteel and J. P. Draayer, Phys. Lett. **B124** (1983) 281.
G. Rosensteel and J. P. Draayer, Phys. Lett. **B125** (1983) 237.
J. P. Draayer, K. J. Weeks and G. Rosensteel, Nucl. Phys. **A413** (1984) 215.
G. Rosensteel, J. P. Draayer and K. J. Weeks, Nucl. Phys. **A419** (1984) 1.
G. Rosensteel and J. P. Draayer, Nucl. Phys. **A436** (1985) 445.
O. Castaños and J. P. Draayer, Nucl. Phys. **A491** (1989) 349.
J. P. Draayer, Nucl. Phys. **A520** (1990) 259c.
C. Bahri, J. P. Draayer, O. Castaños and G. Rosensteel, Phys. Lett. **B234** (1990) 430.

— Potential energy surfaces

G. Gneuss and W. Griener, Nucl. Phys. **A171** (1971) 449.
P. O. Hess, M. Seiwert, J. Maruhn and W. Greiner, Z. Phys. **A296** (1980) 147.
P. O. Hess, J. Maruhn and W. Griener, J. Phys. G: Nucl. Phys. **7** (1981) 737.
M. Seiwert, J. Maruhn and P. O. Hess, Phys. Rev. C **30** (1984) 1779.
J. P. Draayer, S. C. Park, and O. Castaños, Phys. Rev. Lett. **62** (1989) 20.
O. Castaños, P. Hess, P. Rochford and J. P. Draayer, Phys. Lett. **B277** (1992) 27.

— Vector coherent states

K. T. Hecht, *The Vector Coherent State Method and its Application to Problems of Higher Symmetries* (Springer, Heidelberg, 1987).
R. Le Blanc, J. Carvalho, M. Vassanji, and D. J. Rowe, Nucl. Phys. **A452** (1986) 263.
W.-M. Zhang, D. H. Feng and R. Gilmore, Rev. Mod. Phys. **62** (1990) 867.

REVIEW OF THE SYMPLECTIC MODEL

KRISHNA KUMAR[1]
Physics Department, Tennessee Technological University
Cookeville, Tennessee 38505-5051, USA

and

CYRUS BAKTASH
Physics Division, Oak Ridge National Laboratory
Oak Ridge, Tennessee 37831, USA

and

D. H. FENG
Department Of Physics & Atmospheric Sciences, Drexel University
Philadelphia, Pennsylvania 19104, USA

and

MICHAEL W. GUIDRY and JING YE ZHANG
Department Of Physics, University Of Tennessee
Knoxville, Tennessee 37996-1200, USA

ABSTRACT

A summary of the report of the Review Committee For the Symplectic Model, presented at the Workshop On Nuclear Structure Models, is given.

1. Attractive Features Of The Symplectic Model

The Symplectic Model, discussed at the Workshop by Jerry Draayer[1], has a number of attractive features which are listed below. The committee also raised a number of questions about the model which are discussed in the next section.

• Problems associated with the strong spin-orbit coupling are minimized by employing a pseudo-spin representation of the oscillator shell model where the usual quantum numbers (N, l, j, m) are replaced by $(\tilde{N} = N - 1;\ \tilde{l} = l + 1 = \tilde{N}, \tilde{N} - 2, ...;\ \tilde{j} = j = \tilde{l} \mp 1/2;\ \tilde{m} = m)$.

• Numerical problems connected with the mixing of major shells of the same parity are minimized in such a representation. Thus, it is possible to include mixing in a large configuration space at a low numerical cost.

• Powerful group theoretical methods are employed to minimize the numerical cost of describing collective bands arising from correlated motions of a large number of nucleons.

• Basic invariants in the group theory are related to the corresponding basic invariants in the Copenhagen approach and one to one relations are found between

[1] spokesperson for the review committee

the group theory quantum numbers (λ, μ) and the shape variables (β, γ). This provides an important link between two seemingly different languages.

• A deformed mean field, for instance that of the Nilsson Model, is utilized to determine the best (λ, μ) representation for a given nucleus.

• Each (λ, μ) representation generates a set of collective bands which can be identified with the rotational bands built on the ground state, the $K = 2$ (γ-vibrational) band-head, and higher-K bands (K=4, 6, ...).

2. Questions About The Symplectic Model

The committee raised the following questions about the model:

• Why are the excited $K^{\Pi} = 0^{+}$ (or the β-vibrational) bands missing from the model?

• Is the approximation μ (Nilsson) $= 2D/C = 0.5$ good enough for odd-A nuclei?

• Is the group symmetry chain used in the actual calculations?

• How close to the Dynamical Symmetry are the final solutions? In other words, if the remaining terms like the spin- orbit part not included in the basis need to be diagonalized in a large basis, then what is the advantage of using the symplectic basis?

• Is the Surface-Delta-Interaction good if the three-body and four-body interactions are used to push the β-band up?

• If the Nilsson model is used as input (for determining the basic λ, μ), then what is the advantage over Deformed-Mean-Field theories, especially those used for collective bands with no local (for each nucleus) fitting parameters[2,3,4]? Also, since a potential energy function is not minimized to determine the ground state deformation, how can one trust the model for a case for which no such calculation is available? A deformation value determined from the experimental B(E2) value (when available!) may not be reliable since such a value is non- zero even for the spherical nuclei. Also, the value of γ is seldom available from the B(E2) values.

• What is the connection to Superdeformation?

• Since a number of model parameters are determined by fitting the data for *each* nucleus, how can the model have any predictive property, especially for nuclei far from the line of stability for which no data are available?

1. J. P. Draayer, in *Proc. Workshop On Nuclear Structure Models, Oak Ridge, TN, USA, March 14-24, 1992*, ed. W. Nazarewicz (World Scientific, Singapore, in press).
2. K. Kumar, *Prog. Part. Nucl. Phys.* **9** (1983) 347.
3. P. H. Heenen, in *Proc. Workshop On Nuclear Structure Models, Oak Ridge, TN, USA, March 14-24, 1992*, ed. W. Nazarewicz (World Scientific, Singapore, in press).
4. J. F. Berger, in *Proc. Workshop On Nuclear Structure Models, Oak Ridge, TN, USA, March 14-24, 1992*, ed. W. Nazarewicz (World Scientific, Singapore, in press).

Rebuttal

J. P. DRAAYER
Department of Physics and Astronomy
Louisiana State University, Baton Rouge, LA 70803

The evaluation team identified what they considered to be "attractive features of the symplectic model" and also raised "questions about the symplectic model" which indicate that some of the basic concepts underpinning the scheme and its pseudo-spin extension are *not* understood, even by researchers working in the field. We will attempt in this response to answer the questions raised by the evaluation team in the intended constructive *modus operandi* of the workshop. Not all of the questions have straightforward answers, and in those cases where this is the case we will indicate the enhancements required for the theory to be able to address the more difficult issues. In the final paragraph of this response we will take the liberty of expanding the evaluation team's list of questions by raising two additional issues we know about that require further study.

The reviewers identified six "attractive features of the symplectic model" and we shall first comment on these to set the stage for our consideration in the next section of the evaluation team's concerns about the model.

- The transformation from the normal-spin to the pseudo-spin concept, which is applicable for nuclei with $A \gtrsim 100$, allows one to perform calculations in a frame of reference where the self-interaction, spin-orbit coupling term (which can mix spatial symmetries and as a consequence, when large, gives preference to a jj-coupling scheme over an LS-coupling scenario) is small. A further consequence of good pseudo-spin symmetry, which follows if the self-interaction, orbit-orbit splitting is also small relative to the major shell separation distance, is that the Elliott SU(3) scheme with its simplicity and elegance can be applied to the pseudo-space dynamics. A full appreciation for the importance of the pseudo-space/spin concept as introduced within this framework, and in contrast with the way pseudo-spin is introduced and used in other theories, requires a thorough understanding of the SU(3) scheme.

- The symplectic model is a multi-$\hbar\omega$ extension of the SU(3) scheme. It allows one to select out of a huge (infinite in principle) model space special states which display the coherence manifest in collective rotations. The group theory that is required to make this scheme a practical tool for carrying out shell-model calculations has been under development over a number of years. User friendly computer codes for calculating SU(3) coupling and recoupling coefficients are now public domain software.

- The pseudo-SU(3)/Sp(3,R) scheme allows one to give a shell-model description of correlated motion among nucleons that results in rotational bands. A point to be emphasized in this regard is that many-particle wavefunctions, with all the required exact symmetries (fixed particle number, total angular momentum, etc.) can be written down, and that calculated Hamiltonian eigenstates can be used to test observables in heavy deformed nuclei, a region that heretofore has been left to the realm of mean-field and collective model theorists.

- As the review team correctly points out, the mapping between invariants of SU(3) and the rotor establishes an important direct connection between the shell-model and the Copenhagen picture of collective rotation in nuclei. The connection has a solid group theoretical underpinning. But it does more because it allows one to define and compare the collective-model concept of a potential energy surface with its shell-model analog. In doing this one learns of fundamental differences in the two theories and about the role particle statistics plays in determining nuclear structure — a feature the collective model can only include in an approximate way.

- The Nilsson model has been used as a guide in determining the important shell-model configurations. While the reviewers seemed to find this reassuring, we consider it to be a method of only marginal value. It is used because we presently are not able to perform a full search of the available shell-model configurations to determine which of the large but finite number of representations should actually dominate in low-lying eigenstates. For the latter we need faster routines and guidance from mean-field theorist who are experts in seeking out such minima.

- Each (λ,μ) representation, which in an SU(3) theory labels a $0\hbar\omega$ irrep and in its multi-$\hbar\omega$ Sp(3,R) extension the symplectic bandhead, defines a set of rotational sequences. The angular momentum content of a particular (λ,μ) can be determined by a very simple rule: $K=\min(\lambda,\mu)$, $\min(\lambda,\mu)-2$, ..., 1 or 0 where if K=0 we have that $L = \lambda+\mu, \lambda+\mu-2, \lambda+\mu-4, ..., 1$ or 0 while for $K\neq 0$ we get $L=K, K+1, ..., \lambda+\mu-K$. The K=0 series of the leading (λ,μ) representation forms the ground-state rotational band. There are normally $K\neq 0$ bands ($\min(\lambda,\mu) \neq 0$) associated with the ground band with bandheads (L=K) lying at some positive excitation energy. This structure is the same as that of a quantum rotor. The next representation, which usually corresponds to a slightly less deformed configuration and therefore lies somewhat higher in energy, defines another similar sequence. The K=0 member of this second series is the beta-vibrational band of the collective model. The next (λ,μ) corresponds to a double-beta vibrational state, and so on. The gamma band of the collective model is actually the K=2 configuration of the leading (λ,μ) symmetry. Here there is an important difference with the collective model picture: a prolate quantum droplet cannot rotate about its symmetry axis but nuclei can because prolateness does not necessarily imply a corresponding featureless internal structure.

The reviewers formulated eight "questions about the symplectic model" that we will now address. Some may be covered, at least partially, in the discussion given above, but for completeness they will also be addressed in this paragraph.

- Why are the excited $K^\pi=0^+$ (or the β-vibrational) bands missing from the model? They are not missing — this has been explained numerous times and in various ways, nevertheless we will try again. The problem lies in attempts to superimposed the dynamical symmetry concept usually associated with the interacting boson model and its derivative, the fermion dynamical symmetry model, on the symplectic model. The symplectic model, as we define and use it, is the shell-model augmented with multi-$\hbar\omega$ correlations generated by the quadrupole $[\sim r^2 Y_{2\mu}(\theta,\phi)]$ operator. Each $0\hbar\omega$ SU(3) shell-model irrep defines a symplectic representation. Since there is only one $L^\pi=0^+$ state in any SU(3) irrep (assuming both λ and μ are even) the β-band has to be associated with a non-leading SU(3) irrep. Our dynamics is *not* restricted to a single symplectic irrep — to do so would rule out the β-band which is wrong physics and simply not done. The fact that the so-called β-band excitations are associated with a changes in λ and μ can also be seen directly from the $(\beta,\gamma) \leftrightarrow (\lambda,\mu)$ mapping.

  As explained above, the excitation energy of the so-called β-band is determined primarily by the quadrupole-quadrupole part of the interaction. Mixing between the β-band and the ground band occurs through interactions like pairing that lie outside the symplectic symmetry algebra. Since the theory includes these interactions, the symplectic scheme departs from the interacting boson model and fermion dynamical symmetry model concept of a dynamical symmetry. While a single irrep suffices to describe one collective rotational structure, the β-band is a second collective structure (the ground band being the first) in nuclei and this calls for a second symplectic irrep. Our dynamical symmetry is the group of the shell model — augmented with symplectic excitations — and hence has in it the $L^\pi=0^+$ states of the shell model, including the ground state, the β-vibrational bandhead, and all others.

- Is $\mu=2D/C=0.5$ good enough for odd-A nuclei? First of all we note that results for even-even nuclei are the same whether $\mu=0.5$ or $\mu\approx0.5$ because the residual nucleon-nucleon interaction strongly favors $\tilde{S}=0$ configurations and the matrix elements of an $\tilde{S}=1$ operator (which $\tilde{l}\cdot\tilde{s}$ is) between states with $\tilde{S}=0$ vanish identically by angular momentum selection. Setting $\mu=0.5$ uses this result to simplify the calculation, but it is not a necessary condition for pseudo-spin symmetry to be good. Recall that in the ds-shell case low-lying states of even-even nuclei are S=0 configurations — the S>0 states lie higher in energy. Even though the spin-orbit interaction is quite strong in the ds-shell case there is relatively little mixing among configurations with different S values. The effect of $\mu\neq0.5$ enters into the *off-diagonal* couplings in the Hamiltonian matrix — couplings between favored ($\tilde{S}=0$) and unfavored ($\tilde{S}>0$) configurations.

  In odd-A nuclei the $\tilde{l}\cdot\tilde{s}$ interaction make a *diagonal* contribution because low-lying configurations have $\tilde{S}=1/2$ so the angular momentum selection rule does not apply. But it is well-known from the particle-plus-rotor model, which treats odd-A nuclei as a unpaired nucleon coupled to an even-even core, that the $l{\cdot}s$ (as well as the l^2) term is essential in getting the right spectral features. The same is true within the framework of the pseudo-SU(3) model and within its pseudo-Sp(3,R) extension. The fact the $\mu\neq0.5$ is very important in determining spectral features of odd-A nuclei. Furthermore, the theory predicts a difference in the results for $\mu<0.5$ (odd neutron) and $\mu>0.5$ (odd proton). Just as in the even-even case, however, the most symmetric spatial symmetry is still predicted to dominate the low-lying spectrum. In this case that will be an $\tilde{S}=1/2$ configuration with non-zero diagonal $\tilde{l}\cdot\tilde{s}$ matrix elements. The case of an odd-A system treated within the framework of the pseudo-SU(3)/Sp(3,R) scheme was the thesis project of Husney Naqvi (LSU 8/92 Ph.D.).

- Is the symmetry chain actually used? Absolutely! While the Nilsson level scheme can be used to determine which spatial [f] symmetries are favored and the most deformed (λ,μ) representation in each of these, a full knowledge of the $[f] \to (\lambda,\mu)$ branching is required to insure that this choice is optimal for the system under consideration. The $[f] \to (\lambda,\mu)$ branching is required whenever more than one symplectic irrep is taken into account, as in calculations which include pairing. The $[f] \to (\lambda,\mu)$ branching is given in tabular form in a bound but unpublished volume by Draayer and Leschber (*Branching Rules for Unitary Symmetries of the Nuclear Shell Model*, 1987). There is also a software package available that generates this decomposition for any particular case.

  In fairness we should admit that much more work needs to be done on this topic. As interest in applications of the theory grows, it will be important to be able to make fast systematic searches of regions in configurations space to pick out dominant configurations for realistic Hamiltonians. Our need is to be able to do what mean-field theorists can do, only using many-particle wavefunctions of the symplectic scheme rather than determinant forms which ignore exact symmetries of the Hamiltonian, relegating their recovery to after-the-fact projections. This is a symplectic model binding-energy project that awaits an interested and ambitious young researcher.

- How close to the dynamical symmetry limit are the final results and is there a real advantage to the scheme if ultimately one must diagonalize in a larger space to incorporate residual interactions like the spin-orbit term? This is a good question! It requires care in answering because the answer reflects a philosophy concerning how nuclear structure physics can be done best. We always choose to express the Hamiltonian as an expansion over a set of Hamiltonian-like operators (Q·Q, P, etc.) with the most important of these first in the series and the least important last. Statistical spectroscopy methods can be used to assign quantitative measures to the qualitative adjectives ("most" and "least") modifying important in the previous sentence. Having done this, one selects a scheme (coordinate system) that brings as many of the most important terms into diagonal form as possible. (In principle this philosophy underlies all algebraic theories, though we know of no other one which starts with a realistic interactions and actually carries out such a decomposition to verify the assumptions of the model at this elementary level.)

The next step in the procedure is to use as basis states for a larger calculation a few (far less than all) of the lowest eigenstates of the simpler Hamiltonian which respects the selected dynamical symmetry. In our case we firmly believe the symmetry of choice is the symplectic model as Q·Q clearly dominates in regions of strong deformation. The effect of interactions left out of the simpler picture can be assessed by using a large number (~100?) of eigenstates of the symmetry preserving Hamiltonian as basis states for calculating eigenstates of the realistic Hamiltonian, including the symmetry breaking terms in its expansion. We are to a point in the development of symplectic model procedures that it is now possible to implement a procedure of this type. The main problem, of course, lies in evaluating reduced matrix elements of non-generator forms between basis states that are given in terms of the dynamical symmetry group chain. This requires a working knowledge of the appropriate Wigner-Racah algebra which in the case of the symplectic model is SU(3). All the required coefficients (coupling *and* recoupling) for SU(3) are available through codes developed in the early 1970's by Akiyama and Draayer. It also requires a very tedious determination of selected unreduced matrix elements from which the reduced one can be extracted via a generalized Wigner-Eckart Theorem factorization procedure. Fortunately, this too is now available through the effort of Chairul Bahri, a student workshop participant.

An alternative procedure to the one outline above would use a Lanczos procedure to produce (hopefully) equivalent results. Unfortunately, when we tried to implement a symmetry adapted Lanczos procedure we found it to be much to efficient in spanning the full shell-model space. After only a few iterations (3-5, typically) the Hamiltonian matrix becomes dense and one loses the advantages of the algebraic method. (The Lanczos procedure is highly efficient at finding small contributions to an eigenstate, which is a necessary attribute for it to be of general use.) Nonetheless, we do use a block Lanczos procedure for diagonalizing our matrices, which are typically on the order of 5000 by 5000, but only after having generate them via more conventional shell-model methods adapted to the special symmetries of the symplectic model. As currently done, each symplectic band calculation takes about 100 hours of CPU time on an IBM 3090/6J using its full vector capabilities. As one would like to do 100 or so of these single-irrep calculations and then combine ten or so eigenvectors from each in a larger mixed symmetry calculation, computational constraints place such an ambitious program currently out of reach. However, as we learn more about the connection between SU(3) and the rotor and how vibrations can be folded into the picture, there may be ways to streamline the first of these processes, using among other things the power of parallel processing. We appear to be close to being able to carry out realistic shell-model calculations for heavy deformed nuclei.

- Is the surface-delta interaction good if the three-body (X_3) and four-body (X_4) interactions are used to push the beta band up? First of all the X_3 and X_4 operators have nothing to do with the placement of the beta band. The beta band is positioned by the quadrupole-quadrupole interaction or the quadrupole-quadrupole part of a more realistic interaction form. [Recall that $\langle Q\cdot Q\rangle_{L=0} \sim \beta^2$ because $Q\cdot Q = 4C_2\text{-}3L^2$ and $\beta^2 \sim (C_2+3)$.] By construction the matrix elements of the X_3 and X_4 operators *vanish* in L=0 states. Furthermore, since they are built solely out of generators of SU(3), they do not couple different (λ,μ) representations. We repeat, the X_3 and X_4 operators are very special three and four-body forms — together with L^2, the X_3 and X_4 operators can be used to reproduce the dynamics of a rotor within an irreducible representation of SU(3). Additionally, while the $H_{ROT} \leftrightarrow H_{SU3}$ mapping may appear to be a trivial result it most certainly is not — the X_3 and X_4 operators allow one to reproduce the rotor dynamics without affecting the system in any other direct way. So the answer is very simple: X_3 and X_4 have nothing whatsoever to do with whether the surface-delta interaction is good or whether it is not good.

- Since the Nilsson scheme is used to determine the leading configuration, what if any are the advantages of the symplectic scheme over mean-field calculations? In this case I would like to respond by posing a series of questions for which I hope the answers are clear to the reader: what must be done in mean-field calculations to insure exact symmetries like total angular momentum and a fixed number of particles are conserved; do mean-field theories produce solutions that can be used to calculate electromagnetic transition strengths, electron scattering form factors, beta decay widths, particle transfer amplitudes, etc.; is there reasonable hope for moving from results of a mean-field theory to a more fundamental understanding of the nucleon-nucleon interaction; and so on? This response should not be construed as a put-down of mean-field calculations as they can do a number of things more quickly and with less restrictions than we can. The shell-model is the most fundamental of all nuclear structure theories. The only question is whether or not it can be rendered tractable. We believe algebraic methods developed over the past two decades have brought us much closer to the realization of a tractable shell-model theory. The symplectic model enjoys the advantage of being guided by established successes of simpler phenomenological models.

- What is the connection to superdeformation? Good question; no (simple) answer. Having established a connection between the shell-model and the collective model concept of a potential energy surface, it is natural to speculate that second (super) and third (hyper) deformed minima result from competition between shell and subshell closure effects. Within the framework of the symplectic model these minima are determined primarily by the Pauli Principle, modified by gains in the binding energy due to the (deformation inducing) residual interaction. While it is rather difficult to argue with such a simple picture, the devising of a microscopic theory which gives quantitative results reproducing it has escaped us! There may be a connection between good pseudo-spin symmetry and identical bands — this is under consideration — but it is premature to offer a more definitive response to a very wide-open question concerning superdeformation.

- Since the parameters of the model are fit to data, what if any is the predictive power of the theory? First let me say that the theory as presently implemented in even-even nuclei uses four parameters: χ, a, b, and c. The χ value measures the strength of the quadrupole-quadrupole interaction and this in turn determines the amount of vertical (multi-$\hbar\omega$) mixing that occurs as well as interband E2 transition strengths. The a, b and c parameters multiplying respectively L^2, X_3 and X_4 can be determined from the excitation energies of the first two excited 2^+ states in the case of a prolate geometry (specifying a prolate shape counts as a constraint) or by the lowest few states in the ground and gamma-bands for the general case of a triaxial configuration. Recall that the $H_{ROT} \leftrightarrow H_{SU3}$ mapping is one-to-one so the a, b and c values are fixed once the moments of inertia of a quantum rotor that fits the observed spectrum is specified.

While we admit to four parameters, we should hasten to add that an analytic result is available for χ, which we have found to be accurate to within about 10% for applications in the rare earth and actinide regions. Furthermore, as pointed out above, the a,b and c follow once the moments of inertia of a rotor fit to the experimental spectrum are specified. It should therefore come as no surprise that our code rarely needs more than three iterations to find best-fit parameter values. In each case we have considered the final parameter values lie close to the starting ones determined quite literally by a simple back-of-the-envelop calculation, using as input the ground band $B(E2{:}2^+ \to 0^+)$ strength and the excitation energies of the $E_{2_1^+}$ and $E_{2_2^+}$ states. If the system display triaxiality additional work is necessary but even in this case a prolate assumption usually gives good starting values for the parameters. Regarding the matter of predictive power of the theory, it is difficult to say much as one would like first to have available a systematic study of variations in the four parameters as a function of (Z,N) and A, or equivalently in the $B(E2{:}\ 2^+ \to 0^+)$ strengths and the inertia parameter of a best-fit rotor as a function of the same set. Because of the close similarity of the symplectic and collective model pictures, however, one does not expect such a study to produce and surprises — χ will vary smoothly as a function of A, since a multiplies L^2 its value will decrease with increasing deformation reflecting a corresponding drop in the excitation energy of the first excited 2^+ level, and so on.

Responding to the evaluation team's questions has been an instructive exercise for us and we thrust we have been able to address their concerns without being overly defensive and without over-selling the symplectic model. One questions that was not raised which we feel should have been asked because it represents our number one concern, deals with the question of the intruder levels. Thus far in applications of the symplectic model, the nucleons assigned to intruder orbitals have been treated as spectators, adding at most a renormalization factor proportional to the total number of particles divided the number considered to be spectroscopically active. Yet in mean-field studies these same intruder orbitals are championed as drivers of the deformation, particularly when attempting to understand the structure of superdeformed bands. Both pictures cannot be right! Can this be tested, are there direct experimental indicators favoring one interpretation over the other, what must be done to include intruder levels in the symplectic scheme?

Another challenge we must face is how to handle particle-hole configurations. The symplectic model builds in multi-$\hbar\omega$ correlations but ignores low-lying collective particle-hole excitations whose existence is well-known, for example in ^{16}O and ^{40}Ca. There is also clear evidence suggesting that these particle-hole structures, which tend to lie at high excitation energy (6.06 Mev in ^{16}O) in doubly closed-shell nuclei, move downward as the number of valence particles increases. As a consequence particle-hole configurations are expected to mix strongly into the ground and other 0-$\hbar\omega$ rotational bands. Particle-hole configurations are themselves multi-$\hbar\omega$ excitations, but different from and orthogonal to those of the symplectic model. They too can be organized into irreps of (λ,μ) with the most deformed of these expected to be the ones that contribute most strongly through mixing in the ground state domain. The symplectic extension of a n_{ph}-$\hbar\omega$ SU(3) irreps is generated in exactly the same way as for 0-$\hbar\omega$ irreps. Each distinct n_{ph}-$\hbar\omega$ SU(3) irrep defines the bandhead of an independent symplectic irrep. So the shell-model picture expands as it must to a collection of symplectic bands, organized into n_{ph}-$\hbar\omega$ sets with positive or negative parity for n_{ph} even (including 0) or odd, respectively. Superdeformation needs to be addressed within this context.

THE NUCLEAR SYMPLECTIC MODEL

DAVID J. ROWE
Department of Physics, University of Toronto,
Toronto, Ont. M5S 1A7, Canada

ABSTRACT

An overview and status report is given on the symplectic model, the symplectic shell model and various approximations to these models. Particular emphasis is given to the physical content of the symplectic model, the kinds of phenomena it can describe and its relationship to other models. It is shown that the basic model gives rise to lowlying rotational bands and giant monopole and giant quadrupole resonances built upon them. The lowlying vibrational states of the collective model, e.g., beta vibrations, come from mixing of irreducible symplectic representations which arise, for example, from the spin-orbit and pairing interactions. It is shown how the spin degrees of freedom can be accommodated within the model and that the symplectic shell model is, in principle, an exact model. The deficiencies of the approach and some problems that need to be addressed are also discussed.

1. Introduction

In his talk, Jerry Draayer represented the symplectic model as an extension of Elliott's SU(3) model [1] to include the giant resonance degrees of freedom. He then went on to show that the pseudo-SU(3) model [2] can similarly be extended to a pseudo-symplectic model. My talk will be complementary to his. I shall focus initially on the symplectic model as a *microscopically realizable* collective model; i.e., a collective model that has both a macroscopic and a microscopic interpretation. In its macroscopic interpretation, it is a version of the Bohr-Mottelson collective model [3] which includes vorticity, as well as irrotational-flow, degrees of freedom. In its microscopic interpretation, it is a submodel of the shell model which includes SU(3) plus giant resonance degrees of freedom.

The symplectic model was proposed by Rosensteel and Rowe [4, 5] as a solution to the problem of how to express the Bohr-Mottelson collective model in microscopic terms. What was wanted was a version of the model that would be compatible with the microscopic structure of the nucleus as a system of interacting neutrons and protons. The desired collective model should be a submodel of the nuclear shell model. If could one find such a collective model, one could hope to derive it from microscopic considerations.

The early approaches to this problem attempted to change the variables of the shell model from many-nucleon coordinates to collective and intrinsic coordinates. Much progress was made, for example, by Villars [6], Goshen and Lipkin [7], Scheid and Greiner [8], Zickendraht [9] and Dzyublik *et al.* [10]. A practical solution was finally achieved by the use of algebraic methods and dynamical symmetry.

Some of the most important early contributions to the development of the algebraic approach to the problem were by Tomonaga [11], Elliott [1], Cusson [12], Ui [13], Weaver and Biedenharn [14, 15, 16], Vanagas [17] and Raychev [18].

The basic idea of the algebraic approach is to express the collective model in terms of a dynamical algebra that has a microscopic realization; i.e., the observables of the algebra should be expressible in terms of nucleon position and momentum coordinates. The collective model then becomes a submodel of the shell model. Moreover, since the Hilbert space of such a collective model carries an irreducible representation (irrep) of the dynamical algebra, it can be embedded as a subspace of the shell model Hilbert space by finding an equivalent shell model irrep of the collective model's dynamical algebra. The general philosphy of the algebraic approach was nicely outlined by Larry Biedenharn at the 15'th Solvay Conference on Physics [19].

2. The important steps

2.1. The choice of suitable shape variables

It was recognized quite early that, if progress was to be made, one needed better collective coordinates than the surface shape coordinates used in the Bohr-Mottelson model. A suitable set of shape observables is given by the multipole moments of the nuclear density distribution. For example, the quadrupole moments, $Q_{2\nu}$, for a nucleus with a density distribution $\rho(r,\theta,\varphi)$, are given by

$$Q_{2\nu} = \int r^2 Y_{2\nu}(\theta,\varphi)\rho(r,\theta,\varphi)\, dv \tag{1}$$

and, for a distribution of point particles by

$$Q_{2\nu} = \sum_n r_n^2 Y_{2\nu}(\theta_n,\varphi_n). \tag{2}$$

To have a common framework that applies to both a hydrodynamic nucleus and a many-nucleon nucleus, it is convenient to divide the nuclear mass distribution up into equal mass volume elements, as shown schematically in Fig. 1. The n'th element can then be assigned position and momentum coordinates (x_{ni}, p_{ni}) with $i = 1, 2, 3$ corresponding to the three Cartesian (x, y, z) directions. We then introduce Cartesian quadrupole moments (the trace of which is the monopole moment, in spherical tensor notation)

$$Q_{ij} = \sum_n x_{ni} x_{nj}. \tag{3}$$

Thus, in the hydrodynamic model, n labels a small volume element of the nuclear mass distribution and, in the microscopic model, n labels a nucleon. The quadrupole moments, Q_{ij}, are therefore suitable, microscopically realizable, collective coordinates.

2.2. The collective flows

In order that the collective model should have a dynamical algebra, its momentum observables should be the infinitesimal generators of collective flows corre-

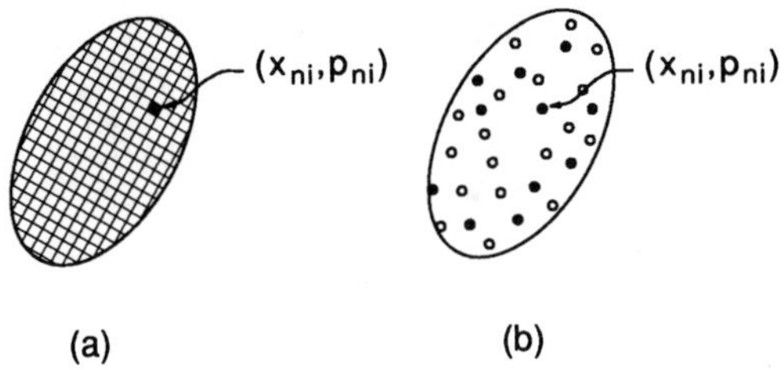

Fig. 1. Assignment of position and momentum coordinates: (a) to an element of the nuclear matter distribution and (b) to a nucleon. The index n labels the element (or nucleon) and i indexes the three Cartesian coordinate axes.

sponding to Lie group transformations. For rotations, the group of rotations, with infinitesimal generators given by the angular momentum operators, would seem to be appropriate. What should one choose for vibrations?

The answer was given by Cusson [12]. Suppose (x, y, z) are Cartesian coordinates of a point in the nucleus relative to the principal axes frame of the quadrupole tensor; i.e., the frame in which $Q_{ij} = Q_i\delta_{ij}$. Then, for example, vibrational flows which leave the principal axes invariant are given by scale transformations in which $(x, y, z) \to (ax, by, cz)$. To have vibrations about any axis, Cusson proposed that one consider the general linear transformations in which the coordinates (x_{ni}) of an element of nuclear matter (e.g., a nucleon) transform according to the equation

$$x_{ni} \to \sum_j x_{nj} g_{ji}, \tag{4}$$

where the matrix (g_{ji}) is an element of the general linear group GL(3,R) of real 3×3 matrices.

Some of the flows corresponding to general linear transformations are illustrated in Fig. 2. They include vibrations, rigid flow rotations, vortex rotations and irrotational flow rotations. Rigid flow rotations are rotations, generated by the angular momentum operators, in which all elements of the fluid rotate as if they were frozen in place relative to one another. Vortex rotation is a rotation in which the fluid circulates but leaves the shape fixed in space. Irrotational flow rotation is a combination of rigid rotation and vortex rotation in opposite directions such that the elements of the fluid move the minimum amount possible in accommodating the fluid's rotating shape.

The infinitesimal generators of general linear flows are the operators

$$P_{ij} = \sum_n x_{in} p_{jn}. \tag{5}$$

These operators span the Lie algebra, gl(3,R), of the general linear group GL(3,R). They are not canonical to the quadrupole operators Q_{ij}. However, they satisfy the

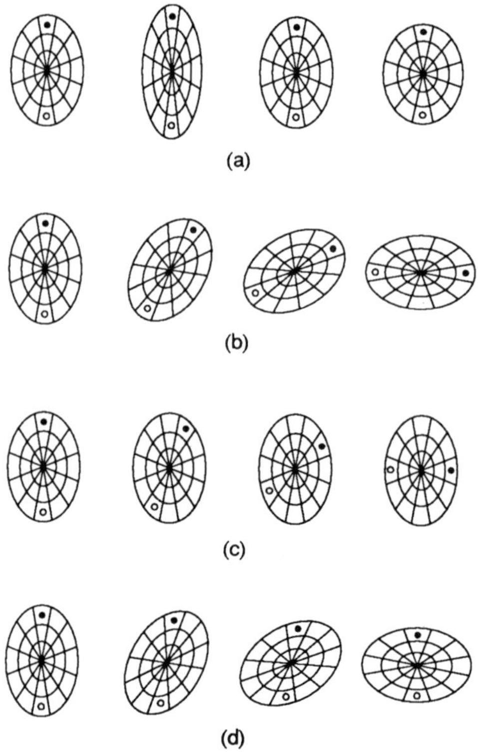

Fig. 2. General linear flows: (a) vibration, (b) rigid rotation, (c) vortex rotation and (d) irrotational flow rotation.

commutation relations

$$[P_{ij}, Q_{kl}] = -\mathrm{i}\hbar(\delta_{jk}Q_{il} + \delta_{jl}Q_{ik}). \tag{6}$$

Thus, the combined set of operators $\{Q_{ij}, P_{ij}\}$ closes on a Lie algebra. It is the Lie algebra of a group, CM(3); the *collective motion* group in three dimensions. The group CM(3) was identified by Weaver, Biedenharn and Cusson [16] as a suitable dynamical group for a microscopically realizable model of nuclear collective motions.

2.3. The kinetic energy

Independently of the above considerations, Zickendraht [9] and Dzyublik *et al.* [10] decomposed the many nucleon kinetic energy $\sum_{ni} p_{in}^2/2m$ into collective and intrinsic parts. The same decomposition was later derived by Gulshani and Rowe [20] with the CM(3) model specifically in mind. Thus, the decomposition corresponded to making a change of variables from many nucleon coordinates to quadrupole plus general linear collective coordinates and complementary intrinsic coordinates. The result was a decomposition of the form

$$T = \frac{1}{2m}\sum_{ni} p_{in}^2 = T_{\mathrm{IRR}} + T_{\mathrm{VORT}} + T_{\mathrm{COUP}} + T_{\mathrm{INTR}}, \tag{7}$$

where T_{IRR} is an irrotational flow kinetic energy, T_{VORT} is a kinetic energy associated with vorticity and T_{COUP} couples the irrotational and vortex flows.

Such considerations of the kinetic energy showed two things:

(i) The kinetic energy strongly couples irrotational and vorticity degrees of freedom. Thus, one can understand why nuclear moments of inertia are not those of the irrotational flow model.

(ii) The kinetic energy does not conserve vortex spin. In other words, the kinetic energy mixes CM(3) irreps.

3. The phenomenological symplectic model

To come to terms with the fact that the kinetic energy mixes CM(3) irreps, Rosensteel and Rowe [4] proposed that one add the kinetic energy operator to the cm(3) observables to form a larger Lie algebra. They found that the smallest Lie algebra that contains the kinetic energy operator as well as all of cm(3) is the Lie algebra sp(3,R) of the non-compact symplectic group Sp(3,R). It is the Lie algebra spanned by the observables $\{Q_{ij}, P_{ij}, K_{ij}\}$, where

$$\begin{aligned} Q_{ij} &= \sum_n x_{in}x_{jn} \\ P_{ij} &= \frac{1}{2}\sum_n (x_{in}p_{jn} + p_{jn}x_{in}) \\ K_{ij} &= \sum_n p_{in}p_{jn} \,. \end{aligned} \tag{8}$$

Whereas the group GL(3,R) is a group of linear canonical point transformations in which

$$x_{in} \to \sum_j x_{jn}g_{ji}\,, \quad p_{in} \to \sum_j p_{jn}g^{-1}_{ji}\,, \tag{9}$$

the group Sp(3,R) is the group of all linear canonical transformations of the phase-space coordinates in which

$$(x_{in}, p_{in}) \to \sum_j (x_{jn}, p_{jn}) \begin{pmatrix} a_{ij} & b_{ij} \\ c_{ij} & d_{ij} \end{pmatrix}. \tag{10}$$

The Lie algebra sp(3,R) is a rank 3 semi-simple Lie algebra.[1] Thus, it has very nice properties with (discrete series) irreps given by raising operators from a lowest weight state. In this respect, it differs from cm(3) which has an Abelian

[1] We denote a symplectic Lie algebra of rank n by sp(n,R). In doing so we follow the convention favoured by mathematicians. Many physicists, on the other hand, use the notation sp(2n,R).

ideal (namely the subalgebra spanned by the quadrupole operators) and whose representation theory, as a consequence, is much more complicated.

The discrete series irreps of sp(3,R) are all known [21]. They are labelled by the quantum numbers $N(\lambda,\mu)$ of a lowest weight state. The precise meaning of these quantum numbers is given in the following section. For the moment, one need only note that the lowest weight state corresponds (in a deformed basis [22]) to the intrinsic state of a Bohr-Mottelson rotational band and the quantum numbers are related to the mean square radius and mean intrinsic quadrupole moments of the intrinsic state by

$$\begin{aligned} \langle\sum_n r_n^2\rangle &\approx \frac{\hbar}{m\omega}N \\ \langle\sum_n r_n^2 Y_{20}\rangle &\approx \frac{\hbar}{m\omega}2(2\lambda+\mu) \\ \langle\sum_n r_n^2 Y_{22}\rangle &\approx \frac{\hbar}{m\omega}2\sqrt{\frac{3}{2}}\,\mu, \end{aligned} \tag{11}$$

where $\sqrt{\hbar/m\omega}$ is a harmonic oscillator unit of length.

The matrix elements of the sp(3,R) operators were determined numerically by Rosensteel [23]. Analytical expressions were given for the so-called closed-shell irreps by Castaños, Chacon and Moshinsky [24]. A general method, which gives analytical matrix elements whenever there are no missing labels, and a simple algorithm, for cases where there are, is given generally by Vector Coherent State theory [25, 26]. In fact, VCS theory was invented specifically for this purpose .

The Lie algebra sp(3,R) is a dynamical algebra for collective model Hamiltonians of the type

$$H = T + V(Q), \tag{12}$$

where

$$T = \frac{1}{2m}\sum_i K_{ii} \tag{13}$$

and $V(Q)$ is a polynomial in the quadrupole moments. For example, $V(Q)$ might be a polynomial of the form

$$V(Q) = aQ_0 + b(Q_2\times Q_2)_0 + c(Q_2\times Q_2\times Q_2)_0 + \cdots. \tag{14}$$

Then, with $a = \frac{1}{2}m\omega^2$ and $Q_0 = \sum_i Q_{ii}$,

$$T + aQ_0 = \frac{1}{2m}\sum_{in} p_{in}^2 + \frac{1}{2}m\omega^2\sum_{in} x_{in}^2 \tag{15}$$

is the harmonic oscillator shell model Hamiltonian. Moreover, the rotationally invariant potential

$$v(\beta,\gamma) = b(Q_2\times Q_2)_0 + c(Q_2\times Q_2\times Q_2)_0 + \cdots \tag{16}$$

is a collective potential of the kind used, for example, by Gneuss and Greiner [27] in their applications of the generalized Bohr-Mottelson model. An example of a two-parameter potential $v(\beta,\gamma)$ is shown in Fig. 3.

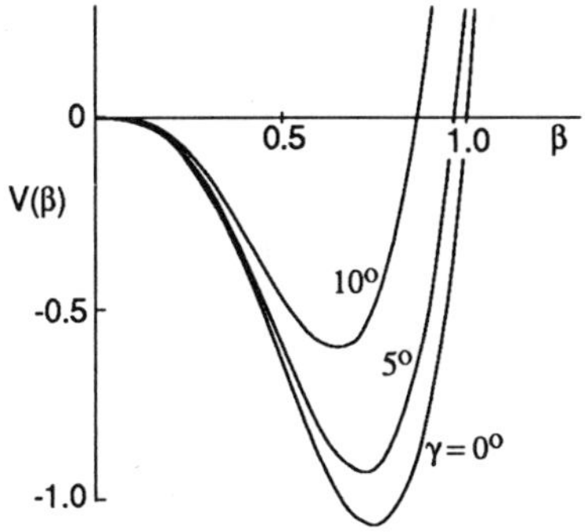

Fig. 3. A two-parameter potential of the type used by Park *et al.* [28] in their symplectic model calculations.

A two parameter potential $v(\beta,\gamma)$ of the type shown was used to calculate some low energy states of a few rare-earth nuclei by Park *et al.* [28] The harmonic oscillator frequency of their calculations was chosen according to the usual formula $\hbar\omega = 41A^{-1/3}$ MeV. The two parameters of the potential $v(\beta,\gamma)$ were chosen such that $v(\beta,\gamma)$ would have a minimum at the observed deformation of the nucleus under investigation and its overall strength should be such that the wave functions of low lying states would have the same deformation; i.e., there would be a self-consistency between the deformation at the minimum of the potential and the deformation of the wave function. The results of their calculations for the lowest four levels of the ground state rotational bands of four deformed nuclei are shown in Fig. 4. It is important to note that no effective charge was used and there was no adjustment of the moment of inertia. The energy levels are just as given by the eigenvalues of the above Hamiltonian with parameters fixed as described.

The fact that no effective charge is needed comes as no surprise. The renormalization of the shell model effective charge arises as a result of coupling of the $0\hbar\omega$ shell model states to the giant quadrupole resonance states. This coupling is included explicitly in the symplectic model because the full quadrupole moment operators are used; they are not restricted to the valence shell space. Thus, a symplectic model calculation includes states from many major oscillator shells. In fact, some 15-20 major shells are needed to obtain convergent results for the eigenstates of the above Hamiltonian. Calculations over so many shells would, of course, be unthinkable without having a dynamical algebraic structure which restricts the states to a single irreducible representation.

It is of considerable interest to enquire why the moment of inertia comes out correctly in spite of the fact that no pairing forces are included in the calculation. A possible interpretation of this result is as follows. Since pair correlations compete with quadrupole deformation correlations, it is reasonable to believe that, if pairing were included, one would obtain states, for a given Hamiltonian, with smaller

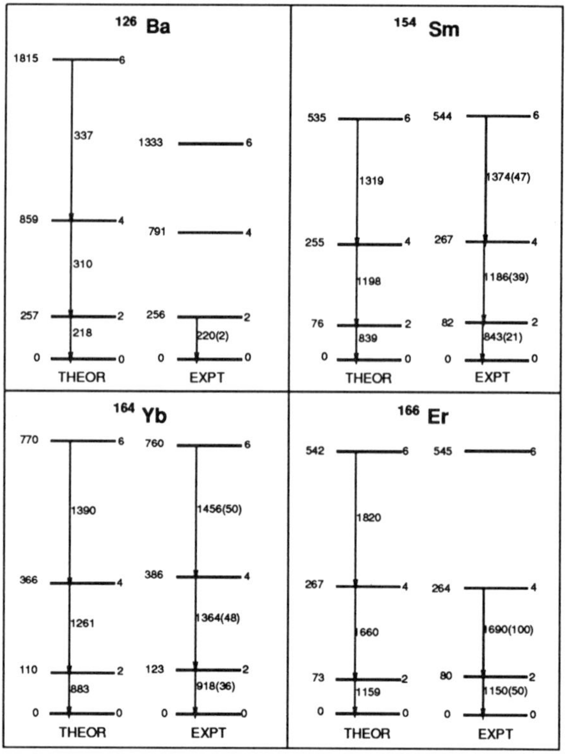

Fig. 4. A comparison of the symplectic model calculations of Park *et al.* [28] with experimentally observed energy levels and E2 tramsition rates.

intrinsic deformations. To get pair-correlated states of the experimentally observed deformation one would then have to increase the deformation producing part of the Hamiltonian. Thus, pairing reduces the moment of inertia of a given band of states by reducing its deformation. However, if the deformation is given correctly, with or without pairing, it is reasonable to expect that one should get correct moments of inertia. The argument is that, if the rotational energy comes from the kinetic energy, there can be no interference terms coming from different sp(3,R) irreps; the kinetic energy is an element of the sp(3,R) Lie algebra and, hence, has vanishing matrix elements between states of different irreps. Thus, even though pairing could result in rotational states with components coming from many sp(3,R) irreps, the kinetic energy spectrum that results is simply the average of the contributions from all the participating irreps. This interpretation needs further investigation because it appears, from the symplectic model calculations we have done, that the rotational energy spectrum does not come entirely from the kinetic energy.

4. The microscopic symplectic model

In the microscopic version of the symplectic model, the Lie algebra is the same sp(3,R) algebra. The elements of the nuclear fluid are specified to be nucleons and the available irreps of the sp(3,R) algebra are restricted to those that occur within the shell model space of appropriately antisymmetrized many-nucleon wave functions.[2]

In constructing irreps, it is useful to form linear combinations of the sp(3,R) elements which have the properties of raising, lowering and a complementary set of operators. Making the usual change of nucleon position and momentum coordinates to harmonic oscillator raising and lowering (boson) operators

$$\begin{aligned} b^\dagger_{in} &= \sqrt{\frac{m\omega}{2\hbar}}(x_{in} - \frac{\mathrm{i}}{m\omega}p_{in}) \\ b_{in} &= \sqrt{\frac{m\omega}{2\hbar}}(x_{in} + \frac{\mathrm{i}}{m\omega}p_{in}), \end{aligned} \tag{17}$$

we find that the desired basis is given by the quadratics in the harmonic oscillator bosons

$$\begin{aligned} A_{ij} &= \sum_n b^\dagger_{in} b^\dagger_{jn} && \text{giant resonance raising operators} \\ C_{ij} &= \sum_n (b^\dagger_{in} b_{jn} + b_{jn} b^\dagger_{in}) && \text{elements of a u(3) subalgebra} \\ B_{ij} &= \sum_n b_{in} b_{jn} && \text{giant resonance lowering operators.} \end{aligned} \tag{18}$$

Thus, the set of sp(3,R) operators is seen to comprise a subset of u(3) operators, a set of giant monopole and giant quadrupole resonance raising operators and a corresponding set of giant resonance lowering operators. Thus, indeed the sp(3,R) Lie algebra can be regarded as an extension of Elliott's u(3) algebra to include the giant resonance vibrational degrees of freedom, as emphasized in Draayer's talk.

A lowest weight state for an sp(3,R) irrep is now defined to be a state $|\varphi_0\rangle$ which satisfies the equations

$$B_{ij}|\varphi_0\rangle = 0 \tag{19}$$

$$C_{ij}|\varphi_0\rangle = 0\,, \quad i \leq j \tag{20}$$

$$C_{ii}|\varphi_0\rangle = N_i|\varphi_0\rangle\,, \tag{21}$$

where (N_1, N_2, N_3) is a triple of integers or half-odd integers. Eqs. (20) and (21) are also the equations that must be satisfied by the highest weight state of a u(3) irrep $\{N_1, N_2, N_3\}$. Thus, an sp(3,R) lowest weight state is a u(3) highest weight state

[2] It is worth noting that it is possible to devise a symplectic model in which the position and momentum coordinates of the nuclear matter distribution are taken to a still more microscopic level (e.g., quarks).

that is annihilated by the giant monopole and quadrupole lowering operators. The quantum numbers $\{N_1, N_2, N_3\}$ are related to the quantum numbers $N(\lambda, \mu)$ by

$$\begin{aligned} N &= N_1 + N_2 + N_3 \\ \lambda &= N_1 - N_2 \\ \mu &= N_2 - N_3 \,. \end{aligned} \tag{22}$$

The labels (λ, μ) are those used by Elliott [1] to define the unique su(3) irrep contained in the u(3) irrep $\{N_1, N_2, N_3\}$.

The first step in applying the symplectic model is to select an irrep appropriate for the states of interest in the nucleus under investigation. It is important to understand that the whole nuclear shell model space, not just the valence shell space, is a direct sum of irreducible sp(3,R) representation spaces. The decomposition of the shell model space into sp(3,R) irreps is made, in principle, by decomposing each harmonic oscillator shell into u(3) irreps. Every u(3) irrep whose states are annihilated by the giant resonance lowering operators of the sp(3,R) algebra then defines an sp(3,R) irrep. Moreover, as shown above, the sp(3,R) quantum numbers are given by the quantum numbers of the associated (lowest weight) u(3) irrep. For a light nucleus, it is appropriate to select the leading irrep, i.e., the irrep based on the su(3) irrep having the largest value of the su(3) Casimir invariant. These are the irreps selected by Elliott [1] in application of the su(3) model to light nuclei. The same irreps are naturally extended to the symplectic model by regarding them as lowest weight u(3) irreps for sp(3,R) irreps. For heavier nuclei, other considerations suggest that sp(3,R) irreps with their lowest weight u(3) irreps in higher shells may be appropriate. We return to this question in section 8.

Having selected an sp(3,R) irrep, it is possible, at least in principle, to diagonalize a microscopic Hamiltonian within the shell model space of the chosen irrep. Many practical problems are involved in carrying out this program. The first is to choose a suitable effective Hamiltonian for a calculation whose active space involves many harmonic oscillator shells. Suitable Hamiltonians are probably of the same type as those used in Hartree-Fock calculations with states covering many shells. The second problem is to calculate matrix elements of the Hamiltonian between basis states of the irrep. This is now a much more complicated problem than for the phenomenological symplectic model. The reason is that an arbitrary microscopic Hamiltonian is not related in any natural way to the elements of the sp(3,R) Lie algebra. Thus, one cannot simply use the abstract knowledge of the irrep to calculate matrix elements. One has to consider the detailed shell model structure of the wave functions. Important advances in the technology of calculating the matrix elements of a microscopic Hamiltonian in an sp(3,R) basis have been made by Arickx *et al.* [29, 30], Filippov, Chopovsky and Vasilevsky [31], Vassanji and Rowe [32], Hecht, Reske and Draayer [33] and Blokhin [34].

Vassanji and I used the generator coordinate method which is well adapted for the purpose. The fundamental principle underlying the method is that an

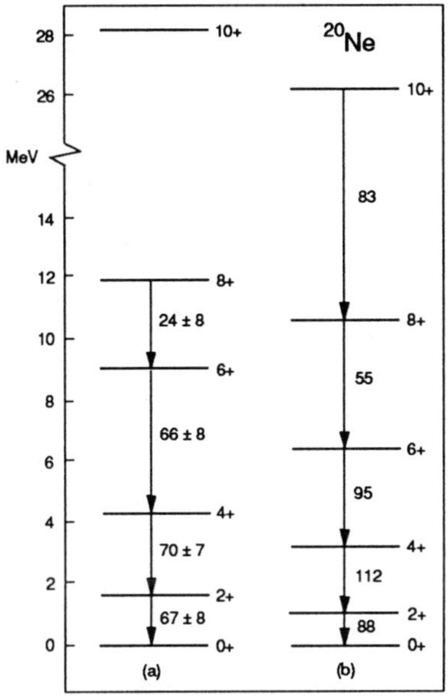

Fig. 5. The low energy spectrum of ^{20}Ne calculated with a Brink-Boeker interaction.

sp(3,R) irrep is spanned by the set of generator coordinate state of the form

$$|\varphi(g)\rangle = \Gamma(g)|\varphi_0\rangle\,, \quad g \in \mathrm{GL}(3,\mathrm{R}) \subset \mathrm{Sp}(3,\mathrm{R})\,, \tag{23}$$

where $|\varphi_0\rangle$ is a lowest weight state for an sp(3,R) irrep and $\Gamma(g)$ defines the transformation corresponding to the Sp(3,R) group element g. The proof that the generator coordinate states $\{|\varphi(g)\rangle; g \in \mathrm{GL}(3,\mathrm{R})\}$ span an sp(3,R) irrep is given along the same lines as the proof that an SU(3) irrep is spanned by the SO(3) rotations of its highest weight state; a property that was used by Elliott [1] in the construction of a basis for irreps of his model.

Since an arbitrary GL(3,R) matrix contains nine independent parameters, a full sp(3,R) model calculation requires the use of nine generator coordinates. While it can be done in principle, this is a non-trivial computational problem. In the calculations of Vassanji and myself, only the six most parameters were included. Some of the results obtained with a Brink-Boeker interaction for the ground state rotational bands of ^{20}Ne and ^{24}Mg are shown in Figs. 5 and 6.

5. Boson and rotor expansions of the symplectic model

It was shown by Rosensteel and Rowe [35] that, in the limit in which the sp(3,R) quantum number N becomes large, the raising and lowering operators of sp(3,R) become proportional to boson (giant resonance phonon) operators. Speci-

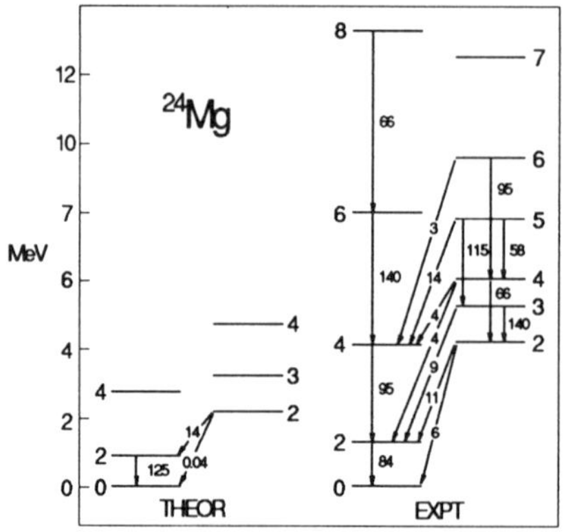

Fig. 6. The low energy spectrum of ^{24}Mg calculated with a Brink-Boeker interaction.

fially they showed that, to leading order in $1/N$,

$$\begin{aligned} A_{ij} &\rightarrow \sqrt{2N/3}\ a^\dagger_{ij} \\ C_{ij} &\rightarrow C_{ij} + (a^\dagger a)_{ij} \\ B_{ij} &\rightarrow \sqrt{2N/3}\ a_{ij}\,, \end{aligned} \tag{24}$$

where a_{ij} and $a^\dagger_{ij}$ satisfy the boson commutation relations

$$[a_{ij}, a^\dagger_{kl}] = \delta_{il}\delta_{jk} + \delta_{ik}\delta_{jl}\,. \tag{25}$$

Thus, the monopole and quadrupole raising operators, defined by

$$A_{l=0} = \frac{1}{\sqrt{6}}\sum_i A_{ii}, \quad A_{l=2,m=0} = \frac{1}{\sqrt{12}}(2A_{11} - A_{22} - A_{33})\,, \tag{26}$$

map to s and d boson operators

$$A_0 \rightarrow \sqrt{2N/3}\ s^\dagger\,, \quad A_{2\nu} \rightarrow \sqrt{2N/3}\ d^\dagger_\nu\,. \tag{27}$$

This limit is called a contraction limit. It has also been considered by other authors, e.g., Biedenharn [36] and Castanños and Draayer [37]. One can easily show [35], by means of eq. (11), that for a nucleus of mass number A, the value of N is given, to a good approximation, by

$$N \approx 0.9A^{4/3}\,. \tag{28}$$

For ^{40}Ca, for example, $N = 118\frac{1}{2}$ while, for ^{166}Er, $N = 811\frac{1}{2}$. Thus, N is a large number for medium and heavy nuclei and the leading order term in the boson expansion of an element of the sp(3,R) algebra is an excellent approximation.

The higher order terms in a boson expansion of the sp(3,R) algebra have been determined in VCS theory [25, 26] and the boson expansion is found to give extraordinarily accurate analytical expressions of the sp(3,R) matrix elements. Thus, to an extremely good approximation, the sp(3,R) Lie algebra can be regarded as a combination of a u(3) Lie algebra and a giant resonance boson algebra. The VCS expansion of ref. [25] can also be used to obtain exact matrix elements by numerical methods.

Now, it has also been known, since the work of Elliott [1] that the matrix elements of the su(3) Lie algebra approach those of the rigid rotor model in the limit in which the su(3) quantum numbers (λ, μ) are large. The reason for this has been explained, by Carvalho [38, 39] as a contraction of the su(3) algebra to a rotor algebra. Specifically, she showed that if one sets $q = \varepsilon Q$, the su(3) commutation relations

$$\begin{aligned} [L, L] &= -\sqrt{2}\, L \\ [L, Q] &= -\sqrt{6}\, Q \\ [Q, Q] &= 3\sqrt{10}\, Q \end{aligned} \tag{29}$$

contract to the commutation relations

$$\begin{aligned} [L, L] &= -\sqrt{2}\, L \\ [L, q] &= -\sqrt{6}\, q \\ [q, q] &= 3\sqrt{10}\varepsilon q \rightarrow 0 , \end{aligned} \tag{30}$$

of the rigid rotor algebra as $\varepsilon \rightarrow 0$. By identifying ε with $1/\langle C_2 \rangle^{1/2}$, where

$$\langle C_2 \rangle = \lambda^2 + \mu^2 + \lambda\mu + 3(\lambda + \mu) \tag{31}$$

is the value of the su(3) Casimir invariant, one sees that the contraction limit is realized as λ and/or μ become large. From experiment [40], one finds that typical values of λ for deformed rare-earth nuclei are in the neighbourhood of $\lambda \approx 60$. Thus, the su(3) to rotor algebra contraction is expected to be very good for heavy deformed nuclei.

The su(3) to rotor algebra contraction was examined further using VCS theory in refs. [41, 42]. We showed that VCS theory gives an exact rotor expansion of the su(3) algebra. Moreover, it gives approximate analytical expressions that are extraordinarily accurate for states of angular momentum $L < (2\lambda + \mu)/4$. Thus, we conclude that, for states of angular momentum $L < 30$, the sp(3,R) algebra contracts to a combination of a rotor algebra, describing the contribution to the spectrum from the particles of the lowest shell, and a boson algebra for the giant resonance degrees of freedom. Thus, the symplectic model contracts for medium and heavy nuclei to a Coupled Rotor Vibrator (CRV) model in which rotational states are described by a u(3) rotor renormalized by the extra quadrupole deformations it induces by polarization of the giant quadrupole resonance vacuum.

6. The Coupled Rotor Vibrator model

The Coupled Rotor Vibrator (CRV) model was proposed by Le Blanc *et al.* [43] and further developed by myself and colleagues [42] as a physically intuitive

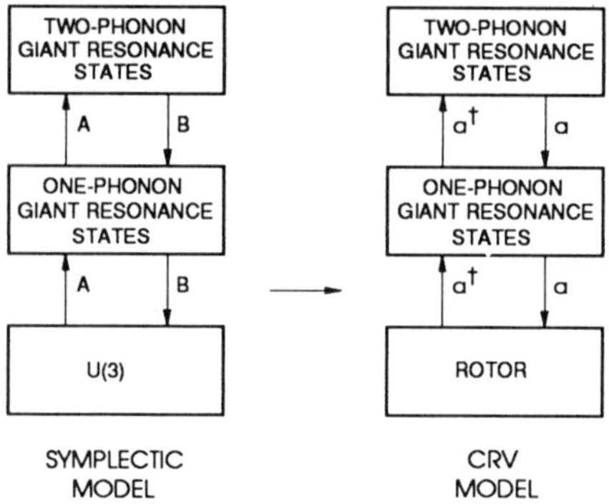

Fig. 7. The symplectic model contracts to a Coupled-Rotor-Vibrator model for large values of N and $2\lambda+\mu$.

limit of the symplectic model that is valid whenever the quantum numbers N and $(2\lambda+\mu)$ of the symplectic model are large. The content of the CRV model, as it relates to the symplectic model, is shown schematically, in Fig. 7.

The CRV model is a simple algebraic model having distinct rotor and vibrator degrees of freedom. The rotor is interpreted as the contraction limit of an Elliott su(3) rotor and the vibrator represents the monopole and quadrupole giant resonance excitations. The CRV algebra is a direct sum of a rotor algebra, spanned by a commuting set of quadrupole operators and angular momentum operators $\{Q_{2\nu}^{\rm rot}, L_k^{\rm rot}\}$, together with the set of operators $\{s, d_\nu, s^\dagger, d_\nu^\dagger, I\}$ of a Heisenberg-Weyl boson algebra ; I is the identity operator. The total angular momentum and quadrupole moments for the CRV model are the sums of their rotor and vibrator components; viz.

$$\begin{aligned} L_k &= L_k^{\rm rot} + \sqrt{10}\,(d^\dagger \times d)_{1k} \\ Q_{2\nu} &= Q_{2\nu}^{\rm rot} + (2N)^{1/2}(d_\nu^\dagger + d_\nu) + \sqrt{2}\,[2(d_\nu^\dagger s + s^\dagger d_\nu - \sqrt{7}\,(d^\dagger \times d)_{2\nu}]\,. \end{aligned} \tag{32}$$

CRV operators corresponding to other observable quantities are similarly given by their sp(3,R) limits.

A simple Hamiltonian for the CRV model is of the form

$$H = H_{\rm intr} + H_{\rm rot} + H_{\rm vib} + H_{\rm coup}\,. \tag{33}$$

$H_{\rm intr}$ is any suitable function of the irrep labels $N(\lambda,\mu)$ which fixes the band-head energies. $H_{\rm rot}$ is a rotor model Hamiltonian which might be of the form

$$H_{\rm rot} = AL(L+1) + BK^2\,. \tag{34}$$

$H_{\rm vib}$ is a suitable boson Hamiltonian for the giant-resonance excitations; e.g.,

$$H_{\rm vib} = \hbar\omega_0 s^\dagger s + \hbar\omega_2 \sum_\nu d_\nu^\dagger d_\nu\,. \tag{35}$$

Finally, a suitable expression for the coupling interaction might be of the form

$$H_{\text{coup}} = \chi Q^{\text{rot}} \cdot (d^{\dagger} + d)\,. \tag{36}$$

The parameters of such a CRV Hamiltonian can be derived in several ways. For example, Leschber, Draayer and Rosensteel [44] have given the mapping between an su(3) Hamiltonian and a rotor Hamiltonian. The frequencies of the giant resonance could be determined from a preliminary microscopic Random Phase approximation (RPA) calculation and the coupling constant χ of the coupling interaction can be estimated on the basis of self-consistency (cf., ref. [45]). A CRV Hamiltonian can also be inferred from a microscopic Hamiltonian using the extension of the RPA, called the partial RPA, designed for deformed nuclei and the way vibrations couple to deformations [46]. The spectrum of such a Hamiltonian was given in ref. [46]. One finds that, to a good approximation (cf., also ref. [22]) that the coupling to the giant resonance degrees of freedom gives ground state rotational bands with an effective quadrupole operator of the form

$$Q^{\text{eff}}_{2\nu} = \frac{2Ae}{A} Q^{\text{rot}}_{2\nu}\,. \tag{37}$$

The important features of the CRV model (and, of course, the symplectic model from which it derives) are:

(i) It shows how the giant resonance vibrational states couple to rotational states to give renormalized effective charges for the low lying SU(3)-rotor bands.

(ii) It shows how the giant resonances themselves acquire a structure from the coupling. In particular, for an axially symmetric nucleus, the giant quadrupole resonance becomes three rotational bands with K quantum numbers given by $K = 0$, 1 and 2. The structure of the giant resonance obtained in this way is very much in accord with what one expects from collective model predictions.

(iii) It predicts a discrete set of rotor bands with intrinsic quadrupole moments given by the set of su(3) irreps in the shell model that are annihilated by the giant resonance lowering operators; i.e., by the set of sp(3,R) irreps in the shell model.

(iv) It shows that, although the symplectic model was constructed to include rotational and shape-vibrational degrees of freedom, the shape vibrations turn out to be those of the high frequency giant resonance type; there are no beta or gamma vibrations explicitly in the model. There are $K = 2$ bands for irreps with $\mu \neq 0$ which one may choose to call gamma bands, albeit with some misuse of terminology. Of course, a nucleus has many (intrinsic) degrees of freeom that are not embraced by either the symplectic model or its CRV limit. Thus, intrinsic excitations (including beta vibrations) are simply regarded as excitations to different irreps. However, since the matrix elements of the quadrupole operator between states of different sp(3,R) irreps are precisely zero, one must have mixing of irreps to explain the existence of excited rotational bands with non-zero E2 transitions to the ground state band.

7. Low energy vibrations

7.1. The symplectic shell model

The symplectic shell model [39] is simply the full shell model (not restricted to the valence space) in a basis that reduces the the subalgebra chain

$$\mathrm{sp}(3,\mathrm{R}) \supset \mathrm{su}(3) \supset \mathrm{so}(3) \ .$$

Some beginnings of calculations in a mixed sp(3,R) basis, with realistic interactions, have been attempted by Hecht, Reske and Draayer [33] and we await results.

However, in spite of the shortage of calculations, the symplectic shell model is still of importance for conceptual and practical purposes. Conceptually, it leads one to understand how correlations associated with degrees of freedom not in the basic symplectic model can be accommodated by mixing irreps. It also shows that a general shell model wave function, given in such a basis, has an immediate interpretation in collective model terms. Such an interpretation is especially simple if the mixed representation wave functions are given in the CRV limit of symplectic model irreps.

First, it should be clearly understood, that the CRV limit of a single sp(3,R) irrep corresponds to a rotor with high frequency giant resonance vibrational degrees of freedom. It has no low frequency vibrational degrees of freedom. Thus, if one factors out the giant resonance degrees of freedom from the model, so that the sp(3,R) model projects to the su(3) model based on its lowest weight u(3) subrepresentation, one is left with su(3) wave functions which, in the limit, correspond to those of a rigid rotor. A rigid rotor may be defined as one having precise values for its intrinsic quadrupole moments; it has no vibrational shape fluctuations and it can have not vibrational excitations. In contrast, a soft rotor has an intrinsic state with a distribution of intrinsic quadrupole moments. Thus, a soft rotor can have low energy excited rotational bands with orthogonal distributions of intrinsic quadrupole moments. In particular, it can exhibit beta vibrational bands. The same is true in the su(3) model. In the absence of band mixing, there can be no non-zero quadrupole matrix elements between states of different su(3) irreps.

7.2. Strong coupling of the spins

A possible source of mixing of sp(3,R) irreps is the spin-orbit interaction which is known to be strong in heavy nuclei. However, to be able to discuss spin-orbit interaction effects, we need to bring the intrinsic spins of the nucleons explicitly into the picture. The way to do this has been given recently by Rowe and de Guise [47] following ideas suggested by Rochford and Rowe [48].

First recall that the inclusion of spin is easy in the rotor model; one simply couples the spin to an intrinsic state of the rotor and generates rotational bands of total angular momentum states which, for odd nuclei, can have half-odd integer values. The reason this is possible for the rotor is that nucleon spin wave functions, of spin S, can be combined with orbital rotational wave functions, having a sequence of L values, to give new rotational bands with sequences of total angular momentum J values. Moreover, the new bands can be constructed in such a way that they separately span irreps of the rotor algebra. The coupling which achieves this result is the familiar strong coupling of the rotor model. These observations are made precise in the language of group representation theory in ref. [47].

A parallel strong coupling of spin wave functions to su(3) wave functions can be defined (cf., ref. [47]) in precisely the same way. However, for su(3), strong coupling cannot, in principle, give new su(3) irreps. This is especially clear for the half-odd integer spin bands because su(3) does not have spinor irreps with half-odd integer angular momentum values. Nevertheless the contraction limit, in which the low angular momentum states of an su(3) irrep behave as though they belonged to a rotational band, implies that strongly coupled spin-su(3) wave functions belong to quasi su(3) irreps which, for practical purposes, may be treated as if they were spinor su(3) irreps; i.e., they have matrix elements which can be inferred from the rotor model and they have neglible quadrupole transition matrix elements between different quasi irreps.

We now consider the mixing of bands (irreps) by the spin-orbit interaction, first in the rotor model and subsequently in the su(3) model. Fig. 8 shows two sets of rotational bands; one set has $K_L = 0$ and $S = 0$ and the others results from strong coupling spin $S = 1$ to a $K_L = 1$ rotational band. The inclusion of the diagonal components of a spin-orbit interaction results in the separation of the various K_J, $S = 1$ bands and stabilizes the strong coupling scheme. If now the two $K_J = 0$ bands, one with $S = 0$ and the other with $S = 1$ are allowed to mix, as a result of the off-diagonal components of the spin-orbit interaction, the mixing can be quite strong. However, the outcome, shown in Fig. 9, is quite remarkable. It reveals that, as a result of the mixing, the two bands repel one another, as one would expect, but each of them retains the characteristics of a rotational band. Because of the mixing there can now be quadrupole transitions between the two bands. However, one finds that the interband transitions are very small. The excited band can be interpreted as a beta vibrational band albeit with a very small transition strength to the states of the ground state band.

The smallness of the beta to ground band transitions can be understood as follows. One finds that the spin-orbit interaction only mixes bands which have very similar intrinsic quadrupole deformations. Consequently the vibrational shape fluctuations of the intrinsic states for each of the two mixed bands remain small; they remain close to their, almost equal, rigid values. The retention of the properties of a rotor band in the face of strong mixing can be understood as a consequence of the *adiabaticity* of the rotational dynamics relative to the "beta-vibrational" dynamics.

Similar results occur in the parallel su(3) situation where the fact that the spin-orbit interaction only mixes states of almost equal quadrupole moments is further clarified. Consider, for example, an axially symmetric $(\lambda, \mu = 0)$ $K_L = 0$ su(3) band of states with spin $S = 0$. The spin-orbit interaction is an su(3) tensor of rank $(\lambda, \mu) = (1, 1)$. Now, a (1,1) tensor can couple $(\lambda, \mu = 0)$ $S = 0$ states to $(\lambda + 1, \mu = 1)$ $S = 1$ states. Thus, one obtains strongly coupled bands of spin-su(3) states very

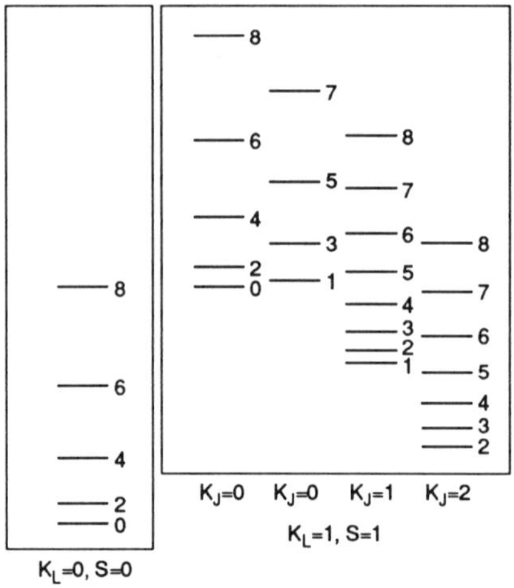

Fig. 8. Two sets of rotational bands; one set has $K_L = 0$ and $S = 0$ and the others results from strong coupling spin $S = 1$ to a $K_L = 1$ rotational band.

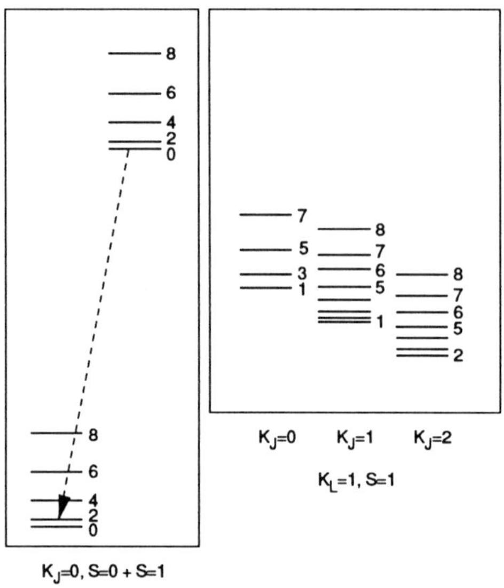

Fig. 9. Strong "adiabatic" mixing of two $K_J = 0$ bands, one with $S = 0$ and the other with $S = 1$, due to a spin-orbit interaction. As a result of the mixing there is a weak $\Delta K = 0$ interband E2 transition between the two mixed bands.

similar to those shown for the rotor. Note, however, that the su(3) bands $(\lambda, 0)$ and $(\lambda+1, 1)$ have very similar intrinsic quadrupole deformations. Thus, we expect to obtain very small transitions between su(3) bands resulting from mixing due to the spin-orbit interaction.

We conjecture that an explanation can be given in this way for why the quadrupole operators of the pseudo-su(3) model [2] should be so close to those of the standard su(3) algebra and how, as a consequence, one is able to transform away the spin-orbit interaction without having to introduce effective quadrupole operators that would mix pseudo-SU(3) irreps (cf., talk by J. P. Draayer).

We also conclude from this analysis that, if beta vibrational bands are observed experimentally with large ground state E2 transitions, then some mixing of irreps by an interaction other than the spin-orbit interaction is probably responsible. A likely candidate is the pairing interaction which could be responsible for relatively large shape fluctuations. However, it is not clear that there is any very compelling experimental evidence to support the concept of a beta vibration as a strong collective mode. The situation for gamma vibrations is even less clear. The problem here is largely one of semantics. One can have a $K = 2$ band belonging to the same su(3) irrep as the $K = 0$ ground band. From the su(3) rotor contraction, it is most natural to associate such a band with a rotation of a triaxial rotor about its z-axis, although one can call it a gamma band if one wishes.

8. The relevant irreps for nuclear rotational bands

In light nuclei, it is appropriate to select the symplectic irrep based on the leading su(3) irrep[3] of the lowest harmonic oscillator shell as being most relevant for the description of the ground state rotational band. However, even in light nuclei, excited bands sometimes come from higher shells. For example, in ^{16}O, the rotational band with its band head at 6.03 MeV is believed to be a 4 particle-4 hole excitation brought low in energy by the energy gained from its deformation. To a first approximation, the ground state of ^{16}O is a closed shell state and spherical. However, excited particle-hole states can be deformed. The remarkable fact is that the deformation energy gained by promoting four particles to a higher shell generates a lower energy state than just promoting two particles. It appears that, if the deformation energy gained by the 4 particle-4 hole state were just 7MeV greater, ^{16}O would be a deformed nucleus. This observation makes it clear that, in heavy nuclei, one should not automatically assume the shell model space spanned by the lowest spherical shell model states to be the most appropriate effective shell model space for the description of deformed nuclei. In fact, that experiment and Nilsson model considerations [40, 22] suggest that, in strongly deformed nuclei, ground state bands come from 10 - 16$\hbar\omega$ shell model configurations.

First observe that the quantum numbers $N(\lambda, \mu)$, which define an sp(3,R) irrep, can be related to the mean square radius and intrinsic quadrupole moments;

[3]The leading su(3) irrep is the one with the largest value of the Casimir invariant.

the relation is given approximately by Eq. (11). Thus, even if experimentally observed rotational states are mixtures of states from different sp(3,R) irreps, one can measure the average values of these quantum numbers. From a consideration of known quadrupole moments and E2 transition strengths, Jarrio *et al.* [40] inferred the implied averages of the quantum numbers $2\lambda+\mu$ and μ for a number of rare-earth nuclei. A summary of their results is given in Table 1. They also

Table 1. Average values of $2\lambda+\mu$ and μ inferred from experiment (from Jarrio *et al.* [40]).

	$\langle 2\lambda+\mu\rangle$		$\langle\mu\rangle$
Nucleus	$K^\pi=0^+$	$K^\pi=2^+$	$2^+\to0^+$
^{160}Er	92–136		
^{162}Er	107	135 ± 44	
^{164}Er	108–155	181 ± 20	10
^{166}Er	136–173	165 ± 14	14–21
^{168}Er	151–183	110–233	7–33
^{170}Er	112–165	151 ± 26	8–17
^{182}W	48–173		15–18
^{184}W	117–159		13
^{186}W	108–158	93 ± 16	15

considered the values of the quantum numbers $N(\lambda,\mu)$ implied by a Nilsson model description of the intrinsic states of the corresponding bands. They found values consistent with those determined experimentally. However, the same Nilsson model wave functions belonged to sp(3,R) representations having lowest weight states in a spherical shell at some 10 - 16$\hbar\omega$ in harmonic oscillator energy above the lowest shell available. Such a result is not surprising if one starts from a deformed shell model (Nilsson model) perspective. It is a direct consequence of the single particle level crossings that occur with increasing deformation on a Nilsson diagram and the fact that often one fills single particle states linked continuously to spherical states from higher shells ahead of those linked to lower ones. Nevertheless, it means that a shell model calculation, restricted to a space of low spherical shell model states, will have essentially zero overlap with any states of the sp(3,R) irreps found necessary in the analysis of Jarrio *et al.* It should be clearly understood that no amount of renormalization of such low lying shell model states by coupling to giant quadrupole resonances can change this fact. This is because the sp(3,R) irreps required have vanishing components in the lowest spherical shells.

A more detailed theoretical analysis of which sp(3,R) irreps are expected to be most relevant for the description of low lying rotational bands has recently been made by Carvalho and Rowe [22]. We addressed the following problem: given a list of sp(3,R) irreps $N(\lambda,\mu)$ that occur in the shell model space of a given nucleus, how can one order the list in decreasing order of importance for the description of the low energy spectrum of the nucleus in question? The first problem is that the list is infinite. Moreover, the number of irreps whose lowest weight states belong to a particular shell grows very rapidly with the harmonic oscillator energy, $N\hbar\omega$, of the shell.

In tackling the problem, we first determined the largest values of $(2\lambda+\mu)$ and the corresponding values of the intrinsic deformations of the rotational bands to which they correspond for all the lowest spherical shells of ^{168}Er up to $26\hbar\omega$. The results are listed in Table 2. We found that the experimental deformation of $\epsilon \approx 0.32$ for ^{168}Er and the corresponding average value of $(2\lambda+\mu)$ (in the range 151-183) could not be obtained with an sp(3,R) irrep whose lowest weight state belongs to a spherical shell of less than $10\hbar\omega$ above that of the lowest shell. Note that, because of level crossings due to the spin-orbit interaction, a standard spherical shell model calculation would be in the $4\hbar\omega$ space.

Table 2. The quantum numbers $N(\lambda,\mu)$ of the sp(3,R) irreps having the largest values of $2\lambda+\mu$ for a given value of N. With removal of the centre of mass contribution of 3/2, the N quantum number is given by $N = 812.5+n$. (from ref. [22]).

n	0	2	4	6	8	10	12
$2\lambda+\mu$	68	96	124	146	168	190	212
λ	30	40	52	60	70	82	96
μ	8	16	20	26	28	26	20
ϵ	0.12	0.17	0.22	0.26	0.30	0.33	0.37
γ	11.88	16.6	16.2	17.9	16.9	14.1	9.9
n	14	16	18	20	22	24	26
$2\lambda+\mu$	228	244	260	270	280	290	297
λ	104	114	126	128	134	136	140
μ	20	16	8	14	12	18	16
ϵ	0.39	0.42	0.44	0.46	0.47	0.49	0.50
γ	9.2	6.9	3.38	5.5	4.6	6.6	5.6

We then enquired what sp(3,R) irrep the Nilsson model would predict to be appropriate for ^{168}Er. For each value of the Nilsson deformation parameter, we determined the Nilsson model energy (the sum of the Nilsson single particle energies) as a function of the possible N values. Note that the value of N depends on the choice of which Nilsson single particle states one chooses to occupy. Not surpisingly, one finds, for each value of the Nilsson deformation parameter, a minimum energy (many-particle) state with a deformation quite close to that of the Nilsson Hamiltonian from which it was generated. However, precise self-consistency is only obtained for special values of the Nilsson deformation parameter. The Nilsson model energies of states is plotted, in Fig. 10, as a function of n, where $N = N_0 + n$ and N_0 is the smallest possible value of N, consistent with the Pauli principle, and N is evaluated for the asymptotic component of each Nilsson model state. Note that N is a good quantum number for asymptotic Nilsson model states. The results show that, for a value of the deformation parameter $\epsilon_N = 0.3$ in the Nilsson model Hamiltonian, the asymptotic component of the lowest energy Nilsson model state belongs to a $14\hbar\omega$

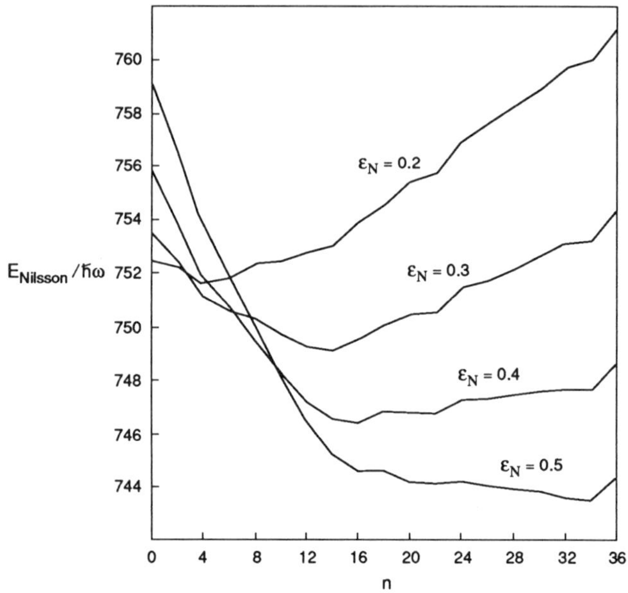

Fig. 10. Nilsson model energies of many particle states as a function of n, corresponding to differently occupied single particle states, for different values of the deformation parameter ϵ_N in the Nilsson model Hamiltonian (from ref. [22]).

spherical shell model configuration.

It is of interest to observe also that, for a Nilsson deformation parameter $\epsilon_N = 0.5$, one gets a $34\hbar\omega$ state coming lowest in energy. We do not find this convincing evidence for the occurence of states of the corresponding sp(3,R) irrep in the low energy spectrum of ^{168}Er. However, it does indicate that there should be no difficulty in explaining the appearance of superdeformations in rare-earth nuclei. Of course, predicting just where and when superdeformed states should be found is a more difficult task and one for which a much more sophisticated approach is needed. We suggest that an ordering of sp(3,R) irreps by means of Hartree-Fock-like methods would be appropriate and we are pursuing this possibility.

9. Concluding remarks

I have given a report of the current status of the symplectic model approach to the microscopic theory of nuclear collective states as it is being pursued at Toronto. I hope to have shown that collective rotational flows are reasonably interpreted in the model as strongly coupled irrotational and vortex-spin flows and that they can be realized in microscopic terms as su(3) rotations renormalized by coupling to the giant quadrupole resonance degrees of freedom. The validity of this interpretation is being pursued by predicting electron scattering form factors in the model, to see if they can probe the vortex content of rotational flows. A description of low energy vibrational bands emerges in the model as the result of mixing sp(3,R) irreps due to non-collective interactions. Moreover, it is possible to obtain results consistent with the predictions of the standard Bohr-Mottelson collective model but without the necessity of introducing specific beta and gamma vibrational degrees of freedom.

It is found that phenomenological calculations within an irrep are weakly dependent on the choice of irrep. This is because quadrupole matrix elements depend more or less linearly on the representation labels. However, to get irreps with "enough" deformation, it is necessary to go to rather high lying spherical shell model configurations to find the lowest weight states for the appropriate sp(3,R) irreps. This result is consistent with the Nilsson model interpretation of rotational bands.

Finally, microscopic calculations within the shell model space of a single or several sp(3,R) irreps is, in principle, possible. Calculations with realistic forces are already underway using a generator coordinate basis for an sp(3,R) irrep. However, there is some way to go before realistic calculations in a mixed sp(3,R) basis are feasible.

What remains to be done? The following are some of the projects we have in mind:

(i) Find out if having "correct" shell-model wave functions matters. Suppose we have the wrong sp(3,R) irrep, albeit one with almost the right representation labels. Will it still be possible to fit the experimental data? As far as the energy spectrum, quadrupole moments and E2 transitions are concerned, the answer is almost certainly "yes". This is interesting because it means that one can get correct results with wave functions that are literally orthogonal to the correct ones. What about other observable quantities? Are they sensitive to the detailed character, or the particular shell model wave functions that span different but equivalent, or nearly equivalent, irreps? I surmise that long range collective phenomena will be insensitive to the representation but that detailed short range correlation effects may be sensitive.

(ii) Calculate electron scattering form factors with wave functions from different irreps to see if there are significant differences and if coherent mixing of irreps, due, for example, to pairing, lead to special coherent effects.

(iii) Do calculations of the Hartree-Fock type with realistic forces to order the sp(3,R) irreps contained in the shell model space of a given nucleus according to decreasing importance as far as their contributions to the low energy spectrum of the nucleus is concerned. A given nucleus has an infinite number of sp(3,R) irreps available to it. In the absence of mixing, each gives rise to a rotor band with giant resonance excitations built upon it. Thus, the symplectic shell model predicts the existence of large (in fact, infinite) numbers of different bands all with different deformations and different moments of inertia. One should therefore expect to see many different deformations appearing in the spectrum of a given nucleus. If the bands occurred in the order in which their lowest weight states appeared in the shell model, i.e., ordered by the values of $N\hbar\omega$, then one might expect to see the deformations of rotational bands slowly varying with excitation energy. However, since our experience, already in light nuclei, is that the highly deformed bands coming from irreps with lowest weight states in higher shells often fall below those of lower shells, we should expect to see a variety of excited bands with deformations different from those of the ground state band. Bands from higher shells in the low energy spectrum are often referred to as "intruders". However, our experience suggests that all the low lying bands might be regarded as intruders in this sense. If a particular band has a very different structure from nearby bands because it comes from different shells, we should expect it to mix rather less than might otherwise be the case with the result that is would be particularly rigid. Thus, superdeformed bands have a natural interpretation in the symplectic model.

(iv) Finally, one of the outstanding problems for the development of the microscopic theory of nuclear collective states is to discover how to include pairing correlations in deformed nuclei. We are aware that pairing can be included in a mean field theory by using the BCS, or Hartree-Bogolyubov, approximation. However, we are seeking a general theory that is not restricted to independent particle, or independent quasiparticle, states. The theory of pair correlated spherical nuclei is in relatively good shape. This is because there are algebraic methods (based on extensions of Kerman's quasi spin algebra) which can realistically describe them. However, the algebraic structures for pairing and quadrupole collective motions do not appear to be amenable to combination. In other words, the smallest Lie algebras which contain the symplectic algebra and the quasispin algebra are very large and not very useful.

10. References

1. J. P. Elliott, *Proc. Roy. Soc.* **A245** (1958) 128, 562.
2. A. Arima, M. Harvey and K. Shimizu, *Phys. Lett.* **B30** (1969) 517; K. T. Hecht and A. Adler, *Nucl. Phys.* **A137** (1969) 129.
3. A. Bohr, *Kgl. Danske. Videnskab. Selskab Mat. Fys. Medd.* **26** (1952) no. 14; A. Bohr and B. R. Mottelson, *Kgl. Danske. Videnskab. Selskab Mat. Fys. Medd.* **27** (1953) no. 16.
4. G. Rosensteel and D. J. Rowe, *Phys. Rev. Lett.* **38** (1977) 10.
5. G. Rosensteel and D. J. Rowe, *Ann. Phys., N.Y.* **126** (1980) 343.
6. F. M. H. Villars, *Nucl. Phys.* **3** (1957) 240.

7. S. Goshen and H. J. Lipkin, *Ann. Phys., N.Y.* **6** (1959) 301.
8. W. Scheid and W. Greiner, *Ann. Phys., N.Y.* **48** (1968) 493.
9. W. Zickendraht, *J. Math. Phys.* **12** (1971) 1663.
10. A. Y. Dzyublik, V. I. Ovcharenko, A. I. Steshenko and G. F. Filippov, *Sov. J. Nucl. Phys.* **15** (1972) 487.
11. S. Tomonaga, *Prog. Theor. Phys.* **13** (1955) 467, 482.
12. R. Y. Cusson, *Nucl. Phys.* **A114** (1968) 289.
13. H. Ui, *Prog. Theor. Phys.* **44** (1970) 153.
14. L. Weaver and L. C. Biedenharn, *Phys. Lett.* **32B** (1970) 326.
15. L. Weaver and L. C. Biedenharn, *Nucl. Phys.* **A185** (1972) 1.
16. L. Weaver, L. C. Biedenharn and R.Y. Cusson, *Ann. Phys., N.Y.* **77** (1973) 250.
17. V. V. Vanagas, *Algebraic Methods in Nuclear Theory* (Vilnius: Mintis, 1971).
18. P. Raychev, *I. Nucl. Phys.* **16** (1972) 1171; *C. R. Bulg. Acad. Sci.* **25** (1972) 1503.
19. L. C. Biedenharn, *Symmetry Properties of Nuclei*, in the *Proceedings of the 15'th Solvay Conference on Physics* (1970).
20. P. Gulshani and D. J. Rowe, *Can. J. Phys.* **54** (1976) 970.
21. G. Rosensteel and D. J. Rowe, *Int. J. Theor. Phys.* **16** (1977) 63.
22. J. Carvalho and D. J. Rowe, *The Shell Model Distribution of Nuclear Shapes,* to be published in the proceedings of the NATO Advanced Workshop at Cargése Jun. 3-7, 1991 *Nuclear Shapes and Nuclear Structure at Low Excitation Energies,* ed. M. Vergnes (Plenum); University of Toronto preprint (to be published, 1992).
23. G. Rosensteel, *J. Math. Phys.* **21** (1980) 924.
24. O. Castaños, E. Chacon and M. Moshinsky, *J. Math. Phys.* **25** (1984) 1211.
25. D. J. Rowe, *J. Math. Phys.* **25** (1984) 2662.
26. D. J. Rowe, G. Rosensteel and R. Carr, *J. Phys. A: Math. Gen.* **17** (1984) L399.
27. G. Gneuss and W. Greiner, *Nucl. Phys.* **A171** (1971) 449.
28. P. Park, J. Carvalho, M. Vassanji, D. J. Rowe and G. Rosensteel, *Nucl. Phys.* **A414** (1984) 93.
29. F. Arickx, *Nucl. Phys.* **A268** (1976) 347.
30. F. Arickx, J. Broeckhove and E. Deumens *Nucl. Phys.* **A318** (1979) 269; **A377** (1982) 121.
31. G. F. Filippov, L. L. Chopovsky and V. S. Vasilevsky, *Nucl. Phys.* **A388** (1982) 47.
32. M. G. Vassanji and D. J. Rowe, *Phys. Lett.* **127B** (1983) 1; *Nucl. Phys.* **A426**

(1984) 205.

33. E. J. Reske, *Ph.D. thesis.* (University of Michigan, 1984); K. T. Hecht and J. P. Draayer, private communication.
34. A. L. Blokhin, *Collective Nuclear Potential within the Microscopic Sp(6,R) Model,* Kiev preprint ITP-91-74E (1992).
35. G. Rosensteel and D. J. Rowe, *Phys. Rev. Lett.* **47** (1981) 223; D. J. Rowe and G. Rosensteel, *Phys. Rev.* **C25** (1982) 3236.
36. L. C. Biedenharn, *Kinam* **4** (1982) 3.
37. O. Castaños and J. P. Draayer, *Nucl. Phys.* **A491** (1989) 349.
38. J. Carvalho, *Ph.D. thesis* (University of Toronto,1984).
39. J. Carvalho, R. Le Blanc, M. Vassanji, D. J. Rowe and J. B. McGrory, *Nucl. Phys.* **A452** (1986) 240.
40. M. Jarrio, J. L. Wood and D. J. Rowe, *Nucl. Phys.* **A528** (1991) 409.
41. D. J. Rowe, R. Le Blanc and J. Repka, *J. Phys. A: Math. Gen.* **22** (1989) L309.
42. D. J. Rowe, M. G. Vassanji and J. Carvalho, *Nucl. Phys.* **A504** (1989) 76.
43. R. Le Blanc, J. Carvalho and D. J. Rowe, *Phys. Lett.* **B140** (1984) 155; R. Le Blanc, J. Carvalho, M. G. Vassanji and D. J. Rowe, *Nucl. Phys.* **A452** (1986) 263.
44. Y. Leschber and J. P. Draayer, *Phys. Rev.* **C33** (1986) 749; Y. Leschber, J. P. Draayer and G. Rosensteel, *J. Phys. G: Nucl. Phys.* **12** (1986) L179.
45. D. J. Rowe, *Phys. Rev.* **162** (1967) 866.
46. D. J. Rowe, P. Rochford and R. Le Blanc, *Nucl. Phys.* **A464** (1987) 39.
47. D. J. Rowe and H. de Guise, *Strong Coupling and Quasispinor Representations of the SU(3)-Rotor Model* (Nucl. Phys., in press).
48. P. Rochford and D. J. Rowe, *Phys. Lett.* **B210** (1988) 5.

GROUP 3

Collective Models

MICROSCOPIC THEORY OF NUCLEAR COLLECTIVE DYNAMICS

- Present Status of Self-consistent Collective Coordinate Method -

Fumihiko SAKATA

Institute for Nuclear Study, University of Tokyo, Tanasi, Tokyo 188, JAPAN

ABSTRACT

The classical-level theory and full quantum theory of nuclear collective dynamics are discussed. The nonlinear dynamics between the collective and single-particle excitation modes of motion specific for the finite, self-sustained and self-organizing system as the nucleus is discussed within the TDHF theory, the basic equation of which is formally equivalent to the Hamilton's canonical equations of motion in the classical nonlinear dynamical system. An importance to relate the structure of the TDHF symplectic manifold with an inexhaustible rich structure of the classical phase space in the nonlinear system is stressed. A full quantum theory of nuclear collective dynamics proposed under a dictation of what has been developed in the classical-level TDHF theory is discussed. It is shown that the proposed quantum theory enables us to explore exceeding complexity of the Hilbert space. It is discussed that a resonant denominator known as a source of the extraordinary rich structure of the phase space trajectories, also plays a decisive role in generating a rich structure of the quantum Hilbert space.

1. Introduction

Owing to the recent experimental and theoretical studies on the nuclear structure physics, it is generally recognized that there hold many types of shell models with different symmetries in one nucleus. In the last decade, the most exciting development of the nuclear structure physics has been an establishment of the rotating shell model as well as the superdeformed shell model in rapidly rotating nuclei. It is really astonishing that there exists a set of characteristic collective states which are well separated from the thousands of neighboring excited states and are described by a simple rotating shell model. The microscopic justification of various shell models is given by the self-consistent mean-field theory called the Hartree-Fock (HF) theory. It describes how a stable mean-field is produced as a result of the nucleonic motion which are in turn travelling in the resultant stable mean-field self-consistently. The shape of the mean-field is specified by a set of collective parameters and its single-particle states are also characterized by these parameters.

Aiming at studying the global aspect of the mean-field, one usually emploies the constrained (C) HF theory specified by a set of mean-values of collective operators. Based on the CHF theory, it is generally considered that the mean-field is stabilized by a pronounced shell gap structure in the single-particle states and there are many level crossing or psuedocrossing points at the barrier region between two local minima of the collective potential. However, it is still an open question what are happening in between two local minima with different symmetries, and what kinds of single-particle dynamics are responsible for the collective motion which moves around in a fairly wide area covering two local minima. Consequently, it is an important task to explore the dynamical interrelationship between the large-amplitude collective motion and single-particle modes of motion. To this end, one has to employ the time-dependent (TD) HF theory by going beyond the static mean-field theories like the HF or CHF theories.

As is well known, the basic equation of the TDHF theory is formally equivalent to the classical Hamilton's equations of motion with many degrees of freedom. Here, one should remember a recent enormous progress in the general theory of nonlinear dynamical system which has been developed in the last few decades [1,2]. It has been clarified that there exists exceedingly rich structure in the classical phase space. In the large-amplitude region far from the potential minimum, there exist various motions with different topologies, which are not expected from the shape of the potential. It is thus naturally expected that there might be an inexhaustible rich structure in the TDHF-manifold like in the classical phase space.

It has been a cornerstone in the nonlinear dynamics to understand infinite series of island structure of the regular motion and an appearance of stochastic sea in between these islands. Understanding an exceedingly rich structure of the phase space, one usually introduces the *most natural coordinate system* for a given trajectory by using the theory of canonical transformation. It gives a good starting point for studying the phase space structure and for getting the fixed points near vicinity of the given trajectory. In §2, we discuss the classical-level theory of nuclear collective dynamics which is capable of defining *the most optimal dynamical coordinate system* for a given TDHF trajectory. Starting with a brief introduction of the self-consistent collective coordinate method, we discuss how to explore the large-amplitude collective motion in connection with the single-particle dynamics. In §3, we discuss the quantum theory of nuclear collective dynamics.

2. Classical-Level Theory of Nuclear Collective Dynamics

2.1. Self-consistent Collective Coordinate (SCC) Method

Aiming at treating the non-linear dynamics specifically for the nuclear system, we start with the TDHF equation

$$\delta < \phi(C,C^*)|\hat{H} - i\frac{\partial}{\partial t}|\phi(C,C^*)> = 0 ,$$

$$\hat{H} = \Sigma T_{\alpha\alpha}\hat{c}^{\dagger}_{\alpha}\hat{c}_{\alpha} + \frac{1}{4}\Sigma V_{\alpha\beta,\gamma\delta}\hat{c}^{\dagger}_{\alpha}\hat{c}^{\dagger}_{\beta}\hat{c}_{\delta}\hat{c}_{\gamma} \,. \tag{2.1}$$

Here the single Slater-determinant is expressed as

$$|\phi(C,C^*)\rangle \equiv e^{i\hat{F}(k)}|\phi_0\rangle \,, \; \hat{F}(k) \equiv \sum_{\mu i}\left(K_{\mu i}\hat{a}^{\dagger}_{\mu}\hat{b}^{\dagger}_{i} + K^{*}_{\mu i}\hat{b}_{i}\hat{a}_{\mu}\right), \tag{2.2}$$

where $|\phi_0\rangle$ denotes HF state satisfying

$$\delta\langle\phi_0|\hat{H}|\phi_0\rangle = 0, \;\; \hat{a}_{\mu}|\phi_0\rangle = \hat{b}_i|\phi_0\rangle = 0, \;\; \hat{c}_{\alpha} = \begin{cases} \hat{a}_{\mu} \;;\; \text{for particle states} \\ b_i \;;\; \text{for hole states} \end{cases} . \tag{2.3}$$

The infinitesimal variation in Eq.(2.1) is expressed by introducing the following one-body operators

$$\hat{O}^{\dagger}_{\mu i}(k) \equiv \frac{\partial e^{i\hat{F}(k)}}{\partial C_{\mu i}} \cdot e^{-i\hat{F}(k)} \,, \;\; \hat{O}_{\mu i}(k) \equiv -\frac{\partial e^{i\hat{F}(k)}}{\partial C^{*}_{\mu i}} \cdot e^{-i\hat{F}(k)} \,, \tag{2.4}$$

where, C and C* are related to K and K* through

$$C^{\dagger} \equiv -iK^{\dagger}\frac{\sin\sqrt{KK^{\dagger}}}{\sqrt{KK^{\dagger}}} \;\;,\;\; C \equiv i\frac{\sin\sqrt{KK^{\dagger}}}{\sqrt{KK^{\dagger}}}K \;\;. \tag{2.5}$$

The operators in Eq.(2.4) satisfy the following relations

$$\langle\phi_0|e^{-i\hat{F}(k)}\left[\hat{O}_j(k), \hat{O}^{\dagger}_i(k)\right]e^{i\hat{F}(k)}|\phi_0\rangle = \delta_{ij} \;\;,$$

$$\langle\phi_0|e^{-i\hat{F}(k)}\left[\hat{O}_j(k), \hat{O}_i(k)\right]e^{i\hat{F}(k)}|\phi_0\rangle = 0 \,. \tag{2.6}$$

Here and hereafter, the double indexes (μi) specifying the particle-hole pair are also expressed by the single indexes (j). By using relation (2.6), Eq.(2.1) results in the classical canonical equations of motion with K degrees of freedom

$$i\dot{C}_j = \partial H / \partial C^{*}_j \,,\; j = 1,\cdots,K, \;\; H \equiv \langle\phi(C,C^*)|\hat{H}|\phi(C,C^*)\rangle \,, \tag{2.7}$$

where K denotes a number of particle-hole pairs. Equation (2.7) determines the TDHF-trajectory within 2K-dimensional TDHF-manifold M^{2K} .

For the classical dynamical system organized by Eq.(2.4), one may imagine two extreme cases; If an ensemble of trajectories with slightly different initial conditions develops in almost the same way forming a bundle, the system is classified to be in the *collective regime*. In this case, one could introduce a *collective submanifold* Σ^{2L} on which a representative trajectory of the bundle is travelling. On the other hand, the system is classified to be in the *chaotic regime*, if many trajectories with almost the same initial conditions develop quite differently, having lost their initial focussing. In the latter case, one could not introduce any submanifold within M^{2K}. From the above discussion, one may expect that the transition mechanism between "regular" and "chaotic" motions may be studied by first extracting the collective submanifold Σ^{2L} out of M^{2K}. In the collective regime where many trajectories are accumulating on a certain submanifold Σ^{2L}, there must be an optimal coordinate system called the *dynamical canonical coordinate* (DCC) system [3] where the minimum number of relevant (collective) variables (η_a,η_a^*: a=1,...,L<<K) is required in describing Σ^{2L}. The remaining canonical variables (ξ_α,ξ_α^*: α=L+1, ..,K) in the DCC system are called the irrelevant (intrinsic) variables. The coordinate system (C_j,C_j^*) used in describing Eq.(2.1) is called the initial canonical coordinate (ICC) system, because this coordinate system has nothing to do with a specific correlation in the Hamiltonian $H(C,C^*)$.

The extraction of Σ^{2L} is achieved by determining a canonical transformation between the ICC and DCC systems. Since the transformation is canonical, there holds a relation

$$\sum_j \left\{ C_j \frac{\partial C_j^*}{\partial \eta_a^*} - C_j^* \frac{\partial C_j}{\partial \eta_a^*} \right\} = \eta_a + 2i \frac{\partial S}{\partial \eta_a^*}, \quad a = 1,\dots,L,$$

$$\sum_j \left\{ C_j \frac{\partial C_j^*}{\partial \xi_\alpha^*} - C_j^* \frac{\partial C_j}{\partial \xi_\alpha^*} \right\} = \xi_\alpha + 2i \frac{\partial S}{\partial \xi_\alpha^*}, \quad \alpha = L+1,\dots,K, \tag{2.8}$$

where S denotes a generating function of the canonical transformation. To fix the transformation, it is important to introduce some expansion scheme. Since (η,η^*) are supposed to describe the large-amplitude collective motions inside Σ^{2L}, whereas (ξ,ξ^*) describe small fluctuations around Σ^{2L}, it is reasonable to introduce a power series expansion with respect to irrelevant variables. For example,

$$C_j = \sum_n C_j^{(n)} = [C_j] + \sum_\alpha \left\{ \xi_\alpha \left[\frac{\partial C_j}{\partial \xi_\alpha} \right] + \xi_\alpha^* \left[\frac{\partial C_j}{\partial \xi_\alpha^*} \right] \right\} + \cdots, \tag{2.9a}$$

$$H(C,C^*) = \sum_n H^{(n)} = [H] + \sum_\alpha \left\{ \xi_\alpha \left[\frac{\partial H}{\partial \xi_\alpha} \right] + \xi_\alpha^* \left[\frac{\partial H}{\partial \xi_\alpha^*} \right] \right\} + \cdots, \tag{2.9b}$$

where the symbol [g] for any function $g(\eta,\eta^*:\xi,\xi^*)$ denotes a function on the expansion surface Σ^{2L}, i.e. $[g]=g(\eta,\eta^*:\xi=\xi^*=0)$. [H] contains only the relevant variables and is called the collective Hamiltonian H_{coll}.

According to the SCC method [4], the basic equations for defining Σ^{2L}, i.e. the diffeomorphic mapping $(\eta,\eta^*)\rightarrow([C],[C^*])$, are given by the following two requirements

Req.1: There should hold the lowest order canonical variable condition

$$\sum_j \left\{ [C_j] \frac{\partial [C_j^*]}{\partial \eta_a^*} - [C_j^*] \frac{\partial [C_j]}{\partial \eta_a^*} \right\} = \eta_a + 2i \frac{\partial [S]}{\partial \eta_a^*}, \tag{2.10}$$

which is obtained by substituting Eq.(2.9a) for Eq.(2.8).

Req.2: Σ^{2L} should satisfy a stationary condition (*maximal-decoupling condition*) given by $H^{(1)}=0$, which is alternatively expressed as

$$\left[\frac{\partial H}{\partial C_j^*} \right] = \sum_a \left\{ \frac{\partial [H]}{\partial \eta_a^*} \frac{\partial [C_j]}{\partial \eta_a} - \frac{\partial [H]}{\partial \eta_a} \frac{\partial [C_j]}{\partial \eta_a^*} \right\}. \tag{2.11}$$

With the aid of these two requirements together with the "boundary condition of Σ^{2L}" derived from the initial condition of the representative trajectory of the collective motion under discussion, we uniquely determine the functional form of [C] and [C*], i.e. Σ^{2L}. Namely, a separation between the relevant and irrelevant degrees of freedom is performed uniquely.

In order to get the DCC system, we have to further define the higher order expansion coefficients $C_j^{(n)}$ in Eq.(2.9a). By using the higher order canonical variable condition obtained by substituting Eq.(2.9a) for Eq.(2.8), it has been shown [3] that the DCC system is defined in a way compatible with the SCC method.

The above stated SCC method is also expressed by using the one-body operator algebra, i.e. by the single Slater-determinant representation. In this representation, Req.1 is expressed as

$$\delta \langle \phi([C],[C^*]) | \hat{H}' | \phi([C],[C^*]) \rangle = 0 \ , \hat{H}' \equiv \hat{H} - \sum_a \left\{ \frac{\partial [H]}{\partial \eta_a} \hat{O}_a^\dagger + \frac{\partial [H]}{\partial \eta_a^*} \hat{O}_a \right\}, \tag{2.12}$$

where

$$\hat{O}_a^\dagger = \left(\frac{\partial e^{i[\hat{F}]}}{\partial \eta_a} \right) e^{-i[\hat{F}]}, \ \hat{O}_a = -\left(\frac{\partial e^{i[\hat{F}]}}{\partial \eta_a^*} \right) \cdot e^{-i[\hat{F}]} . \tag{2.13}$$

Req.2 is given by

$$\langle \phi([C], [C^*]) | \hat{O}_a^\dagger(\eta, \eta^*) | \phi([C], [C^*]) \rangle = \frac{1}{2} \eta_a^* - i \frac{\partial [S]}{\partial \eta_a}. \tag{2.14}$$

It is clearly recognized from Eq.(2.12) that the SCC method just corresponds to a generalization of the CHF theory. An essential generalization rests on the fact that the CHF theory contains a constraining operator with fixed microscopic structure put in by hand, whereas the SCC equation includes many pairs of dynamical constraining operators which describes the instantaneous collective direction (phase space vector $(\partial[H]/\partial\eta_a, \partial[H]/\partial\eta_a^*)$) chosen by the nucleus itself. Note that the SCC method does not introduce any adiabatic assumption, whereas the CHF theory does.

2.2. Applicability of the Collective Submanifold

A range of an applicability of the SCC method is evaluated by using the *stability* and *separability* conditions. Since the extracted manifold Σ^{2L} is defined so as to satisfy the stationary condition $H^{(1)}=0$ with respect to the irrelevant directions, the stability of Σ^{2L} is evaluated by using $H^{(2)}$. Since $H^{(2)}$ is bi-linear function of irrelevant variables, it is approximately diagonalized by using the RPA-type equation which describes the small-amplitude oscillation toward the irrelevant directions at each point of Σ^{2L}. By using the auxiliary Hamiltonian in Eq.(2.12), the RPA-type equation is expressed as

$$\langle\phi([C],[C^*])\big[\hat{X}_\beta,\big[\hat{H}',\hat{X}^\dagger_\alpha\big]\big]\phi([C],[C^*])\rangle=\omega_\alpha,$$
$$\hat{X}^\dagger_\alpha\equiv\sum_{\mu i}\left\{\phi^\alpha_{\mu i}\hat{a}'^\dagger_\mu\hat{b}'^\dagger_i+\varphi^\alpha_{\mu i}\hat{b}'_i\hat{a}'_\mu\right\},\tag{2.15}$$

where we use new basis operators for $|\phi([C],[C^*]\rangle$ defined by

$$\hat{c}'^\dagger_\alpha\equiv e^{i[\hat{F}]}\hat{c}^\dagger_\alpha e^{-i[\hat{F}]}=\sum\hat{c}^\dagger_\beta D_{\beta\alpha}(\eta,\eta^*);$$
$$\text{i.e. for particle-state }\ \hat{a}'^\dagger_\mu\equiv e^{i[\hat{F}]}\hat{a}^\dagger_\mu e^{-i[\hat{F}]},\quad\text{for hole-state }\ \hat{b}'^\dagger_i\equiv e^{i[\hat{F}]}\hat{b}^\dagger_i e^{-i[\hat{F}]}.\tag{2.16}$$

With the aid of ω_α, the *stability condition* of Σ^{2L} at a point (η,η^*) is given by

$$\omega^2_\alpha(\eta,\eta^*)\le 0\ .\tag{2.17}$$

It has been shown [5] that an introduction of the local RPA modes is justified provided there holds the *separability condition* expressed as

$$\sum_{\mu i}\left\{\phi^\beta_{\mu i}\frac{\partial\phi^{\alpha*}_{\mu i}}{\partial t}-\varphi^\beta_{\mu i}\frac{\partial\varphi^{\alpha*}_{\mu i}}{\partial t}\right\}=0,\quad\sum_{\mu i}\left\{\varphi^\beta_{\mu i}\frac{\partial\phi^{\alpha*}_{\mu i}}{\partial t}-\phi^\beta_{\mu i}\frac{\partial\varphi^{\alpha*}_{\mu i}}{\partial t}\right\}=0\,.\tag{2.18}$$

Note that ϕ and φ just represent $[\partial C_j/\partial\xi_\alpha]$ and $[\partial C_j/\partial\xi^*{}_\alpha]$ in Eq.(2.9a) within the small amplitude approximation. Since the time dependence of correlation amplitudes ϕ and φ only comes from that of the collective amplitudes η and η^*, Eq.(2.18) simply means that the non-collective motion should not strongly depend on η,η^* and/or the collective motion should be quite slow in comparison with the non-collective excitation modes. Therefore, Eq.(2.18) is called the separability condition which justifies a separate treatment between collective and non-collective degrees of freedom.

Equation (2.12) defines a set of single-particle states (Routhian) depending on the collective amplitudes (η,η^*). Equation (2.12) is expressed in a form which explicitly shows how a unitary transformation $D_{\alpha\beta}(\eta,\eta^*)$ in Eq.(2.16) is defined, i.e.

$$\sum_{\beta}\left\{\begin{aligned}&T_{\alpha\beta}+\Gamma_{\alpha\beta}-\sum_{a}\frac{\partial[H]}{\partial\eta_a^*}\left(\frac{\partial D(\eta,\eta^*)}{\partial\eta_a}D^{\dagger}(\eta,\eta^*)\right)_{\alpha\beta}\\&\qquad+\sum_{a}\frac{\partial[H]}{\partial\eta_a}\left(\frac{\partial D(\eta,\eta^*)}{\partial\eta_a^*}D^{\dagger}(\eta,\eta^*)\right)_{\alpha\beta}\end{aligned}\right\}D_{\beta\gamma}(\eta,\eta^*)=E_{\gamma}(\eta,\eta^*)D_{\alpha\gamma}(\eta,\eta^*), \tag{2.19}$$

where

$$[H]=\sum_{\alpha\beta}\left\{T_{\alpha\beta}+\frac{1}{2}\Gamma_{\alpha\beta}(\eta,\eta^*)\rho_{\beta\alpha}(\eta,\eta^*)\right\}\ ,\ \Gamma_{\alpha\beta}(\eta,\eta^*)=\sum_{\gamma\delta}V_{\alpha\gamma,\beta\gamma}\rho_{\delta\gamma}(\eta,\eta^*)\ ,$$

$$\rho_{\alpha\beta}(\eta,\eta^*)=\left\langle\phi([C],[C^*])\right|\hat{C}_{\beta}^{\dagger}\hat{C}_{\alpha}\left|\phi([C],[C^*])\right\rangle=\sum_{i=1}^{N}D_{\alpha i}(\eta,\eta^*)D_{i\beta}^{\dagger}(\eta,\eta^*)\ . \tag{2.20}$$

Equation (2.19) defines a set of single-particle energies $E_\gamma(\eta,\eta^*)$ of new basis operators as well as the unitary transformation $D_{\alpha\beta}(\eta,\eta^*)$. Note that the breaking of the stability condition just corresponds to an appearance of the crossing of the single-particle energies $E_\gamma(\eta,\eta^*)$.

With the aid of both the stability and separability conditions, one may divide the collective submanifold Σ^{2L} into three characteristic regions.

	Collective Region	Dissipative Region	Chaotic Region
Stab. Cond.	$\omega^2>>0$	$\omega^2>0$	$\omega^2\leq 0$
Sepa. Cond.	well separated	weakly separated	strongly mixed .

In the collective region, a TDHF trajectory starting from the vicinity of Σ^{2L} is expected to stay always in a tiny region near Σ^{2L}. In the dissipative region, a TDHF trajectory found near Σ^{2L} at certain instant may oscillate toward the irrelevant direction (perpendicular to the submanifold) with a fairly large amount of deviation from Σ^{2L}. In the chaotic region, the trajectory occasionally travelling near Σ^{2L} does not come back in its neighborhood.

The above classification of the extracted manifold into three characteristic regions is too coarse to examine an inexhaustibly rich structure of the TDHF manifold. In such a finite, self-sustained system as the nucleus, the separability and stability conditions for Σ^{2L} have to be examined carefully for individual irrelevant excitation modes. Since one may naturally expect not only many level-crossing points on Σ^{2L} indicated by the stability condition, but also many tiny regions where the separability condition does not hold, a detailed study of the single-particle dynamics on what are happening in the "dissipative region" will be the central theme for further development of the nuclear collective dynamics.

2.3. *New Stable Mean-Field Associated with the Resonant Condition*

In §2.1, the whole system is divided into the relevant and irrelevant variables by means of the SCC method. However the basic equation of the SCC method has singularity when there holds a resonant condition between relevant and irrelevant degrees of freedom. In this case, a number of relevant variables should be enlarged so as to explicitly include the secular coupling terms into the collective submanifold.

An existence of resonances is known to be a source of an exceedingly rich structure of the phase space in the nonlinear classical theory. In order to illuminate an essential effects of the resonance, let us consider the following ideal situation; There are two degrees of freedom which satisfy resonant relation and are decoupled from the other degrees of freedom. (η_1, η_1^*) describe collecive motion under discussion and (η_2, η_2^*) describe non-collective motion. In this case the Hamiltonian is given by

$$\langle \phi([C],[C^*]) | \hat{H} | \phi([C],[C^*]) \rangle$$
$$\cong H = \omega_1 \eta_1^* \eta_1 + \omega_2 \eta_2^* \eta_2 + \Sigma_n V^{(n)}(J_1,J_2) \left\{ \left(\eta_1^*\right)^{ns} \left(\eta_2\right)^{nr} + \text{h.c.} \right\}. \tag{2.21}$$

Here action and angle variables are defined by

$$\eta_1 = \sqrt{J_1} e^{i\varphi_1},\ \eta_1^* = \sqrt{J_1} e^{-i\varphi_1},\ \eta_2 = \sqrt{J_2} e^{i\varphi_2},\ \eta_2^* = \sqrt{J_2} e^{-i\varphi_2}\ . \tag{2.22}$$

To make the discussion simple, we consider only secular coupling terms in Eq.(2.21). However, an expression in Eq.(2.21) is general enough because the Hamiltonian near the resonant point can be locally transformed into

$$H = H_0 + H_1;$$
$$H_0 = H_0(J_1,J_2)\ ,\ H_1 = \Sigma V^{(n)}(J_1,J_2) \left\{ e^{-in(s\varphi_1 - r\varphi_2)} + \text{h.c.} \right\}, \tag{2.23}$$

by means of the Birkhoff-Gustavson normal-form perturbation theory [6]. The coordinate system after applying the Birkhoff-Gustavson perturbation theory is called the DCC system with the *maximal integrable form representation* [7], because the angle dependence of the Hamiltonian is eliminated as much as possible. In Eq. (2.21) and in Eq.(2.23) , we assume that there holds a resonant condition

$$\frac{\omega_1}{\omega_2} \cong \frac{r}{s} \quad \text{and} \quad \frac{\partial H_0}{\partial J_1} : \frac{\partial H_0}{\partial J_2} \cong r:s\ , \ \text{(r and s are prime numbers)}, \tag{2.24}$$

respectively.

In the classical nonlinear dynamical theory, the resonant Hamiltonian in Eq.(2.21) is treated by introducing a set of resonant coordinates defined by

$$\theta_1 = \varphi_1,\ \theta_2 = r\varphi_1 - s\varphi_2 \ \ ; I_1 = J_1 + \frac{s}{r}J_2 \ \ , I_2 = \frac{1}{r}J_2. \tag{2.25}$$

By using resonant coordinates in Eq.(2.25), the Hamiltonian in Eq.(2.21) is expressed as

$$\mathcal{H}(I_1,I_2,\theta_1,\theta_2) = \omega_1 I_1 + (r\omega_2 - s\omega_1)I_2 + \Sigma_n \tilde{V}^{(n)}(I_1,I_2)(e^{in\theta_2} + \text{h.c.})\ , \tag{2.26}$$

which does not depend on θ_1, indicating an existence of constant of motion I_1. With the aid of the Hamiltonian in Eq.(2.26), one may get a fixed point by

$$\left.\frac{\partial\mathcal{H}(I_1,I_2,\theta_1,\theta_2)}{\partial\theta_2}\right|_{I_2 = I_2^{(0)},\theta_2 = \theta_2^{(0)}} = \left.\frac{\partial\mathcal{H}(I_1,I_2,\theta_1,\theta_2)}{\partial I_2}\right|_{I_2 = I_2^{(0)},\theta_2 = \theta_2^{(0)}} = 0. \tag{2.27}$$

Using the one-body operators in Eq.(2.13), one may introduce the following one-body operators

$$\hat{\varphi}_a \equiv \frac{\partial e^{i[\hat{F}]}}{\partial J_a} e^{-i[\hat{F}]} = \frac{1}{2}\left(\frac{\hat{O}_a^\dagger}{\eta_a^*} - \frac{\hat{O}_a}{\eta_a}\right),\ \hat{J}_a \equiv \frac{\partial e^{i[\hat{F}]}}{\partial\varphi_a} e^{-i[\hat{F}]} = i\left(\eta_a^*\hat{O}_a^\dagger + \eta_a^*\hat{O}_a\right). \tag{2.28}$$

As is easily proved by using Eq.(2.14), the one-body operators in Eq.(2.28) satisfy

$$\langle\phi_0|e^{-i\hat{F}}\left[\hat{J}_a,\hat{\varphi}_b\right]e^{i\hat{F}}|\phi_0\rangle = i\delta_{ab}\quad . \tag{2.29}$$

Let us introduce the following operators by using the unitary transformation which corresponds to the canonical transformation in Eq.(2.25);

$$\hat{\Theta}_1 = \hat{\varphi}_1 + \frac{s}{r}\hat{\varphi}_2,\ \hat{\Theta}_2 = -s\varphi_1 + r\hat{\varphi}_2\ ;\ \hat{I}_1 = \hat{J}_1 + \frac{s}{r}\hat{J}_2,\ \hat{I}_2 = \frac{1}{r}\hat{J}_2. \tag{2.30}$$

Equation (2.27) clearly states that there exists a stable mean field

$$\left|I_2^{(0)},\theta_2^{(0)};I_1,\theta_1\right\rangle = \left|[C]^{(0)},[C*]^{(0)}\right\rangle\ ;\ [C]^{(0)} = f(I_2^{(0)},\theta_2^{(0)};I_1,\theta_1), \tag{2.31}$$

where $[C]^{(0)}$ is defined by

$$\frac{\partial}{\partial I_2}\left.\langle\phi([C],[C^*])|\hat{H}|\phi([C],[C^*])\rangle\right|_{I_1^{(0)},\theta_1^{(0)}} = \left.\langle\phi([C],[C^*])|[\hat{H},\hat{\Theta}_2]|\phi([C],[C^*])\rangle\right|_{I_1^{(0)},\theta_1^{(0)}} = 0\ ,$$

$$\frac{\partial}{\partial\theta_2}\left.\langle\phi([C],[C^*])|\hat{H}|\phi([C],[C^*])\rangle\right|_{I_1^{(0)},\theta_1^{(0)}} = \left.\langle\phi([C],[C^*])|[\hat{H},\hat{I}_2]|\phi([C],[C^*])\rangle\right|_{I_1^{(0)},\theta_1^{(0)}} = 0\ , \tag{2.32}$$

with arbitral given values of I_1, irrespective of the value θ_1. Note that this stable mean field (corresponding to the fixed point) can't be reached by the conventional HF theory.

3. Quantum Theory of Nuclear Collective Dynamics

3.1. *Basic Idea of Quantum Theory*

In the classical theory, an introduction of the DCC system associated with Σ^{2L} plays a central role in characterizing the integrability of the system. In a case of the quantum system, we discuss an importance of introducing the concept of *dynamical representation* (DR) [8] associated with a collective subspace under discussion. Let us start with the boson system described by K-kinds of boson operators

$$\{B_j, B_j^\dagger;\ j=1,\dots,K\} \quad . \tag{3.1}$$

Since the even-number many-fermion system is known to be transformed into the boson system by means of the boson expansion theory, it is general enough to formulate a quantum theory of nuclear collective dynamics within the boson system. The representation using $(B,B^\dagger)$ will be called an *initial representation* (IR), because it has nothing to do with a specific correlation characteristic to the Hamiltonian $H(B^\dagger,B)$. The boson Hilbert space $\mathcal{V}^{2K}$ is spanned by

$$|n_1,\dots,n_K) \equiv \prod_{j=1}^{K} \frac{1}{\sqrt{n_j}} (B_j^\dagger)^{n_j} \,|\, 0) \ , \tag{3.2}$$

where $|0)$ denotes the vacuum of the boson operators $B^\dagger$ and B, i.e. $B_j|0) = 0$. The dynamics of the system is governed by the Heisenberg equation of motion

$$i\dot{B}_j^\dagger(t) = \left[B_j^\dagger(t),\, H(B^\dagger,B) \right] \ , \tag{3.3}$$

where the time-dependent operator $B^\dagger{}_j(t)$ is defined as

$$B_j^\dagger(t) \equiv e^{iH(B^\dagger,B)t} B_j^\dagger e^{-iH(B^\dagger,B)t} \ . \tag{3.4}$$

Let us consider an ideal situation where a set of collective excited states constitutes an invariant subspace (, i.e. an ideal collective subspace) $\mathcal{V}^{2L}$ (L<< K) of the Hamiltonian. In this case, there must be an optimal representation called the DR where the minimum number of relevant operators is required in describing $\mathcal{V}^{2L}$. Let

$$\{\alpha_j, \alpha_j^\dagger;\ j=1,\dots,L\} \ , \tag{3.5a}$$

be a set of relevant (collective) operators necessary for spanning $\mathcal{V}^{2L}$ and let

$$\{\beta_j, \beta_j^\dagger;\ j=L+1,\dots,K\} \ , \tag{3.5b}$$

be the remaining operators called the irrelevant (intrinsic) operators in the DR. In the DR, the dynamics are still governed by $H(B^\dagger,B)$. Namely, there holds

$$i\dot{\alpha}_j^\dagger(t) = [\alpha_j^\dagger(t)\ , H(B^\dagger,B)]\ ; \qquad j=1,..L,$$

$$i\dot{\beta}_j^\dagger(t) = [\beta_j^\dagger(t)\ , H(B^\dagger,B)]\ ; \qquad j = L+1,\ldots K, \tag{3.6}$$

where the time-dependent operators in the DR are given by

$$\alpha_j^\dagger(t) \equiv e^{iH(B^\dagger,B)t}\alpha_j^\dagger e^{-iH(B^\dagger,B)t}, \quad \beta_j^\dagger(t) \equiv e^{iH(B^\dagger,B)t}\beta_j^\dagger e^{-iH(B^\dagger,B)t}\ . \tag{3.7}$$

In the DR, the Hamiltonian is expressed as

$$\begin{aligned} H(B^\dagger,B) &= e^{F(\alpha^\dagger,\alpha;\beta^\dagger,\beta)}H(\alpha^\dagger,\alpha;\beta^\dagger,\beta)e^{-F(\alpha^\dagger,\alpha;\beta^\dagger,\beta)} \\ &= \mathcal{H}_{coll}(\alpha^\dagger,\alpha) + \mathcal{H}_{irr}(\beta^\dagger,\beta) + \mathcal{H}_{coupl}(\alpha^\dagger,\alpha,\beta^\dagger,\beta)\ . \end{aligned} \tag{3.8}$$

Here $\exp\{F(B^\dagger,B)\}=\exp\{F(\alpha^\dagger,\alpha,\beta^\dagger,\beta)\}$ denotes the nonlinear unitary transformation between IR and DR;

$$\begin{aligned} \alpha_j^\dagger &= e^{-F(B^\dagger,B)}B_j^\dagger e^{F(B^\dagger,B)}\ , \ j = 1,\ldots,L\ ; \\ \beta_j^\dagger &= e^{-F(B^\dagger,B)}B_j^\dagger e^{F(B^\dagger,B)}\ , \ j = L+1,\ldots K\ . \end{aligned} \tag{3.9}$$

The dynamical condition specifying the invariant subspace is obviously given by the *complete decoupling condition* between the relevant and irrelevant operators;

$$\text{(A - I)} \quad \mathcal{H}_{coupl}(\alpha^\dagger,\alpha,\beta^\dagger,\beta) = 0\ . \tag{3.10}$$

With the aid of condition (3.10), the time-dependent operators in Eq.(3.7) are reduced into

$$\alpha_j^\dagger(t) = e^{i\mathcal{H}_{coll}t}\alpha_j^\dagger e^{-i\mathcal{H}_{coll}t}\ , \quad \beta_j^\dagger(t) = e^{i\mathcal{H}_{irr}t}\beta_j^\dagger e^{-i\mathcal{H}_{irr}t}. \tag{3.11}$$

satisfying

$$\text{(A - II)} \quad \begin{aligned} i\dot{\alpha}_j^\dagger(t) &= [\alpha_j^\dagger(t), \mathcal{H}_{coll}(\alpha^\dagger,\alpha)]\ ; \quad j = 1,\ldots,L\ , \\ i\dot{\beta}_j^\dagger(t) &= [\beta_j^\dagger(t), \mathcal{H}_{irr}(\beta^\dagger,\beta)]\ ; \ j = L+1,\ldots,K. \end{aligned} \tag{3.12}$$

Namely, the relevant and irrelevant operators are governed separately by $\mathcal{H}_{coll}$ and $\mathcal{H}_{irr}$, which satisfy the following commutability condition

$$\text{(A - III)} \quad [\mathcal{H}_{coll}(\alpha^\dagger,\alpha), \mathcal{H}_{irr}(\beta^\dagger,\beta)] = 0\ . \tag{3.13}$$

Equations (A-I), (A-II) and (A-III) constitute a set of basic relations characterizing the invariant subspace $\mathcal{V}^{2L}$ of the Hamiltonian. From the above discussion, it is clear that an introduction of the DR for a given invariant subspace $\mathcal{V}^{2L}$ is decisive so as to manifest an invariant property of the Hamiltonian.

3.2. Basic Equations for Determining the Dynamical Representation

Let us discuss the general case where the complete decoupling condition (3.10) does not necessarily hold. Namely, we intend to extract a set of collective states which form an approximate invariant subspace of the Hamiltonian. To this aim, we start with the resultant DR where the collective subspace D^{2L}: $\{|m_1,m_2,...m_L>\}$ is expressed as

$$|m_1,m_2,...,m_L\rangle = \prod_{j=1}^{L}\frac{1}{\sqrt{m_j!}}\left(\alpha_j^\dagger\right)^{m_j}|0\rangle , \tag{3.14}$$

where $|0\rangle\rangle=\exp\{-F(B^\dagger, B)\}|0)$. In order to get the DR, i.e. to determine a generating function $F(B^\dagger,B)=-F^\dagger(B^\dagger,B)$ of the unitary transformation in Eq.(3.9), we have to discuss how to characterize $F(B^\dagger,B)$.

We start with introducing the following new notation for the vacuum state;

$$|0> = |0>>_R \otimes|0>>_I , \tag{3.15}$$

which satisfy $\alpha_j|0>>_R=0$ and $\beta_j|0>>_I=0$. In this notation, the total Hilbert space is expressed as a product space whose basis states are represented as

$$\begin{aligned} &|m_1,m_2,...,m_L;n_{L+1},n_{L+2},...,n_K> \\ &= |m_1,m_2,...,m_L>>_R \otimes|n_{L+1},n_{L+2},...,n_K>>_I \\ &= \prod_{j=1}^{L}\frac{1}{\sqrt{m_j}}\left(\alpha_j^\dagger\right)^{m_j}|0>>_R \otimes \prod_{j=L+1}^{K}\frac{1}{\sqrt{n_j}}\left(\beta_j^\dagger\right)^{n_j}|0>>_I . \end{aligned} \tag{3.16}$$

With the aid of this notation, the Hamiltonian is decomposed into

$$\begin{aligned} H(B^\dagger,B) &= e^{F(\alpha^\dagger,\alpha;\beta^\dagger,\beta)}H(\alpha^\dagger,\alpha;\beta^\dagger,\beta)\, e^{-F(\alpha^\dagger,\alpha;\beta^\dagger,\beta)} \\ &= \mathcal{H}_{coll}(\alpha^\dagger,\alpha) + \Delta\mathcal{H} , \end{aligned} \tag{3.17}$$

where

$$\mathcal{H}_{coll}(\alpha^\dagger,\alpha) = {}_I<<0|\, H(B^\dagger,B)\, |0>>_I . \tag{3.18}$$

It is obvious that $\mathcal{H}_{coll}$ satisfies

$$\left[\mathcal{H}_{coll} , \beta_j^\dagger\right] = 0 . \tag{3.19}$$

In order to get the basic equations specifying the most optimum representation, we have to employ some expansion scheme; Since the relevant operators are supposed to describe the large-amplitude collective motion within D^{2L} whereas the irrelevant operators describe the small-fluctuation around D^{2L}, it is reasonable to introduce a power series expansion with respect to the irrelevant boson operators. For example, $\Delta\mathcal{H}$ is expanded as

$$\Delta\mathcal{H} = \sum_{n=1}\mathcal{H}(n),$$

$$\mathcal{H}(1) = \sum_{j=L+1}\left\{\beta_j^\dagger\, {}_I\langle\langle 0|[\beta_j,\, H(B^\dagger,B)]|0\rangle\rangle_I + h.c. \right\},$$

$$\mathcal{H}(2) = \sum_{j,k=L+1}\beta_j^\dagger\, {}_I\langle\langle 0|[\beta_j,[H(B,B),\, \beta^\dagger{}_k]]|0\rangle\rangle_I\beta_k$$

$$+\sum_{j,k=L+1}\mathcal{N}_{jk}^{-1}\left\{\beta_j^\dagger\beta_k^\dagger{}_I\langle\langle 0|[\beta_k\beta_j, H(B^\dagger,B)]|0\rangle\rangle_I + \text{h.c.}\right\} \text{ etc.} \tag{3.20}$$

Here, the normalization constant is given by

$$\mathcal{N}_{jk} = {}_I\langle\langle 0|\beta_k\beta_j\beta_j^\dagger\beta_k^\dagger|0\rangle\rangle_I \ . \tag{3.21}$$

Within the first order expansion in $\Delta\mathcal{H}$, the time-dependent operators in Eq.(3.7) are expressed as

$$\begin{aligned}\alpha_j^\dagger(t) &\equiv e^{iH(B^\dagger,B)t}\alpha_j^\dagger e^{-iH(B^\dagger,B)t}\\ &\approx e^{i\mathcal{H}_{coll}t}\alpha_j^\dagger e^{-i\mathcal{H}_{coll}t}\\ &\quad + [\int_0^1 d\lambda\, e^{\lambda(i\mathcal{H}_{coll}t)} i\Delta\mathcal{H}\, t\, e^{-\lambda(i\mathcal{H}_{coll}t)},\ e^{i\mathcal{H}_{coll}t}\alpha_j^\dagger e^{-i\mathcal{H}_{coll}t}] ,\end{aligned} \tag{3.22a}$$

$$\begin{aligned}\beta_j^\dagger(t) &\equiv e^{iH(B^\dagger,B)t}\beta_j^\dagger e^{-iH(B^\dagger,B)t}\\ &\approx \beta_j^\dagger + [\int_0^1 d\lambda\, e^{\lambda(i\mathcal{H}_{coll}t)} i\Delta\mathcal{H}\, t\, e^{-\lambda(i\mathcal{H}_{coll}t)},\ \beta_j^\dagger] \ .\end{aligned} \tag{3.22b}$$

If there holds the maximal-decoupling condition

$$\text{(B-I)} \qquad \mathcal{H}(1) = 0 \ , \tag{3.23}$$

the relevant time-dependent operators in Eq.(3.22a) satisfy the following equation of motion given as

$$\text{(B-IIa)} \qquad i\frac{d}{dt}\alpha_j^\dagger(t) = [\alpha_j^\dagger(t),\ \mathcal{H}_{coll}] \ , \tag{3.24}$$

within the approximation of neglecting the second order effects with respect to the irrelevant operators. Further more, the irrelevant time-dependent operator in Eq.(3.22b) satisfy

$$\text{(B-IIb)} \qquad i\frac{d}{dt}\beta_j^\dagger(t) = [\beta_j^\dagger(t),\ \mathcal{H}(2)] \ , \tag{3.25}$$

provided that there holds the separability condition expressed as

$$\text{(B-III)} \qquad [\mathcal{H}_{coll},\ \mathcal{H}(2)] = 0 \ . \tag{3.26}$$

By comparing Eqs.(B-I),(B-II) and (B-III) with Eqs.(A-I),(A-II) and (A-III), it is easily seen that the maximal-decoupling condition (3.23) and the separability condition (3.26) just correspond to the complete-decoupling (3.10) and the commutability condition (3.13), respectively. Equations (B-I), (B-II) and (B-III) constitute the basic equations of quantum theory of nuclear collective dynamics [8].

3.3. Dynamical Representation and Validity of Collective Subspace D^{2L}

It has been shown [8] that the maximal-decoupling condition (B-I) gives necessary conditions for determining the functional form of $F(B^\dagger,B)$ which specifies the unitary transformation between IR and DR.

According to the quantum theory of nuclear collective dynamics, the generating function $F(B^\dagger,B)$ specifying the DR is determined so as to satisfy the maximal decoupling condition (B-I), under the boundary condition that the relevant and irrelevant operators are reduced to

$$\lim(\alpha_j^\dagger) = B_j^\dagger \ , \ \text{for } j = 1,\dots,L; \ \lim(\beta_j^\dagger) = B_j^\dagger \ , \ \text{for } j = L+1,\dots,K \ , \tag{3.27}$$

in the small-amplitude region, respectively. This boundary condition is only one physical input of the theory in specifying the collectivity under discussion.

We are now at the position to discuss the validity of D^{2L} which is spanned by the basis states defined by Eq.(3.14) in terms of the relevant operators thus obtained. After getting an explicit expression of $F(B^\dagger,B)$, we have a concrete forms of $\mathcal{H}_{coll}$ and $\Delta\mathcal{H}$. With the aid of an explicit expression of $\mathcal{H}_{coll}$, we have a set of collective excited states by diagonalizing $\mathcal{H}_{coll}$ within D^{2L};

$$\mathcal{H}_{coll}|\mu\rangle\rangle_R = E_\mu|\mu\rangle\rangle_R, \quad |\mu\rangle\rangle_R \in D^{2L}:\{|m_1,m_2,\dots,m_L\rangle\rangle_R\}. \tag{3.28}$$

With the aid of the collective excited states $|\mu\rangle\rangle_R$ thus introduced, the separability condition (3.26) is expressed as

$$_R\langle\langle\mu| \, [\mathcal{H}_{coll},\mathcal{H}(2)] \, |\mu'\rangle\rangle_R = 0, \quad \mu \neq \mu'. \tag{3.29}$$

Equation (3.29) states that the collective states $|\mu\rangle\rangle_R$ defined by $\mathcal{H}_{coll}$ should not be disturbed under the effect of $\mathcal{H}(2)$ and gives the first condition to evaluate the validity of collective states.

If there holds the separability condition (3.29), it is easily derived that $\mathcal{H}(2)$ has only diagonal terms with respect to collective states, i.e. $\mathcal{H}(2)$ is expressed as

$$\mathcal{H}(2) = \sum_\mu \mathcal{H}_\mu(2) \ ,$$

$$\mathcal{H}_\mu(2) = \sum_{j,k\geq L+1} \mathcal{N}_{jk}^{-1}\left\{\beta_j^\dagger\beta_k^\dagger|\mu\rangle\langle\mu|\cdot f_\mu^{jk} + \text{h.c.}\right\} + \sum_{j,k\geq L+1} \beta_j^\dagger|\mu\rangle\cdot g_\mu^{jk}\cdot\langle\mu|\beta_k \ . \tag{3.30}$$

Here,

$$|\mu\rangle \equiv |\mu\rangle\rangle_R \otimes |0\rangle\rangle_I,$$

$$f_\mu^{jk} = \langle\mu|[\beta_k\beta_j, H(B,B)]|\mu\rangle, \qquad g_\mu^{jk} = \langle\mu|[\beta_j,[H(B,B),\beta_k^\dagger]]|\mu\rangle. \tag{3.31}$$

Introducing a general coherent state for the irrelevant part of the wave function defined by

$$|\mu,\gamma\rangle \equiv |\mu\rangle\rangle_R \ \otimes \ e^{\sum_{j=L+1}\left(\gamma_j\beta_j^\dagger - \gamma_j^*\beta_j\right)}|0\rangle\rangle_I \ , \tag{3.32}$$

we have

$$\langle\mu,\gamma| \, \mathcal{H}_\mu(2) \, |\mu,\gamma\rangle = \frac{1}{2}e^{-\sum_j\gamma_j^*\gamma_j}(\gamma^*,\gamma)S_\mu\begin{pmatrix}\gamma\\ \gamma^*\end{pmatrix} \ , \tag{3.33}$$

where

$$S_\mu \equiv \begin{pmatrix} g_\mu^{jk} & f_\mu^{jk} \\ f_\mu^{jk^*} & g_\mu^{jk} \end{pmatrix} \quad . \tag{3.34}$$

Even though the matrix S_μ in Eq.(3.34) has the same structure as the usual stability matrix appearing in the RPA theory, it should be noticed that S_μ has only components between irrelevant degrees of freedom. With the aid of the stability matrix S_μ, one may study whether or not the collective state $|\mu\rangle\rangle_R$ defined by Eq.(3.28), or more generally, the collective subspace D^{2L} is stable with respect to small variations toward the irrelevant directions. Namely, the stability condition is given by

$$S_\mu \geq 0 \,, \quad \text{for each collective state } |\mu\rangle\rangle_R \,, \tag{3.35}$$

which is the second condition to evaluate the validity of collective states.

Here, it should be stated that the number of relevant operators 2L necessary for describing the collective states under consideration is determined in such a way that the corresponding subspace D^{2L} should be the smallest subspace where both the stability and separability conditions are satisfied.

3.4. How to Evaluate Stochasticity of Individual Eigenfunctions

In the previous subsection, we have discussed how to estimate the validity of D^{2L}. Namely, our discussion is limited to the collective states $|\mu\rangle\rangle_R \otimes |n{=}0\rangle\rangle_I$ with no intrinsic excitations. In order to introduce the chaoticity of individual eigenfunctions, we have to treat the eigenstates situated at any part of the boson Hilbert space. In this subsection, we will discuss how to evaluate the property of an arbitrary eigenstate by extending the separability and stability conditions to the whole Hilbert space.

mn-Representation

Aiming at extending the separability and stability conditions given by Eqs. (3.29) and (3.35) into the whole Hilbert space, we will hereafter consider the simplest case with L=1 and K=2, i.e. the system described by one relevant $\alpha^\dagger$ and one irrelevant $\beta^\dagger$ boson operators. The extension to the general case is trivial. In this case, the physical boson space $\mathbb{1}$ corresponding to the fermion space is described as

$$\mathbb{1} : \{|m,n\rangle \,;\, m+n \leq N\} \,, \tag{3.36}$$

where

$$|m,n\rangle = \frac{1}{\sqrt{m!}} \frac{1}{\sqrt{n!}} (\alpha^\dagger)^m (\beta^\dagger)^n |0\rangle \,. \tag{3.37}$$

The projection operator $\boldsymbol{1}$ onto $\mathbb{1}$ is expressed as

$$\mathbf{1} = \sum_{m,n(m+n\leq N)} |m,n\rangle\langle m,n| \,. \tag{3.38}$$

The representation using the basis states {|m,n>} is called the mn-representation. As is easily recognized, the mn-representation has a sense in every part of the physical boson space, provided the complete decoupling condition (3.10) is satisfied.

μn-Representation

Let us divide the physical boson space $\mathcal{L}$ into multi-layer of subspace $\{\mathcal{D}_n;\ n=0, 1,\dots,N\}$. Here $\mathcal{D}_n$ consists of a set of basis states having the same irrelevant boson number n, i.e.

$$\mathcal{D}_n: \{\, |m,n\rangle \ ;\ m = 0,1,\dots,N-n \,\} \,. \tag{3.39}$$

Here, $\mathcal{D}_{n=0}$ just corresponds to the collective subspace D^{2L} discussed in the previous subsection. The projection operator $\mathcal{P}(n)$ onto $\mathcal{D}_n$ is given by

$$\mathcal{P}(n) = \sum_{m=0}^{N-n} |m,n\rangle\langle m,n| \,, \qquad \mathbf{1} = \sum_{n=0}^{N} \mathcal{P}(n) \,. \tag{3.40}$$

By using the projection operator $\mathcal{P}(n)$, $\Delta\mathcal{H}$ in Eq.(3.17) is divided into two parts

$$\mathbf{1}\Delta\mathcal{H}\mathbf{1} = \Delta\mathcal{H}^{dia} + \Delta\mathcal{H}^{off-dia} \,,$$

$$\Delta\mathcal{H}^{dia} = \sum_n \mathcal{P}(n)\Delta\mathcal{H}\,\mathcal{P}(n) \,, \quad \Delta\mathcal{H}^{off-dia} = \sum_{n,n'(n\neq n')} \mathcal{P}(n)\Delta\mathcal{H}\,\mathcal{P}(n') \,. \tag{3.41}$$

Here, we briefly discuss characteristic roles of each part of the Hamiltonian. As is clearly understood from Eq. (3.19), $\mathcal{H}_{coll}$ commutes with any irrelevant boson operators by definition. Namely, it does not violate the irrelevant quantum number n. It is also the case for the part $\Delta\mathcal{H}^{dia}$ in Eq.(3.41), because it interacts only between two states with the same irrelevant boson number n. Namely, $\mathcal{H}_{coll}$ and $\Delta\mathcal{H}^{dia}$ conserve the irrelevant quantum number n. On the other hand, $\Delta\mathcal{H}^{off-dia}$ plays a role to violate the irrelevant quantum number n.

Let us extend the concept of collectivity μ defined in Eq.(3.28) to the whole physical boson space. This task is achieved by diagonalizing $H(B^\dagger,B)$ within the subspace $\mathcal{D}_n$ with arbitrary n,

$$\mathcal{P}(n)H(B^\dagger,B)\mathcal{P}(n)|\mu,n\rangle = \mathcal{P}(n)\{\mathcal{H}_{coll} + \Delta\mathcal{H}^{dia}\}\mathcal{P}(n)|\mu,n\rangle = E_\mu^{(n)}\,|\mu,n\rangle. \tag{3.42}$$

Here, the quantum number μ coincides with the definition in Eq.(3.28) in a case with n=0 and specifies the relevant part of the states in order of their energy eigenvalues;

$$\mu = 0,1,\dots,N-n, \qquad \text{with } E^{(n)}_{\mu=0} \leq E^{(n)}_{\mu=1} \leq \ ,\dots, \leq E^{(n)}_{\mu=N-n} \,. \tag{3.43}$$

With the aid of a set of basis states |μ,n> thus obtained, the subspace $\mathcal{D}_n$ is alternatively expressed as

$$\mathcal{D}_n : \{\, |\mu,n\rangle \;\; ; \mu = 0,1,\ldots,N-n \,\} \; . \tag{3.44}$$

The representation using the basis states $|\mu,n\rangle$ is called the μn-representation. Here, it should be mentioned that this representation is the best one may choose in incorporating the dynamical correlations among the relevant bosons into the basis states under the restriction of conserving the number of quantum numbers.

$\mu\nu$-Representation

Since the relevant parts of the basis states $|\mu,n\rangle$ and $|\mu,n'\rangle$ with $n \neq n'$ are expected to have almost the same structure, one may divide the physical boson space $\mathcal{I}$ into a set of vertical-layers of subspaces $\{\mathcal{R}_\mu \;; \mu=0,1,\ldots,N\}$

$$\mathcal{R}_\mu : \{\, |\mu,n\rangle \;\; ; n = 0,1,\ldots,N-\mu \,\} \; . \tag{3.45}$$

The projection operator onto $\mathcal{R}_\mu$ is expressed as

$$\mathcal{P}(\mu) = \sum_{n=0}^{N-\mu} |\mu,n\rangle\langle\mu,n| \;\; , \;\; \mathbf{1} = \sum_{\mu=1}^{N} \mathcal{P}(\mu). \tag{3.46}$$

By means of $\mathcal{P}(\mu)$, the Hamiltonian is expressed as

$$\mathbf{1}\,H(B^\dagger,B)\,\mathbf{1} = \sum_{\mu=0}^{N}\sum_{n=0}^{N-\mu} |\mu,n\rangle\, E_\mu^{(n)}\, \langle\mu,n| + \Delta\mathcal{H}^{off-dia} \; , \tag{3.47}$$

where $\Delta\mathcal{H}^{off-dia}$ splits into two parts

$$\Delta\mathcal{H}^{off-dia} = \mathcal{H}^{[stab]} + \mathcal{H}^{[sepa]} \; ,$$

$$\mathcal{H}^{[stab]} = \sum_{\mu} \mathcal{H}_\mu^{[stab]} = \Sigma_\mu \; \mathcal{P}(\mu)\Delta\mathcal{H}\,\mathcal{P}(\mu) \; ,$$

$$\mathcal{H}^{[sepa]} = \Sigma_{\mu,\mu'(\mu\neq\mu')}\mathcal{H}_{\mu,\mu'}^{[sepa]} = \Sigma_{\mu,\mu'(\mu\neq\mu')}\mathcal{P}(\mu)\Delta\mathcal{H}\,\mathcal{P}(\mu') \; . \tag{3.48}$$

In the expression of Eq.(3.48), it is directly recognized that a part $\mathcal{H}^{[stab]}$ conserves the relevant quantum number μ with violation of the irrelevant quantum number n. For a case with n=0, $\mathcal{H}^{[stab]}$ acts to induce an instability of the collective subspace $D^{2L} = \mathcal{D}_{n=0}$ (see Eq. (3.33) where the diagonal component of the second order Hamiltonian $\mathcal{H}_\mu(2)$ is discussed). On the other hand, a part $\mathcal{H}^{[sepa]}$ violates both the relevant quantum number μ and irrelevant quantum number n. From the above discussion, it becomes clear that the dynamical effects coming from $\mathcal{H}^{[stab]}$ can be incorporated into the basis vectors by diagonalizing $H(B^\dagger,B)$ within the subspace $\mathcal{R}_\mu$, i.e.

$$\mathcal{P}(\mu)H(B^\dagger,B)\mathcal{P}(\mu)\,|\mu,\nu\rangle = E_{\mu,\nu}|\mu,\nu\rangle \;\; , \;\;\; |\mu,\nu\rangle \in \mathcal{R}_\mu . \tag{3.49}$$

Since there holds

$$\mathcal{P}(\mu)H(B^\dagger,B)\mathcal{P}(\mu) = \mathcal{P}_\mu\left\{\mathcal{H}_{coll} + \Delta\mathcal{H}^{dia} + \mathcal{H}_\mu^{[stab]}\right\}\mathcal{P}_\mu , \tag{3.50}$$

the dynamical correlations contained in $H(B^\dagger,B)$ except for a part $\mathcal{H}^{[sepa]}$ are successfully incorporated into the basis states $|\mu,\nu>$. By using $|\mu,\nu>$, the subspace $\mathcal{R}_\mu$ is spanned as

$$\mathcal{R}_\mu: \ \{|\mu,\nu\rangle;\nu = 0,1,\dots,N-\mu\}. \tag{3.51}$$

The representation using the basis states $|\mu,\nu>$ is called the $\mu\nu$-representation. As is easily understood from the above prescription, the $\mu\nu$-representation is the best representation which one may choose in incorporating the correlations into the basis states as much as possible under a restriction of conserving the number of quantum numbers. In this sense, the $\mu\nu$-representation is nothing but a quantum analogue of the DCC system with the *maximum integrable form representation* in Eq.(2.23). The validity of $\mu\nu$-representation is studied by only the separability condition given by

$$\mathcal{H}_{\mu,\mu'}^{[sepa]} = 0 , \ (\mu \neq \mu') , \tag{3.52}$$

because the $\mu\nu$-representation is defined so as to satisfy the stability condition

$$\langle\mu\nu|\,\mathcal{H}_\mu^{[stab]}\,|\mu\nu'\rangle = 0 , \quad \text{for all } \mu \text{ with } \nu \neq \nu'. \tag{3.53}$$

Equations (3.52) and (3.53) are natural extension of the separability and stability conditions for the collective subspace D^{2L} given by Eqs.(3.29) and (3.35) to the whole Hilbert space, respectively.

Definition of Quantum Chaos

Let $\{|\, i >\}$ be a set of exact eigenstates of the total Hamiltonian $H(B,B^\dagger)$

$$\mathbf{1}\,H(B^\dagger,B)\,\mathbf{1}\,|i\rangle = E_i\,|i\rangle , \quad |i\rangle \in \mathbf{1} : \{|m,n\rangle \,;\, m+n \leq N\} . \tag{3.54}$$

From the above discussion, it is clear that the exact eigenstate $|\, i >$ is no more specified by one of the (μ,ν)-basis states when the separability condition (3.52) is violated. Namely the exact eigenstate is not specified by K-kinds of quantum numbers any more but by smaller number of quantum numbers in accordance with the violation of the separability condition. In other words, there occurs a *dissolution of quantum numbers* when the separability condition does not hold. In the classical theory, a dissolution of integrability is induced by the violation of the separability condition (2.18) and simply means an onset of chaotic motion. In the quantum system, therefore, it is reasonable to state that a dissolution of quantum numbers may indicate an onset of "quantum chaos". In accordance with the above discussion of the quantum chaos, we may classify the exact eigenstates $\{|\, i >\}$ into three characteristic cases with the aid of the (μ,ν)-basis states;

Quantum Integrable Motion ; If one finds one of the (μ,ν)-basis states for a given eigenstate $|i>$ satisfying $|<\mu,\nu|i>|^2 \approx 1$, then $|i>$ is classified to be in "quantum integrable motion" because it is specified by K-kinds of quantum numbers (μ,ν) associated with the DR.

Quantum Dissipative Motion; If $|i>$ is described perturbatively starting from the most appropriate (μ,ν)-basis state, then it is specified to be in "quantum dissipative motion" because it is still characterized by K-kinds of asymptotic quantum numbers.

Quantum Chaotic Motion ; If $|i>$ is not described perturbatively starting from any (μ,ν)-basis states, then it is regarded to be in "quantum chaotic motion" because $|i>$ is not characterized by K-kinds of quantum numbers any more.

In order to express the chaoticity of each eigenstate, let us expand the eigen state $|i>$ in terms of (μ,ν)-basis states as

$$|i>= \Sigma_{\mu\nu} a^i_{\mu\nu} |\mu\nu> \ . \tag{3.55}$$

Looking at the distribution of square of expansion coefficients $|a_{\mu\nu}{}^i|^2$, one may recognize whether the eigenstate $|i>$ belongs to the quantum integrable, dissipative or chaotic regime. A figure expressing $|a_{\mu\nu}{}^i|^2$ is called the *fractional parentage plot.*

3.5. Resonant Quantum Number

Let us discuss what happens as the collective excited states develops into the large-amplitude region. Here it should be noticed that the division of the Hilbert space into three characteristic regimes discussed in the previous subsection may be too rough to explore an exceedingly rich structure of the Hilbert space. As discussed in §2, and is known in the classical nonlinear dynamics, an appearance of the resonant denominator is a source of an inexhaustible complexity of topology. Let us discuss what happens when the quantum system satisfies the resonant condition. There appears resonant denominator when one solves the maximal decoupling condition (3.23) by using the perturbative expansion method. It also appears in evaluating the effects coming from $\mathcal{H}^{[sepa]}$ perturbatively on the basis of the $\mu\nu$-representation when the unperturbed energies of the $E_{\mu\nu}$ in Eq.(3.49) degenerate accidentally.

In the following, we discuss an effect of resonances by using the mn-representation. Although the resonance condition depends on the emploied representation, the following discussion is applicable for any representation. We restrict ourselves to a simple system with two degrees of freedom. The unperturbed Hamiltonian of the boson system with two degrees of freedom is given by

$$H_0 = \omega_1 B_1^\dagger B_1 + \omega_2 B_2^\dagger B_2 \quad , \tag{3.56}$$

and the mn-representation of the system is given by introducing the following basis states;

$$|m,n\rangle = \frac{1}{\sqrt{m!}}\frac{1}{\sqrt{n!}} (B_1^\dagger)^m (B_2^\dagger)^n |0\rangle \quad . \tag{3.57}$$

Here m and n are the quantum numbers of the "oscillator quanta" which are eigenvalues of the following two commutable operators,

$$N_1 = B_1^\dagger B_1 \quad , N_2 = B_2^\dagger B_2 \quad . \tag{3.58}$$

The resonance condition for the harmonic Hamiltonian is expressed as

$$k_1\omega_1 - k_2\omega_2 = 0 \ , \ \ k_1 \text{ and } k_2 \text{ are prime numbers.} \tag{3.59}$$

If the resonance condition (3.59) is satisfied, one may introduce new quantum number $\mathcal{N}$ called "resonance oscillator quanta" which is specified by the following operator

$$\mathcal{N} = k_2 N_1 + k_1 N_2 \quad . \tag{3.60}$$

(Compare I_1 in Eq.(2.25)). To make the following discussion simple, we hereafter treat a special case with k_1=2 and k_2=1. However the generalization for any value of k_1 and k_2 is straight forward. By making use of the resonance oscillator quanta, one may introduce following subspace

$$\mathcal{D}^{\mathcal{N}} : \left\{ |\mathcal{N},0\rangle, |\mathcal{N}-2,1\rangle, |\mathcal{N}-4,2\rangle, \cdots, \left|0,\frac{\mathcal{N}}{2}\right\rangle \text{ or } \left|1,\frac{\mathcal{N}-1}{2}\right\rangle \right\}, \tag{3.61}$$

whose elements have the same quantum number $\mathcal{N}$. It is also noted that the states in $\mathcal{D}^{\mathcal{N}}$ have the same unperturbed energy $E_0=\omega_1\mathcal{N}$.

Now let us discuss another new quantum number specifying the states when the resonance condition (3.59) is satisfied. For this aim, we consider a simple case with $\mathcal{N}$=2;

$$\mathcal{D}^{\mathcal{N}=2} \ : \ \{|2,0\rangle, \ |0,1\rangle\} \equiv \left\{ \begin{pmatrix} 1 \\ 0 \end{pmatrix}, \begin{pmatrix} 0 \\ 1 \end{pmatrix} \right\} \quad . \tag{3.62}$$

With the aid of the last notation in Eq.(3.62), the Hamiltonian in the subspace $\mathcal{D}^{\mathcal{N}=2}$ is represented as

$$\mathcal{H}^{\mathcal{N}=2} = \alpha E + \beta T, \quad E = \begin{pmatrix} 1 & 0 \\ 0 & -1 \end{pmatrix}, T = \begin{pmatrix} 0 & 1 \\ 1 & 0 \end{pmatrix} \quad . \tag{3.63}$$

The first term denotes the unperturbed harmonic Hamiltonian and α denotes a degree of violation from complete degeneracy. If two unperturbed states are completely degenerate, there holds a resonance condition α=0. The second term expresses an anharmonic interaction

$$H_1 = \frac{\beta}{\sqrt{2}} (B_1^\dagger B_1^\dagger B_2 + B_2^\dagger B_1 B_1) \quad , \tag{3.64}$$

within the subspace $\mathcal{D}^{\mathcal{N}=2}$. Apparently, a type of the anharmonic Hamiltonian acting between two degenerate unperturbed states depends on k_1 and k_2 in the resonance condition (3.59). Now let us consider the resonant case $\alpha=0$. In this case, one may introduce a set of new basis states

$$\mathcal{R}_+ = \frac{1}{\sqrt{2}}\begin{pmatrix}1\\1\end{pmatrix} \ , \ \mathcal{R}_- = \frac{1}{\sqrt{2}}\begin{pmatrix}-1\\1\end{pmatrix} , \tag{3.65}$$

which satisfy

$$T\mathcal{R}_\pm = t_\pm \mathcal{R}_\pm \quad , \quad t_\pm = \pm 1 . \tag{3.66}$$

Here it is worthwhile to mention that there exists a new representation $\mathcal{R}_\pm$ called "resonant representation" which diagonalizes the anharmonic Hamiltonian. Since the eigenstates $\mathcal{R}_\pm$ of the Hamiltonian in Eq.(3.63) with $\alpha=0$ are still specified by two kinds of quantum number ($\mathcal{N}$, $t_\pm$) called "local resonance quanta" (instead of the original global two oscillator quanta (m,n)), there does not occur any dissolution in the number of the good quantum numbers. Namely, there still exists new regular quantum states specified by the anharmonic effects which just induces a well-known divergent difficulty in the perturbative treatment based on the mn-representation.

An extremely important character of a pair of resonance quanta ($\mathcal{N}$,$t_\pm$) rests on its local property. If there exists two states whose unperturbative energies are accidentally degenerate, one may always introduce a corresponding local subspace $\mathcal{D}^{\mathcal{N}=2}$ spanned by these two states, irrespective of the other part of the Hilbert space. Then one may further introduce local operator T expressing anharmonic coupling between these two state. Namely, the local isolated resonance quanta is defined in a tiny region of the Hilbert space. This local property is very similar to the single resonance of the classical system.

The above discussion can be extended to the non-degenerate case. If the subspace $\mathcal{D}^{\mathcal{N}=2}$ is an approximate invariant subspace of the total Hamiltonian, the eigenstates of the Hamiltonian in Eq.(3.63) has a sense irrespective of the values of α and β. The eigenvalues and eigenstates of Eq.(3.63) are given by

$$\mathcal{H}^{\mathcal{N}=2}U_\pm = \Omega_\pm U_\pm \ , \ \Omega_\pm = \pm\sqrt{\alpha^2+\beta^2} \ ,$$

$$U_+ = \frac{1}{\sqrt{2}}\frac{1}{\sqrt{(\alpha^2+\beta^2)+\alpha\sqrt{\alpha^2+\beta^2}}}\begin{pmatrix}\alpha+\sqrt{\alpha^2+\beta^2}\\ \beta\end{pmatrix} \quad ,$$

$$U_- = \frac{1}{\sqrt{2}}\frac{1}{\sqrt{(\alpha^2+\beta^2)+\alpha\sqrt{\alpha^2+\beta^2}}}\begin{pmatrix}-\beta\\ \alpha+\sqrt{\alpha^2+\beta^2}\end{pmatrix} \quad . \tag{3.67}$$

As is easily justified, $U_\pm$ reduce to two states in Eq.(3.62) in a case with $\beta=0$, whereas they reduce to $\mathcal{R}_\pm$ in a case with $\alpha=0$. As discussed above, the global oscillator quanta (m,n) become good quantum numbers in a case with β=0 and the local resonance quanta ($\mathcal{N}$,t) become good quantum numbers in a case with α=0. More generally, there always exist two good quantum numbers ($\mathcal{N}$,Ω), provided the subspace $\mathcal{D}^{\mathcal{N}=2}$ is an approximate invariant subspace of the total Hamiltonian. The above discussion may be summarized by stating that, as far as the subspace $\mathcal{D}^{\mathcal{N}=2}$ is an invariant subspace of the total Hamiltonian, there exist regular quantum states specified by a set of local quantum numbers ($\mathcal{N}$,Ω), even in a case where the global quantum numbers (m,n) does not remain to be good quantum numbers any more. Note that this situation is very similar to an existence of a resonant action variable in the classical-level theory discussed in §2.3.

In this article, we have discussed on the present status of the microscopic theories of nuclear collective dynamics which provide us with a powerful method for questing an exceedingly rich structure of the TDHF manifold and the many-fermion space of states. In both cases, the perturbative treatment plays an important role; an introduction of the DCC system and DR is decisive in understanding the property of the TDHF-trajectory and exact eigenstates, respectively, which is very similat to the essential role of the most natural coordinate system in understanding the structure of the phase space of the nonlinear dynamical system. In the TDHF manifold, one may expect many DCC systems which correspond to various characteristic trajectories. One may also expect many DR in the quantum system which exhibits a variety of collective excitation modes. An investigation on dynamical interrelationship between various DCC systems and between many DR in terms of primary, secondary and higher order resonances would provide us with fruitful fields of nuclear structure physics.

References

1 A.J. Lichtenberg and M.A. Lieberman, Applied Mathematical Sciences 38 (Springer, Heidelberg 1983).

2 B.V. Chirikov, Phys. Rep. 52 (1979),263.

3 F. Sakata, M. Matsuo, T. Marumori and Y. Zhuo, Ann. of Phys. 194 (1989) 30

4 T. Marumori, T. Maskawa, F. Sakata and A. Kuriyama, Prog. Theor.Phys. 64 (1980) 1294

5 F. Sakata, T. Marumori, Y. Hashimoto and T. Une, Prog. Theor. Phys.70 (1983) 424.

6 G.D. Berkhoff, Dynamical Systems (Am. Math. Soc. 1927, New York) Vol.4

F.G. Gustavson, Astron.J. 71 (1966) 670

7 T. Marumori, Y. Hashimoto, K. Iwasawa and F. Sakata, Univ. of Tsukuba Preprint NT-Rep. No.2

Y. Hashimoto, T, Marumori and F. Sakata, Prog. Theor. Phys. 78 (1987) 1364.

8 F. Sakata, T. Marumori and M. Ogura, Prog. Theor. Phys. 76 (1986) 400

DISCUSSION OF THE PRESENTATION BY F. SAKATA

J. MARUHN, M. R. STRAYER, P. QUENTIN, V. ZELEVINSKY

Out of the many points made in the talk, the discussion focussed on three crucial points:

- the influence of chaos theory on the definition of collective coordinates,
- the relation to other microscopic foundations for collective models, and
- the possibility for practical applications.

For the first problem, the principal question is that of the interpretation of the concepts of "collective regime" vs. "chaotic regime". The classical phase space employed is that given by the very high-dimensional coefficients determining a Slater determinant via a particle-hole expansion in terms of a fixed reference Slater determinant. Now it appears that collective motion can be chaotic, and single-particle motion can be ordered, so that the juxtaposition of these two cases may not be justified. It appears rather that the "chaotic regime" is characterized by a development into very complicated particle-hole excitations similar to thermalization. This led to the longstanding problem of whether TDHF describes thermalization or because it involves a single pure wave function should be regarded as a system with entropy remaining at zero.

The final agreement was that the concept of chaos does not play an essential role in the definition of the collective coordinates; they are rather determined by the condition of decoupling from the internal coordinates.

This of course also raises the question of what the principal advantages and differences are of this method compared to the traditional theories such as ATDHF, and how the notions of nonlinear mechanics can be exploited in practice. The discussion focused on the example of the TDHF solution in the three-level problem, in which there is a small number of "classical" degrees of freedom. There was general agreement that the examples of Poincaré plots and similar methods of nonlinear classical dynamics were very intriguing, but that it must be seen whether anything comparable will be possible in practical applications in nuclear physics, where the number of degrees of freedom is exceedingly large.

The results of the discussion may be thus be summarized as follows:

- For the existence and actual definition of collective degrees of freedom the concepts presented are quite similar although not identical to ATDHF with self-consistent collective coordinates, although differing in detail .
- The "chaotic regime" of motion may correspond to thermalization, although it is not clear to what extent thermalization can be describe by a single Slater-determinant. The real thermalization comes from collision-like processes which are completely outside the TDHF approximation. Then it is not clear what is the physical meaning of "chaotic phenomena" within the TDHF: maybe it is incorrect to take into account such random modes and to neglect collisions at the same time. If one goes beyond TDHF, the results (the "chaotic" component) will change drastically. It is known from macroscopic kinetic theory that only smooth components of kinetic equations (hydrodynamic modes and so on) make sense in a collisionless approximation. Here the situation may be similar.
- The analysis of TDHF solutions in terms of classical dynamics appears to provide a new and very interesting approach to the field; however it may not be practical except for very simple models. Thus the main future direction of this research will probably be the study of such simple yet interesting models with sufficient relevance to nuclear physics.

REBUTTAL

F. SAKATA
University of Tokyo, Tokyo, Japan

The main problem of the microscopic description of nuclear collective motion is to find the most natural coordinate system for a given TDHF trajectory. A schematic concept of collective regime and chaotic regime is used only to illustrate the situation whether or not it is possible to introduce the collective submanifold described by a few relevant degrees of freedom, and decoupled from the other degrees of freedom. If one likes, one may use the language of regular regime and chaotic regime.

I did not discuss anything about the thermalization. The concept of "chaos" does not necessarily relate to the thermalization. We have discussed this problem by proposing the theory of nuclear master equation which is capable of treating the dynamics displayed by a bundle of TDHF trajectories. Note that the effects of one-body dissipation and that of two-body dissipation should be studied more carefully before drawing some definite conclusion, because the situation strongly depends on the excitation energy. However, this subject is out of the scope of my discussion at ORNL.

What I stressed was how rich structures there are in the TDHF manifold and how one can extract these complex topological structures of the TDHF manifold. In order to understand the microscopic dynamics between collective and single-particle modes of motion, one should explore the dynamics of the time-dependent mean-field, and should learn an enormous progress in the general theory of nonlinear dynamical system which is formally equivalent to the TDHF symplectic manifold. In this sense, the dynamical process how the chaotic motion arises out of the regular motion may play an essential role in describing the dynamical change of the collective motion.

Although the TDHF manifold is described by a huge number of degrees of freedom, they may be divided into two; active small dimensional relevant degrees of freedom and non-active large dimensional irrelevant degrees of freedom. In this sense, an applicability of the SCC method may not be restricted in the schematic model.

With regards to the relation of the ATDHF theory, ATDHF theory intends to extract the collective path within the configuration space, whereas the SCC method within the phase space. Note that the structure of the potential energy surface obtained in the ATDHF theory gives a tiny information of the TDHF manifold near the local minimum. On top of that, the dynamical role of the adiabatic v.s. diabatic single-particle states, that of diabatic v.s. nonadiabatic collective potential, etc. are only explored by going beyond the adiabatic assumption. Consequently, the difference between the ATDHF and SCC is decisive.

THE GEOMETRIC COLLECTIVE MODEL

JOACHIM A. MARUHN*
Joint Institute for Heavy Ion Research
Holifield Heavy Ion Research Facility
Oak Ridge, TN 37831, USA

ABSTRACT

The geometric collective model is one of the most widely known and oldest approaches for the description of low-lying states in even-even nuclei. This paper presents an overview of the assumptions made in the model and the various versions actually applied to an interpretation of experimental data.

1. Introduction

The geometric model is one of the oldest models of nuclear structure. The basic idea of a deforming nuclear surface was introduced by Bohr and Wheeler in their theory of fission[1] and developed and applied to nuclear structure in[2,3] leading to its most well-known formulation in the Bohr Hamiltonian[4]. The first systematic application and comparison to experimental data was done by Faessler, Greiner, and Sheline[5]. For many years it was simply *the* collective model. This long history, however, also implies that many things about it are taken for granted, so that it is an interesting subject to subject the model in its various manifestations to a critical examination. I will thus give a brief overview of the various approaches united under this title and focus especially on the assumptions and approximations made, which go beyond the very general idea of a geometric formulation.

2. Basic Definitions

The principal assumption of the geometric collective model is that the state of the nucleus is determined completely by its shape. This is, of course, motivated by the physical picture of an incompressible liquid drop and has been very successful in many formulations of nuclear models, most notably in the field of nuclear fission. But whereas there any parametrization of the nuclear surface can in principle be studied, for the description of low-lying excited states it seems more appropriate to explicitly start with an expansion about the equilibrium configuration of such a drop, the spherical shape, and to use deformation parameters with good angular momentum properties. This motivates the basic definition of the *collective deformation parameters* $\alpha_{2\mu}$ via the expansion of the radius of the nucleus in the direction (θ, ϕ) as

$$R(\theta,\phi) = R_0 \left(1 + \sum_{\lambda=0}^{\infty} \sum_{\mu=-\lambda}^{\lambda} \alpha^*_{\lambda\mu} Y_{\lambda\mu}(\theta,\phi)\right) , \tag{1}$$

where R_0 is the radius of the unperturbed spherical nucleus.

A study of the lowest multipoles in the expansion shows the following well-known properties[6]:

*Permanent address: Institut für Theoretische Physik, Universität Frankfurt, Postfach 11 19 32, W-6000 Frankfurt, Germany

- $\lambda = 0$: the monopole mode does not play a role at low energies because of the incompressibility of nuclear matter. However, it can be used to correct the nuclear volume for deviations introduced by the other multipole deformations; one finds

$$\alpha_{00} = -\frac{1}{\sqrt{4\pi}} \sum_{\mu} |\alpha_{2\mu}|^2 \tag{2}$$

for the quadrupole.

- $\lambda = 1$: to lowest order the dipole mode just describes a shift in the center of mass and thus generates only spurious motions; for larger values of this deformation, though, it also leads to a shape deformation which may be quite interesting for applications; I will return to this point later.

- $\lambda = 2$: this is the well-known quadrupole deformation tensor, which has been used successfully to describe a large number of excited states and will be the focus of attention in the following.

To build a model Lagrangian and, later, quantized Hamiltonian based on the set of coordinates $\alpha_{2\mu} = (-1)^{\mu}\alpha_{2-\mu}$ some assumptions have to be made about the dynamics as well. The previously mentioned one, that the state of the nucleus depend only on its surface shape, i. e., on the $\alpha_{2\mu}$, can be fulfilled in the adiabatic limit or, if necessary, also in the diabatic one. For dynamics we have to assume in addition that the kinetic energy should involve only the collective velocities $\dot{\alpha}_{2\mu}$ together with possibly deformation-dependent mass parameters, and not more complicated ingredients like memory effects.

3. The Intrinsic System

The quadrupole deformation parameters describe both the deviation of the nuclear shape from a sphere and its orientation in space. To separate these two, one may use a transformation to the principal axis like for a rigid body frame given by a rotation

$$\alpha_{2\mu} = \sum_{\nu} \mathcal{D}^{(2)*}_{\mu\nu}(\theta_1, \theta_2, \theta_3)\, a_{2\nu}, \tag{3}$$

where the tensor $a_{2\nu}$ in the principal axis frame has to fulfil $a_{21} = a_{2-1} = 0$. Since $a_{22} = a_{2-2}$, we thus replace the five degrees of freedom contained in the $\alpha_{2\mu}$ by the new set of five coordinates θ_1, θ_2, θ_3, a_{20}, and a_{22}. Clearly the Euler angles describe the orientation in space, while a_{20} and a_{22} indicate the axially symmetric elongation and the triaxiality, respectively.

An alternative definition of the principal-axis or intrinsic system is the introduction of polar coordinates in the (a_{20}, a_{22})-plane[4] in terms of

$$a_{20} = \beta \cos\gamma, \qquad a_{22} = \frac{1}{\sqrt{2}}\beta \sin\gamma. \tag{4}$$

These two coordinate systems naturally also lead to two principal possibilities for setting up the nuclear models: in the laboratory frame, one may construct the terms in the Hamiltonian by explicit coupling of the angular momenta to zero, while in the principal axis frame all terms may only depend on β and γ, with the kinetic energy also involving the time derivatives of the Euler angles.

4. Quantization

One of the principal problems in a theory which deals with such abstract coordinates is that of quantization. The only coordinates for which quantization can be done trivially are the cartesian coordinates of a particle, and this can be generalized to curvilinear coordinates by the standard methods of differential geometry. For the collective surface deformation coordinates, however, quantization can only be performed by analogy and should be founded in principle on a microscopic interpretation. In practice, however, two methods are used:

1. Harmonic oscillator quantization consisting essentially in the postulation of the commutation relations
$$[\alpha_{2\mu}, \pi^*_{2\mu'}] = i\hbar\delta_{\mu\mu'}, \tag{5}$$
where the momentum operator $\pi_{2\mu}$ can either be introduced *ad hoc* or, for the case of the simple harmonic oscillator itself is found to be given by $\pi_{2\mu} = (-1)^\mu \dot{\alpha}_{2-\mu}$. This method is applied in the case of the spherical vibrator and, with a *ad hoc* introduction of the momentum operator, in the Gneuss-Greiner Hamiltonian.

2. Pauli-Podolski quantization[7] is based on the interpretation of the kinetic energy expression as a curvilinear line element related to cartesian — and thus easily quantizable — coordinates. The kinetic energy operator is then given by the curvilinear Laplacian operator and contains the mass parameters even in the derivatives. The rotation-vibration model and the general numerical model use this method.

The justification for these procedures should be seen first of all in the very successful interpretation of experimental data, but also in the intensive ensuing work — discussed in other articles in this volume — to derive boson operators from microscopic degrees of freedom.

5. The Geometric Interpretation

It is important to realize that formulating a Hamiltonian in terms of the quadrupole degrees of freedom does not necessarily imply the actual use of the underlying geometric ideas, especially if the Hamiltonian is then used in a phenomenological way to interpret data by fitting parameters. The Hamiltonian essentially reflects the existence of a quadrupole degree of freedom and the degree of expansion used for the potential and kinetic energies. There is nothing in it to distinguish the surface vibrational bosons introduced here from, say, the more abstract bosons of the IBM[8,9] . In fact, one can show that the Hamiltonian can be subjected to a large number of canonic transformations without any change in its spectrum.

The place where the geometric interpretation really has to be used is the construction of the transition operators. The quadrupole operator in its simplest form is given microscopically by

$$Q_{2\mu} = e\sum_{i=1}^{Z} r_i^2 Y_{2\mu}(\Omega_i), \tag{6}$$

the sum being over all protons. This can be translated into a collective operator by utilizing the assumption of homogeneous charge distribution:

$$\begin{aligned} Q_{2\mu} &= \frac{3Ze}{4\pi R_0^2}\int \frac{R^5(\Omega)}{5} Y_{2\mu}(\Omega)\mathrm{d}\Omega \\ &= \frac{3Ze}{4\pi R_0^2}\left(\alpha_{2\mu} - \frac{10}{\sqrt{70\pi}}[\alpha\times\alpha]_{2\mu}\right). \end{aligned} \tag{7}$$

Note that here the coefficients are completely determined! While one has some leeway in adjusting the radius parameter in the first one, the admixture of the second order term is predicted uniquely. In practice one finds that in most cases this expression works quite well, but fitting the second-order coefficient does improve the agreement with experiment.

If the geometric model is interpreted in its more general form as a parametrization of a set of possibly quite complicated internal states, one is of course justified in taking free parameters in the quadrupole operator.

6. Hierarchy of Models

Practical applications of the geometric model have been done in a variety of formulations. We discuss them working upwards from the simple analytically solvable special cases to the general numerical versions.

6.1 The Spherical Vibrator

This model is based on the simplest conceivable Hamiltonian fulfilling all the symmetry requirements,

$$H = \frac{\sqrt{5}}{2B}[\pi \times \pi]_0 + \frac{\sqrt{5}C}{2}[\alpha \times \alpha]_0. \tag{8}$$

The solutions are the well-known quadrupole phonons, and the excitatioon spectrum shows the equidistant structure with the multiplet sequence (0^+) — (2^+) — $(0^+,2^+,4^+)$ — $(0^+,2^+,3^+,4^+,6^+)$ and so on. The partial presence of these level sequences in spherical nuclei provided considerable insight, although it must be admitted that not many nuclei can be interpreted in this way and that the quantitative description of the transition probabilities is not convincing. These failures may be ascribed, however, to the excessive symmetry present in the Hamiltonian, which leads to large changes in matrix elements even for small deviations from the pure harmonic oscillator. The main application of this Hamiltonian is thus rather as providing a convenient basis set for expansions of more complex models, for which the extensive mathematical results available are quite useful[10–12].

6.2 The Rotation-Vibration Model

In this model[4,5,6] an axially symmetric ground state at $(\beta = \beta_0, \gamma = 0)$is assumed and the potential is expanded around this minimum in the body-fixed system; introducing a small deviation coordinate $\xi = \beta - \beta_0$ we can write

$$V(\beta,\gamma) = \tfrac{1}{2}C\xi^2 + D\gamma^2. \tag{9}$$

It is an interesting point in this model that the kinetic energy is taken to be that of the spherical vibrator of Eq.8 transformed into the intrinsic system and then quantized according to the Pauli prescription. This leads to a Hamiltonian of the form

$$\begin{aligned}
H &= H_{\rm vib} + H_{\rm rot} + H_{\rm vib-rot} \\
H_{\rm rot} &= \frac{J^2 - J_3^2}{2\Theta_0} \\
H_{\rm vib} &= \frac{-\hbar^2}{2B}\frac{\partial^2}{\partial\xi^2} + \frac{C}{2}\xi^2 - \frac{\hbar^2}{4B}\frac{\partial^2}{\partial\gamma^2} + +D\gamma^2 + \frac{J_z^2 - \hbar^2}{16B\gamma^2} \\
H_{\rm rot-vib} &= -\frac{J^2 - J_z^2}{2\Theta_0}\frac{2\xi}{\beta_0} - \frac{J_+^2 - J_-^2}{2\Theta_0}\frac{2\sqrt{6}}{3}\frac{\xi}{\beta_0}
\end{aligned} \tag{10}$$

This Hamiltonian has been extremely successful in understanding the structure of rotational nuclei and concepts such as the β-band or γ-band are familiar to anybody working in nuclear structure physics. For that reason some of the assumptions made in the model are often forgotten.

An interesting feature of this Hamiltonian is that only one mass parameter appears for both vibration types and also determines the moment of inertia, since $\Theta_0 = 3B\beta_0^2$. This is of course due to the simple kinetic energy assumed and certainly is not expected to hold in general. In fact, in the cranking model the moment of inertia and the vibrational mass are quite unrelated. The rotation-vibration coupling is also a consequence of this simple assumption and for that reason has not been used with much success. Another shortcoming of the model is that the harmonic potential in the γ-direction may be less good an approximation than for β, since the periodicity in γ requires a rapid turning over.

Despite these criticisms, the model has brought immense insight into such fundamental properties as the presence of even or odd angular momenta in the various bands, the coupling of the γ-vibrations with rotations about the symmetry axis of the nucleus, and the selection rules for transitions between the bands.

6.3 The Wilets-Jean Model

That there is a third analytically solvable special case had been generally forgotten until the IBM renewed interest in γ-unstable nuclei through its O(6) limit, which was first thought to have no counterpart in the geometric model. The Wilets-Jean model[13] assumes a γ-independent potential of the form

$$V(\beta,\gamma) = \frac{1}{8}C\left(\beta^2 + \frac{\beta_0^2}{\beta^2} - 2\beta_0^2\right) - D, \tag{11}$$

which has circle of degenerate minima at β_0. Using again the kinetic energy of the harmonic vibrator, the eigenstates are given by

$$E(n_\beta,\lambda) = \sqrt{\frac{C}{B}}(n_\beta + \alpha_\lambda = \tfrac{1}{2}) - \tfrac{1}{4}C\beta_0^2 - D \tag{12}$$

with

$$\alpha_\lambda = \tfrac{1}{4}\sqrt{BC\beta_0^4 + (2\lambda+3)^2}. \tag{13}$$

The quantum number λ describes the excitation along the γ-direction.

The model predicts a characteristic sequence of multiplets (0^+) — (2^+) — $(2^+,4^+)$ — $(0^+,3^+,4^+,6^+)$ and so on, with λ varying, and repeated for each value of n_β and has been applied with some success[14].

6.4 The Gneuss-Greiner Model

While the experimental data shows features of all the limiting cases discussed, it is also clear that many nuclei show transitional behaviour in the sense of interpolating between the features of two of them. The Gneuss-Greiner Hamiltonian[15,16] provides a much more general formulation for the potential energy surface via an expansion up to sixth order:

$$\begin{aligned} H \;=\; & \frac{1}{2B}[\pi \times \pi]_0 + \frac{P}{3}[[\pi \times \alpha] \times \pi]_0 \\ & + C_2[\alpha \times \alpha]_0 + C_3[[\alpha \times \alpha]_2 \times \alpha]_0 \\ & + C_4[\alpha \times \alpha]_0^2 + C_5[\alpha \times \alpha]_0[[\alpha \times \alpha]_2 \times \alpha]_0 \\ & + C_6[[\alpha \times \alpha]_2 \times \alpha]_0^2 + D_6[\alpha \times \alpha]_0^3 \end{aligned} \tag{14}$$

It is interesting to note that while all terms up to sixth order were included in the potential, the kinetic energy is quite simple. There are three reasons for this:

1. While microscopical calculations show strong fluctuations in the mass parameter, they may not be so important if the wave function is concentrated near a single minimum, so that a first order deviation may be sufficient. The potential, however, needs this amount of flexibility to be able to produce minima with adjustable properties.

2. Adding as much flexibility to the kinetic energy would increase the number of parameters too much, making it impossible to fit them from experimental data.

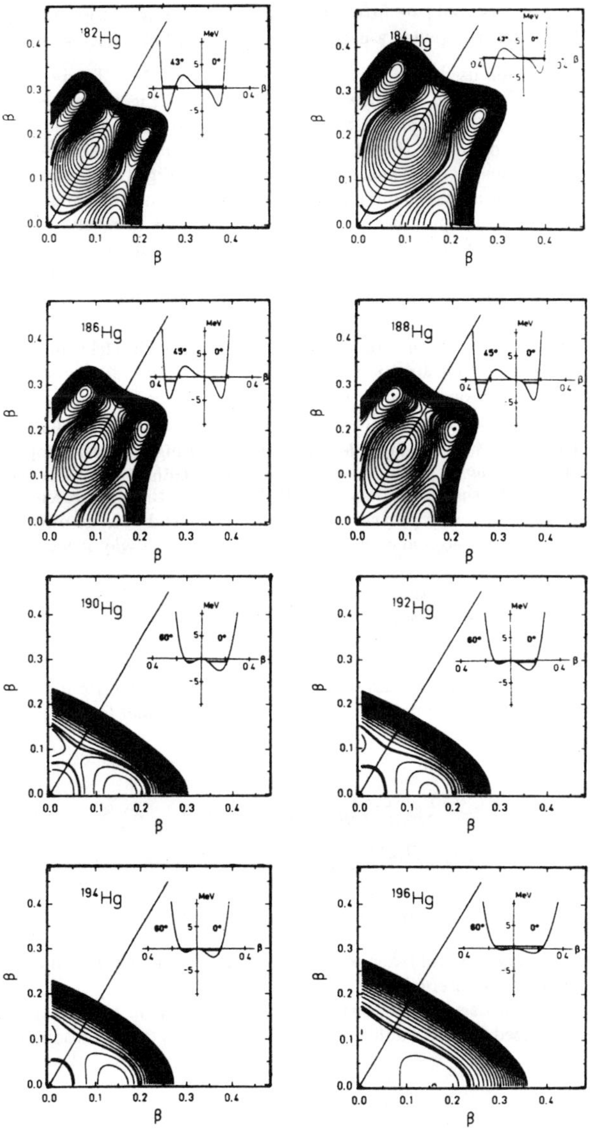

Figure 1: Potential energy surfaces of some Mercury isotopes based on a fit to experimental data with the Gneuss-Greiner Hamiltonian. The cuts in the inset indicate the quantitative behaviour of the potential along the angular directions indicated, which for the isotopes with isomeric minimum intersect the minima.

3. Even if the parameters could be fitted, a complicated kinetic energy would defeat the prime purpose of the model: to provide a useful visualisation of the collective structure.

The model has been used in many applications, e. g. in [17–19]. To give an impression of the practical application of the model, we show the fitted potential energy surfaces for some Mercury isotopes[19]. In the lighter of these isotopes there is a band built on an excited 0^+ state believed to reside in a secondary minimum. At the time the fit was done, it had been seen in $^{182-188}$Hg, and this is clearly reflected in the potential energy surfaces shown in Fig. 1. The triaxial minimum slowly moves up with mass number in this chain; for the heavier isotopes it is not present anymore, because the experimental data do not require it. Any interpretation of the fit results must go hand in hand with an examination of the information content of the experimental data. Thus the location of the minima in the (β,γ)-plane can be definitely established only if it really affects the data: otherwise the location obtained from the fit will just reflect a minor improvement in the fit indicative of the limitations of the parametrization.

This also seems to indicate that the predictive power of the model is quite limited. However, in my opinion its potential in that direction has not yet been exhausted: a prediction of neighboring elements in an isotope of isotone chain seems quite natural given a sequence of similar potential energy surfaces. In our case this would allow the prediction of progressively higher-lying minima in the Mercury isotopes based on the lighter ones; because this prediction would include effects like interactions between levels, the fit would also go beyond a straightforward interpolation of the experimental data. Unfortunately, such a prediction is difficult to do within the present parametrization of the collective potentials, which uses polynomial parameters instead of more physical quantities such as the locations of the minima etc., and this is probably the reason why it has not been attempted up to now.

Finally I would like to mention that the code for fitting the Gneuss-Greiner Hamiltonian to experimental dfata is available in a published and documented form[20].

6.5 General Numerical Formulation

In its most general form the classical Lagrangian of the geometric model can contain arbitrary functions of the deformations both for the mass parameters and for the potential[21–24].

$$
\begin{aligned}
L &= T - V \\
&= \frac{B_{\beta\beta}(\beta,\gamma)}{2}\dot{\beta}^2 + B_{\beta\gamma}(\beta,\gamma)\beta\dot{\beta}\dot{\gamma} \\
&\quad + \frac{B_{\gamma\gamma}(\beta,\gamma)}{2}\beta^2\dot{\gamma}^2 + \frac{1}{2}\sum_{k=x,y,z}\Theta_k(\beta,\gamma)\omega_k^2 \\
&\quad - V(\beta,\gamma).
\end{aligned} \tag{15}
$$

Note that the kinetic energy was still taken to be local in time, requiring the adiabatic approximation or a similar uniqueness of the internal state.

This *ansatz* contains seven functions, which are arbitrary up to certain symmetry requirements. It can be quantized by the Pauli method and this leads to a Schrödinger equation of the form

$$
\begin{aligned}
&\left\{ \left(I(I+1) - K^2\right)\left(\frac{1}{2\Theta_x} + \frac{1}{2\Theta_y}\right) + \frac{K^2}{2\Theta_z} \right. \\
&\qquad \left. + T_{\text{vib}} + \frac{2}{\hbar^2}\left[V(\beta,\gamma) - E_{I\alpha}\right]\right\}\Phi_{I\alpha K} \\
= &\left(\frac{1}{4\Theta_y} - \frac{1}{4\Theta_x}\right)\left(c_{I,K+2}\Phi_{I\alpha K+2} + c_{I,K-2}\Phi_{I\alpha K-2}\right).
\end{aligned} \tag{16}
$$

Here T_{vib} is a complicated differential operator in β and γ and the wave function is decomposed according to

$$\Psi_{I\alpha M}(\alpha_{2\mu}) = \sum_{K=-I}^{I} \Phi_{I\alpha K}(\beta,\gamma)\mathcal{D}^{I*}_{MK}(\theta_1,\theta_2.\theta_3). \tag{17}$$

The notation should be obvious and the quantum numbers are similar to those of the rotation-vibration model. The equations thus form a coupled set of differential equations in the (β,γ) plane and can be solved with numerical techniques; in the recent developments done at Frankfurt the finite element method was used.

The question remains as to how to determine the unknown functions. Clearly a fit to experimental data would require a parametrization in terms of simple functions and thus not go much beyond the Gneuss-Greiner Hamiltonian. The real value of this formulation therefore lies in the possibility of using input from other models such as Nilsson-Strutinsky calculations or the symplectic model[25].

7. Magnetic Properties

The geometric model does not easily lend itself to describing magnetic properties of nuclei. In the spirit of the model, one has to assume that protons and neutrons are homogeneously distributed and move in the same way, so that the magnetic dipole operator becomes proportional to the angular momentum operator and does not produce any transitions. This lead to the development of extended models with a distinction made between proton and neutron deformations[26–28] .

8. Acknowledgments

The author is indebted to W. Greiner, P. O. Hess, D. Troltenier, and M. Rosenstock for a long-lasting fruitful collaboration in various aspects of the geometric collective model.

I would also like to express my gratitude to the Joint Institute for Heavy Ion Physics for the hospitality extended to him. The Joint Institute for Heavy Ion Physics has as member institutions the University of Tennessee, Vanderbilt University, and the Oak Ridge National Laboratory; it is supported by the members and by the Department of Energy through Contract Number DE-FG05-87ER40361 with the University of Tennessee.

6. References

1. N. Bohr and J. A. Wheeler, Phys. Rev. **56** (1939) 426.
2. S. Flügge, Ann. Physik **39** (1941) 373.
3. M. Fierz, Helv. Phys. Acta **16** (1943) 365.
4. A. Bohr, Kgl. Danske Videnskab. Selsk. Mat.-Fys. Medd. **26** (1952) 14.
5. A. Faessler, W. Greiner, and R. K. Sheline, Nucl. Phys. **70** (1965) 33.
6. J. Eisenberg and W. Greiner, *Nuclear Theory 1: Collective Models*, 3rd edition (North Holland, Amsterdam 1972).
7. D. Podolski and W. Pauli, Phys. Rev. **32** (1928) 812.
8. D. Janssen, R. V. Jolos, and F. Dönau, Nucl. Phys. **A224** (1974) 93.
9. A. Arima and F. Iachello, Phys. Lett. **57B** (1975) 39; Phys. Rev. Lett. **35** (1975) 1069.
10. A. Gozdz and S. Szpikowski, Nucl. Phys. **A349** (1980) 359.

11. T. M. Corrigan, F. J. Margetan, and S. A. Williams, Phys. Rev. C**14** (1976) 2279.
12. E. Chacón, M. Moshinsky, and R. T. Sharp, J. Math. Phys. **17** (1976) 668.
13. L. Wilets and M. Jean, Phys. Rev. **102** (1956) 788.
14. S. G. Rohozinski, J. Srebrny, K. Horbaczewska, Z. Physik **268** (1974) 401.
15. G. Gneuss, U. Mosel, and W. Greiner, Phys. Lett. **30B** (1969) 397.
16. G. Gneuss and W. Greiner, Nucl. Phys. **A171** (1971) 449.
17. M. Sedlmayer, Ph. D. Thesis, University of Frankfurt 1976.
18. P. O. Hess, M. Seiwert, J. A. Maruhn, and W. Greiner, Z. Physik **A296** (1980) 147.
19. D. Troltenier, J. A. Maruhn, W. Greiner, V. Velazquez-Aguilar, P. O. Hess, and J. H. Hamilton, Z. Physik **A338** (1991) 261.
20. D. Troltenier, J. A. Maruhn, and P. O. Hess, in: *Computational Nuclear Physics I*, ed. K. Langanke, J. A. Maruhn, and S. E. Koonin (Springer-Verlag, Berlin, Heidelberg, and New York 1991).
21. K. Kumar and M. Baranger, Nucl. Phys. **A92** (1967) 608.
22. K. Kumar, B. Remaud, P. Aguer, J. S. Vaagan, A. C. Rester, R. Foucher, and J. H. Hamilton, Phys. Rev. C**16** (1977) 1235.
23. S. G. Rohozinski, J. Dobaczewski, B. Nerlo-Pomorska, K. Pomorski, and J. Srebrny, Nucl. Phys. **A292** (1977) 66.
24. D. Troltenier, Ph. D. Thesis, University of Frankfurt, 1992.
25. O. Castaños, P. O. Hess, J. P. Draayer, and P. Rochford, Nucl. Phys. **A524** (1991) 469.
26. W. Greiner, Phys. Rev. Lett. **14** (1965) 599.
27. V. Maruhn-Rezvani, W. Greiner, and J. A. Maruhn, Phys. Lett. **57B** (1975) 109.
28. A. Arima, T. Otsuka, F. Iachello, and I. Talmi, Phys. Lett. **66B** (1977) 20; T. Otsuka, A. Arima, F. Iachello, and I. Talmi, Phys. Lett. **76B** (1978) 139.

GROUP 4

Shell Correction Method

THE SHELL-CORRECTION METHOD

MATTHIAS BRACK
Institut für Theoretische Physik, Universität Regensburg
D-8400 Regensburg, Germany

ABSTRACT

We review the basic assumptions and definitions of Strutinsky's shell-correction method. Its derivation from Hartree-Fock and extended selfconsistent microscopical theories is presented and numerical tests based on Hartree-Fock results are summarized. A detailed discussion of the Strutinsky energy averaging procedure, its validity and its relation to other averaging methods is given. We finally mention some recent applications of the Strutinsky method, also outside nuclear physics.

1. Introduction

Strutinsky's shell-correction method[1] (SCM) grew out of the necessity to correct the wrong average behavior of the nuclear deformation energy which in the Nilsson model[2] was taken to be (proportional to) the sum of the lowest occupied single-particle energies ϵ_i of the deformed shell-model potential:

$$E_{sp}(def) = \sum_{i=1}^{A} \epsilon_i(def). \tag{1}$$

It had been realized for some time[3] that $E_{sp}(def)$ does not behave well at large deformations, in particular in attempts to calculate fission barriers. This is mainly due to the fact that Eq. (1) is not the correct expression for the binding (or deformation) energy of a self-saturated fermion system. Furthermore, the spectrum ϵ_i obtained in phenomenological shell-model potentials can approximately describe the real single-particle excitations only near the Fermi energy λ; far below or above λ the energies ϵ_i have little or no physical meaning.

This led Strutinsky[1] to *renormalize* the wrong average part of $E_{sp}(def)$ to that of the empirically known liquid drop model[5] (LDM) energy $E_{LDM}(def)$, by defining the total (deformation) energy to be

$$E_{tot}(Z, N, def) = E_{LDM}(Z, N, def) + \delta E_p(Z, def) + \delta E_n(N, def), \tag{2}$$

where the so-called *shell-correction* energy $\delta E_q(N_q, def)$ for each kind of nucleons ($q = p, n;\ N_p = Z,\ N_n = N$) is defined to be the difference between the corresponding sum E_{sp} and its average:

$$\delta E_q(N_q, def) = \sum_{i=1}^{N_q} \epsilon_i^q(def) - \left\langle \sum_{i=1}^{N_q} \epsilon_i^q(def) \right\rangle. \tag{3}$$

The averaging $\langle \ \rangle$ in Eq. (3) is constructed in such a way that it eliminates the shell effects in E_{sp} without affecting its average part. Strutinsky designed for this a numerical averaging procedure which will be discussed in detail in Sect. 3. As a result, δE_q only depends on the single-particle levels ϵ_i^q near the Fermi energy λ.

The SCM has led to a considerable success in explaining for the first time the nature of the fission isomers[4, 1] and, more generally, in calculating nuclear masses and fission barriers[6, 7, 8] (see also Sect. 4 for more recent applications). In Eq. (2), the average energy E_{LDM} is given in terms of a set of empirical LDM[5] (or droplet model[29]) parameters whereas the shell-corrections δE_q are determined from the levels of phenomenological (deformed) shell-model potentials such as the Nilsson model[2] or other finite-range potentials[7, 8]. The SCM thus gives both empirical nuclear models their balanced role. That this method is more than an *ad hoc* renormalization, but can be solidly based upon selfconsistent Hartree-Fock or Landau theory, will be discussed in the following Sect. 2.

2. Strutinsky's Energy Theorem

We shall in this section discuss the foundation of the SCM upon selfconsistent microscopic theories using effective nucleon-nucleon interactions[1, 7, 9]. For simplicity, we treat here the nucleus as if it had only N nucleons of one kind. The extension to Z protons and N neutrons as used in Eq. (2) is straightforward.

2.1. Derivation within Hartree-Fock Theory

We start from the Hartree-Fock (HF) energy for N fermions interacting through a 2-body force $\hat{V}$:

$$E_{HF} = E_{HF}[\rho] = \mathrm{tr}\,(\mathcal{T}\rho) + \frac{1}{2}\mathrm{tr}\,(\rho\,\mathrm{tr}\,(\mathcal{V}\rho)). \tag{4}$$

Here ρ is the one-body density matrix written in terms of any single-particle basis $|\alpha\rangle$ and the occupied HF states $|i\rangle$

$$\rho_{\alpha\beta} = \sum_{i=1}^{N} \langle\alpha|i\rangle\langle i|\beta\rangle = \sum_i \langle\alpha|i\rangle\langle i|\beta\rangle n_i^{HF}; \tag{5}$$

n_i^{HF} are the HF occupation numbers

$$n_i^{HF} = \begin{cases} 1 & \text{for } \epsilon_i < \lambda, \\ 0 & \text{for } \epsilon_i > \lambda. \end{cases} \tag{6}$$

(Finite temperature and pairing effects will be discussed in Sect. 2.2 below.) In Eq. (4), $\mathcal{T}$ and $\mathcal{V}$ are the one-body and antisymmetrized two-body matrix elements of the kinetic energy and the two-body interaction $\hat{V}$, respectively:

$$\mathcal{T}_{\alpha\beta} = \langle\alpha|\hat{T}|\beta\rangle; \quad \mathcal{V}_{\alpha\beta,\gamma\delta} = \langle\alpha\beta|\hat{V}|\overline{\gamma\delta}\rangle. \tag{7}$$

We now define an *average* density matrix in terms of some smoothed occupation numbers $\tilde{n}_i$ to be defined in Sect. 3.4 below:

$$\tilde{\rho}_{\alpha\beta} = \sum_i \langle\alpha|i\rangle\langle i|\beta\rangle \tilde{n}_i \tag{8}$$

which contains only average information without any shell effects. Next we write

$$\rho = \tilde{\rho} + \delta\rho \tag{9}$$

and Taylor expand the HF energy Eq. (4) around $\tilde{\rho}$:

$$E_{HF}[\rho] = E_{HF}[\tilde{\rho}] + \mathrm{tr}\,(\frac{\partial E_{HF}}{\partial \rho}\Big|_{\tilde{\rho}} \delta\rho) + ... \tag{10}$$

By construction, the first term in Eq. (10) contains no shell effects, i.e. it represents the average part of the HF energy which is smooth as a function of particle number(s) and deformation:

$$E_{HF}[\tilde{\rho}] = \widetilde{E}_{HF}. \tag{11}$$

The second term in Eq. (10) contains all shell effects up to first order in the oscillating part of the density matrix, $\delta\rho$, which is a small quantity. Note now that the partial derivative

$$\frac{\partial E_{HF}[\rho]}{\partial \rho_{\beta\alpha}} = \langle\alpha|\hat{H}_{HF}[\rho]|\beta\rangle = (\mathcal{H}_{HF}[\rho])_{\alpha\beta} \tag{12}$$

is just the matrix element of the HF one-body operator containing the selfconsistent mean field. (This holds also for density-dependent effective interactions!) We thus can rewrite the first-order term in Eq. (10) as

$$\delta_1 E = \mathrm{tr}\,(\frac{\partial E_{HF}}{\partial \rho}\Big|_{\tilde{\rho}} \delta\rho) = \mathrm{tr}\,(\mathcal{H}_{HF}[\tilde{\rho}]\,\delta\rho) = \sum_i \hat{\epsilon}_i \delta n_i = \sum_{i=1}^{N} \hat{\epsilon}_i - \sum_i \hat{\epsilon}_i \tilde{n}_i, \tag{13}$$

where $\hat{\epsilon}_i$ are the eigenenergies of the *average HF field* $\hat{V}_{HF}[\tilde{\rho}]$ which also varies smoothly as a function of particle number(s) and deformation:

$$\hat{H}_{HF}[\tilde{\rho}]\hat{\varphi}_i = \left\{\hat{T} + \hat{V}_{HF}[\tilde{\rho}]\right\}\hat{\varphi}_i = \hat{\epsilon}_i \hat{\varphi}_i. \tag{14}$$

We thus arrive at the so-called[10] *Strutinsky energy theorem*:

$$E_{HF} = \widetilde{E}_{HF} + \delta_1 E + \mathcal{O}[(\delta\rho)^2]. \tag{15}$$

It states that the HF energy of an interacting system of fermions can be written as a smooth part $\widetilde{E}_{HF}$ and an oscillating part $\delta_1 E$. Most importantly, it states that the shell-correction energy $\delta_1 E$ contains *all* contributions of first order in the oscillating part $\delta\rho$ of the density matrix and can be written in the simple form on the r.h.s. of Eq. (13) in terms of the eigenvalues $\hat{\epsilon}_i$ of a smoothed average field.

In the practical shell-correction approach, the average HF energy $\widetilde{E}_{HF}$ is replaced by the phenomenological LDM energy E_{LDM} and the shell-correction $\delta_1 E$ is evaluated in terms of the eigenenergies of a phenomenological shell-model potential. It remains to be checked by explicit HF calculations, to which extent these replacements can be justified and to which extent the higher-order terms $\mathcal{O}[\delta\rho]$ in Eq. (15) can be neglected. Empirically, we know that $\widetilde{E}_{HF}$ is of the order of $\sim$ 1 - 2 GeV, whereas $\delta_1 E$ is only $\sim$ 10 - 15 MeV at most; there are thus good reasons to hope that Eq. (10) and thus Eq. (15) converge fast. (Corresponding numerical tests will be discussed in Sect. 2.3 below.)

Note that in the realistic case $\delta_1 E$ is a sum of neutron and proton contributions, as displayed in Eq. (2), which are obtained from the individual neutron and proton single-particle spectra. This does not mean that – at least in a selfconsistent HF calculation – these two contributions are not coupled through the dependence of the neutron mean field on the proton density and *vice versa*. Such a coupling does, however, not exist in the phenomenological approach where the shell-model potentials for neutrons and protons are fitted independently and are only correlated in a trivial way through their radii which are given in terms of the total nucleon number $A = N + Z$.

It has been repeatedly overlooked by practitioners of the SCM (and still is by some!) that there is ***no double counting*** of the potential energy contributions in $\delta_1 E$: due to the variational way (12) of deriving the average HF field $\tilde{V}_{HF}[\tilde{\rho}]$, the factor 1/2 in Eq. (4) is doubled and $\delta_1 E$ is just the oscillating part of the sum of occupied eigenvalues $\hat{\epsilon}_i$ of the average potential. This holds also for density-dependent effective interactions, since the so-called rearrangement contributions[10] will be taken care of automatically when the variation in Eq. (12) is correctly performed[11].

Another critique[12, 13] of the energy theorem concerned the validity of Eq. (15) as a function of deformation: the way in which it was derived above, it holds only for equilibrium deformations where E_{HF} has a (local) minimum. The SCM, however, parameterizes the deformation dependence through both the LDM energy and the shell-model potentials and is used at arbitrary deformations. The answer[9, 14, 15, 16] lies in a careful inclusion of a *constraint* in the HF equation[17] whose contributions to the first-order shell correction $\delta_1 E$ in the ideal case can be cancelled by a suitable choice of the deformation dependence of the shell-model potential and in practical cases numerically turn out to be negligible (see Fig. 2 below).

2.2. Extensions beyond Hartree-Fock

The derivation of the Strutinsky energy theorem Eq. (15) is not restricted to the pure HF case. In fact, it is a widely exploited property of the variational ground state of a correlated fermion system that to first order in small changes δn_i of the occupation numbers, the change in the ground-state energy can be written as

$$\delta E = \sum_i e_i \delta n_i, \tag{16}$$

where e_i are the single-particle, or more generally, the quasiparticle energies of the system. This is the basis of Landau theory[18] and has been used explicitly to show that the Strutinsky method can also be applied within the framework of the Landau-Migdal Fermi liquid theory[11]. Similarly, the energy theorem (15) has been extended to HF-Bogolyubov theory in order to include pairing and temperature effects selfconsistently[16, 19]. Eq. (16) can also be used to derive the famous Koopmans theorem from the density functional approach[20], and a "force theorem" has been derived in solid state theory along very similar lines[21].

We refer to an extensive presentation on the Strutinsky method and its foundation from the HF-Bogolyubov approximation at finite temperature[16], where also numerical tests are reviewed.

2.3. Numerical Tests of the Energy Theorem

At the time when Strutinsky derived Eq. (15) and formulated the shell-correction approach, no HF results with realistic effective nucleon-nucleon interactions were available to test the convergence of the expansion in Eq. (10) and the applicability of phenomenological liquid-drop and shell models. The SCM thus relied on its success in yielding nuclear masses and fission barriers in good overall agreement with experiment.

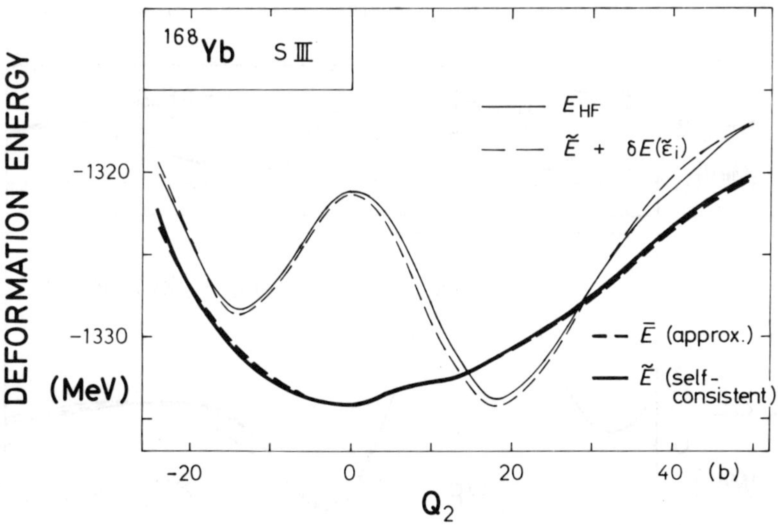

Fig. 1. HF results[26] for the deformation energy of ^{168}Yb (see text for details).

With the successful development of Skyrme's effective nuclear interactions [22, 23], HF calculations for heavy deformed nuclei became possible in the early seventies[17, 24] (see also the presentation of J.-F. Berger for HF calculations with Gogny's finite-range force[25]). This allowed to perform extensive numerical tests [26, 27] of the convergence in Eq. (10) and to compare the phenomenological LDM and shell-model results to selfconsistent HF results of ground-state masses and deformation energies.

We refer to a detailed presentation[16] of these tests using various types of Skyrme interactions and give here only some typical illustrations and a summary of the most important results. The main idea is to compute first the fully microscopical HF energy Eq. (4), then to use the numerically Strutinsky-averaged density matrix (8) to obtain the averaged energy (11) and the averaged HF potential appearing in Eq. (14) from which the first-order shell-correction (13) is extracted, and finally to investigate the difference

$$\delta_2 E = E_{HF} - \widetilde{E}_{HF} - \delta_1 E \tag{17}$$

which gives the sum of all higher-order terms in the expansion (10).

Figure 1 shows the deformation energy of ^{168}Yb as a function of the total mass quadrupole moment Q_2, obtained with the Skyrme III interaction[28] and a quadratic constraint[17] in the HF equation. The thin solid line is the HF energy E_{HF} (4), the heavy lines are obtained by including the average density matrix $\widetilde{\rho}$ (8) either perturbatively in the last HF iteration (dashed line) or iteratively to obtain a *selfconsistent* average HF energy $\widetilde{E}_{HF}$ (solid line). The thin dashed line gives the "Strutinsky approximation" to the HF energy, i.e. the sum of $\widetilde{E}_{HF}$ and $\delta_1 E$.

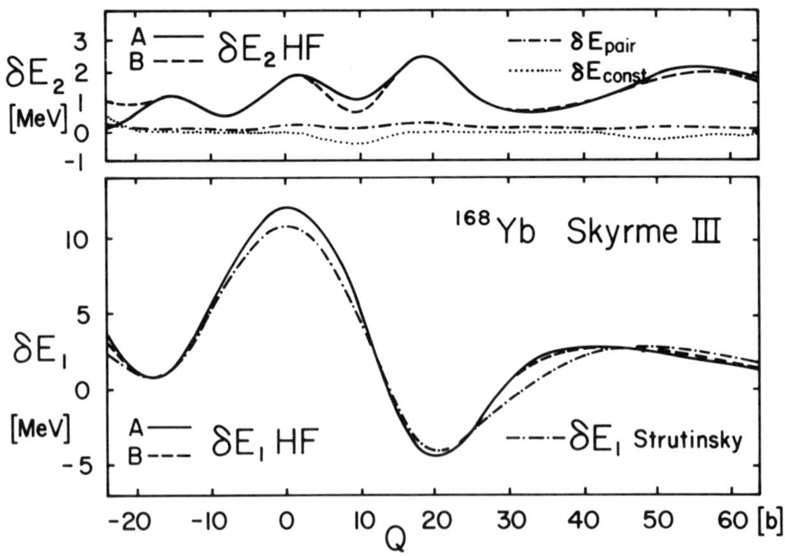

Fig. 2. First-order and sum of higher-order shell-corrections extracted from the HF results[26] shown in Fig. 1.

In the lower part of Figure 2, the shell correction $\delta_1 E$ extracted from the HF results is compared to that from a Woods-Saxon type shell-model potential. In the upper part, the quantity $\delta_2 E$ in Eq. (17) is shown along with the specific (and very small) contributions from pairing correlations and the constraint energy. (See

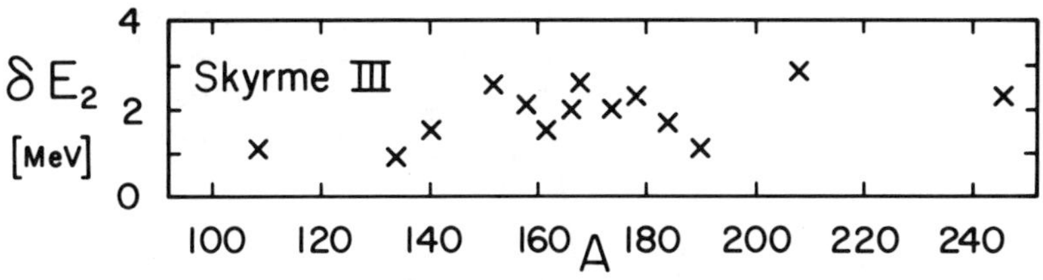

Fig. 3. Sum of higher-order shell-corrections extracted from HF results[26] for a series of nuclear ground states.

ref.[26] for details.) Figure 3 shows the same quantity $\delta_2 E$ for the ground states of a series of nuclei.

The results of these and similar tests can be summarized as follows:

(i) The sum of all second and higher-order corrections defined by Eq. (17) is small; apart from an average value of ~ 2 MeV (which in practice will be renormalized into the average LDM energy, it fluctuates by less than $\pm \sim 1$ MeV for nuclei with $A \gtrsim 50$.

(ii) $\delta_2 E$ is smallest when the averaging (through the smooth occupation numbers $\tilde{n}_i$) is performed iteratively, thus implying a selfconsistency between the average energy $\widetilde{E}$ (usually represented by the LDM) and the average potential $\widetilde{V}$ (usually represented by the shell-model potential).

(iii) $\widetilde{E}_{HF}$ behaves exactly like a LDM energy: it is smooth as a function of both deformation and nucleon numbers and has its minimum at the spherical shape.

(iv) The first-order shell-correction energy $\delta_1 E$ extracted from the HF results is close (usually within less than ~ 1 MeV at all deformations) to the δE obtained in the deformed phenomenological shell-model potentials.

(v) For small nuclei with $A \lesssim 50$, the quantity $\delta_2 E$ can be considerably larger than for heavier systems. It depends rather critically on the inclusion of pairing correlations.

To illustrate the last point, we show in Figure 4 the quantities $\delta_1 E$ and $\delta_2 E$ obtained as above for the nucleus ^{40}Ca. Depending on the strength of the average pairing gap $\widetilde{\Delta}$ (see Ref.[7] for its definition and use), the magnitude of these quantities can vary by a factor of two or more and give a much slower convergence of Eq. (10) than that found for heavier nuclei.

The above results check the validity of the Strutinsky method under *ideal conditions*, i.e. when starting from consistent average (LDM type) energies and mean field (shell-model type) potentials derived from the same effective nuclear interaction. In practice, the two models are fitted independently to different experimental data and no such consistency can be guaranteed. This may lead to larger discrepancies and uncertainties than what could be concluded from the above. A well-studied example is the so-called "Pb anomaly"[7, 30] (see Refs.[31, 32, 33] for

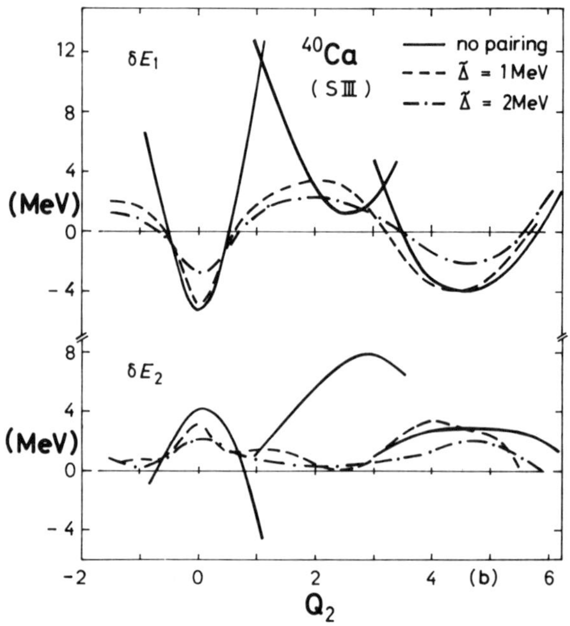

Fig. 4. Same as Fig. 2 for ^{40}Ca using two strengths of the paring gap[27].

its detailed discussion). In general, no reliable results should be expected for nuclei with $A < 40 - 50$.

3. Strutinsky's Energy Averaging Method

The aim of Strutinsky's energy averaging method[1] is to extract the fluctuating part of the sum (1) of occupied single-particle energies in a unique way. In the present section we shall discuss this technique in detail.

In the seventies, a series of alternatives to the Strutinsky energy averaging method has been proposed of which the extended Thomas-Fermi (ETF) model[34], the asymptotic $N \to \infty$ expansion[35, 36], a numerical N-averaging[37, 32], and a "temperature method"[38] were the most promising ones. All of them have some inherent uncertainties of $\lesssim$ 1 MeV in determining the average s.p. energy sum for heavy nuclei and of up to $\sim$ 2 MeV for light nuclei, thus of the same order as the plateau uncertainties of the energy averaging to be discussed in Sect. 3.3 below. We shall in Sect. 3.5 discuss briefly the ETF model, which has been the most successful of these alternatives, and refer to Ref.[31] for an evaluation of the other methods.

3.1. Smoothing Functions and Curvature Corrections

We consider one kind of N particles in a given potential with energy spectrum $\{\epsilon_i\}$. Sums over i are understood to include all degeneracies of the spectrum.

Let us first introduce the exact quantum-mechanical level density

$$g(E) = \sum_i \delta(E - \epsilon_i). \tag{18}$$

The single-particle energy sum E_{sp} can be given by the integral

$$E_{sp} = \sum_{i=1}^{N} \epsilon_i = \int_{-\infty}^{\lambda} E\, g(E)\, dE, \tag{19}$$

where the Fermi energy λ is fixed by the particle number:

$$N = \int_{-\infty}^{\lambda} g(E)\, dE. \tag{20}$$

The basic idea is to assume that $g(E)$ can be written as a sum of a smooth part $g_0(E)$ and an oscillating part $\delta g(E)$:

$$g(E) = g_0(E) + \delta g(E). \tag{21}$$

It has, in fact, be shown by Gutzwiller[39] in the so-called "periodic orbit theory" that this is always possible: $\delta g(E)$ can be expressed in terms of a sum over all classical periodic trajectories of a particle in the given potential (see also Refs.[40, 41, 42]), whereas the average level density $g_0(E)$ may be obtained from the (extended) Thomas-Fermi model (see Sect. 3.5 below). Except for a few model potentials, the two parts in Eq. (21) cannot be determined analytically. Therefore one must resort to numerical methods to determine them approximately.

Strutinsky's method[1] for extracting $g_0(E)$ as accurately as possible by means of a numerical energy averaging of the spectrum $\{\epsilon_i\}$ may be presented in the following way[43]. We start defining an *averaging function* $f(x)$ which is assumed to be analytical, positive, normalized to unity, and, to simplify matters, symmetric in x with a maximum at $x = 0$ (see Figure 5). In a first step, we then define an averaged level density $\tilde{g}_0(E, \gamma)$ by folding $g(E)$ Eq. (18) with f over a range γ:

$$\tilde{g}_0(E, \gamma) = \frac{1}{\gamma} \int_{-\infty}^{+\infty} g(E')\, f\left(\frac{E - E'}{\gamma}\right) dE' = \frac{1}{\gamma} \sum_i f\left(\frac{E - \epsilon_i}{\gamma}\right). \tag{22}$$

This achieves the following result: If the oscillating part $\delta g(E)$ of the level density is dominated by a fundamental period $\hbar\omega$ and its harmonics (like in a spherically symmetric harmonic oscillator potential, see Sect. 3.2 below), the function $\tilde{g}_0$ Eq. (22) will be smooth as soon as $\gamma \gtrsim \hbar\omega$. It will, however, depend on γ except if $g_0(E)$ is a linear function of energy (this is so due to the symmetry and the normalization of f). Furthermore, if $g_0(E)$ is a *polynomial* of degree n in energy, then $\tilde{g}_0$ will also be a polynomial of degree n (with γ-dependent coefficients for $n \geq 2$). In order to get rid of this γ dependence, Strutinsky introduced a *curvature correction*[1]. (This name was chosen because $g_0(E)$ in general is not linear, i.e. it has a nonzero curvature.) To show how this correction works, we first rewrite the folding integral in Eq. (22)

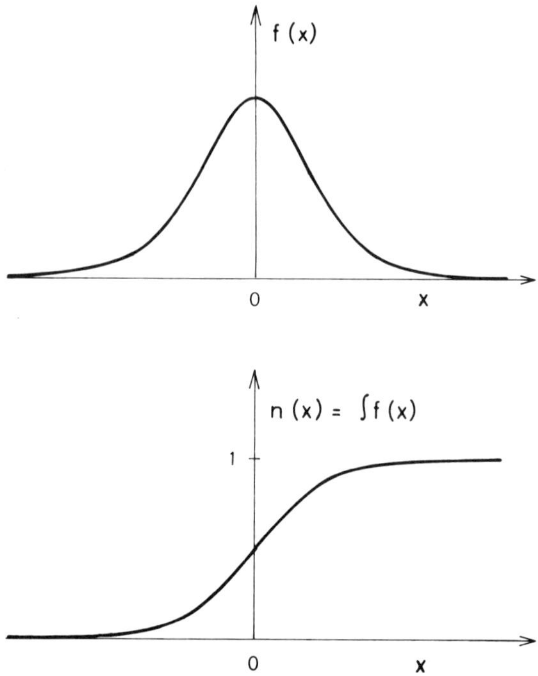

Fig. 5. Upper part: Schematic plot of a smoothing function $f(x)$. Lower part: First integral $n(x)$ of the smoothing function.

as the result of an operator $\widehat{G}_0$, which can be expanded into an infinite sum of derivative operators, acting on $g(E)$:

$$\widetilde{g}_0(E,\gamma) = \left(\sum_{\mu=0}^{\infty} \frac{c_{2\mu}}{(2\mu)!}\gamma^{2\mu}\frac{d^{2\mu}}{dE^{2\mu}}\right) g(E) = \widehat{G}_0[g(E)], \tag{23}$$

where $c_{2\mu}$ are moments of the smoothing function f:

$$c_m = \int_{-\infty}^{+\infty} x^m f(x)dx. \tag{24}$$

(Only the even moments appear in (23) due to the assumed symmetry of f). We then define the inverse operator $\widehat{G}_0^{-1}$ and expand it correspondingly:

$$\widehat{G}_0^{-1} = \sum_{\mu=0}^{\infty} a_{2\mu}\gamma^{2\mu}\frac{d^{2\mu}}{dE^{2\mu}}. \tag{25}$$

The coefficients $a_{2\mu}$ can always be gained recursively from the $c_{2\mu}$ using $\widehat{G}_0^{-1} \times \widehat{G}_0 \equiv 1$. Next we define the "curvature-correction operator" $\widehat{C}_M$ by truncating the sum in

Eq. (25) at $\mu = M$:

$$\widehat{C}_M = \sum_{\mu=0}^{M} a_{2\mu}\gamma^{2\mu}\frac{d^{2\mu}}{dE^{2\mu}}. \tag{26}$$

We now arrive at the *Strutinsky averaging* operator $\widehat{S}_M$

$$\widehat{S}_M = \widehat{C}_M \times \widehat{G}_0 = \frac{1}{\gamma}\sum_{\mu=0}^{M} a_{2\mu}\gamma^{2\mu}\frac{d^{2\mu}}{dE^{2\mu}}\int_{-\infty}^{+\infty} f\left(\frac{E-E'}{\gamma}\right) dE' \times \tag{27}$$

which by construction leaves any polynomial $P_n(E)$ of degree $n \leq 2M+1$ unchanged, independently of $\gamma > 0$

$$\widehat{S}_M[P_{2M+1}(E)] \equiv P_{2M+1}(E) \qquad \forall\, \gamma > 0, \tag{28}$$

since all missing terms with $\mu > M$ of the truncated inverse operator $\widehat{G}_0^{-1}$ are identically zero.

We thus define the *Strutinsky-averaged level density* for a given level spectrum $\{\epsilon_i\}$ by

$$\widetilde{g}_M(E) = \widehat{S}_M[g(E)] = \frac{1}{\gamma}\sum_i\sum_{\mu=0}^{M} a_{2\mu}\gamma^{2\mu}\frac{d^{2\mu}}{dE^{2\mu}} f\left(\frac{E-\epsilon_i}{\gamma}\right). \tag{29}$$

For clarity we repeat the two main features of $\widetilde{g}_M(E)$:

(i) If the averaging width γ is chosen to be larger than the main distance $\hbar\omega$ of the shells in the spectrum $\{\epsilon_i\}$, the oscillating part $\delta g(E)$ of the exact level density is suppressed and $\widetilde{g}_M(E)$ is a smooth function of energy.

(ii) If the average part $g_0(E)$ of the exact level density is a polynomial of degree $2M+1$ (or less), it is identically reproduced in $\widetilde{g}_M(E)$ for any $\gamma > 0$.

One usually refers to $2M$ as to the order of the curvature corrections. As shown numerically[43], the results for $\widetilde{g}_M(E)$ do not depend on the particular form of the averaging function $f(x)$, as long as it fulfils the above requirements. The standard choice is a Gaussian

$$f(x) = \frac{1}{\sqrt{\pi}}\exp(-x^2) \quad \text{with} \quad \widehat{G}_0 = \exp\left(\frac{\gamma}{2}\frac{d}{dE}\right)^2, \quad a_{2\mu} = \frac{(-1)^\mu}{2^{2\mu}\mu!}. \tag{30}$$

The curvature-corrected average level density $\widetilde{g}_M(E)$ Eq. (29) can then be written as a sum of Gaussians multiplied by a linear combination of Hermite polynomials; this is the usual way to present the Strutinsky energy averaging[1, 6, 7]. Another example for the averaging function is

$$f(x) = \frac{1}{2[\cosh(x)]^2}, \quad \widehat{G}_0 = \frac{\frac{\pi}{2}\hat{d}}{\sin(\frac{\pi}{2}\hat{d})} \; (\hat{d} = \gamma\frac{d}{dE}), \quad a_{2\mu} = \left(\frac{\pi}{2}\right)^{2\mu}\frac{(-1)^\mu}{(2\mu+1)!}. \tag{31}$$

[Note that the integral of $f(x)$ in Eq. (31) is a Fermi function and thus leads to the Fermi occupation numbers of a grand canonical ensemble (cf. Sect. 3.4)]. Further examples of averaging functions and their curvature-correction coefficients $a_{2\mu}$ may be found in Ref.[43].

From $\widetilde{g}_M(E)$ we now define the average part of the single-particle energy sum E_{sp}:

$$\widetilde{E}_{sp} = \left\langle \sum_{i=1}^{N} \epsilon_i \right\rangle = \int_{-\infty}^{\widetilde{\lambda}} E\, \widetilde{g}_M(E)\, dE, \tag{32}$$

where the average Fermi energy $\widetilde{\lambda}$ is given by the particle number integral:

$$N = \int_{-\infty}^{\widetilde{\lambda}} \widetilde{g}_M(E)\, dE. \tag{33}$$

3.2. An Analytical Example

For a spherically symmetric 3-dimensional harmonic oscillator potential, the exact level density can be written as[44]

$$g(E) = \frac{1}{(\hbar\omega)^3} \left[E^2 - \frac{1}{4}(\hbar\omega)^2 \right] \left\{ 1 + 2\sum_{n=1}^{\infty} (-1)^n \cos\left(n \frac{2\pi E}{\hbar\omega} \right) \right\} \Theta(E). \tag{34}$$

We thus know the analytical form of its average part:

$$g_0(E) = \frac{1}{(\hbar\omega)^3} \left[E^2 - \frac{1}{4}(\hbar\omega)^2 \right]. \qquad (E \geq 0) \tag{35}$$

This is a quadratic polynomial, and therefore the Strutinsky averaging with $M = 1$ should reproduce exactly this average part. Inserting Eq. (34) into Eq. (19), one obtains for the smooth part of E_{sp}:

$$E_{sp}^{(0)} = \hbar\omega \left[\frac{1}{4}\left(\frac{\widetilde{\lambda}}{\hbar\omega} \right)^4 - \frac{1}{8}\left(\frac{\widetilde{\lambda}}{\hbar\omega} \right)^2 - \frac{7}{320} \right], \tag{36}$$

where $\widetilde{\lambda}$ is given by the positive root of

$$N = \frac{1}{3}\left(\frac{\widetilde{\lambda}}{\hbar\omega} \right)^3 - \frac{1}{4}\left(\frac{\widetilde{\lambda}}{\hbar\omega} \right). \tag{37}$$

[Note that some contributions to Eq. (36) also come from the lower limit ($E = 0$) of the integral over the oscillating terms in Eq. (34). The expressions (36,37), which easily can be extended to a triaxially deformed harmonic oscillator, are identical to those obtained in the extended Thomas-Fermi model; see Sect. 3.5 below.] Since $E_{sp}^{(0)}$ is a fourth-order polynomial in the Fermi energy, one has to use a fourth-order curvature-correction ($M = 2$) in the Strutinsky averaging to obtain a γ-independent average energy $\widetilde{E}_{sp}$ (32).

As an illustration, we show in Figure 6 the results obtained for the shell-correction energy δE by Strutinsky averaging using the standard Gaussian averaging function (30) for various values of M. For any value $M \geq 2$ one obtains, indeed, a unique value of δE (and thus of $\widetilde{E}_{sp}$) which is constant as soon as $\gamma \gtrsim \hbar\omega$. The upper end of the "plateau" in Fig. 6 is merely due to the finite number of levels used in the

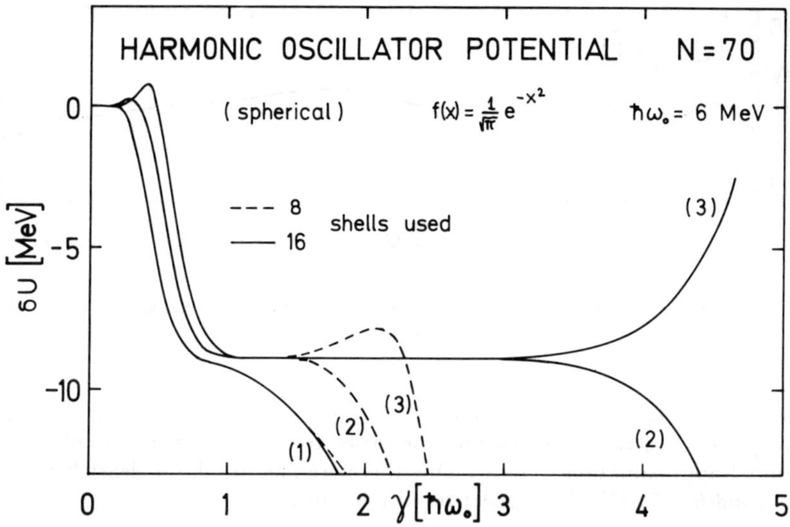

Fig. 6. Shell-correction energy δE versus averaging width γ for 70 fermions in a spherical harmonic oscillator, for various values of M and for two truncations of the energy spectrum[43].

summation. Very similar curves with identical plateau values of δE were obtained in Ref.[43] also with the averaging function in Eq. (31); in fact, it was shown there analytically that the plateau value of $\widetilde{E}_{sp}(\gamma)$ equals $E_{sp}^{(0)}$ Eq. (36) independently of $f(x)$.

The harmonic oscillator represents an ideal case for the Strutinsky averaging. The Nilsson model, for which the method was first applied, also is near to ideal because it is not much different from an oscillator model. In general, however, the average part $g_0(E)$ of the level density is not a polynomial. As a consequence no ideal plateaux like in Fig. 6 can be found. This is particularly so in situations where the local level density is high and δE has a maximum, as e.g. near fission barriers of heavy nuclei[45]. In such cases, a more careful investigation of the plateau in $\widetilde{E}_{sp}(\gamma)$ and its dependence on the correction order $2M$ is necessary.

3.3. The Plateau Condition

When $g_0(E)$ is not a polynomial, $\widetilde{E}_{sp}$ can, in general not be made independent of γ. Some alternatives to the original Strutinsky energy averaging, assuming expansions in non-integer powers of energy, have been investigated[32, 37] but not been able to solve the problem of this γ dependence.

It is, however, still possible to employ the standard averaging method in order to find *a local plateau* in $\widetilde{E}_{sp}(\gamma)$. As we have demonstrated above, this method consists essentially in a polynomial expansion of the average level density. This always works locally, i.e. in a limited energy interval around the Fermi energy

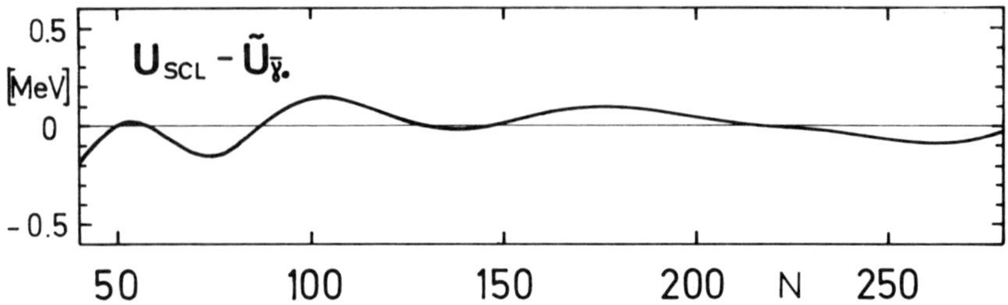

Fig. 7. Difference between semiclassical and Strutinsky-averaged s.p. energy sum in a cubic Hill-Wheeler box[43].

(except at possible – but usually uninteresting – singular points of $g_0(E)$). Hereby the curvature-correction order $2M$ must be varied to find the best degree of the local polynomial approximation to $g_0(E)$. One therefore has to fulfil at least locally the ***plateau condition***[7, 43] with respect to both γ and M:

$$\left.\frac{\partial \widetilde{E}_{sp}(M_0)}{\partial \gamma}\right|_{\gamma_0} = 0; \qquad \left.\frac{\Delta \widetilde{E}_{sp}(\gamma_0)}{\Delta M}\right|_{M_0} = 0. \tag{38}$$

[The plateau condition has also been discussed in connection with the stationarity of a selfconsistently Strutinsky-averaged HF energy[16].] Typically, solutions of (38) can be found with values of the order

$$\gamma_0 \sim (1 - 1.6)\,\hbar\Omega; \qquad M_0 \sim 3 - 8, \tag{39}$$

where $\hbar\Omega$ is the main-shell spacing in the spectrum $\{\epsilon_i\}$.

The uncertainties in finding unique plateau values of $\widetilde{E}_{sp}$ or the shell-correction δE are usually less than ± 1 MeV for nuclei with N or $Z \gtrsim$ 40 - 50. For smaller particle numbers, uncertainties up to $\sim$ 2 MeV or more may be found, in particular in finite-depth potentials as discussed below. (See Ref.[31] for a more detailed summary.)

As an illustration we take the Hill-Wheeler box[46], i.e. an infinite rectangular box potential with side lengths a, b and c subject to the volume conservation condition $abc = L^3$. Its semiclassical average level density is known (see also Ref.[47]):

$$g_0(E) = \frac{\pi}{4}\frac{1}{E_0}\left[\sqrt{E/E_0} - \frac{\pi}{8}(S/2L^2) + \frac{1}{8}(C/L)\sqrt{E_0/E}\right]. \tag{40}$$

Hereby $E_0 = (\pi^2\hbar^2)/(2mL^2)$ is the natural energy unit, $S = 2(ab + ac + bc)$ is the surface and $C = a+b+c$ the "mean curvature" of the box. Since this $g_0(E)$ is not a polynomial in E, the local plateau conditions Eq. (38) must be solved at each energy. In Figure 7 we show the results[43] for the *difference* between the semiclassical average s.p. energy obtained from Eq. (40) by integration and the plateau values of $\widetilde{E}_{sp}$ in the case of the cubic box. The difference is seen to be less than ± 0.2 MeV for all particle numbers $N > 40$, demonstrating that the local polynomial approximation to (40) works very well.

A basic problem is met with the finite-depth potentials frequently used in nuclear physics. Here the spectrum has a continuum for $E > 0$, and the problem is how to treat this continuum. In particular when the Fermi energy comes close to the continuum, i.e. if the separation energy of a nucleon is comparable to or less than the main shell spacing $\hbar\Omega$ (and thus γ), the Strutinsky-averaged results depend rather crucially on the lowest continuum states. In practice, one diagonalizes the potential in a finite harmonic oscillator basis[7, 8]. This leads – for not too large basis sizes – to a set of discrete states with $E > 0$ which approximate the lowest resonances. Including them in the averaging, the plateau conditions (38) can usually be fulfilled reasonably well for not too small nuclei. This has been carefully studied for a realistic spherical Woods-Saxon potential in Ref.[36]. As a rather extreme example where one has to go relatively high correction orders $2M$ to obtain convergence of (38), we show in Figure 8 the neutron shell-correction δE for $N = 170$.

A more rigorous inclusion of the resonances in the continuum by a calculation of the corresponding phase shifts[48] has shown that, indeed, very good plateaux can be found if resonances up to sufficiently high energies $E > 0$ are included.

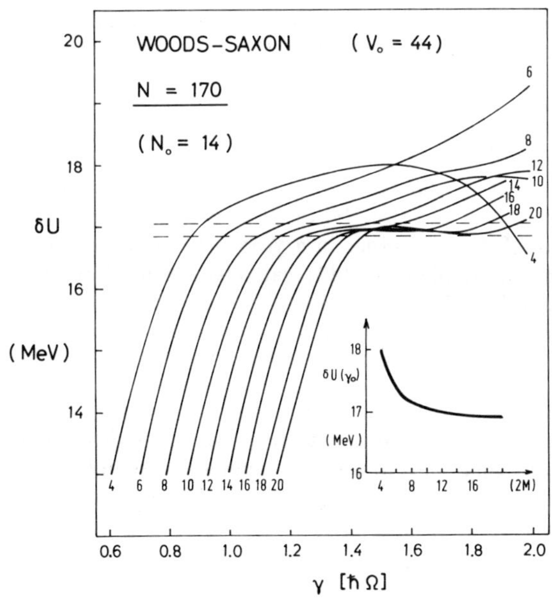

Fig. 8. Shell-correction energy δE versus γ for 170 neutrons in a spherical Woods-Saxon potential for various values of $2M$. The two dashed horizontal lines give the estimated error bars for the optimal plateau value. (Note that the semiclassical ETF energy discussed in Sect. 3.5 lies within these error bars.) The insert shows the plateau values versus curvature-correction order[36].

3.4. Averaged Occupation Numbers

Since Eq. (29) for the Strutinsky-averaged level density can be written in the form

$$\tilde{g}_M(E) = \sum_i \tilde{g}_i(E), \tag{41}$$

it is straightforward to define *averaged occupation numbers*[7, 43] by

$$\tilde{n}_i = \int_{-\infty}^{\tilde{\lambda}} \tilde{g}_i(E)\, dE \tag{42}$$

such that

$$N = \sum_i \tilde{n}_i. \tag{43}$$

[See Figure 5 for a schematic plot of the integral of $f(x)$ appearing in the explicit evaluation of Eq. (42).] It may be shown by algebraic manipulations[43] that, independently of the averaging function $f(x)$, the quantity $\widetilde{E}_{sp}$ in Eq. (32) fulfils the following differential equation:

$$\widetilde{E}_{sp} = \sum_i \epsilon_i \tilde{n}_i + \gamma \frac{d\widetilde{E}_{sp}}{d\gamma}. \tag{44}$$

If the plateau condition (38) can be fulfilled, one can therefore write the shell-correction energy δE in a particularly attractive form:

$$\delta E = \sum_i \epsilon_i (n_i^{HF} - \tilde{n}_i) = \sum_i \epsilon_i \delta n_i. \tag{45}$$

The form (45) of the shell correction shows that it only depend on the s.p. levels near the Fermi energy: far above or below it, the averaged occupation numbers tend towards the HF ones and thus δn_i tends to zero. This is illustrated in Figure 9 for a realistic case of $Z = 94$ protons in a Woods-Saxon potential.

The occupation numbers $\tilde{n}_i$ (42) can also be used to define Strutinsky-averaged nucleon densities[7]

$$\tilde{\rho}(r) = \sum_i |\varphi_i(r)|^2\, \tilde{n}_i \tag{46}$$

or the density matrices $\tilde{\rho}_{\alpha\beta}$ (8) used in Sect. 2.1.

Eq. (44) suggests an interesting analogy[31] of the Strutinsky averaging to the temperature averaging appropriate to a canonical ensemble of (non-interacting) fermions. For the latter, the free energy F is well-known to fulfil the differential equation

$$F = \sum_i \epsilon_i n_i(T) + T\frac{dF}{dT}, \tag{47}$$

where T is the temperature in energy units ($k \equiv 1$), $n_i(T)$ are the Fermi occupation numbers

$$n_i(T) = \frac{1}{1 + \exp\left(\frac{\epsilon_i - \mu}{T}\right)}, \tag{48}$$

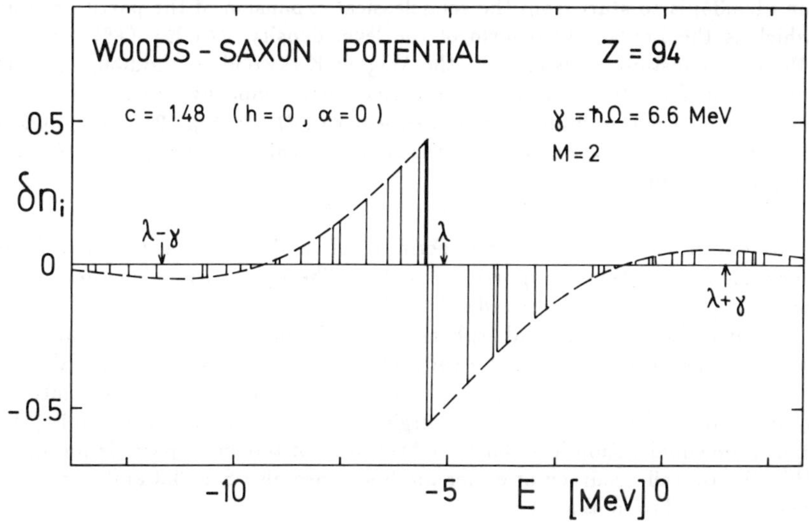

Fig. 9. Occupation number difference $\delta n_i = n_i^{HF} - \widetilde{n}_i$ versus energy for a Woods-Saxon potential corresponding to the $Z = 94$ protons of the Pu fission isomer[7]. The vertical lines indicate the positions of the proton s.p. levels ϵ_i.

μ is the chemical potential and the derivative in Eq. (47) gives the negative entropy S:

$$\frac{dF}{dT} = -S. \tag{49}$$

We thus have the following one-to-one correspondence:

$$\begin{aligned} \widetilde{E}_{sp} &\longleftrightarrow F \\ \gamma &\longleftrightarrow T \\ \frac{d\widetilde{E}_{sp}}{d\gamma} &\longleftrightarrow -S. \end{aligned} \tag{50}$$

This shows that the Strutinsky averaging is very similar to that brought about by a finite temperature – however, with one important difference: it leaves the system unexcited since, due to the curvature corrections (with $M > 0$), the "effective entropy" $-d\widetilde{E}_{sp}/d\gamma$ can be kept to zero by the plateau condition (38).

The close relation of Strutinsky and temperature averaging has been exploited in the so-called "temperature method"[38] for extracting the smooth part of the s.p. energy sum (1) in an alternative way.

3.5. Relation to the Extended Thomas-Fermi Model

Bhaduri and coworkers[34] proposed to use the extended Thomas-Fermi (ETF) model as a semiclassical alternative to Strutinsky averaging. In fact, the close relation between the two methods had been pointed out early[1, 7, 49]. The main idea[34] is to start from the semiclassical expansion of the partition function – which is the Laplace transform of the level density $g(E)$ Eq. (18) – in powers of Planck's constant $\hbar$, using e.g. the Wigner-Kirkwood method[50]. By inverse Laplace transform term by term one arrives at a semiconverging asymptotic expansion of $g(E)$ whose finite part is identical to $g_0(E)$ in Eq. (21). By integration analogously to Eq. (19) one obtains the semiclassical part of E_{sp} in the form of an asymptotic expansion

$$E_{ETF} = E_{TF} + E_2 + E_4 + \ldots.. \tag{51}$$

The first term is the pure Thomas-Fermi result and the terms E_m come from the semiclassical corrections proportional to $\hbar^m$. The series (51) converges very fast; usually the term E_4 is only of order 1 - 2 MeV.

For a three-dimensional (spherical or deformed) harmonic oscillator potential, E_{ETF} has been shown analytically[43, 44] to be identical to the Strutinsky-averaged energy $\widetilde{E}_{sp}$ given by Eq. (32). For realistic Woods-Saxon potentials including spin-orbit interaction, the two energies have been compared carefully[51] and found to be equal within less than ~ 1 MeV for not too small particle numbers, i.e. within the overall accuracy of either method. (See also Sec. 3.3 and Fig. 7 for the case of a cubic box potential.)

The ETF model may also be used in semiclassical density variational calculations[52, 53] to obtain selfconsistent average nuclear properties from a given effective nucleon-nucleon interaction. The resulting binding and deformation energies have been shown[52] to be very close to the Strutinsky-averaged HF energies (11).

In a series of papers[54], the variational ETF model has been combined with the Strutinsky method to develop a microscopically based mass formula which has been successfully applied also to large deformations and nuclei far from β stability.

4. Some Recent Applications

Strutinsky's shell-correction method – by some authors also termed the "macroscopic-microscopic" method – has been extremely successful in calculations of nuclear masses and deformation energies, in particular fission barriers. We refer to the presentation of R. Wyss[55] for some detailed examples and comparisons of the results, also in applications to high-spin physics. The status of fission barrier calculations with the SCM and the HF(B) method in early 1979 was summarized in Ref.[31]. Very recently, a critical assessment of nuclear mass models employing the SCM was given by Möller and Nix[56], covering also the newest results[57] of the ETFSI model by Pearson and coworkers[54].

We shall conclude this little review by mentioning two applications of Strutinsky's ideas in other domains of physics.

Metal clusters have recently obtained much attention – not least because of their striking similarity to nuclei. (See some recent review articles[58, 59] for the detailed literature.) Pronounced shell effects can be observed in mass yields and ionization potentials and suggest that the valence electrons can be described to a good approximation in a mean-field approximation. Since metal clusters can be made neutral, there is no limitation to their size, in contrast to nuclei. Accordingly, one has been able to observe "magic numbers" corresponding to filled major shells up to $N \sim 3000$. Most spectacularly, the so-called "super-shell" structure has been observed[60], which already was obtained theoretically by Balian and Bloch[40] in the level density of a spherical box. For metal clusters, microscopical calculations with a phenomenological Woods-Saxon potential[61] or with the self-consistent jellium model[62] give, indeed, a beating pattern in the level density and the shell-correction energy of metal clusters which is in good agreement with the experimental observations. The Strutinsky energy theorem has been used as a technical means for simplifying the rather time consuming selfconsistent microscopical calculations at finite temperatures[63]. A direct use of the SCM, renormalizing the wrong surface energy of the jellium model to its experimental value, has been made in calculations for deformation energies of large clusters[64]. As an example of the results, we show in Figure 10 the shell-correction energy δE for Na clusters with up to $N = 850$ atoms, evaluated at their axially symmetric ground states. The deepest minima correspond to spherical clusters with atomic numbers 58, 92, 138, 186, 254, 338, etc., whereas the smaller minima marked by arrows correspond to oblate or prolate ground-state shapes. Most of the spherical and many of the deformed magic numbers have been experimentally seen within the statistical error bars of the extracted fluctuating part of the mass yields.

Finally we mention that the Strutinsky energy averaging has been found very useful[65] for extracting the oscillating part $\delta g(E)$ of the quantum-mechanical density

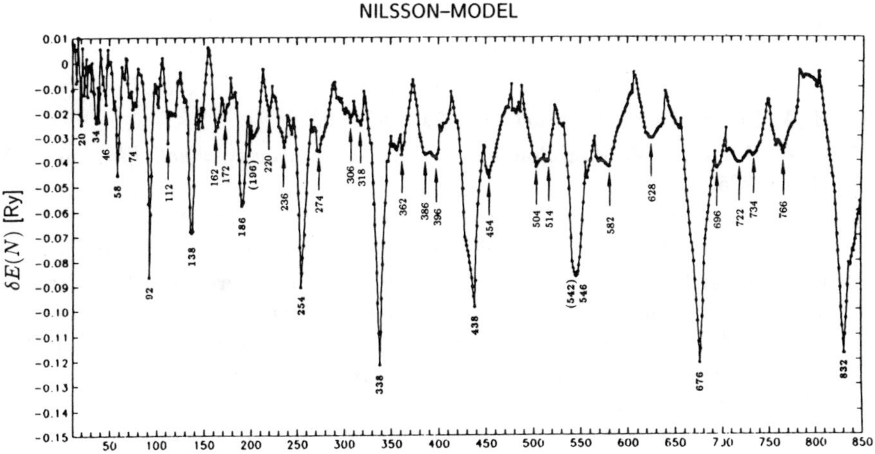

Fig. 10. Shell-correction energies δE of Na clusters versus number N of atoms (or valence electrons) at the corresponding (axially symmetric) ground-state deformations[64].

of states in the Hénon-Heiles potential[66] which has been intensively studied in connection with classical chaos (see, e.g., Ref.[67]). An interesting beating pattern in $\delta g(E)$ was found[65], even at energies where the classical orbits are mostly chaotic. This can be interpreted in terms of Gutzwiller's periodic orbit theory[39] – similarly to the super-shell beating just mentioned above – and demonstrates that long-range order in the energy spectrum can coexist with the short-range correlations that are typical for chaotic behavior.

5. References

1. V. M. Strutinsky, *Yad. Fiz.***3** (1966) 614 [*Sov. J. Nucl. Phys.* **3** (1966) 449]; *Ark Fys.* **36** (1966) 629; *Nucl. Phys.* **A 95** (1967) 420; *ibid.* **A 122** (1968) 1.
2. S. G. Nilsson, *Mat. Fys. Medd. Dan. Vid. Selsk.* **29** (1955) no. 16; B. Mottelson and S. G. Nilsson, *Phys. Rev.* **99** (1955) 1615.
3. S. A. E. Johansson, *Nucl. Phys.* **22** (1962) 529; see also *ibid.* **12** (1959) 449; D. R. Bès and Z. Szymański, *Nucl. Phys.* **28** (1963) 42; C. Gustafson, I.-L. Lamm, B. Nilsson and S. G. Nilsson, *Ark Fys.* **36** (1966) 613.
4. S. M. Polikanov *et al.*, *Soviet Phys. JETP* **15** (1962) 1016.
5. W. D. Myers and W. J. Swiatecki, *Nucl. Phys.* **81** (1966) 1.
6. S. G. Nilsson, C. F. Tsang, A. Sobiczewski, Z. Szymański, S. Wycech, C. Gustafson, I.-L. Lamm, P. Möller and B. Nilsson, *Nucl. Phys.* **A 131** (1969) 1.
7. M. Brack, J. Damgård, A. S. Jensen, H. C. Pauli, V. M. Strutinsky, and C. Y. Wong, *Rev. Mod. Phys.* **44** (1972) 320.
8. M. Bolsterli, E. O. Fiset, J. R. Nix and J. L. Norton, *Phys. Rev.* **C 5** (1972) 1050; J. R. Nix, *Ann. Rev. Nucl. Sci.* **22** (1972) 65.
9. V. M. Strutinsky, *Nucl. Phys.* **A 218** (1974) 169.
10. H. A. Bethe, *Ann. Rev. Nucl. Sci.* **21** (1971) 93.
11. G. G. Bunatian, V. M. Kolomietz, and V. M. Strutinsky, *Nucl. Phys.* **A 188** (1972) 225.
12. L. Wilets, in *Physics and Chemistry of Fission 1969* (IAEA, Vienna, 1969), p. 179; W. H. Bassichis and L. Wilets, *Phys. Rev. Lett.* **27** (1971) 1451.
13. W. H. Bassichis, A. K. Kerman, C. F. Tsang, D. R. Tuerpe and L. Wilets, in *Magic without magic: John Archibald Wheeler, a collection of essays in honor of his 60th birthday*, ed. J. R. Klauder (Freeman, San Francisco, 1972), p. 15.
14. K. Dietrich, in *Nuclear self-consistent fields*, ed. G. Ripka and M. Porneuf (North-Holland, Amsterdam, 1975), p. 373.
15. M. Brack, in *Proceedings of the International School of Nuclear Physics 1974 at Predeal (Romania)*, (Bukarest, 1975).
16. M. Brack, and P. Quentin, *Nucl. Phys.* **A 361** (1981) 35.
17. H. Flocard, P. Quentin, D. Vautherin, M. Vénéroni and A. K. Kerman, *Nucl. Phys.* **A 231** (1974) 176.

18. L. Landau, *Sov. Phys. JETP* **3** (1957) 920.

19. V. M. Kolomietz, *Sov. J. Nucl. Phys.* **18** (1974) 147.

20. J. P. Perdew, in *Density Functional Methods in Physics*, ed. R. M. Dreizler and J. da Providência (Plenum Press, New York, 1985), p. 309.

21. A. R. Mackintosh and O. K. Andersen, in *Electrons at the Fermi Surface*, ed. M. Springford (Cambridge University Press, 1980).

22. T. H. R. Skyrme, *Phil. Mag.* **1** (1956) 1043; *Nucl. Phys.* **9** (1959) 615; D. Vautherin and D. Brink, *Phys. Rev.* **C 5** (1972) 626.

23. For a review on HF calculations with Skyrme (and other) effective interactions, see P. Quentin and H. Flocard, *Ann. Rev. Nucl. Part. Sci.* **28** (1978) 523.

24. D. Vautherin, *Phys. Rev.* **C 7** (1973) 296; H. Flocard, P. Quentin, A. K. Kerman and D. Vautherin, *Nucl. Phys.* **A 203** (1973) 433.

25. J.-F. Berger, these Proceedings.

26. M. Brack and P. Quentin, in *Physics and Chemistry of Fission 1973* (IAEA Vienna, 1974), Vol. I, p. 231.

27. M. Brack and P. Quentin, in *Nuclear self-consistent fields*, ed. G. Ripka and M. Porneuf (North-Holland, 1975), p. 353; *Phys. Lett.* **56B** (1975) 421.

28. M. Beiner, H. Flocard, Nguyen van Giai and P. Quentin, *Nucl. Phys.* **A 238** (1975) 29.

29. W. D. Myers and W. J. Swiatecki, *Ann. Phys. (N.Y.)* **55** (1969) 395.

30. H. C. Pauli, *Physics Reports* **C 7** (1973) 35.

31. M. Brack, in *Physics and Chemistry of Fission 1979* (IAEA Vienna, 1980), Vol. I, p. 227.

32. V. M. Strutinsky, in *Physics and Chemistry of Fission 1979* (IAEA Vienna, 1980), Vol. I, p. 475.

33. E. Werner, K. Dietrich, P. Möller and J. R. Nix, in *Physics and Chemistry of Fission 1979* (IAEA Vienna, 1980), Vol. I, p. 501.

34. R. K. Bhaduri and C. K. Ross, *Phys. Rev. Lett.* **27** (1971) 606; B. K. Jennings, *Nucl. Phys.* **A 207** (1973) 538; B. K. Jennings, *Ann. Phys. (N.Y.)* **84** (1974) 1.

35. A. Bohr and B. Mottelson, *Nuclear Structure II* (Benjamin, New York, 1975).

36. A. Sobiczewski, A. Gyurkovich and M. Brack, *Nucl. Phys.* **A 289** (1977) 346.

37. V. M. Strutinsky and F. A. Ivanyuk, *Nucl. Phys.* **A 255** (1975) 405; F. A. Ivanyuk and V. M. Strutinsky, *Z. Phys.* **A 286** (1978) 291; *ibid.* **A 290** (1979) 107.

38. V. S. Ramamurthy, S. S. Kapoor and S. K. Kataria, *Phys. Rev. Lett.* **25** (1970) 386 and *Phys. Rev.* **C 5** (1972) 1124; R. Bengtsson, *Nucl. Phys.* **A 198** (1972) 591; R. K. Bhaduri and S. Das Gupta, *Phys. Lett.* **47 B** (1973) 129.

39. M. C. Gutzwiller, *J. Math. Phys.* **10** (1969) 1004; *ibid.* **11** (1970) 1791; *ibid.* **12** (1971) 343.
40. R. Balian and C. Bloch, *Ann. Phys. (NY)* **69** (1972) 76.
41. V. M. Strutinsky, *Nukleonika (Poland)* **20** (1975) 679.
42. V. M. Strutinsky, A. G. Magner, S. R. Ofengenden, and T. Døssing, *Z. Phys.* **A 283** (1977) 269.
43. M. Brack, *PhD thesis*, Universität Basel, 1972 (unpublished); see also M. Brack and H. C. Pauli, *Nucl. Phys.* **A 207** (1973) 401.
44. B. K. Jennings, *PhD thesis*, McMaster University, 1976 (unpublished).
45. R. R. Chasman, *Phys. Rev. Lett.* **33** (1974) 544.
46. D. L. Hill and J. A. Wheeler, *Phys. Rev.* **89** (1953) 1102.
47. R. Balian and C. Bloch, *Ann. Phys. (N.Y.)* **60** (1970) 401; *ibid.* **64** (1971) 271.
48. C. K. Ross and R. K. Bhaduri, *Nucl. Phys.* **A 188** (1972) 566.
49. A. S. Tyapin, *Sov. J. Nucl. Phys.* **11** (1970) 401; *ibid.* **14** (1972) 50; D. H. E. Gross, *Phys. Lett.* **42 B** (1972) 41.
50. E. P. Wigner, *Phys. Rev.* **40** (1932) 749; J. G. Kirkwood, *Phys. Rev.* **44** (1933) 31.
51. B. K. Jennings, R. K. Bhaduri and M. Brack, *Phys. Rev. Lett.* **34** (1975) 228; *Nucl. Phys.* **A 253** (1975) 29.
52. M. Brack, C. Guet, and H.-B. Håkansson, *Physics Reports* **123** (1985) 275.
53. J. Treiner and H. Krivine, *Ann. Phys. (N.Y.)* **170** (1986) 406.
54. A. K. Dutta, J.-P. Arcoragi, J. M. Pearson, R. Behrman and F. Tondeur, *Nucl. Phys.* **A 458** (1986) 77; F. Tondeur, A. K. Dutta, J. M. Pearson and R. Behrman, *Nucl. Phys.* **A 470** (1987) 93; J. M. Pearson, Y. Aboussir, A. K. Dutta, R. C. Nayak and M. Farine, *Nucl. Phys.* **A 528** (1991) 1.
55. R. Wyss, these Proceedings.
56. P. Möller and J. R. Nix, in *6th International Conference on Nuclei Far from Stability and 9th International Conference on Atomic Masses and Fundamental Constants (AMCO)*, to be published (Los Alamos Preprint LA-UR-92-2150).
57. Y. Aboussir, J. M. Pearson, A. K. Dutta and F. Tondeur, *Nucl. Phys.* **A** (1992), in print.
58. W. A. de Heer, W. D. Knight, M. Y. Chou, and M. L. Cohen, in *Solid State Physics,* Vol. **40**, edited by H. Ehrenreich and D. Turnbull (Academic, New York, 1987), p. 93.
59. W. A. de Heer, M. Brack, *Rev. Mod. Phys.*, in preparation.
60. J. Pedersen, S. Bjørnholm, J. Borggreen, K. Hansen, T. P. Martin and H. D. Rasmussen, *Nature* **353** (1991) 733.
61. H. Nishioka, K. Hansen and B. R. Mottelson, *Phys. Rev.* **B 42** (1990) 9377.

62. O. Genzken, and M. Brack, *Phys. Rev. Lett.* **67** (1991) 3286.

63. M. Brack, O. Genzken, and K. Hansen, *Z. Phys.* **D 21** (1991) 65.

64. S. Reimann, *Diploma Thesis*, Universität Regensburg (unpublished); S. Reimann and M. Brack, to be published.

65. M. Brack, R. K. Bhaduri, J. Law and M. V. N. Murthy, submitted to *Phys. Rev. Lett.* (McMaster Preprint MCMA-TP-92-03).

66. M. Hénon and C. Heiles, *Astron. J.* **69** (1964) 73.

67. M. Tabor, *Chaos and Integrability in Nonlinear Dynamics* (John Wiley, New York, 1989).

THE SHELL CORRECTION METHOD

Evaluation by Working Group 4

S.A.MOSZKOWSKI
Department of Physics, UCLA, 405 Hilgard Avenue
Los Angeles, CA 90024, USA

D.M.CULLEN
Physics Division, ORNL, Bldg 6000, MS 6371
Oak Ridge, TN 37831, USA

and

J.-F.BERGER
Commissariat à l'Energie Atomique
Service PTN, Centre d'Etudes de Bruyères-le-Châtel
B.P.12, 91680 Bruyères-le-Châtel, France

ABSTRACT

After recalling the principle of the shell correction method based on the Strutinsky energy averaging procedure, different advantages and limitations of this technique are presented. The alternative way of defining average energies based on the extended Thomas-Fermi method is also discussed. A remark concerning the similarity of the average field in metal clusters and in nuclei is given at the end.

1. The Shell Correction Method

The Shell Correction Method (SCM) is essentially a technique for computing the total binding energy of a nucleus in its ground state or as a function of deformation. A very clear presentation of the SCM was given at this Workshop by Matthias Brack. It was divided into three parts:

1.1. Strutinsky Energy Theorem

Strutinsky showed that the total energy E of a nucleus can be split into one part $\tilde{E}$ that varies smoothly with particle number and other global nuclear characteristics, and an oscillatory part δE:

$$E \;=\; \tilde{E} \;+\; \delta E \tag{1}$$

The original theoretical proof of this theorem was given in the mean-field context (Hartree-Fock theory) [1]. The smooth part of the energy is defined by means of averaged occupation numbers $\tilde{n}_i$, and δE is identified as the first order correction to the self-consistent energy.

Numerical tests were first performed using this method with the Skyrme III interaction [2]. They indicate that the second order correction

$$\delta_2 E \;=\; E \;-\; \tilde{E} \;-\; \delta E$$

is always small ($\delta_2 E < 1MeV$), except in "small" systems such as ^{40}Ca. Further tests including a BCS treatment of pairing correlations or employing other effective interactions essentially confirmed this result.

In practice the Strutinsky energy theorem is applied by approximating $\tilde{E}$ with a phenomenological liquid drop model energy and by replacing the self-consistent Hartree-Fock field by a phenomenological shell model potential. The most widely used forms of potentials have been modified harmonic oscillator, Woods-Saxon and folded Yukawa potentials. In this phenomenological context, an efficient way of extracting the shell correction δE is to use the energy averaging technique invented by Strutinsky.

1.2. Strutinsky Energy Average Method

The Strutinsky method firstly consists of deriving a smooth component $\tilde{g}(\epsilon)$ from the one-particle level density

$$g(\epsilon) = \sum_i \delta(\epsilon - \epsilon_i), \tag{2}$$

by convoluting $g(\epsilon)$ with a well-chosen function $f_\gamma(\epsilon)$ having a spreading width γ. The main idea of Strutinsky was to realize that, in order to make the smoothing procedure independent of γ over a broad range of values (the "plateau condition"), $f_\gamma(\epsilon)$ must include so-called "curvature corrections" [3]. This is done by taking for $f_\gamma(\epsilon)$ the product of a spreading function of width γ (usually a gaussian) with a polynomial of order M. The polynomial is chosen in such a way that the first M moments of the $\tilde{g}(\epsilon)$ distribution are identical to those of $g(\epsilon)$.

The shell correction δE is then calculated as

$$\delta E = \sum_i \epsilon_i - \tilde{E}_{sp} \tag{3}$$

where $\tilde{E}_{sp}$ is the averaged total energy computed using $\tilde{g}(\epsilon)$.

1.3. Alternative Methods

Among several alternative methods to the Strutinsky technique, the most interesting one is based on the extended Thomas-Fermi (ETF) method [4]. Its main advantage is to define a microscopic energy functional from which both the smooth part of the total energy and the average nuclear field can be computed . A second advantage of ETF averaging is the possibility of including pairing effects and nuclear temperature in the calculation of smoothed quantities. This clearly cannot be done in the framework of the usual liquid drop model.

2. Discussion

The discussion that follows summarizes the opinion of the authors as well as the many observations and comments made by the Workshop participants during the presentation of the evaluation of the SCM.

Let us first review the main advantages of the traditional SCM, i.e. the method based on the liquid drop model and the Strutinsky energy average method.

i) The most obvious advantage of the SCM is its extreme computing efficiency, especially compared with the usual self-consistent calculations of the Hartree-Fock type;

ii) it is easy to implement, since the ingredients of the method, liquid drop and shell model parameters are known quantities;

iii) it leads to fairly accurate and reliable results (provided the nucleus is not magic or too light, see below). An indication of this is provided by the very similar results derived with different forms of shell model potentials, as well as by comparisons with other approaches of nuclear structure;

iv) as a consequence, very useful, extensive and reliable systematics about nuclear masses, deformations, fission barriers, ... have been derived with the SCM, including both stable and exotic nuclear species.

The shortcomings of the SCM follow from its inherent simplicity:

i) the liquid drop energy and shell model potential depend on many parameters some of them being not very well defined. For instance the curvature coefficient a_c of the liquid drop energy usually is assumed to be zero in mass and fission barrier fits, while a value of the order of 10 MeV is found from self-consistent approaches;

ii) there may exist inconsistencies between the liquid drop description and the shell model one. For instance, the nucleon density distribution computed with the shell model orbitals may differ significantly from the one assumed in the liquid drop model;

iii) the plateau condition may be difficult to fulfill when finite depth potentials (Woods-Saxon, folded Yukawa) are used. The usual way of performing the shell correction in this case is to include in the one-particle density only the quasi-bound states lying in the continuum. However, many of such states seem to be necessary to obtain "good" plateaux;

iv) the SCM is not reliable for nuclei having a particle number A lower than 40 or 50;

v) difficulties are met when the level density at the Fermi energy is low. The shell correction δE is then much too large, for example the so-called "lead anomaly". One possible origin of this discrepancy may be the lack of self-consistency between the shell model potential and the liquid drop in this particular case [5]. Another origin has been suggested by Krishna Kumar [6] who pointed out that the shell correction should in principle include besides the usual term, a component coming from the nucleus kinetic energy. This additional component is large and opposite to the usual one in magic nuclei, resulting in a significantly reduced total shell correction.

Many of these difficulties in the conventional SCM can be corrected by employing the ETF averaging method. Some advantages of the latter have already been given in Section 1.3:

i) the smooth part of the energy and the average nuclear field are evaluated with the same formalism. Hence the consistency between the liquid drop-like description and the average potential is automatically ensured;

ii) since the ETF method can be extended to include pairing and temperature, this form of SCM is more general than the usual one.

One may also add that:

iii) there is no plateau condition to fulfill when ETF is employed;

iv) the equivalence of ETF averaging with the usual Strutinsky procedure has been demonstrated both analytically and numerically in standard cases [7,8].

It is worthwhile to mention that, in spite of its advantages, the ETF method is not completely free from difficulties. Some of them are of a practical nature:

i) ETF averaging clearly is numerically less simple than the conventional SCM;

ii) divergences sometimes appear in integrals involving densities because the latter diverge at the classical turning points.

Other difficulties have a more fundamental character:

iii) the ETF equations cannot be derived from a variational principle as for instance the Hartree-Fock theory, because nucleon densities are not defined at and beyond the classical turning points. As a consequence energies found with ETF generally are too low;

iv) ETF does not adequately describe some effects contributing to the smooth part of the energy. One example is the influence of the Friedel oscillations appearing in the density, near the surface of semi-infinite systems. These oscillations must have an effect on the smooth component of the energy, since the one-particle spectrum is continuous and the shell correction is identically zero. However, the ETF method does not account for Friedel oscillations, and therefore ignore their contribution to the energy. Let us note that this difficulty of describing density oscillations of the Friedel type in a classical or semi-classical context may explain the differences in the estimated curvature coefficient a_c mentioned above.

To complete this discussion, let us finally mention one question that frequently arises in most contemporary nuclear structure applications of the SCM, namely the problem of how to incorporate nuclear excitations in this formalism.

i) One way is to model nuclear excitations by means of a temperature parameter. The ETF method can be employed for performing the energy averaging. This technique clearly is appropriate for high enough excitation energies and/or medium and heavy systems;

ii) individual excitations of the particle-hole type can also be incorporated by first applying the SCM to the ground state and then adding their contributions to the excitation energy;

iii) extending the SCM to collective dynamics on the other hand remains a difficult problem. For instance one may think of how to correct the collective energy of a vibrating or rotating liquid drop for quantum effects. Several attempts in this direction have been made, in particular by Strutinsky [9].
A special case of this situation occurs in cranking calculations. There one tries to obtain the structure and spectrum of high spin rotational bands from a nuclear hamiltonian of the form $h_c = h - \omega j_x$ where j_x is the x-component of the total angular momentum and ω an angular frequency. The SCM allows one to compute the smooth part of h from the smooth part of the density ρ for every value of ω. However, to smooth the second component of h_c it is necessary to be able to define some averaged moment of inertia, which cannot be done unambiguously.

3. Conclusion

To conclude this discussion, one may say that, despite several limitations, the Strutinsky SCM undoubtedly is the most efficient technique ever invented for retrieving or predicting the total energy of nuclear systems. The essential virtue of the method is its simplicity. For this reason it will remain an extremely powerful tool of nuclear structure studies as long as such a simplicity is preserved. For instance, it would be illusory to try to improve the possibilities of the method by introducing many collective parameters. This would lead to much larger computing times, that would be better employed for use in a more fundamental approach. In this respect the ETF formalism certainly represents an interesting way of generalizing the traditional SCM without going to the large size, self-consistent calculations of the Hartree-Fock type. Let us add that the SCM and its extensions can be applied not only to nuclear physics but also to other domains, for instance metal clusters.

4. An additional remark concerning metal clusters

A very interesting point of Matthias Brack's talk was devoted to metal clusters. Since this topic is not directly related to the shell correction method, we just append here a remark based on a recent paper of Cleminger [10], showing the very similar structure between the average field in metal clusters and in nuclei.

It is well known that valence electrons in metal clusters appear to move rather independently in the mean field generated by the background ions and the other electrons. This interpretation followed the experimental discovery of shells and super-shells in these systems. The mean field felt by the electrons can be approximated by a Woods-Saxon potential

$$V(r) = \frac{-U_0}{1+e^{\frac{r-R_0}{a}}}, \tag{4}$$

where R_0 is a characteristic cluster radius proportional to the Wigner-Seitz radius r_S and to the cubic root of the number of valence electrons in the cluster. In Na clusters for example, one has $U_0 = 6.0\ eV$ and the electron Fermi energy is $\epsilon_F = 3.24\ eV$. One notes that $U_0 \simeq 2\epsilon_F$, which can be understood from the virial theorem applied to particles interacting via the Coulomb force ($<U>=-2<T>$).

Now, according to Cleminger's paper, the surface thickness of the potential in metal clusters is approximately

$$w = \frac{3}{8}\lambda_F,$$

where λ_F is the Fermi wavelength. Equating this to the surface thickness

$$w = U_0/\left|\frac{dV}{dr}\right|_{r=R_0} = 4a$$

of the Woods-Saxon potential (4) yields

$$a/\lambda_F = 3/32$$

From this, one concludes that the dimensionless ratio,

$$\chi = \frac{U_0}{\epsilon_F}\cdot\frac{a}{\lambda_F},$$

should always be of the order of $3/16 = 0.1875$. Actual parameters valid for Na clusters in fact yield

$$\chi = \frac{6.0\ eV}{3.24\ eV}\cdot\frac{0.74\ \mathring{A}}{6.81\ \mathring{A}} = 0.201.$$

Evaluating the same ratio in typical nuclei, one obtains

$$\chi = \frac{50.0\ MeV}{37\ MeV}\cdot\frac{0.67\ fm}{4.62\ fm} = 0.196,$$

a value surprisingly close to the one found in Na metal clusters !

References

1. V. M. Strutinsky, *Yad. Fiz.* **3** (1966) 614 [*Sov. J. Nucl. Phys.* **3** (1969) 449].
2. M. Brack and P. Quentin, in *Physics and Chemistry of Fission 1973*, Rochester, (IAEA, Vienna, 1974), Vol.I, p. 231.
3. V. M. Strutinsky, *Nucl. Phys. A* **95** (1967) 420; *ibid.* **122** (1968) 1.
4. R. K. Bhaduri, and C. K. Ross, *Phys. Rev. Lett.* **27** (1971) 606; B. K. Jennings, *Nucl. Phys. A* **207** (1973) 538; M. Brack, C. Guet, and H.-B.Håkansson, *Phys. Rep.* **123** (1985) 275.

5. M. Brack, in *Physics and Chemistry of Fission 1979*, Jülich, (IAEA, Vienna, 1980), Vol.I, p. 227.
6. K. Kumar, *Superheavy Elements* (Adam Hilger, Bristol and New York, 1989), p. 63.
7. M. Brack and H. C. Pauli, *Nucl. Phys. A* **207** (1973) 401.
8. B. K. Jennings, R. K. Bhaduri, and M. Brack, *Phys. Rev. Lett.* **34** (1975) 228; *Nucl. Phys. A* **253** (1975) 29.
9. V. M. Strutinsky, *Nucl. Phys. A* **502** (1989) 67c.
10. K. Cleminger, *Phys. Rev. B* **44** (1991) 12991.

REBUTTAL

M. BRACK
University of Regensburg, Regensburg, Germany

I have nothing to add to the evaluation of Group 4 concerning the shell-correction method itself. However, I want to make a few remarks about the variational ETF method, partially correcting the statements made in point iii) of the 'difficulties of the ETF method'. (References and further details may be found in Ref. [52] of my lectures.)

One should differentiate two versions of the ETF model, commonly running under the same name:

1) The semiclassical expansion (e.g., à la Wigner-Kirkwood) of densities, energy etc. in powers of $\hbar$, such as studied in Refs. [34,44,51], for fermions in a given one-body potential V. Here one finds a divergence of the densities at the classical turning points (see also the 'difficulty ii') – however, only at $T = 0$! (see below).

2) When eliminating the Fermi energy, the potential V and its derivatives from the $\hbar$-expanded densities, one arrives at the functional $\tau_{ETF}[\rho]$ for the kinetic energy density (in which $\hbar$ does not appear explicitly). This functional may now be used in density variational calculations starting from an (effective) two-body interaction (in contrast to the $\hbar$ expansion starting from a one-body potential V). This version I like to call the 'variational ETF method'. Variation of the total energy with respect to the density leads to the so-called 'ETF equations' which therefore do have a solid variational basis, as generally shown in the framework of density functional theory. No divergencies arise here at the classical turning points, for reasons mentioned below. The fact that the functional $\tau_{ETF}[\rho]$ is only approximate may lead to the observed fact that the variational ETF energies often are lower than (averaged) HF energies.

Note that the functional $\tau_{ETF}[\rho]$ is *strictly valid also beyond the classical turning points, even at* $T = 0$*!* (The proof goes over the $T > 0$ extension of the $\hbar$ expansion, where there are no divergencies at the classical turning points, taking a careful limit $T \to 0$ [M. Brack, *Phys. Rev. Lett.* **54** (1985) 851].) Therefore the overbinding of the variational ETF energy has *a priori* nothing to do with the classical turning points. (Perhaps indirectly, as the ETF approximation to the *true* functional $\tau[\rho]$ presumably is worst near and beyond the classical turning points.)

APPLICATIONS TO THE NILSSON STRUTINSKY SCHEME

R. WYSS

Joint Institute for Heavy-Ion Research
Holifield Heavy Ion Research Facility
P.O. Box 2008, Oak Ridge, Tennessee 37831, U.S.A.
and
MSI
Frescativ. 24
S-104 05 Stockholm

ABSTRACT

Applications to the Nilsson-Strutinsky scheme are discussed for selected areas of nuclear structure calculations. Potential energy calculations of shape isomeric states based on two different liquid drop formulas are presented. The importance of bandhead calculations are addressed. The extension of the shell-correction method to high spin is shortly reviewed with examples of shape-coexistence and super deformation.

1. Introduction

Results from recent years nuclear spectroscopy reveal the tremendous richness of nuclear excitations. The shell-correction or Nilsson-Strutinsky (NS) aproach has proven to be a very powerful tool in the analysis and understanding of experimental nuclear structure data. The present paper does not aim to give justice to all calculations based on the Nilsson-Strutinsky aproach, but to show a few selected examples of general interest. For a recent review, the reader is referred to ref. 1. The foundations of the shell-correction method is well documented in ref. 2, see further the contribution by M. Brack in this issue.

The examples to be discussed are based on the Woods-Saxon (WS) model, developed by the Warszawa group, ref. 3 - 7. The WS potential employes one universal set of parameters for all nuclei and has been used extensively in nuclear structure calculations.

The first section is devoted to some general remarks about the NS scheme. The difference in the potential energy resulting from two different liquid drop formulas is discussed. The treatment of pairing correlations is also addressed. In section two, results of bandhead calculations are presented for some selected nuclei. Finally the extension of the NS scheme to high spin by means of the cranking equation is reviewed and demonstrated with two examples.

2. Potential Energy Calculations

Following the predictions by many authors,[8-15] the discovery of rotational bands built upon superdeformed shapes in the Hg- and Pb region,[16] lead to a renewed interest in potential energy calculations. In connection to fission isomers, similar calculations had been performed quite some time ago (see e.g. ref. 17 and the discussion on Pb-nuclei in ref. 2).

Early calculations were in general based entirely on the NS scheme, whereas the recent development of computer techniques has lead to an increased number of large scale calculations based on selfconsistent Hartree-Fock (HF) method.[14,15,18] In the Nilsson-Strutinsky (NS) aproach, the energy is obtained from the sum of the (macroscopic) liquid-drop energy and the (microscopic) Strutinsky shell-correction.

$$E(Z,N,\hat{\beta}) = E_{ld}(Z,N,\hat{\beta}) + \delta E_{shell}(Z,N,\hat{\beta}) + \delta E_{pair}(Z,N,\hat{\beta}), \tag{1}$$

The Energy of the nucleus $E(Z,N,\hat{\beta})$ is minimzed with respect to the deformation parameter $\hat{\beta}$, specifying the potential. The nuclear shape, which is determined by the potential, can be expanded in terms of multipoles. For axial symmetric shapes, the radius of the nucleus can be expressed by the following equation:

$$R(\hat{\beta},A,\theta) = R_{\circ}(A)c(\hat{\beta})\ [1 + \beta_2 Y_{20}(\theta) + \beta_3 Y_{30}(\theta) + \beta_4 Y_{40}(\theta) + ... \tag{2}$$

The constant $c(\hat{\beta})$ assures volume conservation and $R_{\circ} = r_{\circ}A^{1/3}$, $r_{\circ} = 1.2$ fm. Realistic computations have to be restricted in deformation space depending on the physics which is addressed. Calculations involving all odd multipoles of octupole type, $Y_{30}, Y_{31}, ...$ with quadrupole $Y_{20}, Y_{22}...$ have not been feasable up to now. In such cases, the NS apraoch looses its simplicity and selfonsistent HF calculations become more straight forward. On the other hand, since much of the physics lie either in quadrupole-hexadecapole type of deformation or can be restricted to axial symmetric shapes, many important inquieries connected to the shape degree of freedom can be addressed by NS type of calculations. The shape of SD nuclei in the A=150 or A=190 mass-region e.g., is calculated to be axial symmetric[19] and subsequent investigation can be limited accordingly.

The selected deformation space is input to NS type of calculations. The total energy of a given nucleus can be calculated on a grid, where the dimension of the grid is determined by the number of different multipoles to be minimized. If one considers the lowest three even multipoles, the energy is first minimized in each β_4 grid-point with respect to β_6 and in a subsequent step, the two-dimensional energy landscape is minimized at each β_2-value with respect to $\beta_4(\beta_6)$. When even higher order of multipoles or a combination of odd- and even are considered, the gradient method has to be used in order to obtain the lowest minima. It has been shown, that the inclusion of higher order multipoles as β_5 and β_6 play an important roll in lowering the octupole deformed minima in Actinide nuclei [20].

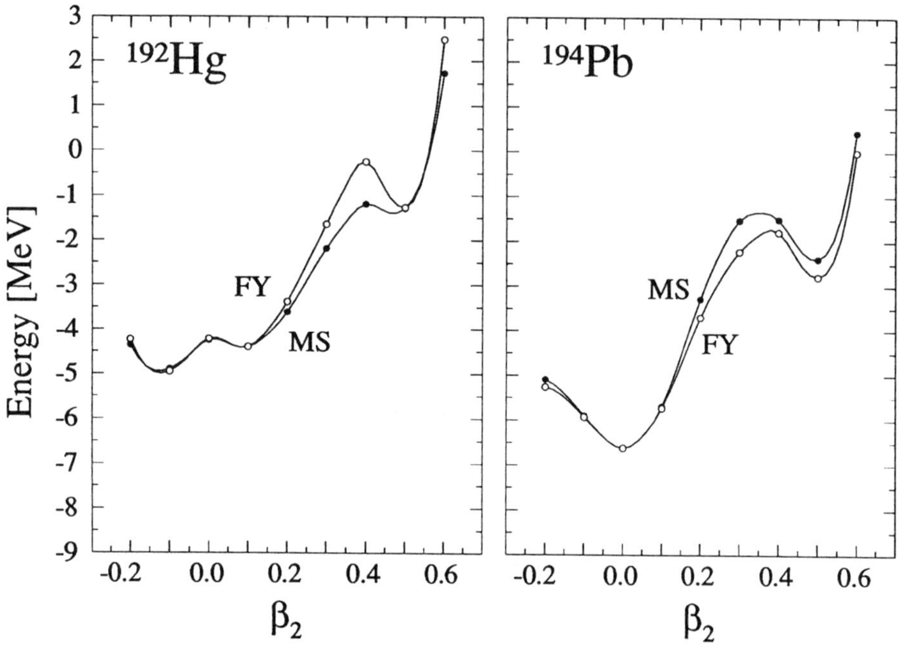

Figure 1: *Potential energy curve for ^{192}Hg and ^{194}Pb. The energy is minimized with respect to β_4 at each β_2 -value. (Calculations from ref. 21.)*

Common formulas for calculation of the liquid drop (LD) energy origin from Myers-Swiatecki(MS) [22,23] or are based on the folded Yukawa plus exponential(FY).[24] The MS-LD formula results in too high fission barriers, which are more accurately calculated by the FY-LD model.[24] Calculations of SD-shapes may therefore be more accurate if the FY-LD formula is chosen. Fig.1 compares results obtained from the two different LD formulas, which are discussed for a serie of isotopes in ref. 21. The calculated equilibrium quadrupole deformations obtained from the two formulas, are very similar in all nuclei considered. As expected, the largest difference occurs in the barrier region, where the level density is high, resulting in many level-crossings and accompanying large fluctuations. Accordingly, from the variation of the barrier height, one can not draw any systematic conclusions. the second minimum is in general always deeper when using the FY-LD formula. The FY-LD formula can describe any kind of shape, whereas the MS liquid drop formula breaks down for shapes with small necks.[24] This is refleced in larger values of the hexadecapole deformation in the second well for the FY-LD formula. The investigation of ref. 21 show, that the excitation energy of the second well in Hg- to Ra-nuclei is strongly model dependent. Results obtained from the Woods-Saxon model agree very well with results from Hartree-Fock calculations with the SKM* force[15] and HFB calculations with the Gogny force[25]. However, NS calculations based on the Nilsson-model[26] or selfconsistent HF calculations involving SIII yield much higher excitation energies.[15] The experimentally observed intensity of the bands suggests the excitation energy rather to be lower than higher, when compared to the WS calculations.[26] For more details, the reader is refered to ref. 21.

Similar to the shell-correction, the pairing correction δE_{pair} is constituted by the difference of the smooth Strutinsky part ($\tilde{E}_{pair}$) and the part obtained from the single particle potential (E_{pair}).

$$\delta E_{pair} = E_{pair} - \tilde{E}_{pair}. \tag{3}$$

The smooth pairing energy $\tilde{E}_{pair}$ has many times been given a constant value [17,27] and is then of no relevance in calculations of the energy surface. The pairing energy E_{pair} can be computed from the BCS theory, assuming e.g. a simple dependence of the monopole pairing force strength G,

$$G_N^Z \cdot A = 19.2 \pm 7.4 \frac{N-Z}{A} \quad [MeV], \tag{4}$$

see e.g. ref. 17. As already discussed in ref. 2, such an aproach has certain drawbacks, since G is not a measurable quantity and the pairing energy depends on the number of levels taken into account in the calculations.

A better etimate of the pairing strength G can be obtained from the average gap method,

$$G(\nu, \hat{\beta}) = [\frac{\sinh(\nu/2\rho\tilde{\Delta})}{\rho}], \tag{5}$$

where ρ is the average level density at the Fermi surface, determined by the Strutinsky smearing procedure, $\tilde{\Delta}$ the average pairing gap, fitted to the odd-even mass difference and ν the number of levels taken into account in the calculations.[28] The average pairing gap $\tilde{\Delta}$, and not G, is now the primary quantity determining the pairing energy, E_{pair}. Similarly, E_{pair} depends only weakly on the number of levels, ν, removing much of the ambiguities present in earlier aproaches. An extensive discussion about the pairing energy and the average gap method can be found in ref. 29.

The average gap method implies that no free parameters are used in the calculations of the pairing energy. However, the connection between odd-even mass-difference and the pairing gap is not very well founded. One can expect additional residual interactions to enter the odd-even mass difference, not only the monopole pairing force. The investigations of ref. 30 show that there is little evidence for BCS-blocking in the moment of inertia of odd-even nuclei. Similarly, in the fit of odd-even mass differences, one can not find any evidence for reduced pairing correlations in odd-even nuclei.[29] A microscopic understanding of the pairing interaction and its relation to the odd-even mass difference is still lacking.

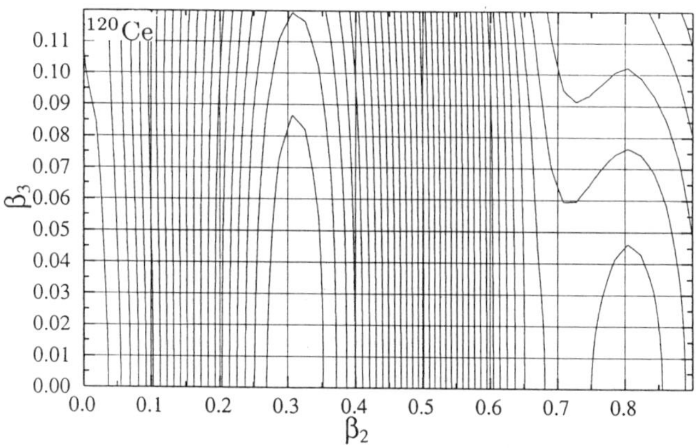

Figure 2: *Potential energy landscape for the doubly magic hyperdeformed nucleus* ^{120}Ce. *At each* β_2, β_3 *point, the energy is minimized with respect to* $\beta_4, \beta_5, \beta_6, \beta_7, \beta_8$ *and* β_{10}. *The energy difference between contour lines is 250keV. The rotational band built upon the hyperdeformed minimum is calculated to cross the normal deformed states around spin* $\approx 55\hbar$.

The differences in the total energy, resulting from either use of the pairing strength G (eq. 4,5), has been addressed for the case of ^{192}Po.[21] In the calculations of ref. 27, where the pairing energy is calculated according to eq.(4), the Po isotopes exhibit a third well around $\beta_2 \approx 0.8$ in adddition to the second well at $\beta_2 \approx 0.5$. The third well appeares in the NS-WS calculations only as a slight shoulder.[21] (The NS-WS calculations are based on the average gap method, employing the Lipkin-Nogami appraoch.) However, when reducing the pairing strength G, a minimum at hyperdeformed shape occurs also in the NS-WS calculations. The shell-effects are smoothed out by the pairing interaction, as discussed in ref. 2. (For more details, see fig. 18 in ref. 21.) A simplified treatment of the pairing energy might thus easily lead to wrong conclusions concerning additional minima in the nuclear potential energy surface.

The influence of the pairing energy on the total energy can also be found in the investigation by Girod and coworkers.[31] A large number of nuclei (800) are investigated for the occurance of shape isomers at large quadrupole deformation. For selected nuclei, single particle states are obtained from selfconsistent HFB calculations and used in subsequent potential energy calculations employing the Strutinsky method without pairing (SHFB). By comparing results of HFB and SHFB calculations, the authors discuss the accuracy of the NS aproach. However, the difference between fully selfconsistent HFB calculations and the former type (SHFB) can rather be traced back to the lack of the pairing correction δE_{pair} than to the accuracy of the Strutinsky method.

The calculations of the pairing energy can be improved further by the extension to the Lipkin-Nogami method, which to first order corrects for the particle number violating terms in the BCS equation.[32,33] The Lipkin-Nogami method introduces an additional parameter to the Hamiltonian, $-\lambda_2 N^2$. Additional sets of non-linear equations have to be solved, but the resulting Lipkin-Nogami pairing gap never vanishes, not even at the magical shell-gaps. One avoids thus the common problem of the BCS gap-equation, which can break down in regions of low level-density, leading to large fluctuations of the pairing energy.

3. Band-Head energies

The great achievment of the Nilsson model relates partly to the successfull explanation of single particle states in odd-A nuclei. Assuming a certain deformation, the spin and parity of states close to the Fermi surface can be obtained by simple inspection of a Nilsson single particle level diagram. In order to include the shape degree of freedom in a general way, the total energy of specific single particle states as a function of deformation has to be computed.

The extension of the NS aproach to odd-A nuclei can be made in several ways.[35] Since the Strutinsky and liquid-drop energy depends smoothly on N and Z, the shell-correction energy can be obtained in a straight forward manner. For odd-A nuclei, one has to block the specific orbital in the BCS pairing calculations. The calculated equilibrium deformation is in general not depending sensitively on the treatment of the pairing energy. Blocking is therefore not very important in order to obtain the equilibrium deformation. However, in order to accurately compare the energy of different single particle states, an exact treatment of BCS-blocking is essential.

For axial symmetric shapes, the projection of angular momentum on the symmetry axis, Ω, and parity π is a good quantum number by which each state can be classified. Within the macro-micro aproach, the equilibrium shape of each single particle state can be calculated and compared to experiment. The total energy is

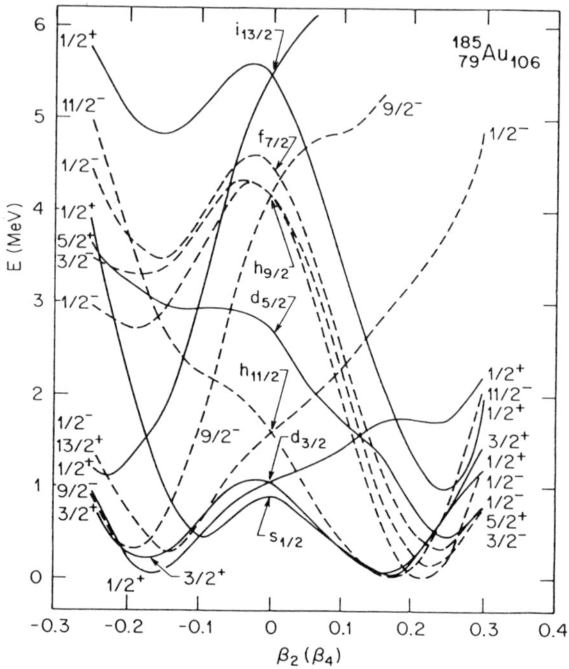

Figure 3: *Deformation-energy curves for various bandheads in* ^{185}Au *as a function of* $\beta_2(\beta_4)$. *Bandheads with positive (negative) parity are shown with solid (dashed) line (Figure from ref. 34).*

obtained in the usual manner as a sum of the liquid drop energy, the Strutinsky shell correction and the pairing correction. In ref. 34, the equilibrium deformation and energy for all single particle states close to the Fermi surface is calculated for ^{185}Au. The Lipkin-Nogami aproach is applied and exact blocking of each state is performed. As is illustrated nicely in fig.3, each configuration minimizes at a specific shape. The envelop of all diabatic single particle configurations form an adiabatic energy-surface of ^{185}Au.

The limits of phenomenological models are addressed in bandhead calculations. Comparison between experimental and calculated bandhead energies serve as a valuable check of the parameters of the potential. Systematic deviations between calculated and experimental levels might give insight on residual interactions, which can not be accounted for by the average field,

Proton bandhead energies have been calculated for Rare Earth nuclei with

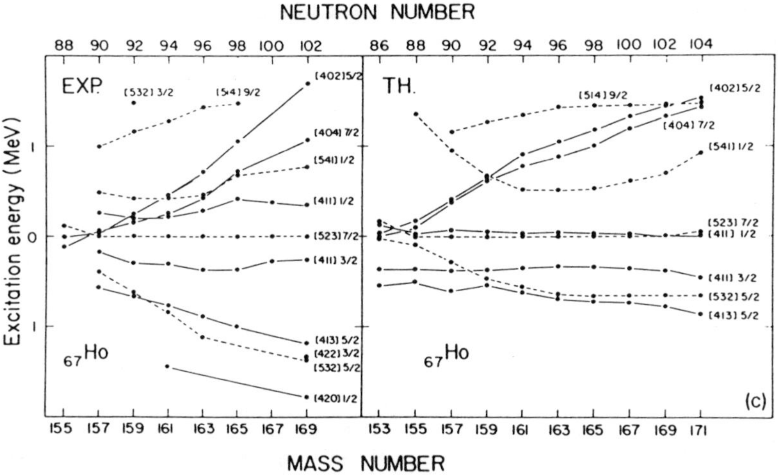

Figure 4: *Comparison of single particle energies in $_{67}Ho$. Note especially the difference in energy between calculated and experimental peudo-spin partner orbitals as the [402]5/2-[404]7/2 and [413]5/2-[411]3/2 (Figure from ref. 36).*

Z=67-79 in ref. 36. The pairing energy is computed by means of the Lipkin-Nogami approach. Interesting conclusions are drawn from the energy of pseudo-spin doublets. The calculated pseudo-spin orbitals lie always closer in energy than the experimental, implying that the pseudo SU(3) symmetry is realized stronger in the single particle potential than in real nuclei.

4. Applications to High Spin Physics - Total Routhian Surface Calculations

The extension of the NS scheme to high spin has been discussed since some time ago (see e.g. ref. 37 - 39). The total energy of the nucleus is expressed in a spin independent term, as discussed previously, and a spin dependent term. The spin dependent term can be calculated from the cranking equation since it correctly reproduces the experimental moment of inertia. The total energy becomes frequency (ω-) dependent and is expressed by the following equation:

$$\begin{aligned} E^{\omega}(Z,N,\hat{\beta}) &= E^{\omega=0}(Z,N,\hat{\beta}) + \\ &\Big[<\Psi^{\omega}|\hat{H}^{\omega}(Z,N,\hat{\beta})|\Psi^{\omega}> -[<\Psi|\hat{H}^{\omega=0}(Z,N,\hat{\beta})|\Psi>\Big]. \end{aligned} \tag{6}$$

The term in brackets describes the contribution to the energy due to rotation and $E^{\omega=0}$ is given by equation (1). The calculations can be performed with or without pairing correlations. If pairing is taken into account, the cranking equation is solved by means of the HFB transformation, which is equivalent to diagonalizing the matrix $\hat{H}^\omega$.

$$H^\omega = \begin{pmatrix} (e^i_{sp} - \lambda)\delta_{ij} - \omega < j_x >_{ij} & \Delta \\ \Delta & (-e^i_{sp} + \lambda)\delta_{ij} + \omega < j_x >_{ij} \end{pmatrix} \tag{7}$$

These kind of calculations are called the Total Routhian Surface (TRS) method, since the Routhian (energy in the rotating frame) is minimized with respect to deformation. The concept and details of the TRS calculations are described in refs. 19,40 and 41. Recent results and discussion of the method can also be found in ref. 42, where the potential energy of 138 nuclei are displayed.

When dealing with high spin states, the pairing equation encounters the common problem of the BCS theory: most times after the first bandcrossing only the trivial solution with $\Delta = 0$ is obtained. This problem can be circumvented, by leaving selfconsistency and prescribing the value of Δ. Calculations with good particle number (particle number projection before variation) show that the obtained Δ-value drops somewhat at each band-crossing and otherwise decreases smoothly with frequency.[43,44] The Δ-value in the TRS-calculations is modeled in a similar manner:

$$\begin{aligned} \Delta^{TRS}(\nu,\hat{\beta},\omega) &= \Delta_0(\nu,\hat{\beta})[1-1/2(\omega/\omega_c)^2] \quad \text{if } \omega \le \omega_c \text{ and} \\ \Delta^{TRS}(\nu,\hat{\beta},\omega) &= 1/2\Delta_0(\nu,\hat{\beta})(\omega/\omega_c)^2 \quad \text{if } \omega > \omega_c \end{aligned} \tag{8}$$

At zero frequency, the BCS equation is solved resulting in the selconsistent value of Δ_o. At the critical frequency ω_c, Δ^{TRS} becomes equal to $1/2\Delta_o$. The value of the critical frequency ω_c is chosen such as to corespond the frequency, where on the average the pairing gap parameter Δ in calculations with particle number projection (PNP) has dropped to half its value in the massregion considered. In this way, stable solutions are obtained and pairing correlations are taken into account.

The TRS aproach has been very successful in interpreting and predicting low energy nuclear structure phenomena, see refs. 19,45- 49. In many even-even Xe-Ba-Ce nuclei, the gsb splits up into two S-bands. This phenomen has not been given a satisfactory explanation in the frame work of different models.[50]. TRS-calculations carried out for selected Ba-nuclei, exhibit moderately deformed prolate shapes in the ground state. However, at higher frequencies, prolate-oblate shape coexistence occures in the calculations, see fig.??. The change in shape with rotation can be understood by means of aligning quasi-particles. In the mass A=120-130 region, the unique parity subshell is $h_{11/2}$ for both protons and neutrons. Accordingly, one expects neutron and proton $h_{11/2}$ orbitals to align their angular momenta along the rotational axis. Since the Fermi level of protons is low in the shell, whereas

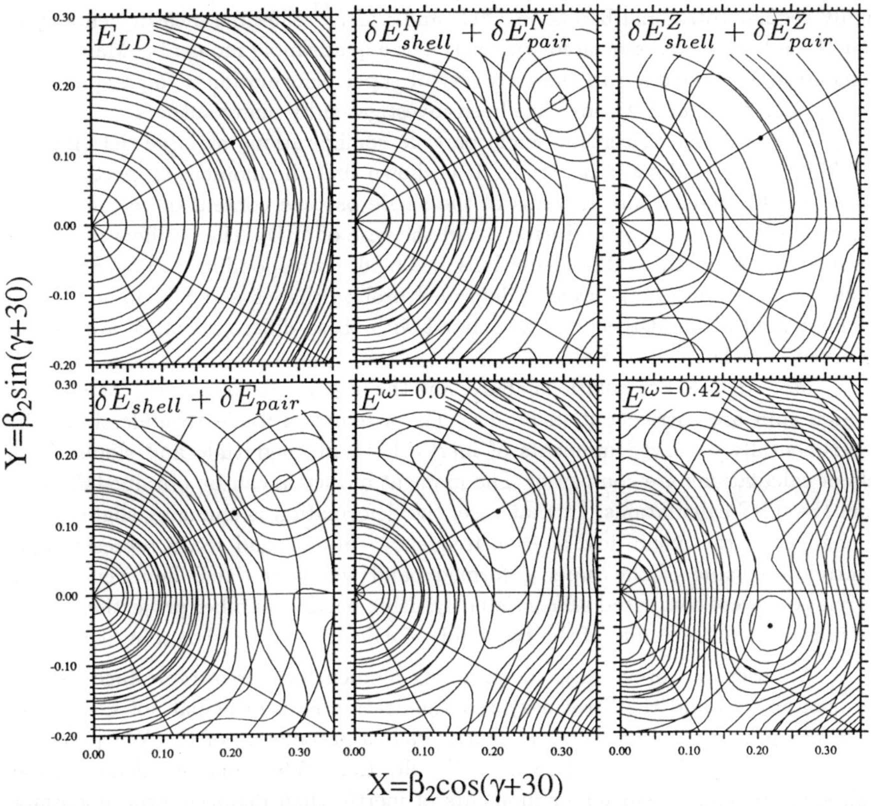

Figure 5: *The composition of the Total Routhian Surface, eq.(1) and (6), for ^{126}Ba. The energy is minimized with respect to the hexadecapole deformation β_4 at each point in the β_2 -γ plain. The 'Lund'-convention is used in the definition of γ-deformation. The energy difference between contour lines is 200keV. The separate contributions of neutrons and protons to the shell energy are displayed. Note how the $\langle \hat{H}^\omega \rangle$-term changes the Routhian landscape with frequency.*

it is above mid-shell for neutrons, the aligned quasi-particles will drive the nucleus to prolate (protons) and oblate (neutrons) shape, respectively. The quasi-particle structure can be extracted from the TRS calculations, showing indeed, that the aligned state at prolate shape is of proton $\pi(h_{11/2})^2$ character and of $\nu(h_{11/2})^2$ at oblate shape (corresponding to the two 12^+ state of the S-band). Investigation along these lines can account for the systematic observations of two S-bands in this mass-region.[48]

Stable super-deformed shapes at high spin are linked to the discovery of highly regular rotational bands, in ^{132}Ce and ^{152}Dy.[51] Lifetime measurments show that these bands are associated with very large quadrupole moments. Many questions arose in connection to the discovery of super deformation at high spin. The feeding and decay of SD states has been investigated by many authors,

The stability of certain single particle configurations at large deformation and high spin is addressed in ref. 52. It is shown that the moment of inertia of SD nuclei can be understood in terms of the single particle content of certain high-j intruder orbitals as $\pi i_{13/2}$ and $\nu j_{15/2}$.[52] In the case of ^{152}Dy, three different kind of minima emerge in the TRS-calculations, see fig.6, corresponding to non collective single particle excitations, weakly deformed prolate shapes and SD shape. Three different kind of bandstructures associated with these different shapes are indeed observed in ^{152}Dy.[51]

More detailed calculations addressing the importance of pairing correlations at high spin are carried out in ref 19. The attempt is made to understand the experimentally observed moments of inertia both in magnitude and slope. Two sets of calculations are compared. The moment of inertia can be obtained directly from the TRS calculations. In that case, the deformation is changing slightly, due to changes in the mean field at each frequency. In a second set, a representative deformation value of the super deformed band was kept fixed, however pairing correlations were computed selfconsistently, using particle number projection before minimization. The results of the two kinds of calculations are shown in fig.6

As discussed previously, the cranking moment of inertia reproduce the experimental ones and the Strutinsky renormalization is in principal not necesseray for a WS potential. However, the central radius parameters of the WS potential is rather large ($r_o^{\nu(\pi)} = 1.35(1.28)$fm). Results from WS cranking calculations at high spin therefore yield larger moments of inertia than deduced experimentally. Comparison with the SD bands in the A=150 mass region show that a Strutinsky renormalization is needed. In order to include the Strutinsky and liquid drop part of the moment of inertia, equation 6 is extended in the following way:

$$E^{\omega}(Z,N,\hat{\beta}) = E^{\omega=0}(Z,N,\hat{\beta}) + \left[<\Psi^{\omega}|\hat{H}^{\omega}(Z,N,\hat{\beta})|\Psi^{\omega}> -[<\Psi^{\omega=0}|\hat{H}^{\omega}(Z,N,\hat{\beta})|\Psi^{\omega=0}>\right]$$
$$-\frac{1}{2}\omega^2\left[\Im_{LD}(A,\hat{\beta}) - \Im_{Strut}(N,Z,\hat{\beta})\right] \quad (9)$$

Investigation of ref. 19 show that the Strutinsky renormaliztion is purely geometrical, i.e. the Strutinsky moment of inertia of the WS-potential is similar to the liquid drop moment of inertia with the corresponding radius. The renormalization can therefore be done just by considering the two liquid drop moments of inertia with different radii.

Calculations of ref. 19 reproduce essentially all moments of inertia of SD-bands in A=150 mass-region and predict favored SD-structures in neighbouring

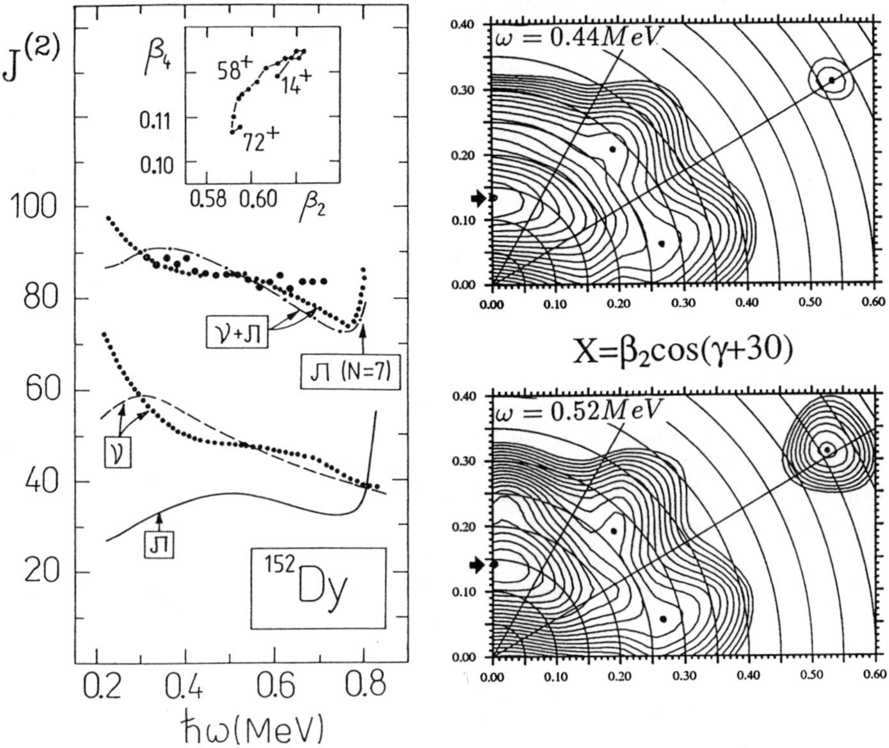

Figure 6: *Dynamical moments of inertia of* ^{152}Dy *- the experimental values are depicted with large dots. The small dots are the results from the TRS calculations and the dash-dotted line show results from pairing selfconsistent PNP-calculation. The shape evolution of the SD band is shown in the inset. The TRS calculations to the right reveal the with spin increasing depth of the SD minima. The non-collective minimum is marked by an arrow. Note also the two normally deformed minima at triaxial shape.*

nuclei. Especially, the occurance of a paired bandcrossing in ^{150}Gd and the importance of pairing correlations is addressed.[19] However, subsequent studies of the SD-bands in the Hg-region show, that many questions in understanding the moments of inertia remain.[16,26,53] Part of it can be related to the description of the pairing interaction in terms of a simple monopole force.[54] In addition, one might probe additional residual interactions at large deformation. The Nilsson orbitals at SD shape origin from three to four different shells, whereas one just deals with one unique parity subshell within the major shell at normal deformation.

Shell structure is related to periodic motion in a potential and thus a general phenomen in a many-body quantum system. In the rotating nucleus, the Cramers degeneracy between time-reversed states is lifted and one expects the shell-structure to be less pronounced. However, certain gaps in the shell structure remain up to highest angular momentum and even new shell-gaps open up in the rotating nucleus.[19]

5. Conclusions

Calculations of potential energy surfaces based on the shell-correction method has proven to be highly accurate. Single particle states in odd-A nuclei can be calculated in a similar manner and be understood in terms of corresponding 'WS'-states and core-polarization due to the odd particle. The NS scheme is still the most frequently used model addressing the structure of high spin states. Its success relates partly to the simple interpretation of rotational spectra in terms of the collective degree of freedom - shape - and single particle excitations - occupation and roll of certain Nilsson orbits. There is a natural feedback from experiment to adjustment of the single particle levels and the 'ultimate' single particle potential has not been obtained yet. The goal is to describe as many areas of nuclear structure with as few parameters as possible. The present WS potential has certainly contributed in this direction. At a certain point, one aproaches the limits of the model and encounters important residual interaction, which are not possible to describe within the mean field. Especially the recent development in high spin physics, with a multitude of different bands and the intriguing properties of the moment of inertias in many SD nuclei, rise serious questions whether the NS sheme is able to account for these new phenomena. Since there are still many ways of improving the NS scheme, future will tell to what extent it will continue to play a dominant roll in nuclear structure theory.

6. References

1. S. Åberg, H. Flocard and W. Nazarewicz, Ann. Rev. Nucl. Part. Sci. 40 (1990) 439.
2. M. Brack, J. Damgård, A.S. Jensen, H.C. Pauli, V.M. Strutinsky and C. Y. Wong, Rev. Mod. Phys. **44** (1972) 320.
3. S. Ćwiok, J. Dudek, W. Nazarewicz, J. Skalski and T. Werner, Comput. Phys. Commun. **46** (1987) 379.
4. J. Dudek and T. Werner, J. Phys. **G4** (1978) 1543.
5. J. Dudek, A. Majhofer, J. Skalski, T. Werner, S. Ćwiok , W. Nazarewicz, J. Phys. **G5** (1979) 1359.

6. J. Dudek, Z. Szymański, T. Werner, Phys. Rev. **C23** (1981) 920.

7. W. Nazarewicz, J. Dudek, R. Bengtsson, T. Bengtsson and I. Ragnarsson, Nucl. Phys. **A435** (1985) 397.

8. E.M. Rastopchin, G.N. Smirenkin and V.V. Pashkevich, *Proc. Workshop on Microscopic Theories of Superdeformation in Heavy Nuclei at Low Spin*, Lyon, May 1990; V.V. Pashkevich, Preprint JINR (Dubna) P4-4383 (1969).

9. C.F. Tsang and S.G. Nilsson, Nucl. Phys. **A140** (1970) 275.

10. U. Götz, H.C. Pauli, K. Adler, K. Junker, Nucl. Phys. **A192** (1972) 1.

11. M. Caillian, J. Letessier, H. Flocard, P. Quentin, Phys. Lett. **46B** (1973) 11.

12. S. Åberg, I. Ragnarsson, T. Bengtsson and R.K. Sheline, Nucl. Phys. **A391** (1982) 327.

13. R.R. Chasman, Phys. Lett. **219B** (1989) 227.

14. M. Girod, J.P. Delaroche, D. Gogny and J.F. Berger, Phys. Rev. Lett. **62** (1989) 2452.

15. P. Bonche, P. Quentin, S.J. Krieger, M.S. Weiss, J. Meyer, M. Meyer, N. Redon, H. Flocard and P.-H. Heenen, Nucl. Phys. **A500** (1989) 308.

16. R.V.F. Janssens and T.L. Khoo, Ann. Rev. Nucl. Part. Sci. **41** (1991) 321.

17. S.G. Nilsson, C.F. Tsang, A. Sobiczewski, Z. Szymański, S. Wycech, C. Gustafsson, I.-L. Lamm, P. Möller, B. Nilsson, Nucl. Phys. **A131** (1969) 1.

18. M. Meyer, N. Redon, P. Quentin and J. Libert, Phys. Rev. C **45** (1992) 233.

19. W. Nazarewicz, R. Wyss, A. Johnson, Nucl. Phys. **A503** (1989) 285.

20. A. Sobiczewski, Z. Patyk, S. Ćwiok and P. Rozmej, Nucl. Phys. **A485** (1988) 16.

21. W. Satuła, S. Ćwiok, W. Nazarewicz, R. Wyss and A. Johnson, Nucl. Phys. **A529** (1991) 289.

22. W.D. Myers, W.J. Światecki, Ark. Fys. **36** (1967) 343.

23. W.D. Myers, W.J. Światecki, Ann. Phys. (NY) **84** (1969) 395.

24. H.J. Krappe, J.R. Nix,, A.J. Sierk, Phys. Rev. **C20** (1979) 992.

25. J.P. Delaroche, M. Girod, J. Libert and I. Deloncle, Phys. Lett. **232B** (1989) 145.

26. M.A. Riley, D.M. Cullen, A. Alderson, I. Ali, T. Bengtsson, M.A. Bentley, A.M. Bruce, P. Fallon, P.D. Forsyth, F. Hanna, S.M. Mullins, W. Nazarewicz, M. Poynter, J.W. Roberts, J.F. Sharpey–Schafer, J. Simpson, G. Sletten, P.J. Twin, R. Wodsworth, R. Wyss, Nucl. Phys. **A512** (1990) 178.

27. I. Ragnarsson, R.K. Sheline, Phys. Scr. **29** (1984) 385.

28. D.G. Madland, R. Nix, Nucl. Phys. **A476** (1988) 1.

29. P. Möller and J.R. Nix, Nucl. Phys. **A536** (1992) 20.

30. C. Baktash, J.D. Garrett, D. Winchell, and A. Smith, Phys. Rev. Lett. (submitted).

31. M. Girod, J.P. Delaroche and J.F. Berger, Phys. Rev. **C38** (1988) 1519.

32. Y. Nogami, Phys. Rev. **134** (1964) B313.

33. H.C. Pradhan, Y. Nogami, J. Law, Nucl. Phys. **A201** (1973) 357.

34. J.A. Larabee, M.P. Carpenter, L.L. Riedinger, L.H. Courtney, J.C. Waddington, V.P. Janzen, W. Nazarewicz, J.Y. Zhang, R. Bengtsson and G.A. Leander, Phys. Lett. **169B** (1986) 21.

35. K. Heyde, P. Van Isacker, M. Waroquier, J.L. Wood and R.M. Meyer, Phys. Rep. **102** (1983) 291.

36. W. Nazarewicz, M.A. Riley and J.D. Garrett, Nucl. Phys. **A512** (1990) 61.

37. J. Krumlinde and Z. Szymański, Phys. Lett. **36B** (1971) 157.

38. M. Brack and B.K. Jennings, Nucl. Phys **A258** (1976) 264.

39. R. Bengtsson, S.E. Larsson, G.A. Leander, P. Möller, S.G. Nilsson, S. Åberg and Z. Szymański, Phys. Lett. **57B** (1975) 301.

40. W. Nazarewicz, G.A. Leander and J. Dudek, Nucl. Phys. **A467** (1987) 437.

41. R.Wyss and R. Bengtsson, in preparation.

42. T. Werner and J. Dudek, At. Data Nucl. Data Tables **50** (1992)179 .

43. P. Ring and P. Schuck, *The Nuclear Many-Body Problem* (Springer-Verlag, 1980).

44. D.R. Bés, R.A. Broglia, J. Dudek, W. Nazarewicz and Z. Szymański, Ann. Phys. **182** (1988) 237.

45. W. Nazarewicz and T. Werner, in *Nuclear Structure of the Zirconium Region*, ed. by J. Eberth, R.A. Meyer and K. Sistemich, (Springer-Verlag, 1988), p. 277.

46. R. Wyss, J. Nyberg, A. Johnson, R. Bengtsson and W. Nazarewicz, Phys. Lett. **215B** (1988) 211.

47. R. Bengtsson, T. Bengtsson, J. Dudek, G.A. Leander, W. Nazarewicz and J.-y. Zhang, Phys. Lett. **183B** (1987) 1.

48. R.Wyss A. Granderath, W. Lieberz, R. Bengtsson, P. von Brentano, A. Dewald, A. Gelberg, A. Gizon, J. Gizon, S. Harrisopulos, A. Johnson, W. Nazarewicz, J. Nyberg and K. Schiffer Nucl. Phys. **A505** (1989) 337.

49. R.Wyss, W. Satuła, W. Nazarewicz and A. Johnson, Nucl. Phys. **A511** (1990) 324.

50. K. Schiffer, A. Dewald, A. Gelberg, R. Reinhardt, K.O. Zell, Sun Xiangfu and P. von Brentano Nucl. Phys. **A458** (1986) 337.

51. P.J. Nolan, P.J. Twin, Ann. Rev. of Nucl. and Part. Sci. **38** (1988) 226.

52. T. Bengtsson, S. Åberg and I. Ragnarsson, Phys. Lett. **208B** (1988) 39.

53. M.W. Drigert, M.P. Carpenter, R.V.F. Janssens, E.F. Moore, I. Ahmad, P. Fernandez, U. Garg, W. Reviol, D. Ye, I.G. Bearden, Ph. Benet, P.J. Daly and R. Wyss, Nucl. Phys **A530** (1991) 452.

54. W. Nazarewicz, Proc. *XXV Zakopane School on Physics, Selected Topics in Nuclear Structure*, ed. by J. Styczeń and Z. Stachura, (World Scientific Publ. 1990) vol. 2, p. 53.

GROUP 5

Fermion Dynamical Symmetry Model

FERMION DYNAMICAL SYMMETRY MODEL

Da Hsuan Feng
Department of Physics and Atmospheric Science, Drexel University
Philadelphia, Pennsylvania 19104

Mike W. Guidry
Department of Physics and Astronomy, University of Tennessee
Knoxville, Tennessee 37996

and

Cheng–Li Wu
Department of Physics and Astronomy, University of Tennessee
Knoxville, Tennessee 37996

The Editors refer the Reader to the published literature on this subject.

THE FERMION DYNAMIC SYMMETRY MODEL: WORKING GROUP REPORT

MICHAEL W. KIRSON
Department of Nuclear Physics
Weizmann Institute of Science
76100 Rehovot, Israel

WORKING GROUP
Bruce Barrett, University of Arizona
Jerry Draayer, Louisiana State University
Martin Jarrio, Georgia Institute of Technology
Michael Kirson, Weizmann Institute of Science
Karl Schmid, Universität Tübingen
Sudha Swaminathan, University of Michigan

ABSTRACT

The Fermion Dynamic Symmetry Model is very briefly reviewed, its pros and cons are mentioned, and some questions are posed concerning its basis and applications.

1. Summary of the model

The Fermion Dynamic Symmetry Model (FDSM) is based on the nuclear shell model, and can be applied in a single major nuclear shell or in several major shells. The underlying idea is a change of single-particle basis which permits some of the nucleon degrees of freedom to be frozen in a convenient manner, thus reducing the number of active degrees of freedom to a manageable number. This reduction is achieved within a natural algebraic framework which greatly simplifies the application of the model.

In terms of the usual spherical shell-model particle creation operators $a^\dagger_{jm}$, the basis change is defined with the aid of Clebsch-Gordan coefficients, as follows:

$$b^\dagger_{k,m_k,i,m_i} = \sum_{jm} (km_k im_i|jm) a^\dagger_{jm}$$

thus separating the single-particle angular momentum j into two components, k and i, analogous to LS coupling. From the antisymmetry of the fermion operators, a two-particle state coupled to total pseudo-orbital angular momentum K and total pseudo-spin angular momentum I, these in turn coupled to total angular momentum J, must have the sum $K+I$ even. If K or I is coupled to zero, the corresponding pseudo angular momentum becoming frozen, the other is restricted to even values. So if $K=0$ and $i=3/2$, the only allowed values of I (and hence of J) for the fermion pair are 0 and 2. Similarly, if $I = 0$ and $k = 1$, the only allowed values of K (and hence of J) are 0 and 2. This technique, invented more than a decade ago by Ginocchio, supplies a natural basis in which there exists a decoupled space of S (J=0) and D (J=2) fermion pairs.

In order to preserve parity, the above basis change is applied only to the normal-parity states of a given major shell. These can be separated into groups with either $i = 3/2$ or $k = 1$ in an essentially unique manner. The remaining, abnormal-parity state in the shell is necessarily assigned $k = 0$, $i = j$, and its pair states accommodate all even values from 0 to $2j$-1.

Each of the two ways of generating a fermion S-D space by freezing certain angular momentum degrees of freedom has a natural algebraic structure. For i=3/2 (K=0), the governing algebra is so(8), while for $k = 1$ ($I = 0$) it is sp(6). The so(8) algebra has an so(6) subalgebra which exactly mimics the so(6) of the interacting-boson model (IBM), including its so(5) - so(3) chain of subalgebras; a "vibrational" subalgebra which closely resembles the su(5) of the IBM; and an so(7) subalgebra which is not found in the IBM framework. Similarly, the sp(6) algebra has an su(3) subalgebra very like that of the IBM and a "vibrational" subalgebra resembling the su(5) of the IBM. The only convenient algebra for the abnormal-parity state is the familiar seniority algebra, essentially an su(2) of S pairs. There is no algebraically decoupled abnormal-parity D pair.

The most general number-conserving, rotation, parity and time-reversal invariant one and two-body fermion Hamiltonian built of $K = 0$ and $I = 0$ operators is called the general FDSM Hamiltonian, and it has an exact eigenspace of S and D pairs. For appropriate choices of parameters, it will also have su(3), so(6), etc., dynamic symmetry. Although the state labelling in the so(6) case is the same in FDSM as in IBM, a distinction arises in the su(3) case, where the labels are the same for states allowed by both models but FDSM does not allow realisation of certain states permitted by IBM. This is referred to as the dynamic Pauli effect, since it arises from the fermion nature of the basic building blocks. It prevents the realisation of the lowest-energy, ground-state irreducible representations of IBM in the FDSM once the number of normal-parity particles exceeds 1/3 of the total capacity of the normal-parity orbitals. This produces dramatic effects in the su(3) dynamic symmetry limit.

2. The pros and cons of the model

The working group felt that the FDSM has certain clear advantages and certain apparent drawbacks. The former, once listed, require very little further discussion, since their desirable qualities are self-evident. The latter need to be expanded on to some extent. Pros and cons will be listed here in brief, and the questions arising from the latter will be gone into in further detail in the subsequent section.

2.1 Advantages

- the FDSM is a fermion model, explicitly involving fermionic degrees of freedom; no bosonisation is needed
- the model is, in principle, as complete as the shell model; it contains all the fermion degrees of freedom, which can be brought into active play by suitable "breaking" of fermion pairs
- although a fermion model, the FDSM contains the familiar symmetry limits of the phenomenologically very successful IBM: the su(3) rotational limit, the so(6) gamma-unstable rotor limit, and an "su(5)"-like vibrational limit
- the FDSM has an exactly-decoupled $S-D$ fermion pair space, when the Hamiltonian is limited to I and K scalar operators; this is an exact fermion implementation of the truncation implied by the success of the IBM
- the model has a simple algebraic structure, allowing the use of the powerful tools of group theory even when the symmetries of the model are broken
- neutrons and protons are explicitly, and separately, treated in the FDSM; this important nucleon degree of freedom is fully included in the model, and protons and neutrons are permitted to behave differently

2.2 Disadvantages

- in the simplest, most tractable form of the FDSM, in which pairs are not broken, the abnormal-parity state plays the role of a particle sink; particles in this state act as spectators, locked in zero-coupled pairs
- the structure of the pairs in the FDSM is purely "kinematic", determined by the requirements of the algebraic truncation and completely independent of the dynamics implicit in the Hamiltonian parameters
- similarly, the underlying symmetry of the model is imposed by the desire to have a decoupled $S-D$ fermion pair space, and is not manifestly required by the dynamics of the fermions
- the IBM symmetries imbedded in the FDSM are not common to a single realisation, but are shared between the $i=3/2$ option (which does include so(6), but not su(3)) and the $k=1$ option (which includes su(3) but not

so(6)); the Casten-Warner triangle of IBM is replaced by a divided "pentangle" in FDSM and there is no smooth transition between the deformed rotor and the gamma-unstable rotor.

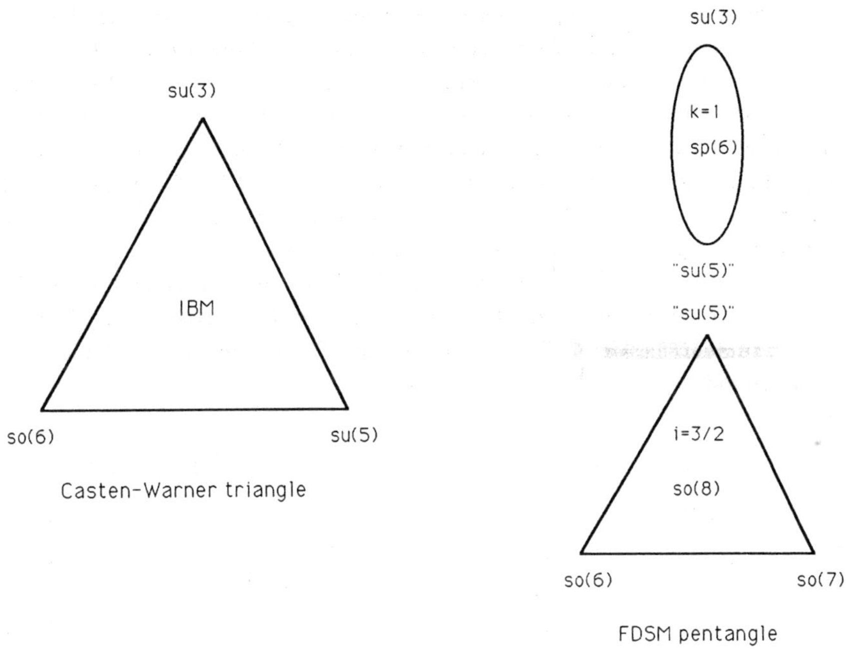

3. Questions concerning the FDSM

The disadvantages noted briefly above raise certain questions concerning the underlying basis of the FDSM, the manner of its application to real nuclei and the significance of its predictions. These were presented in the oral report of this working group and discussed at some length in the workshop plenary. Some attempt will be made to include in the following written presentation the highlights of the plenary discussion.

3.1 Truncation of the shell-model space

The shell model cannot be directly applied well away from closed shells because the size of the model space rapidly becomes intractable. The IBM clearly indicated that truncation to monopole and quadrupole degrees of freedom permits reproduction of the important low-energy collective behaviour of nuclei despite the radical reduction of the model space. The FDSM is an explicit fermion realisation of this truncation to an S-D space. It is clearly of fundamental importance and interest to understand whether this drastic truncation is a property of the nuclear dynamics or a simple computational convenience. The $k-i$ basis of the FDSM could be optimal for one of two reasons: (1) the corresponding $S-D$ space may decouple when the exact shell-model Hamiltonian H is used, or (2) the renormalisation of the effective Hamiltonian required to compensate for forced truncation to a non-decoupled $S-D$ space may be sufficiently mild to be accommodated by a simple FDSM form. If the former is the case, then the FDSM is required by the dynamics of the nucleus; if the latter is the case, then the FDSM is a very convenient way to cope with the explosion of the shell-model basis; if neither is the case, then the FDSM is simply not a useful model. These questions should be testable.

The two possible origins of the truncation may be represented by the following Hamiltonian matrices, where H is the "exact" shell-model Hamiltonian in the full shell-model space:

$$H = \begin{array}{c} \\ SD \\ \\ \neq SD \end{array} \begin{array}{c} \begin{array}{cc} SD & \neq SD \end{array} \\ \begin{pmatrix} FDSM & \vdots & 0 \\ \cdots & \vdots & \cdots \\ 0 & \vdots & rest \end{pmatrix} \end{array} \qquad PH = P \begin{array}{c} \\ SD \\ \\ \neq SD \end{array} \begin{array}{c} \begin{array}{cc} SD & \neq SD \end{array} \\ \begin{pmatrix} FDSM & \vdots & \neq 0 \\ \cdots & \vdots & \cdots \\ \neq 0 & \vdots & rest \end{pmatrix} \end{array} = H_{SD}$$

automatic decoupling

imposed truncation
(P projects on SD subspace)

If the correct description is that of automatic decoupling, so that the SD space generated by the $k-i$ transformation is an approximate eigenspace of the true nuclear Hamiltonian, this should be relatively easily testable. The full shell-model Hamiltonian can be transcribed in terms of $K-I$ tensors of well-defined rank. The scalar part ($K{=}0{=}I$) does decouple automatically, so the decoupling scenario could be confirmed by the vanishing of the non-scalar part, or by its relative unimportance. An estimate of the importance of the non-scalar part of the Hamiltonian could be obtained by use of statistical spectroscopy measures. These could also be interpreted in terms of the adequacy of the no-broken-pair space in accommodating the dynamics of the full shell-model Hamiltonian.

Should the tests advocated here indicate, as would seem to be very likely, that the FDSM truncated SD space does not reflect exact shell-model dynamics but represents a convenient ad hoc truncation of the huge shell-model space, then the question arises as to whether the associated renormalisation of the effective Hamiltonian can be accommodated in the FDSM framework. The simple projection onto the SD space illustrated in the sketch above has to be supplemented by effects arising from multiple virtual scattering into the non-SD space via the non-zero coupling terms. Such a renormalisation always exists, but if the resulting effective Hamiltonian is very complex (contains significant many-body terms) it is not useful. Tests of the simplicity of the effective interaction could be empirical or phenomenological. An empirical test would involve constructing the exact Hamiltonian matrix in few-pair cases, truncating the space to the SD space, and generating the exact effective Hamiltonian by well-established methods of effective interaction theory. The number dependence of the results would indicate the feasibility of a simple two-body representation, perhaps with smoothly varying coefficients. A phenomenological test would involve establishing numerical values for the parameters of an FDSM Hamiltonian which would reproduce to a good approximation a wide range of experimental data on nuclei. This could determine whether an effective FDSM Hamiltonian exists, even if it cannot be precisely derived from the shell-model Hamiltonian, just as the shell-model effective interaction itself was established to exist long before it could be derived, even approximately, from the underlying nucleon-nucleon force.

3.2 Structure and occupancy of pairs

The FDSM has two types of S pairs, one for normal-parity states and one for the abnormal-parity state, and one D pair, for the normal-parity states, in the unbroken-pair limit. The creation operator for the normal-parity S pair is

$$\mathrm{S}^\dagger = \left[\sum_j (2j+1)\right]^{-1/2} \sum_{jm} (-)^{j-m} a^\dagger_{jm} a^\dagger_{j-m}$$

where the sums are over the normal-parity states. This is the $K=0, I=0$ pair generated by the $k-i$ transformed single-particle creation operator. It is an equal-weight sum over all time-reversed pairs in normal-parity states, and reflects no dynamics. The creation operator for the abnormal-parity S pair, denoted $\mathcal{S}^\dagger$, is defined in exactly the same way but without the sums over j, which is restricted to a single value. The only way the dynamics can influence the structure of the fermion S-pair in this model is by defining the relative weight of the abnormal and normal parity parts of a generalised pair generated by a mixing term of the form

$S^\dagger\mathcal{S} + \mathcal{S}^\dagger S$ in the FDSM Hamiltonian. The model contains such a term, but it does complicate the handling of the calculations. Even with such a term, the model can only accommodate an unequal sharing of S-pair strength between normal-parity states by breaking pairs and admixing terms of higher heritage. (The heritage is a generalisation of the seniority, and is written $h = u + v$, where u is the number of non-SD-pair particles in normal parity states and v is the number of non-S-pair particles in abnormal-parity states. This v is the usual seniority for the abnormal-parity state.)

The D pairs, restricted in the symmetry limit to normal parity orbitals, have an equally rigidly determined internal structure, with the relative weight of different j, j' quadrupole pairs fixed by a 6-j symbol. The abnormal parity quadrupole pair cannot be mixed into the normal-parity D pair except by breaking the seniority v, which then should free all abnormal-parity pair multipoles. It was argued that there is only one quadrupole component for the abnormal-parity state, and an order of magnitude more quadrupole components in the general normal-parity quadrupole pair, so that the omission of the abnormal-parity D term could be expected to be a minor error. This would still leave the rigid, geometrically-determined internal structure of the normal-parity pairs a troublesome feature.

In most applications reported to date, the relative occupancy of normal and abnormal parity single-particle orbitals in a given nucleus has been determined by considerations extraneous to the FDSM. Use has been made of filling of Nilsson-model orbitals or of empirically-measured spectroscopic factors to determine the number of particles in the abnormal-parity orbital. Here, again, it would be desirable to allow the dynamics of the model to determine this parameter. An S-pair mixing term of the sort mentioned above would mix states with different abnormal-parity occupancies, which would probably be a more realistic situation. The present practice of having the number of abnormal-parity particles a constant of the motion seems artificial.

3.3 Heritage breaking

As discussed above, the assumption of heritage zero (pure SD) states has some discomforting features, especially the rigid "kinematic" structure of the fermion pairs. This might argue for heritage breaking as a way to improve the model. However, the further one goes from the heritage-zero symmetry limit, the more numerically complicated the calculations become. There is thus a conflict here between computational convenience and physical arguments. A further point in favour of the breaking of the heritage quantum number is that it is required by the non-degeneracy of the single-particle energies within a major shell. The FDSM can permit different single-particle energies for normal and abnormal parity states, and for normal parity states belonging to different values of the frozen pseudo angular momentum. (For instance, the $0h_{9/2}, 1f_{7/2}$ and $1f_{5/2}$ states, as $k = 1$, $i = 7/2$ states, must be degenerate, but can have different energy from the $2p_{3/2}$ and $2p_{1/2}$ states, as degenerate $k = 1, i = 1/2$ states, and from the abnormal-parity $0i_{13/2}$ state.) Non-degeneracy of different j's for a fixed k, i breaks the heritage.

Broken heritage states are essential to handle the backbending phenomenon, in which the presence of two different seniorities, the normal-parity u and the abnormal-parity v, is helpful in dealing with multiple pair breakings. But this would seem to indicate that broken-heritage states would lie higher in the spectrum than the collective states usually treated by IBM, say. Their effect on the low-lying spectrum could then be perturbatively small. However, the FDSM contains the fermion degrees of freedom, so it can also produce the observed low-lying negative-parity states in the collective spectrum. These are $h = 2(u = 1, v = 1)$, broken-heritage states and occur low in energy, so it is not clear that the breaking of heritage will not have significant effects on low-lying positive-parity states. It is also relevant that the $h > 0$ states have a rather high degeneracy, so additional interactions are needed to avoid an unphysically high density of low-lying negative-parity states.

In a similar vein, it is far from clear how to treat transition operators in the FDSM. Given the explicit fermion structure of the pairs, a given operator, such as the familiar $r^2Y^{(2)}$ quadrupole operator, can be exactly transcribed in terms of $k - i$ tensors and will generally not be a K or I scalar. Should the quadrupole operator used in the model be this microscopic operator, or a renormalised shell-model operator, or a generator of one of the limiting symmetries, or a mixture of generators? This connects to the truncation question presented in section 3.1, since the correct approach will evidently depend on whether the FDSM reflects an exact decoupling or a suitably renormalised truncation. Again, a phenomenological fit to data within a well-defined FDSM framework could be convincing.

3.4 Dynamic Pauli effect

One of the most dramatic claims of the FDSM is that the dynamic Pauli effect, which excludes certain su(3) irreducible representations once the number of particles in normal-parity orbitals exceeds one third of their total capacity, has significant empirical consequences which are borne out by data. This claim produced a considerable amount of fairly heated discussion. The degree of support afforded by the data appears to be open to interpretation.

In the su(3) limit of the IBM, the lowest irreducible representation is always (2N,0), where N is the number of bosons, always identified with the total number of valence pairs of nucleons. In the same limit of the FDSM, the lowest irreducible representation is (2N,0) only up to 1/3 filling of the shell, and (2N-4,2) thereafter, where N is the number of pairs of nucleons in normal-parity orbitals. The values of N in the two models differ, in general, in a given nucleus because of the occupancy of abnormal-parity states, but the difference is small enough to make no noticeable difference to the spectra of rotational nuclei as long as (2N,0) is the ground-state representation. There is a single K=0 band in the (2N,0) representation; one K=0 and one K=2 band in the (2N-4,2) representation; one K=0, one K=2 and one K=4 band in the (2N-8,4) representation. Conventionally, the lowest K=0 band is referred to as the ground band, the lowest K=2 band as the gamma band, the lowest excited K=0 band as the beta band. Bands within a representation are degenerate with one another, while different representations are significantly split in energy. (The K quantum number identifying a band is the "old-fashioned" intrinsic angular momentum projection, and should not be confused with the total pseudo orbital angular momentum of the FDSM.)

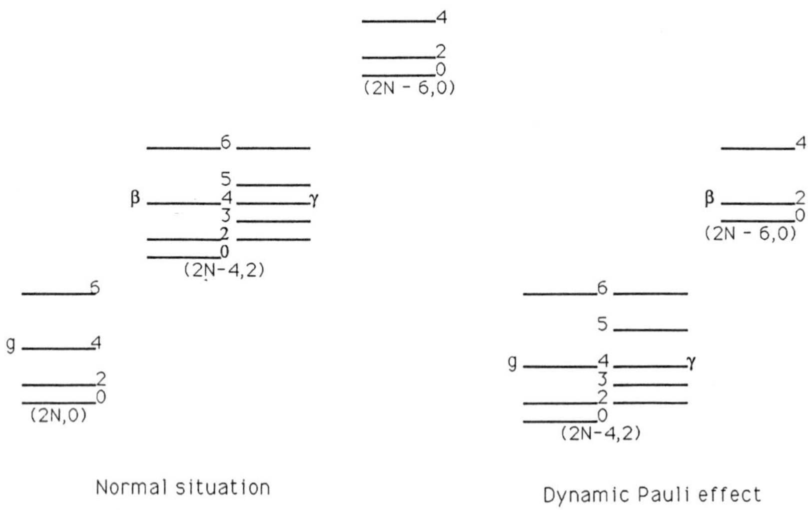

When the (2N,0) is the ground band, the gamma bandhead is significantly higher in energy than the first excited 2^+ state, is very close in energy to the beta bandhead, and has a strongly suppressed quadrupole transition to the ground state (since they belong to different irreducible representations). When the K=0 part of (2N-4,2) becomes the ground band, the gamma bandhead is degenerate with the first excited 2^+ state, is significantly lower in energy than the beta bandhead, and can have a noticeable quadrupole transition to the ground state, since they are in the same irreducible representation. (It should be noted that, if the beta and gamma bands are generated by intrinsic states which are quadrupole eigenstates, the quadrupole transition between them can also be suppressed.)

So far, the dynamic Pauli effect is claimed to be manifested by a fall-off of the gamma bandhead excitation energy as a function of neutron number for a few heavy rotational nuclei. It is not clear that the data really carry such a signature, and the other differences alluded to here have not been looked for. A serious, broad and systematic confrontation with the data is clearly called for. It should also be checked how strongly the dynamic Pauli signature would be expected to persist away from the exact su(3) limit.

4. Summary

The FDSM has a number of attractive features and a number of troublesome attributes. There would seem to be several avenues of investigation which invite further study in order to establish whether the model has any firm basis in nuclear reality. Certainly, a systematic fit to a wide range of data with a small set of empirical parameters would go quite some way towards establishing the model's phenomenological respectability. Failing that, the perceived artificiality of the structure of its fundamental fermion pairs and of the symmetry of its Hamiltonian is likely to continue to generate serious doubts concerning its applicability to real nuclei.

GROUP 6

Kerman–Klein Scheme

THREE DECADES OF THE KERMAN-KLEIN METHOD IN QUANTUM MECHANICS, IN QUANTUM FIELD THEORY, AND IN NUCLEAR PHYSICS

ABRAHAM KLEIN AND NIELS R. WALET
Physics Department, University of Pennsylvania
Philadelphia, PA 19104-6396, USA

ABSTRACT

Though introduced thirty years ago as a method of restoring the broken symmetry of the mean field solutions of the nuclear many-body problem, it has become clear in the interim from a host of theoretical studies and applications that the Kerman-Klein method is a general method of solving problems in the quantum mechanics of particles and fields by the application of Heisenberg's matrix mechanics. To illustrate this thesis, the basic ideas and applications to particle quantum mechanics and to collective motion in low energy nuclear physics are described in some detail, and many of the other applications and associated theoretical developments are discussed briefly.

1. Introduction

It has been a decade since we last reviewed[1] the subject matter of these lectures. In that review we took a more-or-less historical approach to the development of our ideas. From the pedagogical point of view we now consider that this was a disasterous way to proceed, and together with several other circumstances, for which we are at least partly responsible, may account for the fact that the generality and wide applicability of these ideas does not seem to be appreciated; in any event they have certainly not been utilized to the extent that the writers feels they should.

The ideas of which we speak were introduced into nuclear physics as a method of restoring the broken symmetries associated with mean field solutions to the nuclear many-body problem (see the cited review for the original references). In retrospect, it is now clear that the implementation of the basic notions for the case of a realistic nuclear interaction involves subtleties associated with the need to satisfy the Pauli exclusion principle that make this one of the more difficult applications of a very general and basically simple approach to solving problems in quantum mechanics. Unfortunately, in the earliest publications, when interest may have been at a maximum, we did not stress the distinction between the fundamental approach and possible difficulties inherent in specific applications. Indeed by continuing to emphasize these difficulties in later work[2], we have contributed to a view, held at least by the authors (Peter Ring, private communication) of one of the most important textbooks in the field of nuclear structure[3] that the Kerman-Klein(K-K) method is inherently inconsistent.

The first aim of these lectures will be, therefore, to show that, properly defined, the K-K method is simply a version of Heisenberg's matrix mechanics, and therefore to assert that it is inconsistent is tantamount to asserting that wave mechanics is inconsistent because we can't solve Schrödinger's equation for every problem. To make this point, we start, in sec. 2, by describing and illustrating the approach for problems in elementary quantum mechnics, in particular for a general anharmonic oscillator[4,5]. There we show how matrix mechanics may be used in a new way. Instead of diagonalizing the hamiltonian in a fixed basis, we solve the equations of motion and the commutation relations in a basis (to be determined self-consistently) in which the hamiltonian is *already diagonal* in a chosen subspace. This replaces the solution of a linear problem in a large space by the solution of a much smaller set of non-linear equations that characterize a relatively small space of true eigenstates of the system.

The fundamental problem is how to decouple the space of interest from the remaining states. Some of the basic notions involved are discussed in sec. 2 for the simple problem treated there. The problems encountered in decoupling for realistic nuclear applications are more difficult. Nevertheless, several methods have been developed that have proved successful in practice for a number of problems and models, several of which are described somewhat sketchily in secs. 3-5; these all involve the properties of even nuclei. In sec. 6, we describe our current viewpoint with respect to the treatment of odd nuclei by the principles under review. All of the above have to do with nuclear structure at low excitation. In sec. 7 we point out that the K-K method can also be used to study collective properties in the semi-classical domain, where, as an example, the total angular momentum of the system may be large compared to any single-particle value. Finally in sec. 8, we discuss a number of special topics. We first dispose of the question of consistency of the K-K method. We then point out how the method solves the problem of the redundancy of individual and collective coordinates. We follow that with a brief account of the variational properties of the method. We conclude with some observations on a new theory of effective interactions, research that is still partially connected to our main theme.

2. Quantum mechanics, low energy approximation

We study the problem defined by the hamiltonian

$$H = \frac{1}{2}p^2 + V(x), \tag{1}$$

by means of the operator equations of motion,

$$[x, H] = ip, \tag{2}$$

$$[ip, H] = \frac{dV}{dx} = V', \tag{3}$$

derived with the help of the commutation relation

$$[x, p] = i, \quad (\hbar = 1). \tag{4}$$

We apply these equations to a system with a large number of bound states, since this is the type of application for which the method to be described has been developed most fully. In the representation in which H is diagonal with eigenvectors $|n\rangle$, eigenvalues E_n, and where, for example, we write $x_{mn} = \langle m|x|n\rangle$, the equations of motion yield the matrix elements

$$(E_n - E_m)x_{mn} = ip_{mn}, \tag{5}$$

$$(E_n - E_m)ip_{mn} = (V')_{mn}, \tag{6}$$

whereas the commutation relations become, with the help of the completeness relation (we use the summation convention)

$$[x, p]_{mn} = x_{mn'}p_{n'n} - p_{mn'}x_{n'n} = i\delta_{mn}. \tag{7}$$

In principle these equations can be solved for all the matrix elements of the fundamental operators, x and p, and for the energy differences. This assertion has been verified explicitly for some of the problems of one-dimensional quantum mechanics that are exactly solvable analytically[5], though for cases other than the simple harmonic oscillator, this requires that we obtain equations for those functions of the coordinate operator that occur in the force $V'(x)$. The general case, where numerical methods are required, will be discussed below. To complete the scheme, we need a formula for the eigenvalues themselves. These can be computed as expectation values, using completeness explicitly for kinetic energy,

$$E_n = H_{nn} =< n|H|n >= \sum_{n'} \frac{1}{2}|p_{nn'}|^2 + < n|V(x)|n > . \tag{8}$$

In theory, we need only a single one of equations for the energy eigenvalues, since the differences are determined from the equations of motion themselves. In practice, the additional conditions can be used as consistency checks.

The above is simply a formulation of Heisenberg's matrix mechanics. In this formulation, we have treated both the dynamics (equations of motion) and the kinematics (commutation relations) on an equal footing. We have assumed that we know the matrix elements of the basic operators in the representation in which the hamiltonian is diagonal and we evalulate the matrix elements of a product of elementary operators by using the completeness relation. What do we have to offer in the way of a new *non-perturbative* method of dealing with these equations? The general philosophy of our approach is that instead of solving a large-dimensional linear problem, we shall solve non-linear equations for a much smaller number of variables. In fact, we have developed calculational schemes for two different regimes, the low energy algorithm and the semi-classical algorithm. We shall follow the plan of starting with the low energy scheme for the problem at hand and then follow through to show how this scheme may be generalized to the many body problem. (This is because the bulk of the applications have been of this kind.) We shall then turn to the the problem of semi-classical approximation within the current framework.

For the low energy approximation, it is convenient to assume that the potential energy, $V(x)$, is either given as a polynomial in the coordinates or else can be approximated by such a polynomial. To be specific, let us choose for simplicity

$$V(x) = \frac{1}{2}x^2 + \frac{1}{4}\lambda x^4. \tag{9}$$

With the potential as chosen, in the first application of the equation of motion, (6), we would naturally study the equation connecting the state $|0\rangle$ to the state $|1\rangle$, and thus encounter the matrix element

$$(x^3)_{01} = x_{0n}x_{nn'}x_{n'1}. \tag{10}$$

The possibility of obtaining a viable scheme of calculation depends on the rapid convergence of the sums over intermediate states. This will be true provided that

$$x_{n\,n+1} \ll x_{n\,n+3} \ll \dots \,. \tag{11}$$

If (11) is satisfied, a similar ordering will hold for the matrix elements of the momentum operator. In the present instance it is very easy to justify these assumptions on *a priori* grounds by considering the nodal structure of the wave functions, as well as by arguments intrinsic to the matrix method. However, one can also take the point of view that (11) is a guess, whose self-consistency can be checked *a posteriori*. If we write the general matrix element as $x_{n\,n+1+2k}$, what we find in practice when we solve the resulting scheme is that for every increase in the value of k, the *hierarchy* number, by one unit, there is a corresponding decrease in magnitude of the matrix element by roughly an order of magnitude. This occurs independently of the strength of the anharmonicity. The scheme is thus non-perturbative, but is framed in terms of quantities that can, nevertheless, be characterized as zeroth, first, etc. order.

To define a definite sequence of successively more accurate approximations, we must add the notion of a closure approximation. For example, to lowest order we have

$$(x^3)_{01} \cong x_{01}x_{10}x_{01} + x_{01}x_{12}x_{21}. \tag{12}$$

This equation makes it evident that there is no exact decoupling of lower states from higher ones, since $x_{12} \sim x_{01}$, but rather decoupling must be forced. Otherwise we must now study the equation for the matrix element x_{12}, and this will lead to coupling to still higher amplitudes. We obtain a closed set of equations by assuming that

$$x_{12} = \kappa x_{01}, \tag{13}$$

where $\kappa = \sqrt{2}$ in the harmonic limit, though this is not necessarily the best choice. With a definite choice for κ, we obtain three equations, two equations of motion and the ground-state expectation value of the commutator, for three unknowns, x_{01}, p_{01}, and $E_1 - E_0$. It turns out that the results are accurate at the level of a few percent for any value of the coupling constant λ that occurs in eq. (9) for the potential energy.

For the simple problem under discussion, it is trivial to carry out a much more elaborate calculation, starting from the ground state, and cutting off at reasonably large values of n. There is always a closure approximation to be made, but very soon the matrix elements and energy differences referring to the lowest states have converged, numerically speaking, to values independent of where and precisely how the closure is made. The truth of these assertions can be verified by examining the results reported both for energies and for matrix elements in Li's paper[4]. A small sample of results for the ground state energy calculated for a range of coupling strengths λ and by a sequence of successively more accurate approximations, identified simply as $\nu = 1...4$ is reproduced in Table 1.

ν	$\lambda = 0.01$	$\lambda = 0.10$	$\lambda = 1.00$	$\lambda = 10.0$
1	0.50186114	0.51752077	0.62401642	1.02352050
2	0.50185891	0.51736489	0.62093221	1.00920933
3	0.50185891	0.51736485	0.62092704	1.00917042
4	0.50185891	0.51736485	0.62092703	1.00917032

Table 1: Ground state energies of the oscillator for selected values, λ, of the quartic anharmonicity and for a series of approximations of increasing complexity, labeled by an integer, ν, showing convergence to the exact value. Refer to the text for more details.

An important aspect of the equation of motion approach, as we have described it thus far, is that there is an associated variational formulation. In order not to interrupt the flow that is aimed primarily at building the ideas of the low-energy algorithm, we have relegated a technical discussion of the variational approach (or approaches) to sec. 8.

Though some work has been done on problems of more than one dimension, we have not published anything along these lines. However, some of the applications to so-called exactly solvable models of nuclear physics, i. e. models that that can be solved "exactly" on a computer, are approximately equivalent to problems of elementary quantum mechanics in two or more dimensions, and we intend to deal with these problems below as the next simplest class of problems. To be able to treat this kind of problem in a practical way, we have to do two things: (i) introduce a hierarchy of matrix elements just as we did for the most elementary problem, and (ii) devise an algorithm where zero order quantities are treated most accurately, first order quantities next most accurately, etc. In fact, these elements have been unvarying features of all so-called nuclear applications, and therefore will be described below.

3. Algebraic many-particle models, low energy approximation

We turn next to a series of models, that were introduced into nuclear physics for the following reasons: (i) By present standards they are trivial to study numerically, and in limiting cases even analytical results are known. (ii) In the best cases, such as the pairing model, they contain an essential piece of the physics that determines the low energy behavior of the nucleus as a many-particle system. The aim of the present brief interlude is describe for such simple cases how the Kerman-Klein method may be applied to physics within an irreducible representation (irrep) of a Lie algebra. Though in the work to be reported we deal directly with the Lie algebra, all of these models can be boson-mapped[6], and thus the connection with the previous section rendered closer. We shall therefore not repeat the discussion of those aspects of the method, especially the dynamics, where the ideas are essentially the same in both cases. We shall however discuss briefly a new aspect of the kinematics. Afterwards, we shall describe an application to the pairing hamiltonian[7].

For the Lie algebraic framework, let $\{X_\rho\}$ be the generators of the Lie algebra. In practice, these are always constructed as bilinear forms of single-fermion operators and thus only antisymmetric irreps of the associated algebra occur. The hamiltonian, H, chosen as a polynomial in the X_ρ, can then be viewed as an approximate shell-model hamiltonian. The elements of the method parallel those described in sec. 2 - equations of motion, commutation relations, and evaluation of matrix elements of the hamiltonian and other operators. The only additional but vital element is the specification of the representation by means of the Casimir invariants in relations of the form

$$\langle K|C|K'\rangle = \delta_{K,K'}C', \tag{14}$$

where C is one of the Casimir invariants, $|K\rangle$ is one of the eigenvectors of the hamiltonian, and C' is its eigenvalue for the irrep of interest. It is these numbers, simple to compute, that guarantee that we are describing fermions, since in this approach, explicit, antisymmetric wave functions never intervene. Nevertheless by including these constraints in the scheme, we can come as close as we wish to satisfying the Pauli principle. (It turns out that this is not the only method of satisfying the Pauli principle, and in the application described in the next section, another method was utilized.)

To illustrate this approach we show some results from the work of Carlos Dasso[7] on the two-shell pairing problem (solutions were also obtained for three shells). As described for the anharmonic oscillator, in order to decouple the low-energy space from the rest of the vector space we need the twin concepts of a hierarchy of matrix elements and closure approximations. We describe the form these assume for the pairing model.

The basic operators whose matrix elements define the scheme are either occupation number operators or pair operators for a given shell. Linear combinations of the elementary pair operators form either a Cooper or superconducting pair, of which the ground state is overwhelmingly composed, or else one of the less correlated pairs, referred to as broken pairs. The low-lying states can be characterized approximately as states of zero, one, ... broken pairs. A matrix element of an elementary operator has hiearchy number ν, if the operator connects states differing by ν broken pairs. In general the convergence factor associated with unit increase of hierarchy number is $1/\Omega^{\frac{1}{2}}$, where Ω is the average pair degeneracy of a shell. These assumptions, which can be argued on *a priori* grounds and justified *a posteriori*, allow us to pick out the

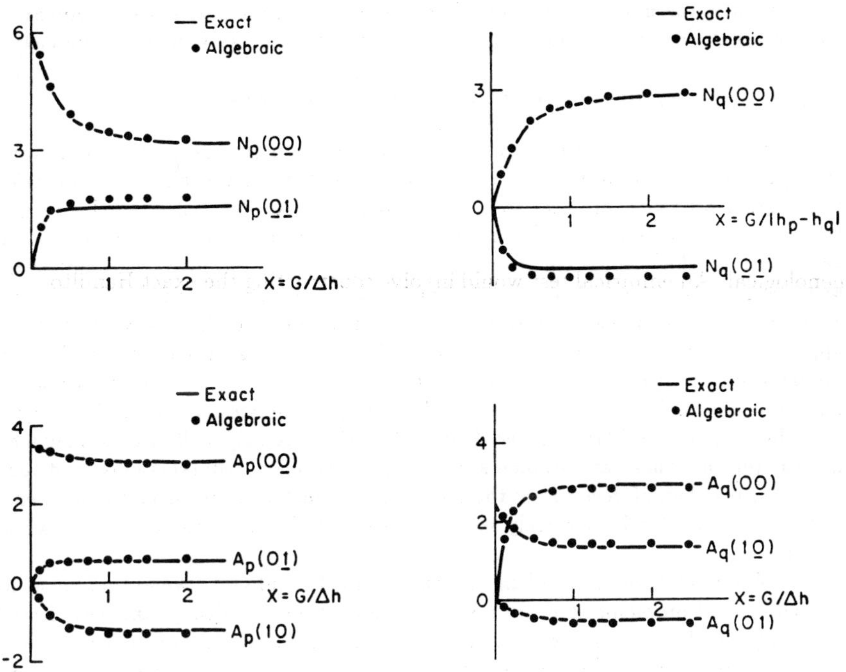

Figure 1: A comparison of some exact matrix elements in the two-level pairing problem to the approximation discussed above. For details refer to the text.

dominant terms of any dynamical or kinematical equation and to define a sequence of increasingly accurate approximations.

The closure argument consists of several pieces, that we can characterize as horizontal and vertical, respectively, and uses the idea that matrix elements that belong to a given level of hierarchy, mainly $\nu = 0$ and $\nu = 1$, are slowly varying functions both of the total particle number (horizontal closure) and of the number of broken pairs in each state (vertical closure). In lowest approximation these assumptions are known as the BCS and no-blocking approximations, respectively. But the method includes a procedure for successive improvement of these approximations.

Since the algebraic structure is SU(2)×SU(2), we use the quadratic Casimir invariants of each of these components to enforce the Pauli principle. As stated before, this is the major difference between this method and that described in the previous section.

In Fig. 1, we reproduce results for several of the matrix elements of the occupation number operators, N_a, and of the pair operators, A_a. The two levels are designated as $a = p, q$, with pair degeneracies $\Omega_p = \Omega_q = 4$ and single-particle energies, $h_p = 0$, $h_q = 2$MeV. The state-labels refer to the ground state and to the first excited state, the underline to particle number $N = 6$, the pair matrix elements also connecting N=8. In this figure, the pairing strength, G, was varied continuously.

4. Low energy properties of the even Ni isotopes

In the method outlined in the previous section for the pairing hamiltonian, the approach was to start with a formulation that preserved all the symmetries of the original problem. In order to obtain a solution that also respected these symmetries, we were forced, on practical grounds, to start the numerical procedures with symmetry-breaking approximations and subsequently to improve the calculations so as to build accurate solutions of the symmetry preserving structures. This approach gave extremely accurate solutions for all the simplified models to which they were applied. In retrospect, this is not too surprising, since these models can all be boson-mapped onto problems almost equivalent to problems of a few coupled oscillators in quantum mechanics, and for these problems at low energies, versions of the method described for one oscillator in sec. 2 should work extremely well. In the language of nuclear physics, these are problems involving collective coordinates only, and though there is a problem of decoupling the low energy collective states from the high energy collective ones, this is relatively easy to do in non-linear formulations, as we have shown.

When we turn to problems of realistic nuclear physics, we encounter at least two additional problems that are related to each other technically, and arise from the fact that a realistic microscopic model has both collective and non-collective degrees of of freedom. The first problem is to decouple these two sets of degrees of freedom to the extent possible, but also to describe how we might successively include the coupling where this is dictated by the physics. The second problem is how to take into account the restrictions imposed by the Pauli principle.

The relation between these problems can be understood qualitatively. Any shell model hamiltonian is a polynomial in the generators of a suitably defined Lie algebra. In a simplified model, all the generators refer to collective degrees of freedom. In a realistic model only a subset of the generators can refer to collective degrees of freedom. However, these generators do not form a closed subalgebra, at least not in our approach, where the collective degrees of freedom are determined by the hamiltonian itself and not imposed from the outside. It has been argued that nevertheless closure can be obtained approximately in a suitable subspace by means of algebraic arguments suggested by the simpler models. Recall from previous discussion that there are three algebraic elements that must be considered in any calculation, the equations of motion, the commutation relations of the algebra, and the Casimir invariants that specify the representation space, i.e., that enforce the Pauli principle.

We have shown repeatedly that it is relatively easy to decouple for the equations of motion and for the commutation relations because we can confine our study to a large degree to the matrix elements of collective operators alone. The intellectual challenge resides in how to deal with the Pauli principle, where this cannot be done so readily. We have suggested two ways of dealing with this problem, both of which have been shown, by example, to work in the vibrational domain, i. e., the domain of the broken pair approximation. We believe that the same ideas can be formulated for the pseudo-symplectic shell model, but nothing has been done in this direction. The first method[8] consists in showing that the quadratic Casimir invariant of the full shell-model algebra can be written approximately as a sum of two separately conserved invariants, one referring to the collective subspace and the other to the rest of the space, and that this division, which is a natural generalization of seniority conservation, works well as long as the nucleus in question has a stable spherical shape. Below, we shall use a second method, appropriate when we introduce the Bogoliubov-Valatin transformation into our hamiltonian and thus attack the problem somewhat more conventionally than the method applied to the pairing hamiltonian. In this method, we use the Pauli principle to solve algebraically for the quasi-particle multipole operators in terms of the quasi-particle pair operators as an expansion where collective and non-collective terms can easily be distinguished. Owing to space limitations, we cannot enter into the details of either method here.

We now describe the application of our ideas to the Ni isotopes[9,10], that represents the farthest we have gotten in the application of the algebraic ideas to realistic nuclei. The starting hamiltonian for this study is

$$\begin{aligned} H &= H_{sp} + H_p + H_{Q-Q} \\ &= \sum_a h_a N_a - G \sum_{ab} \sqrt{\Omega_a \Omega_a} A^{(0)\dagger}(a) A^{(0)\dagger}(b) \\ &\quad -\frac{1}{2}\chi \sum_q (Q_q^\dagger Q_q)_{\text{modified}}, \end{aligned} \tag{15}$$

where

$$Q_q^\dagger = \sum_{\alpha\beta} (\alpha|r^2 Y_q^2(\hat{r}|\beta) a_\alpha^\dagger a_\beta \tag{16}$$

is the quadrupole operator associated with the valence shell. The subscript *modified* means that pairing matrix elements have been removed from the operator. This was found to be necessary because of the large and unphysical effect of such matrix elements when the level degeneracies are small, as they are for Ni$(2p_{1/2}, 2p_{3/2}, 1f_{5/2})$. Ultimately other modifications were called for in the hamiltonian.

We outline the method used to study the spectrum of the hamiltonian for a nucleus with a quadrupole phonon spectrum and a "superconducting" ground state.

(i) We start with a Bogoliubov-Valatin transformation to quasiparticle operators denoted by a tilde,

$$a_\alpha^\dagger = u_\alpha \tilde{a}_\alpha^\dagger + s_\alpha v_\alpha \tilde{a}_\alpha, \tag{17}$$

$$u_\alpha^2 + v_\alpha^2 = 1, \tag{18}$$

where the coefficients u_α, v_α will be determined by a method which is an improvement over the usual BCS procedure. Under this transformation the pair and multipole operators $A_M^{(J)\dagger}$ and $B_M^{(J)}$ undergo linear transformations which convert them into linear combinations of the corresponding quasiparticle operators $\tilde{A}_M^{(J)\dagger}$ and $\tilde{B}_M^{(J)}$. With these changes H becomes (symbolically)

$$H = \tilde{H}(u, v, \tilde{A}, \tilde{A}^\dagger, \tilde{B}), \tag{19}$$

where only the operators for $J = 0$ and $J = 2$ intervene. To insure number conservation, on the average, and also to damp number fluctuations, we work with the auxiliary hamiltonian (Lipkin-Nogami method)

$$H' = H - \lambda_1 N - \lambda_2 N^2. \tag{20}$$

(ii) From the resulting operator we drop all terms containing $\tilde{A}^{(0)\dagger}(a)$ or $\tilde{A}^{(0)}(a)$ (or both). This corresponds to the omission of pairing vibrations from the present considerations.

(iii) For the treatment of the Pauli principle, as stated above, we choose the method that involves the elimination of multipole operators from the hamiltonian. The expression we need, written in the angular momentum coupled representation and truncated at fourth order is

$$\begin{aligned}\tilde{B}_M^{(J)}(ab) = {} & (2/\hat{J}) \sum_{1,2} (-1)^{J_2 - M_2} y(J\bar{2}1) \tilde{A}_1^\dagger \tilde{A}_2 \\ & -(4/3\hat{J}) \sum_{1\ldots 6} (-1)^{J_3 + J_4 - M_3 - M_4} y(J61) y(65\bar{3}) y(5\bar{4}2) \tilde{A}_1^\dagger \tilde{A}_2^\dagger \tilde{A}_3 \tilde{A}_4,\end{aligned} \tag{21}$$

where

$$y(123) = \hat{J}_1 \hat{J}_2 \hat{J}_3 \delta_{b_1 b_2} \delta_{a_1 a_3} \delta_{a_2 b_3} (-1)^{j_{a_1} + j_{b_1} + M_1} \begin{pmatrix} J_1 & J_2 & J_3 \\ -M_1 & M_2 & M_3 \end{pmatrix} \begin{Bmatrix} J_1 & J_2 & J_3 \\ j_{a_2} & j_{a_3} & j_{b_1} \end{Bmatrix}, \tag{22}$$

in terms of $3-j$ and $6-j$ symbols, $\hat{J} = (2J+1)^{1/2}$, $\bar{2}$ means the set $(a_2, b_2, J_2 - M_2)$, and the J in $y(J\bar{2}1)$, e.g., represents (J, M, a, b). In eq. (21), terms with six or more operators have been neglected.

(iv) Substituting eq. (21) in H' and into the algebra yields, symbolically again,

$$H' = H'(u, v, \lambda_1, \lambda_2, \tilde{A}^{(2)}, \tilde{A}^{(2)\dagger}), \tag{23}$$

$$[\tilde{A}, \tilde{A}^\dagger] = 1 + c\tilde{A}^\dagger \tilde{A} + \ldots, \tag{24}$$

and c represents an appropriate set of constants.

We now outline the calculational procedure designed to utilize the tools and ideas described above. The main point, not to be obscured by the details that follow, is that it represents a non-linear generalization of the standard BCS plus QRPA approach and includes an improved treatment of number conservation as well. The first step is to evaluate the ground state expectation value of H'. In the associated sum rule evaluation, besides the ground state, there intervene the one broken-pair or one phonon 2+ states (of which there will be five in this calculation = number of independent one particle quadrupole operators) that will be called $2, 2', 2''...$ and the two phonon 2+ states $2_2, 2_2', ...$. Of the latter we shall retain only the collective one. The matrix elements that appear in this calculation of $\langle 0|H'|0\rangle$ up to the two phonon or second layer of states are (symbolically)

$$X = \langle 0|A^{(2)}|2_1\rangle, \quad \text{Tamm} - \text{Dancoff}, \tag{25}$$

$$Y = \langle 0|A^{(2)\dagger}|2_1\rangle, \quad \text{ground state correlation}, \tag{26}$$

$$\theta = \langle 2_1|A^{(2)}|2_1'\rangle, \quad \text{quadrupole matrix}, \tag{27}$$

$$\Phi = \langle 0|A^{(2)}|2_2\rangle, \quad \text{crossover matrix}, \tag{28}$$

where we have indicated the usual designation of these amplitudes. Thus we have

$$F_0 \equiv \langle 0|H'|0\rangle = F_0(u, v, \lambda, X, Y, \theta, \Phi). \tag{29}$$

(v) To determine the quantities $u, v, \lambda_{1,2}$, as well as the amplitudes just defined, we employ various means - variational, equations of motion, and commutation relations. For example, from the conditions

$$\delta F_0/\delta u_a = \delta F_0/\delta v_a = 0 \tag{30}$$

and number conservation, we obtain a generalized form of the BCS theory which depends on the unknown amplitudes (25)-(28) in addition to the usual zero order BCS quantities. From the equations of motion for $\tilde{A}^{(2)}$, namely $[\tilde{A}^{(2)}, H']$, we obtain equations for the first order X and Y amplitudes, as well as for the second order amplitudes θ and Φ. From the commutation relations (24), we obtain normalization conditions, as well as alternative equations for computing the θ and Φ amplitudes.

(vi) To obtain a solution, we proceeded as follows: Setting $X = Y = \theta = \Phi = 0$, we solved the BCS-like equations for the $u_a, v_a, \lambda_{1,2}$. These values were then inserted into the equations for the X,Y which constitute a non-linear version of QRPA. The latter equations were then solved, neglecting the θ, Φ amplitudes, and the solutions normalized. Finally the θ and Φ amplitudes were calculated by perturbation theory. This constitutes a single cycle of the calculation, and provides new input for a second cycle. The steps were then repeated until convergence was achieved.

Some results of applying this theory to the isotopes of Ni will be described below. There is no reason that the calculation cannot be extended in many directions: inclusion of the two phonon states, odd nuclei, doubly open nuclei, larger shell model space.

Comparison of the results of the calculation with an exact shell model calculation for the same hamiltonian shows that the method yields accurate solutions. A comparison of theory with experiment, however, shows that the Q-Q interaction is inadequate to describe the four non-collective one-phonon states, implying too compressed a spectrum. We proposed a rough-and-ready means of removing this essential disagreement: Insofar as the equations of motion have the QRPA form, the various excited states can be written in terms of phonon creation operators ($i = 1, ...5$),

$$|2q(i)>= O^{\dagger}_{2q,i}|0>, \tag{31}$$

where $i = 1$ is the collective state. We then find that an additional term in the hamiltonian of the form

$$\delta H = \sum_{q,i} \delta\omega_i(1-\delta_{i,1})O^{\dagger}_{2q,i}O_{2q,i}, \tag{32}$$

can be used to correct the deficiency noted above. In practice we used values $\delta\omega_i = \delta\omega =$ constant.

^{60}Ni	Present work	MSDI	ASDI	Experiment
	$\chi' = 0.40, \delta\omega = 1.17)$			
ω_1 [MeV]	1.33	1.58	1.46	1.33
ω_2	3.03	2.96	2.76	3.12
ω_3	3.15	3.10	2.94	3.27
ω_4	3.39	3.33	3.34	3.39
ω_5	3.66	3.51	3.59	3.87
$\omega_{1,1}$	-	2.13	2.22	2.16
e_n	1.59	1.64	1.57	
B(E2)$[e^2 \cdot fm^4]$				
$2_1 \rightarrow 0_g$	186	186	186	186
$2_2 \rightarrow 0_g$	0.003	0.293	0.899	$0.75^{+0.52}_{-0.31}$
$2_3 \rightarrow 0_g$	0.710	0.110	0.129	$1.1^{+0.7}_{-0.4}$
$2_4 \rightarrow 0_g$	4.37	0.055	0.510	$0.49^{+0.34}_{-0.22}$
$2_5 \rightarrow 0_g$	4.83	0.786	0.167	$0.47^{+0.31}_{-0.20}$
$2_{1,1} \rightarrow 0_g$	1.68	3.19	0.001	< 1.5
$Q(2_1^+)[e \cdot f_m^2]$	-9.63	-16.8	-12.3	3 ± 7
$\frac{B(E2; 2_{1,1} \rightarrow 0_g)}{B(E2; 2_{1,1} \rightarrow 2_1)}$	-	0.020	5×10^5	$3.2^{+0.7}_{-0.5} \times 10^{-3}$

Table 2: Various results obtained for ^{60}Ni by the algebraic method are compared with shell model calculations carried out with the MSDI and ASDI. See text for definitions.

Some results obtained with the altered hamiltonian are then given in Tables 2 and 3, where they are compared with experiment and with two standard shell model calculations. In these tables ω_i are the one-phonon energies, $\omega_{1,1}$ the excitation of the two-phonon 2+ state, e_n the effective charge. In addition to energies, various B(E2) values and the quadrupole moment of the state 2_1+ have been computed. The overall agreement - with a hamiltonian that has hardly been chosen in an optimum way - appears to be of comparable quality when compared with the results obtained with

^{62}Ni	Present work	MSDI	ASDI	Experiment
	$(\chi' = 0.5, \delta\omega = 0.999)$			
ω_1 MeV	1.22	1.44	1.42	1.17
ω_2	3.00	2.88	2.75	3.06
ω_3	3.12	2.95	3.00	3.16
ω_4	3.54	3.25	3.17	3.26
ω_5	3.67	3.44	3.49	3.27
$\omega_{\underline{1.1}}$	-	2.28	2.22	2.30
e_n	1.48	1.48	1.43	-
$B(E2)[e^2 \cdot fm^4]$				
$2_1 \rightarrow 0_g$	179	179	179	179
$2_2 \rightarrow 0_g$	7.27	3.51	1.20	-
$2_3 \rightarrow 0_g$	0.200	0.621	0.851	
$2_4 \rightarrow 0_g$	8.50	0.395	0.683	
$2_5 \rightarrow 0_g$	3.18	0.018	0.073	
$2_{1,1} \rightarrow 0_g$	6.43	4.91	0.013	6.4 ± 1.9
$Q(2_1^+)[e \cdot fm^2]$	-6.78	-10.2	1.44	5 ± 12

Table 3: Same as Table 2, but results are for ^{62}Ni.

the modified surface delta interaction (MSDI) and with the adjusted surface delta interaction (ASDI). In any event the validity of the approximation scheme does not appear to be in doubt. Much work could be done in improving the hamiltonian and in extending the calculation.

5. Low energy approximation, deformed nuclei

Though the problem of restoring rotational invariance to the description of deformed nuclei was the problem that led to the invention of the K-K method, our own contributions to the solution of practical problems[1] has lagged behind those of several other groups of workers, in particular behind the efforts of Belyaev and Zelevinsky[11,12], who reinvented the K-K method, and Mikhailov and Nadjakov and their collaborators[13], who did have distinct ideas, that can nevertheless be connected mathematically to consequences of K-K. We list what has been done:

(i) Derivation of the self-consistent cranking model both at low and at high angular momentum. For the former case this means the derivation of the self-consistent version of the cranking formula for the moment of inertia, but going somewhat beyond that in including rotational corrections to the ground-state correlation energy.

(ii) Formal derivation of the phenomenology of the Bohr-Mottelson model.

(iii) As a generalization of (i), microscopic derivation of the Bohr-Mottelson model in the rigid rotor, harmonic vibration limit (adiabatic limit). Systematic perturbative treatment of non-adiabatic corrections to this limit. For more details of this work, we refer to Zelevinsky's report in this volume.

To the present day, the most satisfactory microscopic treatment of the low energy properties of deformed nuclei remains the Kumar-Baranger (K-B) theory[14] and its more modern developments in the hands of Kumar[15], who has gone in the direction of simplifying the theory but expanding drastically the size of the shell-model space in which calculations are carried out, and of Berger, Girod, and Gogny[16], who have improved the original program by taking more realistic interactions and including more collective coordinates. However, these theories are not self-consistent in the sense of the K-K theory. Since 1983, we have been studying the theory of large amplitude collective motion. The original stimulus for this work was the same for us as it apparently was for Baranger, who pioneered in this effort more than a decade earlier, namely, to generalize the K-B theory. We list the accomplishments of our program, which are certainly non-negligible overall, but are still disapointingly short of where we wish to be, in the light of the original aims of the work. For a more detailed account of this work, we refer to the separate article by Walet in this volume.

(i) The main part of a theory of large amplitude collective motion, whether fully self-consistent or otherwise, consists in the extraction of a classical collective hamiltonian, for example a Bohr hamiltonian, from time-dependent mean field theory, followed by the requantization. One fundamental task is to embed this classical theory into a suitable quantum framework. We tried this first at the beginning of our program in 1983, using the K-K method, and succeeded sufficiently to get the program off the ground[17]. Within the last year, however, we have become dissatisfied with the original papers, for two distinct reasons. The first reason is a very practical one, in that neither our early work nor anyone else's provided a systematic method of calculating quantum corrections as a series in $1/N$, where N is the number of degrees of freedom that participate in the collective motion. In an application still in progress, to the nucleus ^{28}Si, where N is not that large, it was considered important to include leading order quantum corrections. A formalism has been developed and the corrections made[18]. Though not vintage K-K, there is some residuum of the same ideas in this approach.

A second reason for wishing to reexamine the quantum foundations is fueled by the resurgence in interest in the Born-Oppenheimer approximation and its relation to the Berry phase. If one thinks about it, there really is a big difference between the way the "fast" variables are treated in molecular physics and the way in which they are usually treated in nuclear physics. Conceptually, we all know that the nuclear physics problem is much more difficult, in that in contrast to the molecular case, where the separation into slow and fast degrees of freedom is simple enough to explain in an introductory physics course, in the nuclear case the same array of single-particle motions must account for both the fast and for the slow degrees of freedom. As a consequence, most of the technology of the theory of large amplitude collective motion has been developed precisely for effecting this separation. The result is that the collective hamiltonian is a sum of a potential energy and a kinetic energy quadratic in the momenta. But we now know that a careful decoupling between the fast and slow degrees of freedom should yield effective magnetic interactions in the collective part. In work in progress, we are addressing this question for the theory of large amplitude collective motion.

(ii) At the mean field level, the problem of decoupling the slow degrees of freedom can be studied most profitably by mapping the fermion mean field theory onto a problem in hamiltonian dynamics, as has been known to be possible for several decades[19]. In the adiabatic limit, it is recognized that the goal is to find manifolds in configuration space that are exactly or approximately invariant under dynamical evolution of the system. Thus phrased, the problem is relevant for coupled systems with any number of degrees of freedom, and the theory that has been developed has been applied to several problems of interest that are intrinsically simpler than the many-particle problem of nuclear physics. The development of a theory, i. e., a mathematical structure and associated algorithms, and their application to systems with a few degrees of freedom is a relatively closed chapter of our work, of which a review has been given[20].

(iii) The above theory has been translated back into the mean field picture of nuclear physics, and we have tried to apply it to the low-lying spectrum of ^{28}Si, with some, but not yet complete success[18]. We have shown how to decouple four slow degrees of freedom, the angular momentum and one shape coordinate. In the potential energy involving the latter, we have included the leading quantum corrections. And we have solved the Bohr hamiltonian for the resulting problem. We have not yet succeeded in obtaining self-consistent solutions with two shape parameters, but recently we have undertaken an attack on this problem with renewed vigor, in the belief that it was our algorithm rather our physics that was at fault in previous attempts.

In addition to the above program which is still underway, we think that there are alternative dynamical approaches to the problem of deformed nuclei that are fully within the spirit of K-K and that in our view would be interesting to try. We mention two. The first is an extension to deformed nuclei of the method described in sec. 4 for the Ni isotopes, that would start with the transformation of the hamiltonian by means of a general Hartree-Bogoliubov transformation, and has been characterized as a non-linear generalization of the BCS plus QRPA appraoch. A second method would attempt to apply to these nuclei, within the pseudosymplectic formulation, the symmetry preserving techniques shown to be successful for the vibrational regime of the single-j shell model[8]. Just as in the vibrational case, where the essential point was the relation of the full shell-model algebra to an approximate closure of a collective subalgebra generalizing the pairing subalgebra, here the emphasis would be on the relation of the full symplectic algebra to its SU(3) subalgebra.

6. Odd nuclei, core-particle coupling methods

Thus far in this paper we have discussed only the physics of even nuclei. As older readers may know, the original K-K method was derived as a theory of single particle amplitudes that coupled together in a self-consistent way the properties of even and of neighboring odd nuclei. Though unobjectionable in principle, and a perfectly acceptable way of deriving all the know mean-field limits of the nuclear many-body problem, in practice it proved cumbersome to carry out numerical investigations based on algorithms constructed from this formalism. It was just too difficult to control all the elements of self consistency. The methods based on pair amplitudes and Lie algebras have proved to be more promising to the extent to which they have been applied to the same problems. (We refer to this development as the decoupling of the even-nucleus problem form the odd-nucleus one.) The formalism based on amplitudes for single fermion operators is nevertheless alive and well, continuing to play a role, and likely to play an increasingly important role in the future in the description of odd nuclei. For an alternative and more complete account of some of the ideas discussed below, see the report by Semmes in this volume.

The relationship of this formalism as currently used to that which was originally envisaged is similar to the relationship of the Hartree-Fock problem to a phenomenological single-particle model, in that in the former the single-particle potential is to be computed self-consistently, whereas in the latter, the problem is linearized by choosing the single particle potential according to one's best judgment. In our case, what we obtain by giving up full self-consistency are generalized core-particle coupling models.

In chapter 3 of the book by Ring and Shuck, one finds an account of the phenomenological core-particle coupling model for deformed nuclei, with the statement that though this model has proved of great value in the correlation of data, nobody has provided a microscopic foundation for it. In fact, we had published just such a derivation more than a decade earlier[21], based on the K-K formalism. It involves taking a relevant version of the original K-K formulation for deformed nuclei and replacing self-consistency requirements by phenomenological input, as already implied by previous remarks. Here the generalized self-consistent potential is replaced by the sum of a prescribed deformed potential and the rotor hamiltonian. (As an example of what one loses thereby is the ability to describe the change of the self-consistent potential with angular momentum.) With a student, we have undertaken a reexamination of the relationship between the phenomenological and the microscopic model, and extended the range of the investigation to include the basis for the rotating shell model.

The fundamental approach to the core-particle method of describing odd nuclei is also alive and well in an adaptation developed by Dönau and co-workers[22,23]. We shall explain in a very schematic way what they have accomplished. We describe the formalism for a hamiltonian that is the sum of a spherical single-particle part, a pairing part, and a quadrupole-quadrupole interaction, but as far as principle is concerned, the only restiction is that the interaction be given as a sum of pairing and of multipole forces separately. We label the subspace of states of the even nucleus that are included in the model by the quantum numbers I, M_I, s, where s includes all quantum numbers other than the angular momentum, those of the odd nucleus so included by J, M_J, ν, where ν again comprises all quantum numbers other than the angular momentum, and the labels $a, b, \ldots$ will identify single-particle states, except for their magnetic quantum numbers. One starts with the fundamental equation of motion of the original K-K method for the model under discussion. It is a generalized Hartree-Bogoliubov equation for amplitudes that connect odd and neighboring even nuclei by means of a single-particle creation or annihilation operator. After application of the Wigner-Eckart theorem it has the form of an eigenvalue equation

$$\mathcal{E}_{J\nu}\Psi_{J\nu}(aIs) = \sum_{bI's'} \mathcal{H}(aIs|bI's')\Psi_{J\nu}(bI's'), \tag{33}$$

where the eigenvalue, $\mathcal{E}_{J\nu}$ measures the energy of a state in the odd system relative to the ground state of one of the neighboring even systems, and the generalized core-particle hamiltonian, $\mathcal{H}$, has the structure

$$\mathcal{H} = \begin{pmatrix} -\epsilon + \omega - \Gamma & \Delta \\ \Delta & \epsilon + \omega + \Gamma \end{pmatrix}. \tag{34}$$

Here ϵ are the single particle energies measured relative to the chemical potential, Δ is the generalized pairing potential, Γ is the self-consistent field generated by the quadrupole-quadrupole interaction, and the ω are the excitation energies in the even nucleus. As written, the equation explicitly conserves angular momentum but not particle number. But a version that restores particle number conservation can easily be obtained. As an example of the elements of the matrix (34) we have

$$(aI|\Gamma|a'I') = -\frac{1}{2}\kappa(-1)^{j_{a'}+I+J} \begin{Bmatrix} j_a & j_{a'} & 2 \\ I' & I & J \end{Bmatrix} \langle I||Q||I'\rangle q_{j_a j_{a'}}, \tag{35}$$

where κ is the strength of the quadrupole interaction and $\langle I||Q||I'\rangle$ and $q_{j_a j_{a'}}$ are reduced matrix elements of the quadrupole operator for core and particle respectively.

We note in passing that the equation given above implies, in this formalism, the basic dynamical equations of the Lie algebraic method. If we define a "generalized density matrix" by means of the equation

$$\mathcal{R}(aIs|bI's') \equiv \sum_{J\nu} \Psi_{J\nu}(aIs)\Psi^{\dagger}_{J\nu}(bI's'), \tag{36}$$

then from eq. (33) and its hermitian conjugate we derive, in matrix form

$$[\mathcal{R}, \mathcal{H}] = 0. \tag{37}$$

Prior to the introduction of the method under discussion, our approach to the formalism presented schematically above was to insert the physics of a limiting model and thus to treat separately the vibrational and rotational regimes. This seemed particularly necessary if self-consistency was to be achieved, or if we were to derive one of the standard versions of the core-particle coupling model. Dönau and collaborators introduced a new way of dealing with these equations in a semi-phenomenological manner. We shall here describe only the basic elements of their approach:

(i) The potentials and energies that occur in the effective hamiltonian, $\mathcal{H}$, are to be chosen phenomenologically. This linearizes the equations.

(ii) Angular momentum is conserved exactly.

(iii) Particle number also plays an important role, in that a given odd nucleus is compared to neighboring even nuclei on either side in particle number. Though this is not shown in eq. (34), the excitation spectrum of both even partners should intervene in the equations. This suggests that this formalism may be capable of treating odd nuclei even in a transition region between spherical and deformed neighbors. This has turned out to be the case.

(iv) Once the formalism has been linearized, one can treat a much larger space of states than is possible in a self-consistent treatment. This is both an advantage and a disadvantage in that the larger space should improve the accuracy of the treatment, but on the other hand, information about some of the higher lying states may be insufficient to provide reliable phenomenological input concerning these states. We refer to the report of Semmes for further details.

7. Semi-classical approximations; high spin, high energy

In our exposition thus far, we have chosen to emphasize the study of both few-body and many-body problems in the neighborhood of the ground state. For systems with a few degrees of freedom, this is intrinsically a quantum problem. For the many-body problem, in a formulation that preserves all the symmetries of the hamiltonian, we also have an intrinsically quantum problem. However, most of the progress to date has been made by expanding the solution in powers of $1/N$, where N is some measure of the number of degrees of freedom that participate in the collective motion of interest. The leading term is always a mean field solution, that represents a classical or semi-classical limit of the many-body problem. In this sense there is a false dichotomy between the work described previously and the work to be described in this section. There is another sense, however, in which the distinction is completely justified. In the low energy domain, it is possible to obtain solutions without expanding about the mean field solution, as we have already shown for simplified models. On the other hand, for the problems studied in this section, we do not believe that it is possible to proceed except by expanding about a classical or mean field solution. Depending on the problem discussed, the large dimensionless quantity that puts us into the semi-classical regime can be, among others, a large oscillator quantum number (usual WKB approximation), a large value of the angular momentum relative to that carried by a single particle (cranking model, rotating shell model), and the ratio of the periods of the bound state motion to the collision time for heavy ion collisions.

In this account, we have space enough only to advertise the questions that have been addressed. Let us first consider the problems of one or a few degrees of freedom. By studying the equations of sec. 2 in the limit of large quantum number, n, we have given a new derivation of the WKB bound-state quantization condition[24]. A side benefit of this procedure is the provision of a non-perturbative, time-independent method of obtaining solutions of the equations of motion. The formal reasoning has been extended to multi-dimensional systems[25] where it provides a method of solving the equations of motion for non-resonant cases as provided by the famous KAM theorem. Another example of an application to a relatively simple system was the construction of approximate analytic solutions to the triaxial rotor at large values of the angular momentum[26]. Before returning to the discussion of nuclear physics, we point to the existence of a series of papers that can be traced from ref.[27], in which the K-K method was applied to field theories in 1 + 1 dimensions, with soliton solutions representing the requisite mean field starting point. Quite recently, we have returned to this class of applications, but at a more difficult level, namely, an effort to turn the Skyrme model into a full-fledged quantum field theory[28].

Turning back to nuclear structure, our main interest at the moment is in seeing to what extent our approach can provide some insight for problems at high angular momentum. Concerning this problem, we can draw attention to a long history of interest in this subject, centering about two periods. In 1965, Dreizler and one of the authors (AK) produced the first derivation[29] of the self-consistent cranking model from a many-body starting point, but only for the abelian case (rotation about a fixed axis). This work was reexamined and extended to the non-abelian case by Vassanji in his thesis in the late seventies[30]. These results are of very limited interest in the light of the rapidly accumulating data concerning the behavior of real nuclei at high spin, but it should be possible to extend the methods employed in these earlier preliminary works to obtain microscopic versions of the rotating shell model and band-mixing models.

Finally we call attention to some other fragmentary work in the semi-classical domain, a study of peridic solutions within the mean field approximation[31] and several papers on how to relate scattering solutions of TDHF to the quantum theory of scattering from and to definite reaction channels[32].

8. Consistency of the Kerman-Klein theory; treatment of redundant coordinates; variational principle; effective interactions

We now turn to a question that we have assiduously avoided up to now. Is the Kerman-Klein theory inconsistent? Here we shall explain briefly how such a question arose and then try to put it to rest. To illustrate the problem, let us consider a general shell model hamiltonian of the form

$$H = \sum_{\alpha\beta} h_{\alpha\beta} a_\alpha^\dagger a_\beta + \frac{1}{4}\sum_{\alpha\beta\gamma\delta} V_{\alpha\beta\gamma\delta} a_\alpha^\dagger a_\beta^\dagger a_\delta a_\gamma \tag{38}$$

$$= \sum_{\alpha\beta} h'_{\alpha\beta} a_\alpha^\dagger a_\beta + \frac{1}{4}\sum_{\alpha\beta\gamma\delta} V_{\alpha\beta\gamma\delta} a_\alpha^\dagger a_\gamma a_\beta^\dagger a_\delta, \tag{39}$$

where the two forms differ by an obvious rearrangement.

Let $|K\rangle$ denote the subspace of states that we wish to include in a dynamical scheme. To illustrate the type of problem that has given rise to controversy, let us consider the evaluation of a matrix element of the hamiltonian, $\langle K|H|K'\rangle$. The delicate point is associated with the interaction. If we use completeness in the most straightforward unthinking way, we would, in general, obtain a different result from the two forms (38) and (39), one involving matrix elements of operators that change particle number by two units, and the other matrix elements of number-conserving operators; except for the completely unrealistic limit that the space of states $|K\rangle$ is complete, these two decompositions are not equal. The first of the possible choices emphasizes pairing correlations at the expense of multipole correlations, whereas the second does the reverse (but at the same time fails to satisfy the Pauli principle). For this reason, we worried very hard[1,2] to invent more physically motivated but necessarily approximate decompositions that included both kinds of correlations and, depending on the particular choice, did less violence or no violence at all to the Pauli principle. One obvious inconsistency (and depending on the choice made not necessarily the only one) is that if the suggested modifications were applied to a complete space we would run into an overcounting problem. This means that, in contradistinction to the schemes proposed in previous sections of this paper, the schemes now under discussion never have the exact solution as a limit. An exception to this is the case where the interaction is given as a sum of schematic interactions, each of multipole or pairing type, where the factorization to be used in each case is obvious. In practice, one often confines oneself to such a hamiltonian, but it should be emphasized that such a choice really begs the question.

Nevertheless we believe that these schemes for approximate factorization have merit and in the general case represent the physics in the collective subspace much better than any of the straightforward uses of completeness. The strongest support for this assertion is based on the demonstations, known from the earliest days of the theory, that these forms reduce to all the well-known limiting mean field solutions. But, up to now, very little has been done to solve such schemes in a symmetry-preserving approximation for realistic models. For the moment, the so-called inconsistencies are therefore of no consequence whatever, since we have yet to explore fully the physics of symmetry preserving approximations within a collective subspace. Furthermore, to treat the coupling between the collective and the non-collective spaces, we have recently emphasized[18] the availability of systematic expansions in $1/N$ which treat the Pauli principle correctly and are free of contradictions.

To summarize this discussion, what we have found is that for fermion systems with short range interactions, where the hamiltonian has not been schematized, the formal implementation of the K-K method is such that one cannot specify a single symmetry preserving algorithm that both gives good physics for a small space and at the same time will be theoretically correct in the limit that we keep the entire Hilbert space. For the time being and for the forseeable future, this represents no practical limitation for any useful applications. Furthermore, we know of no competing approach to the study of collective motion where this problem does not arise in some form.

We next turn to a topic about which we need not be apologetic. It was owing to a remark by Marumori that we first realized that the K-K method provides a "natural" solution to the problem of redundant coordinates that concerned many authors who tried to find a derivation of the Bohr-Mottelson model in the early days of that model. The problem is simply this: One starts with a hamiltonian expressed in terms of a complete set of single-particle degrees of freedom. One then introduce a small number of collective coordinates. In these early attempts, it was very difficult to develop a formalism in which one introduces explicitly a set of independent coordinates to complement the collective subset. The preferred method was then to work with an overcomplete or redundant set of variables, collective plus original set, and to impose supplementary conditions. This brings problems of its own such as the need to use Dirac's method for quantization of systems with constraints, though this is only a technical complication that one can live with in many instances. In the K-K method, the problem apparently never arises because we work only with the original single-particle degrees of freedom. Collective coordinates enter only in the description of the Hilbert space of states. As the most elementary example, if in the chosen collective subspace $|K\rangle$, the total momentum is included among the labeled subsumed by K, then the coordinate of the center of mass may be introduced by a standard application of transformation theory to the states (fourier transform) The use of such localized states and their properties is essential in some of the approximation schemes developed for the K-K method, but we are never forced to make a transformation of operators. A connection with the Dirac method of quantization may be made through the study of sum rules for the commutation or anti-commutation relations, where it can be seen that the same final results may be obtained from different starting points.

We consider next an attribute of the K-K method that has importance both theoretically and practically, namely the existence of associated variational principles. We illustrate the ideas with the help of the problem of one-dimensional quantum mechanics treated in sec. 2. A natural question, in any event one that we asked ourselves, is whether one can consider the expression for the energy, eq. (8), to be stationary with respect to variation of the $x_{nn'}$ and $p_{nn'}$. There are two objections to such a view: (i) the matrix elements are not all independent. (ii) The same matrix elements appear in the energy functionals for different states. A solution to the second problem, though not the only one, is to form an average of the stationary functionals. Though many averages are possible, the trace is usually the most convenient, as it turns out to be the most natural. Thus we require

$$\delta \sum_n H_{nn} = \delta \,\mathrm{Tr}\, H = 0. \tag{40}$$

A solution to the first problem is to impose all the possible kinematical constraints, namely,

$$\delta [x, p]_{nn'} = 0. \tag{41}$$

Introducing a Lagrange multiplier matrix $(-i)\Lambda_{nn'}$ (Λ is Hermitian), we are thus led to a master variational principle

$$\delta \,\mathrm{Tr}\{H - i\Lambda[x,p]\} = \delta \,\mathrm{Tr}\{H - ip[\Lambda, x]\} = \delta \,\mathrm{Tr}\{H + ix[\Lambda, p]\} = 0, \tag{42}$$

using the cyclic invariance of the trace. Carrying out the variations, now unconstrained, we find the equations

$$p_{nn'} = -i[x, \Lambda]_{nn'} \tag{43}$$

$$(V')_{nn'} = i[p, \Lambda]_{nn'}. \tag{44}$$

Because of the invariance of the trace with respect to choice of basis, the representation $|n\rangle$ is, at this point, arbitrary. The most convenient immediate choice is the one in which the Hermitian operator Λ is diagonal. Comparing with the known equations of motion, we then identify Λ as the hamiltonian. Thus we may rewrite the variational principle (42), using the second form,

$$\delta \operatorname{Tr}\{H - p\dot{x}\} = -\delta \operatorname{Tr} L = 0, \tag{45}$$

where L is the Lagrange operator. The final result is a quantum Hamilton'sprinciple, with unconstrained variation of coordinates and momenta, which has been derived from a Rayleigh-Ritz principle with constraints.

This is not the only form of the variational principle[5], but we stop here, only adding a few remarks about applications. We haved used the variational principle directly in the study of several nuclear models, for example in our work on the single-j model[8]. As an example of a different kind[25], we have indicated that for a system of coupled oscillators, e. g. if we view the quantum numbers of the state $|K\rangle$ to be made up of occupation numbers that have survived from the description of uncoupled oscillators, then this would be the quantum analogue of the concept of survival of the invariant tori that is such an important consequence of the KAM theorem. In the semi-classical limit, these occupation numbers becomes action variables. By fourier transformation, the trace turns into an average over angle variables, or equivalently, a multiple time average of the lagrangian. The final result is an important starting point for the study of such systems.

As a final set of observations, we call attention to work that we have done on the theory of effective interactions[33], still at quite a young stage, that mixes equation of motion techniques with coupled cluster methods. The point is that the theory of collective motion, or at least our approach to it, starts from a hamiltonian with smooth effective interations. It was natural, therefore, to become interested in looking once more at the problem of relating the effective interaction to the "observed" two-body interaction. We have asked the following question. Can we find a smooth, energy-independent interaction that can be used both in the calculation of nuclear binding and in the equations of motion that describe collectivity at low energy? We have proved that this can done provided differences of order $1/N$ can be neglected, where N is the number of particles in the system. We have thus far been unable to give this theory the further attention that it deserves.

9. Acknowledgement

This work was supported in part by the U. S. Department of Energy under grant number 40264-5-25351.

10. References

1. A. Klein, in *Progress in Particle and Nuclear Physics*, edited by D. Wilkinson (Pergamon, London, 1983), Vol. 10, p. 39.
2. A. Klein, *Phys. Rev.* **C30**, 1680 (1984).
3. P. Ring and P. Schuck, *The Nuclear Many Body Problem*, (Springer Verlag, New York, 1980).
4. C. T. Li, A. Klein, and F. Krejs, *Phys. Rev.* **D17**, 2311 (1975).
5. C. T. Li and A. Klein, *Fizika* **22**, 67 (1990).
6. A. Klein and E. R. Marshalek, *Rev. Mod. Phys.* **63**, 375 (1991)
7. C. Dasso, A. Klein, C. Y. Wang-Keiser, and G. J. Dreiss, *Nucl. Phys.* **A205**, 200 (1973).
8. M. Vallières, A. Klein and R. M. Dreizler, *Phys. Rev.* **C7**, 2188 (1973).
9. C. T. Li, P. K. Chattopadhyay, A. Klein, and M. J. Vassanji, *Phys. Rev.* **C19**, 2002 (1979).
10. C. T. Li and A. Klein, *Phys. Rev.* **C19**, 2023 (1979).
11. S. T. Belyaev and V. G. Zelevinsky, *Sov. J. Nucl. Phys.* **16**, 657 (1974).
12. V. G. Zelevinsky, *Proc. 5th Kyoto Summer Institute*, Y. Abe and T. Suzuki, eds. (Kyoto U. 1983), p. 251.
13. D. Karadjov, I. N. Mikhailov, E. Nadjakov, and J. Piperova, *Nucl. Phys.* **A305**, 78 (1978); **A305**, 93 (1978).
14. K. Kumar and M. Baranger, *Nucl. Phys.* **A122**, 273 (1968).
15. K. Kumar, *Nuclear Models and the Search for Unity in Nuclear Physics* (Universitetsforlaget, Oslo).
16. J. F. Berger, M. Girod, and D. Gogny, *Nucl. Phys.* **A502**, 85c (1989).
17. A. Klein, *Nucl. Phys.* **A410**, 74 (1983); **A431**, 90 (1984).
18. N. R. Walet, A. Klein, and G. Do Dang, *Phys. Rev.* **C45**, 249 (1992).
19. E. R. Marshalek and J. Weneser, *Phys. Rev.* **C2**, 1682 (1970).
20. A. Klein, N. R. Walet and G. Do Dang, *Ann. Phys. (N. Y.)***208**, 90 (1991).
21. G. Do Dang, G. J. Dreiss, R. M. Dreizler, A. Klein and Chi-Shiang Wu, *Nucl. Phys.* **A114**, 481 (1968).
22. F. Dönau and S. Frauendorf, *Phys. Lett.* **B71**, 263 (1977).
23. H. Helpi, J. Hattula, A. Lunkko, M. Jäaskelainen, and F. Dönau, *Nucl. Phys.* **A357**, 333 (1981).
24. A. Klein, *J. Math. Phys.* **19**, 292 (1978).
25. A. Klein and C. T. Li, *J. Math. Phys.* **20**, 572 (1979).
26. M. G. Vassanji and A. Klein, *Phys. Rev.* **C19**, 2349 (1979).
27. A. Klein and A. Weldon, *Phys. Rev.* **D17**, 1009 (1978).

28. D. Cebula, A. Klein and N. R. Walet, *J. Phys. G: Nucl. Part. Phys.* **18**, 499 (1992).
29. A. Klein and R. M. Dreizler, *Phys. Rev. Lett.* **15**, 893 (1965).
30. M. G. Vassanji and A. Klein, *Phys. Rev. Lett.* **42**, 436 (1979).
31. A. Klein and A. S. Umar, *Phys. Rev.* **C34**, 1965 (1986).
32. A. Klein and A. S. Umar, *Phys. Rev.* **C35**, 1672 (1987).
33. A. Klein and T. Une, *Phys. Rev.* **C38**, 1886, 1897 (1988).

REVIEW OF THE KERMAN-KLEIN METHOD

P.-H. HEENEN
Université Libre de Bruxelles, Brussels, Belgium

W. SATUŁA
University of Lund, Lund, Sweden

P. SEMMES
Tennessee Technological University, Cookeville, USA

The Kerman–Klein method (KK) is an attempt to formulate in a very general way a set of rules and approximations to treat complex many-body problems. A review of such method must deal with at least two facets of the method: the generality and the consistency of the approximations on the one hand, and, on the other hand, the applicability of the theory to specific physical problems.

The basic ingredients of the KK method are:

- Heisenberg's matrix mechanics
- equations of motion (EOM)
- relations of completeness

Thus, the only important quantities are the **matrix elements** of operators and not the eigenstates themselves. The Heisenberg formulation of quantum mechanics requires solving a large-dimensional set of linear equations. In principle, this can be done explicitly, yielding energy differences and matrix elements of given operators (additionally, the energy scale should be set by solving a single eigenenergy problem). The spirit of the KK method is to replace this linear large-scale problem by a set of non-linear equations for few fundamental (elementary) variables of physical interest and reduce the dimensionality by introducing a *hierarchy* for matrix elements. In other words, one assumes that the matrix elements of interest can be ordered in a descending, rapidly converging sequence. This last step is equivalent to a closure approximation and thus violates the Pauli principle.

The advantage of this prescription is the possibility, at least in principle, to take into account in a controlled, systematic way the nonlinearity effects generated by *small* matrix elements. One should note that the treatment is nonperturbative, with "small" and "large" quantities treated on the same footing. Moreover, in calculations, one has to regard only relevant physical states.

The disadvantage lies in the fact that, in practical cases, a *hierarchy* must be assumed or guessed from experimental data. The procedure thus requires either an experimental input which diminishes the predictive power of the KK method or strong assumptions which, beyond simple models, could be checked only a posteriori. Consequently, for nontrivial problems, one has to elaborate extra selfconsistency conditions to control e.g. the degree of Pauli principle violation, the dependence of final results on the closure approximation procedure, the procedure of decoupling of low-lying and high-lying modes etc.

The application of this scheme to the one-dimensional anharmonic potential works satisfactorily well. However, this example seems to be too simplified to be used as a test-case for consistency of all the assumptions of general scheme.

The difficulties traced previously can be controlled much more easily in algebraic models. This requires new ingredients in the formulation of the KK method, namely, specifying a Lie algebra and the physically relevant irreducible representations. Again, an application of this procedure to a simple pairing Hamiltonian ($SU(2) \times SU(2)$ model) works very well, and provides accurate solutions. However, in this case, both the algebraic structure (all generators of Lie algebra are regarded to be collective) and the *hierarchy* (classified naturally in terms of seniority) seem to be far away from realistic situations.

The generalization to realistic cases seems to be more problematic. Due to the complexity of nuclear spectra, one immediately encounters difficulties connected with the coupling of collective and non-collective degrees of freedom. For practical reasons, one must devise an algorithm which decouples (to some extent) the collective subspace. Furthermore, one must be able to evaluate at least the leading order corrections arising from the coupling of collective and non-collective motions. How to proceed with such problems seems to be an open question.

Up to now, for practical reasons, realistic applications have been limited to a selected subset of generators which do not form a closed algebra. As a consequence, one can trace back how to deal with the problem of violation of the exclusion principle. A method to restore the Pauli principle was applied to Ni nuclei (working in a vibrational limit). It is based on successive elimination of multipole operators from a pairing plus quadrupole Hamiltonian (they are replaced by pair operators) by means of Bogoliubov transformation. However, the final results appear to be quite far away from the experimental data. For example, the E2 strength of $2_2 \to 0_g$ transition in ^{60}Ni is underestimated by 2-3 orders of magnitude, while the $2_4 \to 0_g$ and $2_5 \to 0_g$ transitions in the same nucleus are overestimated by an order of magnitude. It is not clear to what extent these discrepancies are caused by inadequacies of the quadrupole plus pairing Hamiltonian and to what extent by the approximations themselves.

The KK method has also been used to derive a selfconsistent particle-core coupling formalism for odd-mass nuclei, in which the properties of the physical even-even neighbors appear as "cores". Constraints imposed by the Pauli principle are expressed as sum rules derived from the fermion anticommutation relations. Practical calculations in this approach have not been feasible without relaxing some selfconsistency conditions; nonetheless, with judicious approximations to the selfconsistent scheme, various phenomenological particle-core coupling models can be recovered, including the usual particle-rotor model. A crucial step in this derivation is the replacement of the Pauli principle sum rules by a simple normalization condition on the odd-A wave function. At present, the consequences of this replacement and the severity of the approximation remain unclear.

In summary, the Kerman-Klein method remains an interesting and promising approach for describing many-body systems, although the practical applications to realistic nuclear problems remain lacking. Since the method is essentially a version of Heisenberg's matrix mechanics, it is formally correct and practically intractable without added approximation schemes. Devising useful approximations for the realistic applications remains the central problem.

REPLY TO REVIEW OF KERMAN-KLEIN METHOD

ABRAHAM KLEIN AND NIELS R. WALET
Physics Department, University of Pennsylvania
Philadelphia, PA 19104-6396, USA

Though accurate and fair on many counts, it is our opinion, nevertheless, that the preceding critique of the Kerman-Klein (KK) method does not accord sufficient credit to either its power or its accomplishments. For this we accept at least partial blame, since we were forced by space limitations to restrict severely the choice of illustrative examples that we could describe in any detail and appear, as a consequence, to have omitted some important points.

Let us center this discussion largely - though not exclusively - about the classes of applications of the method:
(i) Relatively simple group theoretically based models, including problems of elementary quantum mechanics. From the examples cited in our paper, as well as from more elaborate models not included, models that contain both collective and noncollective degrees of freedom[1,2], we believe we have shown an understanding of how to carry through non-perturbative schemes of calculation based on the division of matrix elements into hierarchies of decreasing collectivity. Some of these models share significant features with realistic nuclear models[1,2].
(ii) Relativistic field theories with soliton solutions. In the very first applications of the KK method to such theories[3], it was already clear that a natural division of matrix elements into hierarchies could be made, and that this yielded the so-called loop expansion for such field theories. For renormalizeable models, this provides a complete and systematic method for studying these theories.
(iii) Realistic nuclear models. Within this class of applications, our critics have correctly identified the weakest point of our defense. But here a little historical perspective is necessary. We have presented the work on the even Ni isotopes as the most impressive we have to offer. This work, completed fifteen years ago, was the last such application attempted, not because it was considered a failure, but simply because of changing interests. It is true that agreement with experiment is not too impressive, but this was the fault primarily of the hamiltonian, not of the method. How do we know this? Because of shortage of space we omitted an account of the part of Li's work in which the results of the KK calculation were compared, favorably, with exact shell-model diagonalizations of the same schematic hamiltonian.

Our overall conclusion, based on this and previous experience, is that as far as the nuclear many-body problem is concerned, the KK method is powerful and "correct", in a sense to be described further below, for any problem for which an identifiable collectivity plays a dominant role. In terms of the standard concepts of collective *vrs.* single-particle degrees of freedom, it has not proved difficult to construct the appropriate hierarchy of matrix elements in every application ever attempted by us. As an early example, we claim that the *concept* (though not the name) of broken pair approximation appeared first in our work on number-conserving methods for vibrational spectra[4]. In every case in which comparison with an exact shell-model calculation was made, the validity of the approximations could be verified.

Nevertheless a valid criticism of our work is that the *number* of realistic applications has been pitifully small, and except for work in progress, nothing has been done for a decade and a half. It is the frustration of this circumstance that has led us to try to promote our ideas anew at this time.

We conclude with a remark that is also germane to the preceding critique. There, it is pointed out, correctly, that for realistic applications, our method will always contain some degree of violation of the Pauli exclusion principle. It is our contention that this is not necessarily a weakness of the approach. In many-body methods based on wave functions, there is a clear separation of kinematics and dynamics. Wave functions that (may) give a poor account of dynamical correlations are, nevertheless, required to satisfy the Pauli principle exactly. In the KK method, kinematics and dynamics are treated on an equal footing via the same type of sum rule approximations. In this approach, to make an error in the Pauli principle is considered no more heinous a crime than to make an error of the same magnitude in the dynamical correlations, where we know that for realistic models, such errors are unavoidable. Again from numerous worked examples, we have gained the distinct impression that far from being a defect of the approach, it may instead be one of its strengths.

In summary, it is our contention that the elements of the KK method are quite complete. What we lack at this time is some systematic effort to verify the truth of the confident assertions with which we have filled this space.

References

1. M. J. Vassanji and A. Klein, Phys. Rev. **C17**, 755 (1978).
2. M. Vallières, R. M. Dreizler, and A. Klein, Phys. Rev. **C7**, 2188 (1973).
3. J. Goldstone and R. Jackiw, Phys. Rev. **D11**, 1486 (1975).
4. G. Do Dang *et al*, Phys. Rev. **172**, 1022 (1968).

SELF-CONSISTENT DETERMINATION OF COLLECTIVE COORDINATES IN THE THEORY OF LARGE AMPLITUDE COLLECTIVE MOTION

NIELS R. WALET
Physics Department, University of Pennsylvania
Philadelphia, PA 19104-6396, USA

ABSTRACT

We discuss the possibility of determining a collective coordinate, or constraining operator, from the properties of the nuclear Hamiltonian only. Our technique applies to the adiabatic limit, and uses a combination of a constrained Hartree-Fock and a local RPA equation that determines the matrix elements of the constraining operator. We show an application of this method to ^{28}Si treated in the *sd* shell.

1. Introduction

In many cases one would like to obtain an approximate solution of the nuclear many body problem by reducing the number of degrees of freedom to a limited set of collective coordinates. One of the oldest of such approaches is the geometric model, which in its original form is based on a parametrization of the nuclear shape in terms of a few multipoles (see the contribution by Maruhn to this workshop for more details). The idea of our work is to extract such coordinates self-consistently from a description of a many-particle system in terms of a mean-field approximation, either Hartree-Fock or, when pairing correlations are important, Hartree-Fock-Bogoliubov. In less ambitious (and usally more practical) approaches one just chooses a set of collective operators, and one determines a collective Hamiltonian, which is then requantized, or one uses the generator coordinate method. Examples of this kind can, e.g., be found in the contributions by Heenen, Berger and Robledo. Of course the first such calculation is the one by Kumar and Baranger[1], where a schematic interaction is used, chosen such that the mass quadrupole operator is the natural choice of collective coordinate.

Still, the problem with these approaches is that the choice of the collective operator is rather arbitrary. One method to bridge this gap, the theory of large amplitude collective motion, utilizes the time-dependent version of mean-field theory to derive a classical Hamiltonian, involving only a few degrees of freedom, that is subsequently requantized. The time-dependent mean-field theory is not treated as an initial value problem, that can be used to generate orbits, but is used as a means to generate a surface that is "close" to a set of orbits. The procedure consists of moving from one point of the collective surface to the next by solving a set of constrained mean-field equations.

The techniques we use may at first sight appear to be very similar to the ones used by the Japanese groups (see the contribution by Sakata). Indeed our work was stimulated in part by the work of Marumori[2]. There are important differences, however: Sakata and coworkers (see his contribution) concentrate on the solution of an initial value problem, and do not make the adiabatic assumption. Matsuyanagi and his collaborators (e.g., Ref.[3]) have gone the farthest in practical applications, but have restricted themselves to anharmonic vibrations.

Stimulated by the realization that there was an open problem, a large number of different approaches to a self-consistent determination of the collective operators have been developed (see Ref.[4] for a complete set of references). Although none can claim total success, some interesting methods have been proposed. One of the most viable of these[5] has been used for heavy ion reactions, but suffers from the limitation that no more than one collective coordinate can be treated. We have proposed our own approach to the problem, which has proven to be successful when applied to toy models[4,6], and have also studied a simple nuclear physics problem[7,8]. The formulation we are currently using has, compared to all previous work in the literature, the large overlap with that of Rowe[9], though the latter has not applied his work to any realistic nuclear physics problem.

2. Formalism

We have used the fact that time-dependent mean-field theories are equivalent to Hamiltonian equations of motion, to separate the problems into two parts. The first is the treatment of decoupling in classical mechanics, which is amply reviewed in[10], and will therefore not be repeated here apart from the most important points in the derivation. The second is the translation of these results into the language of the mean-field theory, which we shall review here with a minimum of details, due to the limited space available to us.

A note about our notation: We use the tensor notation of general relativity, with upper and lower indices. We use a comma to denote a derivative with respect to coordinates, and a semicolon to denote a covariant derivative.

It has been amply documented that the TDHF equation

$$\dot{\rho}_{ph} = \delta W[\rho]/\delta\rho_{ph}, \quad W[\rho] = \rho_{ab}\epsilon_{ba} + \tfrac{1}{2}V_{abcd}\rho_{ca}\rho_{db}, \tag{1}$$

are a disguised from of Hamilton's equations of motion. here we work in a "local basis", i.e., at each point on the manifold of Slater determinants we introduce a set of single particle labels, h for the orbits occupied in that specific Slater determinant, and p for the unoccupied orbits. The number of degrees of freedom of the classical dynamical system is equal to the number of independent $p-h$ pairs used to label the local coordinates. This is especially clear if we introduce the local coordinate ξ and momenta π through the "classical Holstein-Primakoff mapping", i.e., when we define

$$\beta_{ph} = \frac{1}{\sqrt{2}}(\xi^{ph} + i\pi_{ph}), \quad \beta^{\dagger}_{hp} = \frac{1}{\sqrt{2}}(\xi^{ph} - i\pi_{ph}), \tag{2}$$

and

$$\begin{aligned} \rho_{ph} &= (\beta[1-\beta\beta^{\dagger}]^{1/2})_{ph}, \quad \rho_{hp} = ([1-\beta^{\dagger}\beta]^{1/2}\beta^{\dagger})_{hp}, \\ \rho_{hh'} &= \delta_{hh'} - (\beta^{\dagger}\beta)_{hh'}, \quad \rho_{pp'} = (\beta\beta^{\dagger})_{pp'}. \end{aligned} \tag{3}$$

According to (1), the Hartree-Fock energy functional W plays the role of a Hamiltonian. In general this is not a quadratic form in the momenta, so that we will have to truncate to obtain the form desired in our work.

We have used two different algorithms in our work. The first[6] hinges on the fact that we can specify a decoupled path by expanding the self-consistent collective operator in a limited number of basis operators. This may be a good approach to nuclear collective motion described in large model spaces, since it extends the usual constrained mean-field approach, but we found that it fails for the case of ^{28}Si treated in the very limited *sd* model space. Specifically we find that the relevant RPA mode can not be approximated by purely orbital operators (such as the quadrupole operator). For this problem we found it more convenient to use the algorithm discussed in Ref.[7], and repeated here for convenience.

We start from the quadratic approximation to the Hamiltonian,

$$H = \tfrac{1}{2}\pi_{ph} B^{php'h'} \pi_{p'h'} + V(\xi), \tag{4}$$

where

$$B^{php'h'} = \frac{\partial^2 W}{\partial \pi_{ph} \partial \pi_{p'h'}}, \tag{5}$$

which can easily be expressed in terms of the parameters of the nuclear Hamiltonian[7]. The potential energy V is the value of the HF energy W for zero momenta. As discussed at length in our previous work we now look for a point transformation[11] to new coordinates q and momenta p,

$$q^{\mu} = f^{\mu}(\xi), \quad p_{\mu} = g^{ph}_{,\mu} \pi_{ph}, \tag{6}$$

such that our new coordinates show *approximate* decoupling. The reason that we look at approximate as opposed to exact decoupling is that the latter only happens for a small set of exactly solvable models, which are of no interest for realistic calculations.

After the transformations described above we end up with the collective Hamiltonian

$$\bar{H} = \tfrac{1}{2} p_{\mu} \bar{B}^{\mu\nu} p_{\nu} + \bar{V}(q), \tag{7}$$

where the collective potential $\bar{V}$ is obtained by substituting $\xi^{\alpha} = g^{\alpha}(q)$ in V, and the collective mass is calculated from the equation

$$\bar{B}^{\mu\nu} = f^{\mu}_{,ph} B^{php'h'} f^{\nu}_{,p'h'}. \tag{8}$$

In the local harmonic form of our theory (which is generically referred to as the generalized valley approximation) we have to solve a set of two coupled equations. The first expresses that the condition that the force should lie along the collective path. For the case of a single collective coordinate, $Q = q^1$, we find the following condition, using the chain-rule relation for the gradient of the potential energy

$$V_{,ph} = (d\bar{V}/dQ)(dQ/d\xi^{ph}). \tag{9}$$

If we replace the derivative of the collective potential energy by λ and use the fact that $dQ/d\xi^{ph} = f^1_{,ph}$ as well as the fact that the derivative of W is just the Hartree-Fock Hamiltonian, we find that this equation turns into the familiar constrained Hartree-Fock equation

$$\mathcal{H}_{ph} = \lambda f_{,ph}. \tag{10}$$

The second equation of the method is a necessary (but not sufficient) condition for decoupling, that can be obtained from the condition that geometrical forces should also be along the path. (An example of a geometrical force is the centrifugal force.) This turns out to be equivalent to the requirement that the the ph matrix elements of the collective operator be the solution of an RPA like equation, which we usually write in the real form equivalent to Eq. (8.83) in Ref.[12],

$$M^{p'h'}_{ph} f^{\mu}_{,p'h'} = \omega^2_{\mu} f^{\mu}_{,ph}, \tag{11}$$

with

$$M^{p'h'}_{ph} = V_{;php''h''} B^{p''h''p'h'}, \tag{12}$$

and the covariant derivative is defined as

$$V_{;php''h''} = V_{,ph,p'h''} - V_{,p'h'} \Gamma^{p'h'}_{php''h''}. \tag{13}$$

The affine connection Γ is determined in the usual way from the inverse mass B, which for this system is the Riemannian metric. The calculation of the affine connection requires the inverse of the mass, which may be a very expensive computation in a large model space. In the example we shall discuss such a calculation is almost trivial. We have found that the use of an ordinary second derivative instead of the covariant derivative makes only a very small difference in the result of the calculations.

The cranking and local harmonic equations should now be solved self consistently: The constrained mean-field equation (10) determines, for given f, a set of single-particle wave functions labeled by p for empty and h for filled orbits. The local harmonic equation (11) determines f, but depends on the details of the single particle orbitals p and h.

The RPA equation (11) is a linear eigenvalue problem and thus does not fix the scale of the eigenvectors. We propose to make the very convenient choice of normalization

$$f^{\mu}_{,ph} B^{php'h'} f^{\nu}_{,p'h'} = \delta_{\mu\nu}. \tag{14}$$

Since the quantity f_{ph} is symmetric, $f_{ph} = f_{hp}$, and is equal to the derivative $\delta Q/\delta\rho_{ph}$, we find that (there is a factor $\sqrt{2}$ difference between the derivative w.r.t. ρ as compared to that w.r.t. ξ, as explained in Ref.[7])

$$Q = \int f_{,ph} d\xi_{ph} = 2\int f_{ph}\delta\rho_{ph} = \mathrm{Tr}(f\delta\rho) \tag{15}$$

for real $\delta\rho$. One may verify the correctness of this equation for a density independent one-body operator f, where we obtain the familiar relation

$$Q - Q_0 = \int_{\rho_0}^{\rho} \mathrm{Tr}(f\delta\rho) = \mathrm{Tr}(f\int_{\rho_0}^{\rho}\delta\rho) = \mathrm{Tr}(f\rho) - \mathrm{Tr}(f\rho_0). \tag{16}$$

The relation (15) will allow us to evaluate the collective coordinate along the path. Since in the algorithm described below we shall move in small steps along the path, we use a finite difference approximation for the change in Q for such a small step,

$$\Delta Q = \sqrt{2} \sum_{ph} f_{,ph} \Delta \rho_{ph}. \tag{17}$$

Finally we would like to be able to measure the quality of decoupling along the path. This can be obtained by comparing two different forms of the collective mass that can be calculated in the theory; one is the definition of the reciprocal collective mass as a covariant tensor in Eq. (8),

$$\bar{B}^{ij} = f^{i}_{,ph} B^{php'h'} f^{j}_{,p'h'}, \tag{18}$$

and the other can be calculated from a formula for the collective mass itself in terms of the tangents for the path,

$$\breve{B}_{ij} = \frac{\partial \xi^{ph}}{\partial q^{i}} B_{php'h'} \frac{\partial \xi^{p'h'}}{\partial q^{j}}. \tag{19}$$

If we have exact decoupling the second matrix will be the inverse of the first. We use the deviation from this relation as a measure for the quality of the decoupling (K is the number of collective coordinates),

$$D = \sum_{ij} \bar{B}^{ij} \breve{B}_{ji}/K - 1. \tag{20}$$

Due to our choice of normalization for the RPA modes we find that $\bar{B}^{11} = 1$. This can be shown to imply

$$D = \sum_{\mu} (\Delta q^{\mu}/\Delta Q)^{2} - 1, \tag{21}$$

where we have replaced a derivative by a finite difference approximation. The quantities Δq^{μ} are defined in analogy to ΔQ

$$\Delta q^{\mu} = \sqrt{2} \sum_{ph} \Delta \rho_{ph} f^{\mu}_{,ph}. \tag{22}$$

Thus if $\Delta q^{1} \equiv \Delta Q$ is the only non-zero change of coordinate we find that $D = 0$. If we have a large change in any of the non-collective coordinates as we follow the collective path we do not have good decoupling.

3. Algorithm

The algorithmic solution of Eqs. (10, 11) is obtained as follows:

1. Find an extremal point of the Hartree-Fock equations, where $\lambda = 0$.

2. Solve the RPA at this point and calculate f_{ph}. There is no self-consistency requirement for $\lambda = 0$.

3. Move to the next point, using a constrained minimization procedure similar to the one discussed in Ref.[12]. (The constraint is a prescribed change in the collective coordinate.) Use the value of f_{ph} calculated in the previous step.

4. Solve the RPA and find f_{ph}. If this is very close to the previous value of f_{ph}, go to step 3, else continue.

5. Solve the constraint problem again, now without changing the value of the constraint, i.e., without changing the collective coordinate. Use the new *ph* basis as input to step 4.

4. Results

We have applied the algorithm discussed in the previous section to a description of ^{28}Si in the sd-shell, using Kuo's interaction[13], with single-particle energies as given in Ref.[7]. Taking advantage of the fact that isospin is a good symmetry for light nuclei, we require that proton and neutron orbits be occupied with equal probability. This corresponds to considering only the manifold of $T = 0$ states, which constitute the low-energy part of the spectrum of ^{28}Si. This symmetry reduces the effective number of single-particle degrees of freedom from 12 to 6. Furthermore we impose "ellipsoidal" symmetry, i.e., we require the intrinsic nuclear shape to be invariant under a rotation of 180 degrees about any of the three symmetry axes. It is well known (see, e.g.,[14]) that such a symmetry requires that K, the projection of the angular

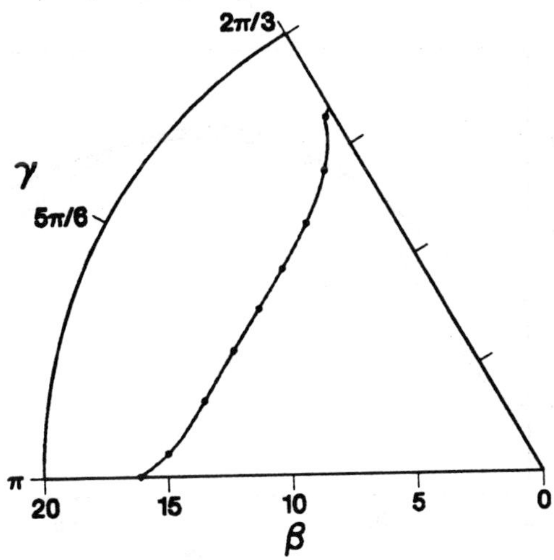

Figure 1: The one-dimensional collective path projected onto the quadrupole deformation $\beta - \gamma$ plane. The oblate minimum is obtained for $\gamma = \pi/2$, corresponding to axial symmetry around the z-axis. The dots indicate the points $Q = i/4$, starting from the oblate minimum where $i = 0$.

momentum on the intrinsic z-axis, be even and also relates wavefunction components with positive K to those with negative K. The requirement of ellipsoidal symmetry is indeed a strong limitation in the implementation of our algorithm, since any reasonable interaction (Kuo, Wildenthal) gives that for ^{28}Si the lowest-energy RPA mode at the Hartree-Fock minimum is a $|\Delta K| = 3$. We thus reject this solution as a possible choice of cranking operator, and choose the lowest even-K solution for this purpose.

The model exhibits a deep deformed minimum on the oblate side built from a Slater determinant in which orbits with $m = \pm 5/2, \pm 3/2, \pm 1/2$ are occupied. There is no stable prolate solution with the same orbits occupied. Further study reveals, however, that there is another minimum with positive quadrupole moment when the orbits $m = \pm 3/2, \pm 1/2, \pm 1/2$ are occupied. Since all axially deformed states within this manifold are orthogonal to all states considered previously, we find that there is no path through the subspace of axially deformed Slater determinants from one state to the other. The lowest mode of the RPA is not an axial ($\Delta K = 0$) mode, however, but a $|\Delta K| = 2$ mode. It therefore seems plausible that there exists a path going through triaxial shapes from one to the other.

We have found the corresponding path. Since each point of the collective path corresponds to a Slater determinant, or equivalently a set of occupied orbits, it is hard to visualize this path. We need projections of the manifold of Slater determinants on some two-dimensional surface in order to represent the path graphically. In Fig. 1 we have drawn a representation of the path, where we plot the quadrupole β and γ parameters along the path, defined as

$$\langle \sqrt{4\pi/5}\, r^2 Y_0^2 \rangle = \beta \cos\gamma, \quad \langle \sqrt{4\pi/5}\, r^2 Y_2^2 \rangle = \beta/\sqrt{2} \sin\gamma. \tag{23}$$

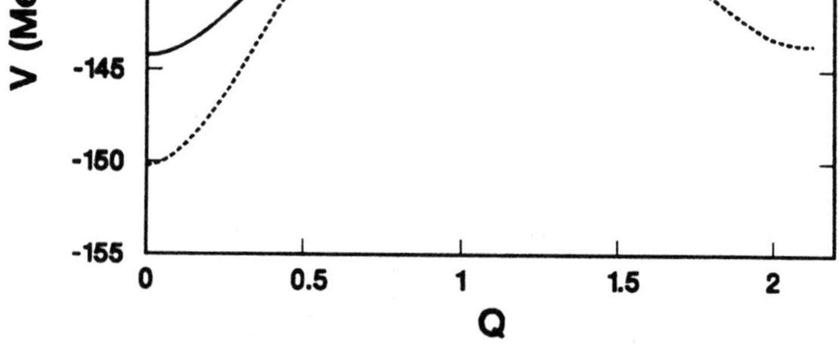

Figure 2: The collective potential energy (solid line). The dashed line shows what happens if we include quantum corrections.

The behavior near the prolate minimum is not as smooth as at all other points, an observation that has important consequences. The difference between the covariant and non-covariant form of the RPA is small, and the resulting curves coincide. This is surprising if one notes that the affine connection does not have small matrix elements.

Using the calculated value of the collective coordinate, we can give the collective potential $\bar{V}$ (Fig. 2) as a function of the collective coordinate Q.

Having obtained the path we can now calculate the decoupling measure D. Using Eq. (21) we find in Fig. 5 that the quantity D is not small everywhere along the path. The largest value of D is found near the prolate minimum, but D also exhibits another "bump" not too far from the oblate minimum. This explains the rapid change in properties of the path in Fig. 1, since the region of rapid change (large curvature) occurs exactly where decoupling is bad. These features seem to indicate that we need to include more than one collective coordinate. To see whether an approach with two coordinates may be able to solve this problem we have separated the contributions $(\Delta q^\mu/\Delta Q)^2$ to D. We have discovered that the second coordinate gives the only important contribution to D. This seems to hold promise for the calculation of a two-dimensional potential energy surface.

To complete the discussion of parameters of the Hamiltonian we have also calculated the moments of inertia. The moments of inertia behave relatively smoothly (although they do not follow the irrotational pattern). Where D has it largest value the moments of inertia seem to change less smoothly.

Though we have found that decoupling is not good everywhere, we shall nevertheless use the Hamiltonian we have constructed, that has the form

$$\mathcal{H} = \tfrac{1}{2}\sum_i \frac{J_i^2}{\mathcal{I}_i} + \tfrac{1}{2}P^2 + \bar{V}(Q), \tag{24}$$

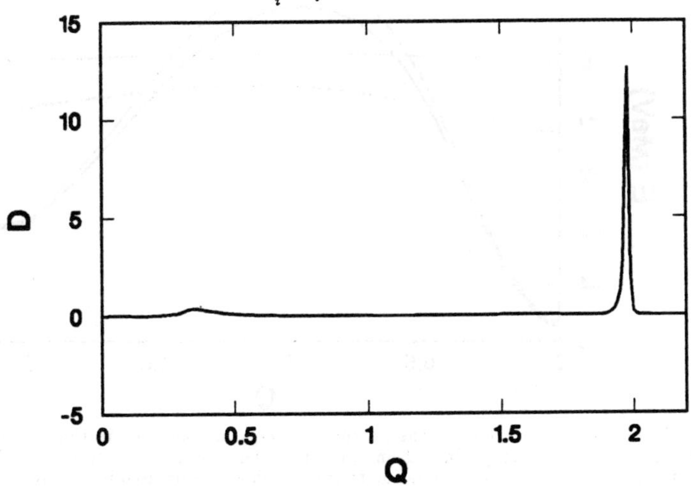

Figure 3: The decoupling measure D.

to generate spectra and compare to the full shell-model calculation in the *sd* model space. To that end we need to requantize the Hamiltonian (24). Since we are working in the intrinsic system we have to be careful. It can be solved easily using the necessary symmetry properties set out a long time ago by Bohr[15].

The ground state can not be very sensitive to the badness of decoupling, since it is very small for $Q > 1$, but the first and other excited 0^+-states have sizable values in the region of bad decoupling. This means that we do not believe that the corresponding eigenvalues are very good approximations to the shell model values. If we look at the spectrum and compare it to the shell model, we indeed see that the ground state rotational band is reproduced quite well. The band built on the oblate minimum (the second 0^+ state), comes at much too low an excitation energy. Again the moment of inertia of that band is more or less correct. The other 0^+-states also come at a lower energy than their shell-model counterparts. The first 3^+ state is found to lie at too high an excitation energy, which is at least partially due to the neglect of odd multipoles in our calculation.

Of course we can calculate the quantum corrections to the potential energy within the current framework. This is discussed in quite some detail in Ref.[8]. The general quantum correction can be unraveled into three distinct pieces, all of which have been plotted in Fig. 4:

1. The quasiboson terms $\frac{1}{2}(\sum_a \omega_a - \mathrm{Tr}\mathcal{A})$, where ω_a are the *position dependent* RPA frequencies, and $\mathcal{A}$ is the corresponding TDA matrix. In the sum over the RPA modes we have excluded the one used to define the collective coordinate.
2. The rotational zero point energy $-\frac{1}{2}\sum_i \langle J_i^2 \rangle / \mathcal{I}_i$.

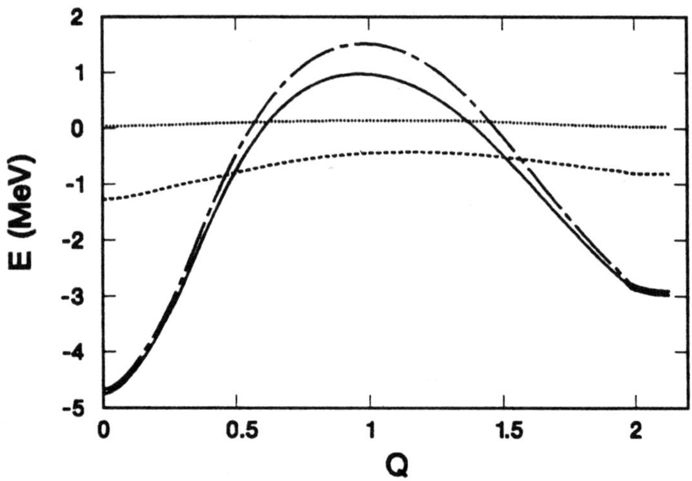

Figure 4: The quantum corrections to the potentials energy, decomposed in the quasiboson result (solid line), the rotational energy (dashed line) and the overcounting correction (dotted line). The dash-dotted line represents the approximation to the quasiboson term used in the GOA by Reinhard and Goeke[16].

3. The overcounting correction, which is due to the fact that a bosonization of the RPA modes is not totally consistent with the Pauli principle.

We have also compared our result to an approximate result obtained in the Gaussian overlap approximation[16], that only involves the collective mode. (This is the expression used by Robledo in his calculation.) As can be seen from Fig. 4 this approximation works well near the minimum, but overestimates the corrections near the saddlepoint, where the collective potential energy is maximal.

5. Acknowledgement

The research reported here was performed in collaboration with A. Klein and G. Do Dang. This work was partially supported by the U.S. Department Of Energy and by the National Science Foundation.

6. References

1. K. Kumar and M. Baranger, *Nucl. Phys.* **A110** (1968) 529; *Nucl. Phys.* **A122** (1968) 273.
2. T. Marumori *et al*, *Prog. Theor. Phys.* **63** (1980) 1576; **64** (1980) 1294.
3. H. Aiba and K. Matsuyanagi, *Prog. Theor. Phys.* **81** (1989) 690.
4. A. Bulgac, A. Klein, N.R. Walet, and G. Do Dang, *Phys. Rev.* **C40** 945 (1989).
5. D. Provoost, F. Grümmer, K. Goeke, and P.–G. Reinhard, *Nucl. Phys.* **A431** (1984) 139.
6. N. R. Walet, A. Klein, G. Do Dang, and A. Bulgac, *Phys. Rev.* **C41** (1990) 318.
7. N. R. Walet, G. Do Dang, and A. Klein, *Phys. Rev.* **C43** (1991) 2254.
8. N. R. Walet, A. Klein, and G. Do Dang, *Phys. Rev.* **C45** (1992) 249.
9. D. J. Rowe and R. Basserman, *Can. J. Phys.* **57** (1977) 112.
10. A. Klein, N.R. Walet, and G. Do Dang, *Ann. Phys. (NY)* **208** (1991) 90.
11. H. Goldstein, *Classical Mechanics* (Addison-Wesley, Reading, 1950), chap. 8.
12. P. Ring and P. Schuck, *The Nuclear Many-Body Problem* (Springer, New York, 1980).
13. T. T. S. Kuo, *Nucl. Phys.* **A103** (1967) 71.
14. J. M. Eisenberg and W. Greiner, *Nuclear Models* (North-Holland, Amsterdam, 1970).
15. A. Bohr, *Mat. Fys. Medd. Dan. Vid. Selsk.* **26** (1952) no. 14.
16. P.–G. Reinhard and K. Goeke, *Rep. Prog. Phys.* **50** (1987) 1.

GENERALIZED DENSITY MATRIX REVISITED

V.G.ZELEVINSKY
The Niels Bohr Institute, Copenhagen 2100, Denmark
and
Budker Institute of Nuclear Physics, Novosibirsk 630090, Russia

ABSTRACT

The review is given of the method of the generalized density matrix (GDM) in the microscopic theory of nuclear structure. It gives a framework capable to incorporate various aspects of interplay of single-particle, collective and chaotic dynamics rather than a specific model. The method is exemplified with the description of a rotational band without external cranking. Then the general operator formulation is explained and the advantages and open problems of the GDM method are discussed. The concluding chapter contains the summary of applications.

1. Introduction

The only practical tool for microscopic studies of many-body systems is the mean field (MF) theory. It has been formulated in numerous modifications which, being different in actual approximations and input parameters, share the common view of a system as composed of independent constituents (quasiparticles) in the MF with properties determined (in principle, self-consistently) by the same constituents. In such a way one considers usually complex atoms, molecules, solids and nuclei.

The symmetry of the resulting MF plays the important role specifying the excitation spectrum and the response function to an external field. These dynamical characteristics can be calculated with the aid of the linear response theory which also exists in a variety of formulations. Roughly speaking, the main idea is to consider the small amplitude self-sustained distortion of the MF as the vibrational mode which can be single-particle or collective, regarding of the number of constituents coherently taking part in formation of the mode. Such approaches can be unified as random phase approximation (RPA). The next natural step should make it possible to describe the large amplitude collective motion (time-dependent MF). Finally, we come to collision-like processes beyond the MF which result in fluctuations and dissipation of simple modes. At high level density such effects dominate resulting in the complicated structure of generic wave functions and quantum chaos.

This handsome hierarchy of phenomena erected on the base of the MF is known to be violated in regions of phase transitions where the MF becomes unstable and the role of fluctuations is decisive. In finite systems like nuclei there is no sharp phase transitions. But the similar situation takes place, for example, in soft transitional nuclei where, even in the ground state, large quantum fluctuations make it necessary to introduce a concept of shape coexistence. In the Fermi liquid theory [1] which is nothing but the regular microscopic way to the MF and RPA, a theorem is proved that there exists a vicinity of the Fermi surface where the single-particle branch of the excitation spectrum is well-defined and determines the response function. This statement is also violated in transitional nuclei where the coupling of quasiparticles to soft phonons modifies completely spectra of odd-A nuclei and analytical properties of the single-particle Green's functions [2].

The most spectacular new element in the description of finite systems is the appearance of specific excitations as collective rotation. In the standard cranking method one substitutes the intrinsic rotational excitation by the motion induced with the help of an external torque. Then the MF is recovered in the body-fixed frame as belonging to a deformed object with fixed orientation. Meanwhile, in the exact quantum formalism, the total nuclear spin would be an integral of motion labeling the stationary states within rotational bands. The intraband transition matrix elements of multipole operators have the same order of magnitude as expectation values, apparently violating the conventional MF hierarchic scheme.

The appropriate extension of the MF theory which allows one to consider problems with strong coupling between single-particle and collective degrees of freedom was suggested by Kerman and Klein [3] and, in different formulation, by S.T.Belyaev and the author [4,5]. The idea was to treat on equal footing all the states connected by strong matrix elements of collective operators. These states form the "band" as a generalization of the usual rotational band. The MF and the single-particle density matrix (SPDM) become the operators within the band whose properties are strongly dependent on the symmetry of the states. Thus, we assume that, to a good approximation, the dynamics is saturated in the space of the band. Coupling to states outside the band can be regularly included as a correction of the next order [6,7].

We called [5] this approach Generalized Density Matrix (GDM) method. Since only collective states were included in the band, now I would prefer the term Collective Density Matrix (CDM). A short review of the development and applications of the CDM-GDM method can be found in [8].

In the recent paper [9] I consider an opposite situation when the dynamics proceeds through a very big number of intermediate states of extremely complicated structure. It is possible to take into account on average the complete set of those states, i.e. consider the dynamics of the full GDM, under simple statistical assumptions. And the circle turns out to be closed: the result is again the MF averaged over a group of typical states at given energy. The MF is generated by the stochastic, or incoherent, dynamics as a smooth and the most stable component reflecting coarse-grained features of multiple interactions in a many-body system. Therefore the MF appears to be the most reasonable candidate for the preferential representation in estimates of degree of complexity of generic wave functions [10].

In what follows I explain main ideas of the CDM approach on the exactly solvable example of an isolated rotational band. The result is the standard Bohr-Mottelson description and the correct value of the moment of inertia obtained without cranking and with exact conservation of the angular momentum. The body-fixed frame and all related concepts appear quite naturally from exact microscopic equations. Then I introduce briefly the formalism of the CDM and GDM and enumerate various directions of application. The problem of chaotic dynamics will be briefly described in the same framework.

2.Rotation without cranking

2.1. What is a collective band

To clarify the main ideas and the machinery of the approach, let us consider an example of an isolated (for instance, ground state) rotational band of a well-deformed nucleus.

By our definition, the rotational band is a set of states of a many-body system which satisfy two conditions:

- *symmetry*: the states within the band are uniquely labeled by the quantum numbers of the $SU(2)$ group (the total spin J and its projection M onto the space-fixed axis z);
- *dynamics*: collective operators, determining the actual configuration of the system in the coordinate (or phase) space, have strong matrix elements between the states belonging to the band and relatively weak ones for the transitions to the states outside the band.

The dynamical condition is rather vague since there is no quantitative definition of "collective" operators. It is intuitively clear that, for example, the quadrupole-type operators like $\sum_a x_i(a)x_j(a)$ should manifest such a property since they characterize not only the shape but also the classical moment of inertia. The point is that in the lowest order we do not need to specify that set of operators: *within the band* all of them will have similar properties defined, up to normalization, by the multipolarity of the operator and geometry of the band. The situation resembles that in the textbook vector model: within the angular momentum multiplet $|JM>$ (J is fixed), all vector operators have the same dependence of matrix elements on the magnetic quantum numbers as matrix elements of the angular momentum $\mathbf{J}$.

2.2. Sum rules and intensity rules

Take, for example, the quadrupole tensor Q_μ constructed from the coordinates of particles. Its components commute with each other, $[Q_\mu, Q_{\mu'}] = 0$. Let us calculate the matrix element $< J_aM_a|\cdots|J_bM_b >$ of this operator equation between the states of the band. The dynamical postulate tells us that we can restrict summation over the intermediate states by the states $|J_cM_c >$ within the same band. The postulate of symmetry defines the dependence of the matrix elements of any tensor operator on the magnetic quantum numbers: according to the Wigner-Eckart theorem,

$$< a|T_{l\mu}|b >= \frac{1}{\sqrt{2J_a+1}}(-)^{\mu} < J_bM_b, l-\mu|J_aM_a > (a||T_l||b). \tag{1}$$

Thus, after the simple angular momentum algebra, the commutator identity leads to the set of equations for the reduced matrix elements $(a||Q||b)$,

$$[1-(-)^L]\sum_{J_c} W(22J_bJ_a; LJ_c)(a||Q||c)(c||Q||b) = 0 \tag{2}$$

where W is the Racah coefficient. Bohr and Mottelson [11] refer to Eq.(2) as the rotational sum rule.

The general solution to Eq.(2) is

$$(a||Q||b) = Q\sqrt{2J_b+1} < J_bK, 20|J_aK > . \tag{3}$$

It contains two constants, the normalization Q and the headband spin K, which should be specified for the case under study. Note that if the quadrupole tensor Q_μ were the Elliott $SU(3)$ quadrupole [12] rather than the ordinary coordinate quadrupole, the $L = 1, J_a = J_b$ component of Eq.(2) would not vanish being proportional to the reduced matrix element $(a||J||a)$ of the angular momentum. Then the solution (3) ceases to be exact. Nevertheless, even in that model Eq.(2) gives a good approximation because the commutator is still small at spins not very close to the band cut-off. In higher orders similar rotational corrections arise as well for the ordinary quadrupole tensor as a result of virtual coupling to the states outside the band. The exact solution of the Elliott model can be also derived with the CDM method [13].

The formal output of Eqs.(1) and (3) has an obvious physical meaning. What we have obtained is nothing but the well known Alaga rule [11] for the E2 transition amplitudes within the rotational band of the axially symmetric rotor. In this approximation K is a good quantum number (later we will see that it coincides with the projection of the body-fixed angular momentum onto the symmetry axis but at this stage we have not yet introduced such a frame). We stress again that the Alaga rule came entirely due to the fact that our band contains each state $|JM >$ only once. For the simplicity, we limit ourselves below by the ground state band of an even-even nucleus, $K = 0$. In this case Eq.(3) shows that the band started from the 0^+ state contains only even spin states. More general results can be found in [4–7].

2.3. Effective operators within the band

Now we can introduce the coordinate representation setting the spherical function $Y_{JM}(\mathbf{n})$ with the even J as a wave function of the state $|JM>$. Any operator inside the band can be expressed in terms of functions of angles $\mathbf{n}$ and differential operators acting onto those functions. In accordance with Eqs.(1) and (2), for the quadrupole tensor we have simply $Q_\mu \sim Y_{2\mu}(\mathbf{n})$, and the normalization constant Q of Eq.(3) is proportional to the intrinsic (static) quadrupole moment of the band.

Up to this moment, we did not use specific features of the dynamics, apart from the existence of the band which is to be confirmed by the microscopic equations of motion. These equations can be written for the pair fermion operators (1 and 2 stand for the complete set of single-particle quantum numbers)

$$A_{l\mu}(12) = [a(\tilde{2})a(1)]_{l\mu}, \tag{4}$$

$$B_{l\mu}(12) = [a^\dagger(2)a(1)]_{l\mu}. \tag{5}$$

For any tensor operator $T_{l\mu}$, it is convenient to form the linear combinations

$$T^{(\pm)}_{l\mu} = \frac{1}{2}\{T_{l\mu} \pm (-)^{l-\mu}T^\dagger_{l-\mu}\}, \tag{6}$$

even ($T^{(+)}$) or odd ($T^{(-)}$) under time reversal. In particular, the particle number is $N \sim B^{(+)}_{00}$ and the total angular momentum is

$$J_\mu = \sum_j \sqrt{j(j+1)(2j+1)}B^{(-)}_{1\mu}(jj) \tag{7}$$

where the sum runs over angular momenta of single-particle spherical orbitals. Now we should write down the hamiltonian H of the system in the spherical representation which explicitly ensures the angular momentum conservation and calculate commutators of H with the operators (4) and (5).

Here we will not display all the details of equations referring to the original paper [4]. We sketch only the main points of reasoning.

Since the commutator with H (time derivative) for a time-even (odd) quantity gives a time-odd (even) one, we can find easily the analytic solution valid in the adiabatic region of low spins where the rotational energy intervals are small as compared to typical energies ($> 2\Delta$) of two-quasiparticle excitations. We assume (and confirm by the results) that the matrix elements of time-even operators like the pairing gap ($\Delta \sim A^{(+)}_{00}$) or the quadrupole moment ($Q_\mu \sim B^{(+)}_{2\mu}$) are large ("static" quantities) whereas those of time-odd operators are comparatively small being induced by rotational effects. Neglecting in the equations of motion for small quantities their time derivatives (second order corrections) we come to the static equations for large quantities. Proceeding in the same way as we did for the derivation of the rotational sume rule, we find the static operators as

$$A^{(+)}_{l\mu}(12) \sim Y_{l\mu}(\mathbf{n})\alpha^{(+)}_l(12), \tag{8}$$

and analogously for $B^{(+)}$, where amplitudes $\alpha_l^{(+)}(12)$ and $\beta_l^{(+)}(12)$ are still to be found. The set of equations for these amplitudes can be partially diagonalized by the transition from the particle-hole or particle-particle angular momentum l to the conserved quantum number m according to

$$\alpha^{(+)}(m;12) = \sum_l (2l+1)(2j_1+1)^{-1/2} < j_2 m, l0|j_1 m > \alpha_l^{(+)}(12). \tag{9}$$

. At this point one can see that the new amplitudes $\alpha^{(+)}(m;12)$ and $\beta^{(+)}(m;12)$ obey the set of equations for the single-particle density matrix (SPDM) [14] where the single-particle space is doubled to take pairing into account. The final diagonalization of the SPDM determines the deformed self-consistent MF, complementary to m quantum numbers of orbitals, single-particle energies and the occupation numbers (quasiparticle vacuum for the ground state band). This picture becomes trivial in the frame where the quantization axis for the single-particle motion [the corresponding conserved quasiparticle angular momentum projection is m, Eq.(9)] is combined with the axis of global orientation used in the spherical functions of Eq.(8). Thus, we obtained the microscopic definition of the body-fixed frame without using any *a-priori* concepts.

The next step of the procedure is the solution of the first order equations (those containing the time derivatives of time-even quantities found at the previous step). Making transformations similar to (8), one can derive the solution for the small quantities in the operator form as (for even l)

$$A_{l\mu}^{(-)} \sim [(\mathbf{J}Y_{l\mu}(\mathbf{n})), \mathbf{J}] \sim [Y_{l\mu}, \mathbf{J}^2] \tag{10}$$

where $\mathbf{J}$ is the angular momentum operator acting onto the angles $\mathbf{n}$ (the representative of the total angular momentum (7) within the band), and the amplitudes $\alpha_l^{(-)}(12)$ are not written explicitly. Eq.(10) implies that all terms of the first order equations including the time derivative, i.e. the commutator of the zero order operator (8) with the hamiltonian of the band, have the same $[J_a(J_a+1)-J_b(J_b+1)]$ dependence on the angular momenta of the bra- and ket-states. It means that, up to the first order, the energy spectrum of the band is $E(\mathbf{J}) \sim \mathbf{J}^2$.

In the single-particle basis of the zero order the small amplitudes $\alpha_l^{(-)}(12)$ are off-diagonal with the selection rules $|\Delta m| = \pm 1$ which correspond to the perturbation proportional to the component of the angular momentum perpendicular to the symmetry axis. This is the Coriolis force induced by the small (rotational) elements of the SPDM. It is easy to show that in the collective space it gives, in addition to (3), off-diagonal corrections to the effective operators with the selection rules $|\Delta K| = \pm 1$. Therefore the formal K quantum number introduced in the solution (3) indeed can be identified with the angular momentum along the symmetry axis. Using the equations of motion one can express all such contributions in terms of the coefficient in the energy spectrum, i.e. the moment of inertia. Finally, substituting the expression for $B_{l=1,\mu}^{(-)}$ into Eq.(7), the requirement of self-consistency for the total angular momentum determines the moment of inertia which coincides with that found in the cranking model [14] including the so called Migdal term [1] caused by the rotational distortion of the MF.

Thus, all the ingredients of the unified model of deformed nuclei (self-consistent nonspherical MF, body-fixed frame, intensity rules, cranking moment of inertia, Coriolis force) are obtained from the pure microscopic theory. The angular momentum is conserved at all stages of calculations. The meaning of all the transformations is strictly defined rather than intuitive; the approximations are accurately formulated which makes it possible to develop the regular way of iterative corrections. There are corrections of two kinds: higher order terms within the band and contributions from the interband coupling (extension of the collective space).

The actual calculations can be performed almost "automatically" after the whole theory is rewritten in terms of the GDM (in this specific problem CDM) as a new operator entity [5].

3. Generalized density matrix

3.1. What is the GDM

The operators (4) and (5) of the previous section can be considered as particular components of the general density operator acting in the total Hilbert space of a system. To simplify formulae, I will write all definitions as if for the system with no pairing; the same expressions are valid for the pairing case if one introduces the "spinor" representation [5] to be able to mix the single particle creation and annihilation operators.

The GDM is the set of the matrix elements

$$R_{12}^{\alpha\beta} \equiv < \alpha | R_{12} | \beta > \equiv < \alpha | a_2^{\dagger} a_1 | \beta > \tag{11}$$

in the total Hilbert space $|\alpha>$. Since the operator R_{12} is at the same time the matrix in single-particle indices, it can be convenient to define the general density operator $\hat{R}$ in the combined, single-particle $\otimes$ global, space $|1;\alpha>$ as

$$R_{12}^{\alpha\beta} = < 1; \alpha | \hat{R} | 2; \beta > . \tag{12}$$

Any one-body operator Q which is a sum of the single-particle operators q over all particles, $Q = \sum_a q_a$, can be expressed as

$$Q = \sum_{12} q_{12} R_{21} = Tr(q\hat{R}) \tag{13}$$

where the trace runs over the single-particle indices only and Q remains an operator in the physical space $|\alpha>$.

The analysis of the mathematical properties of the GDM can be found in Ref.[9]. Here I mention only the very important "conservation law" valid for any

Q of Eq.(13),

$$[Q+q,\hat{R}]=0. \tag{14}$$

Eq.(14) can be checked immediately in the eigenbasis of the operator q where it is reduced to the selection rule $Q_\alpha - Q_\beta = q_2 - q_1$ which is the trivial consequence of the definition (11) of $R^{\alpha\beta}_{12}$. If the operator Q is thought of as a generator of a unitary transformation (it could be natural, for example, for the total linear or angular momentum), Eq.(14) acquires a simple meaning: the GDM $\hat{R}$ is invariant under any operation of such kind in the full space (simultaneous transformation of all single-particle orbitals and the whole system).

3.2. Dynamics and the saturation principle

Eq.(14), along with other similar properties, is fulfilled if the dynamics of the corresponding degree of freedom is saturated within the selected band. Any approximate scheme truncating the physical space can violate it. The standard constrained Hartree-Fock (HF) calculations, where the "band" is restricted by one trial state, try to remedy this shortcoming by the appropriate choice of the MF configuration; fluctuations around this point (matrix elements to the excited states) are taken into account by the RPA. For large amplitude collective motion, the requirement of minimizing the deviations from Eq.(14) ("saturation principle"[15]) gives a recipe for the optimal choice of the collective path. We utilized similar arguments in the previous section demanding that the quadrupole dynamics be saturated within the rotational band; it turned out to be sufficient for deriving the intensity rules.

Let the collective band be selected by physical arguments. Then the generalization of the conventional HF equations can be reached by the following procedure which was illustrated above for the rotational band:

- calculate, for the given microscopic hamiltonian H including the independent particle energy ϵ and the effective (two- or many-body) interaction V, exact equations of motion for the operator $\hat{R}$, Eq.(12);
- take the matrix elements of the equations of motion between the states belonging to the selected band;
- split the interaction term as in the HF approximation and use the saturation assumptions to restrict the summation over intermediate states by those within the band; it gives the nonlinear set of equations for the matrix elements (12);
- use the symmetry properties of the band to extract the information on the dependence of the matrix elements on the quantum numbers.

The result of the first three steps can be written in terms of operators of the combined space as [5]

$$[\hat{R},\hat{h}+\mathcal{H}]=0 \tag{15}$$

where $\hat{h}$ is the self-consistent field operator

$$\hat{h} = h\{\hat{R}\} = \epsilon + Tr(V\hat{R}) \tag{16}$$

(it is written here for the two-body effective interaction V), and $\mathcal{H}$ stands for the unknown collective hamiltonian, acting in the collective space $|\alpha>$ only, whose eigenvalues E_α give the energy spectrum of the band sought for. The commutator in Eq.(15) has to be taken with respect to all the variables, single-particle as well as collective.

In the HF approximation the band is reduced to the ground state $|0>$, the GDM $\hat{R}$ coincides with the SPDM ρ acting in the single-particle space only, $\rho_{12} = R^{00}_{12}$, the collective energy $E_0 =< 0|\mathcal{H}|0>$ becomes a c-number and drops out from the commutator so that the result is the well known form of the MF equation [16]

$$[\rho, h\{\rho\}] = 0. \tag{17}$$

Eq.(17) means that in the eigenbasis $|1)$ of the self-consistent field $h\{\rho\}$ the SPDM is diagonal as well; their eigenvalues are the orbital energies e_1 and occupation numbers n_1, respectively. In the CDM formulation where the band consists of collective states strongly coupled to the single-particle degrees of freedom, Eq.(17) can serve only as the zeroth approximation; the situation is formally similar to that in the particle-core or particle-rotor models.

3.3. Conservation laws

The advantage of the CDM method is manifested by the treatment of conservation laws, broken symmetries, Goldstone modes and etc. It was already demonstrated by the explicit solution for an isolated rotational band at low spins. Now I show how it works in the general case.

Let the additive operator of the type (13), $P = Tr(p\hat{R})$, be a constant of motion. We assume that it is conserved both by the independent particle hamiltonian, $[p, \epsilon] = 0$, and the interaction, $[p_a + p_b, V_{ab}] = 0$. Of course, this conservation law has to be respected by the collective hamiltonian, $[P, \mathcal{H}] = 0$. Using the definition of the generalized self-consistent field (16) it is easy to show that

$$[P + p, h\{\hat{R}\}] = 0, \tag{18}$$

the system is invariant with respect to the simultaneous "shift" of the global and intrinsic (particle) variables. Note that for the GDM $\hat{R}$ the analogous rule (14) is valid for any one-body operator, irrespectively of its conservation.

Given the shift operator P, one can introduce the conjugate collective coordinate X. It is cyclic and does not enter the collective hamiltonian. Then the shift $D(X) = \exp(ipX)$ of the intrinsic coordinate x (conjugate to p) at the collective position X combines both frames. This shift results in the transformations

$$D(X)\hat{R}D(X)^{-1} = \hat{R}', D(X)\hat{h}D(X)^{-1} = \hat{h}', \tag{19}$$

and

$$D(X)(P+p)D(X)^{-1} = P. \tag{20}$$

The point is that the transformed GDM $\hat{R}'$ and nuclear field $\hat{h}'$ do not depend on the cyclic coordinate X (thus, the motion in the c.m. frame does not depend on the position of the center of mass) and therefore, according to Eqs.(14) and (17), commute with the conserved collective momentum:

$$[P, \hat{R}'] = 0, [P, \hat{h}'] = 0. \tag{21}$$

At the same time, the D-transformation creates the recoil term, $DPD^{-1} = P - p$, so that the collective hamiltonian $\mathcal{H}(P)$ acquires the dependence on the intrinsic variables, and instead of (15) one obtains

$$[\hat{R}', \hat{h}' + \mathcal{H}(P-p)] = 0. \tag{22}$$

Here the collective coordinate X is eliminated and the collective momentum P entering the dynamics as a parameter is the conserved quantum number of the global motion.

3.4. CDM (collective dynamics) and GDM (chaotic dynamics)

There are two problems which were not solved in the original papers [3,4].

First, the main Eq.(15) can be satisfied formally by the simultaneous diagonalization of operators $\hat{h} + \mathcal{H}$ and $\hat{R}$ in the combined space. But it is not clear what are the "occupation numbers" (eigenvalues of R). In the usual SPDM approach (17), the ground state energy is minimized over Slater determinants, i.e. the occupation numbers are 1 or 0, $\rho^2 = \rho$. The same recipe is valid in all cases where the collective variables are switched on iteratively as it was, in particular, in the rotational problem of Sect.2. Then the transition SPDM $\Rightarrow$ CDM, $\rho \Rightarrow \hat{R}$, can be treated as a unitary transformation preserving the eigenvalues so that $\hat{R}^2 = \hat{R}$. For the single-particle orbitals, it implies fragmentation due to presence of collective excitations. Similar arguments can be applied for the non-zero temperature case. In some solvable models [17] deviations of the exact result from the one obtained under such a normalization are not negligible. But they are estimated to be small for realistic many-body problems. In general, one should satisfy the whole set of nonlinear kinematical conditions (like the saturation principle (14)) imposed onto the GDM by the requirements of quantum statistics. In reality it could be done in some average sense, see below.

Another shortcoming of the CDM formalism was an *ad hoc* procedure of HF-type splitting of many-operator expressions in the equations of motion following by inserting the set of collective intermediate states in each term. It is appropriate for models with factorizable forces which we used for actual calculations in [4,6,7] where one does not need to include exchange terms. In case of realistic interactions the accuracy of the procedure was questionable. Meanwhile, otherwise one can violate the permutational symmetry which is one of the strict conditions mentioned in the preceding paragraph.

Apparently, the transition from the CDM to the GDM [9] suggests the way of solution. If the complete set of intermediate states is used in the equations of motion, there is no problem of splitting. This set contains collective states, which are to be included in the band under consideration, as well as numerous noncollective states, mostly of very complicated structure. The latter can be taken into account on average. The averaging procedure [9,10] is based on simple statistical assumptions: components of such chaotic wave functions are uncorrelated. In the limit of extreme chaoticity it would lead to the Gaussian orthogonal ensemble. It can be shown that after averaging out contributions of chaotic excited states one comes to the MF equations of correct HF structure. It is important that the correct averaging procedure preserves, although on average, all kinematic and statistic properties of the GDM.

Thus, we come to the non-orthodox view on the MF as a result of smoothing stochastic components of very complicated incoherent dynamics (collisions, level crossings and so on). The universality of chaotic behavior generates the universal features of the MF equations. On the other hand, the MF presents the basis which allows one to compare the degree of complexity of many-body wave functions. Such a "natural" representation is indispensable for studying manifestations of quantum chaos. On the next step we return to collective phenomena which now occur on the chaotic background responsible for fluctuations and dissipation of collective modes.

4. Applications

Below I enumerate various problems, where the CDM-GDM formalism seems to be an adequate instrument, and comment briefly on the characteristic features of specific applications.

4.1. Pairing

This is the simplest example of the broken symmetry. One can, as was indicated earlier, simply double the single-particle space. It corresponds to the BCS-scheme. More refined approach [5] conserves the particle number and extracts the N-dependent corrections and the so called moment of inertia for pair rotation (pair separation energy). In this problem N plays a role of the conserving momentum P in Eq.(18); the corresponding intrinsic operator is the Pauli matrix τ_3 of the "spinor" representation.

4.2. Translational motion

Here the operator P is the ordinary linear momentum $\mathbf{P}$, and its intrinsic counterpart p is the particle momentum operator $\mathbf{p}$. For a Galilean-invariant system (without pairing and velocity-dependent forces), the machinery of the CDM gives the exact prescription for the transformation of the SPDM ρ (for example, the HF solution (17)) into the CDM $\hat{R}$ which keeps the integral of motion $\mathbf{P}$ and therefore can be used in problems where the momentum transfer to a nucleus is essential. This transformation is given by

$$\hat{R} = \exp(-i\mathbf{p}\mathbf{X})\exp[i(\mu/M)\mathbf{P}\mathbf{x}]\rho\exp[-i(\mu/M)\mathbf{P}\mathbf{x}]\exp(i\mathbf{p}\mathbf{X}) \qquad (23)$$

where $\mathbf{x}$ and $\mathbf{X}$ are particle and center of mass coordinates respectively, M is the total mass and μ is the reduced mass of a particle with respect to the rest of a nucleus. The solution (23) is self-consistent giving the total momentum $Tr(\mathbf{p}\hat{R}) = \mathbf{P}$ and the particle number $A = Tr\hat{R} = M/m$ where m is the nucleon mass.

For a system with pairing the problem can be solved approximately within the accuracy A^{-1}.

4.3. Random phase approximation

The collective space of the problem is spanned by the phonon modes, and the RPA is equivalent to the regular expansion of $\hat{R}$ in terms of boson creation and annihilation operators, $A^\dagger$ and A, respectively. In the lowest order the CDM coincides with the SPDM ρ. The corrections to the CDM are linear in phonons,

$$\hat{R} = \rho + \sum_A (\phi_A A^\dagger + \phi_A^\dagger A). \qquad (24)$$

Here ϕ_A is the superposition of quasiparticle excitations forming the given RPA mode with the eigenfrequency ω_A. The complete analysis of the RPA equations and the structure of the biorthogonal set of the eigenmodes can be found in Ref.[8]. For the small amplitude collective motion, this set corresponds to physical excited states. In more complicated problems, it gives the convenient representation where the important correlations beyond the MF are already taken into account.

If the SPDM manifests the spontaneous symmetry breaking, for example, by fixing the collective coordinate X conjugate to the violated integral of motion P, the RPA restores it via the Goldstone mode with the zero frequency. The wave function ϕ_0 for this mode is nothing but the shift of the static SPDM by the momentum p, $\phi_0 = [p, \rho]$. Then it can be shown straightforwardly that the moment of inertia for the Goldstone mode is equal to $Tr(\phi_0 x)$ where the single-particle counterpart x of the cyclic coordinate X satisfies the equation $[x, h\{\rho\}] = \tilde{p}$ and $\tilde{p}$ includes the Migdal term (renormalization of the bare operator p by the interaction). All the calculations are extremely facilitated in the operator form taking matrix elements at the end only.

4.4. Higher order anharmonicity

The most intriguing problem of the transitional nuclei is not solved yet in spite of the vast amount of phenomenological work (various boson models). Here one has to deal with the finite Fermi system in the vicinity of the phase transition to the stable static deformation. The shape of such a nucleus is ill-defined and the RPA quadrupole frequency is abnormally low in the natural scale provided by the pair breaking energy 2Δ. Consequently, the amplitude of collective quadrupole motion is large and corrections to the RPA theory of small vibrations turn out to dominate the whole pattern of low-lying excitations. Moreover, the way to describe anharmonic terms with the aid of the boson expansion technique [18] is not promising since it is necessary to include the response of all non-collective modes which are strongly coupled to the slow changing deformation.

We know that the empirical classification of the low-lying states, at least in even-even nuclei, is the same as in the harmonic vibrator $SU(5)$ scheme although the quantitative predictions of the harmonic approximations are badly violated. It can be understood with the help of the microscopic estimates of the leading anharmonic corrections [10,19] which turn out to be, similar to the Landau theory of the second order phase transitions, of the quartic type. The quartic anharmonicity maintains the $O(5)$ symmetry and permits to label the levels according to the harmonic approximation scheme. The phenomenological model [10,19] with strong quartic anharmonicity works excellent for dozens of soft nuclei. But the actual computation of anharmonic parameters still requires the full machinery of microscopic theory. The main difficulty is caused by admixture of configurations with a different shape due to the large vibrational amplitude.

The reasonable way to perform such a computation would be to select the "band" spanned by the phonon multiplets and to solve the GDM equations within such a space. This program is much more complicated than its rotational prototype of Sect.2 but still appears to be feasible.

4.5. Low spin rotation

In the problem of rotation, the body-fixed frame can be introduced in exact way as it was explained in Sect.1. The formalism of Subsect.3.3 extremely simplifies actual work. The pecularity of the rotational case is the non-abelian character of the corresponding group. Here, instead of the single generator P, we have to deal with three non-commuting components of the angular momentum $\mathbf{J}$ and, instead of the collective coordinate X, with Euler angles ϑ_i. The D-transformation (19) of the single-particle space, $D = D(\vartheta)$, rotates the intrinsic axis to combine them with the global ones. But the transformed quantities $\hat{R}'$ and $\hat{h}'$, commuting with the space-fixed angular momentum $\mathbf{J}$, still can contain the angles through the body-fixed angular momentum components

$$I_\lambda = \sum_\mu D^1_{\mu\lambda}(\vartheta) J_\mu. \tag{25}$$

Under the D-transformation (19), the components of $\mathbf{I}$ acquire the particle angular momentum recoil terms,

$$D(\vartheta)\mathbf{I}D(\vartheta)^{-1} = \mathbf{I} - \mathbf{j}, \tag{26}$$

and the Coriolis force appears in the dynamic equation (22),

$$[\hat{R}', \hat{h}' + \mathcal{H}(A; \mathbf{I} - \mathbf{j})] = 0. \tag{27}$$

Here A stands for collective operators of non-rotational nature like phonon operators of Eq.(24) defined in the body-fixed frame.

Full microscopic calculations of the low spin properties of the ground, beta and gamma rotational bands in deformed rare earth even-even nuclei were performed in Refs. [6,7]. Using pairing and quadrupole-quadrupole forces and enlarging the collective space to include several interacting bands built on the RPA phonons, we have calculated moments of inertia (as a rule, underestimated by the standard cranking calculations), next order terms $\sim \mathbf{J}^4$ in the energy spectra, intraband and interband transition probabilities including non-adiabatic corrections. The agreement with data available at that time was quite reasonable. Unfortunately, the problem of crossing of the rotational bands built on the different MF configurations still was not considered by the CDM method.

The approach to the description of rotation similar to that developed in the Sect.2, was independently suggested in Ref. [20] with the use of the states of odd-A nuclei as in the original Kerman-Klein formulation [3], see also [21].

4.6. High spin rotation

In the asymptotic high-spin region, $I \gg j$, the GDM method allows for the transition to the semiclassical TDHF-cranking approach. In contrast to the standard formulation [22], the corresponding limit of the GDM, Eq.(27), in general should be written in the time-dependent form

$$i\dot{R}'(t) = [h'(t) - \mathbf{j}\mathbf{\Omega}(t), R'(t)] \tag{28}$$

where $R'(t)$ and $h'(t)$ are operators in the intrinsic space only, and the angular velocity is $\mathbf{\Omega} = \partial\mathcal{H}(A;\mathbf{I})/\partial\mathbf{I}$. In general, this three-dimensional TDHF-cranking equation implies non-uniform rotation, and Eq.(28) guarantees the correct equation of motion for the total body-fixed angular momentum $\mathbf{I}(t) = Tr\{\mathbf{j}R'(t)\}$,

$$\dot{\mathbf{I}}(t) = \mathbf{I}(t) \times \mathbf{\Omega}(t). \tag{29}$$

The structure of Eq.(28) and the RPA modes built on the solution, corresponding to the uniform rotation with the body-fixed stationary density matrix $\rho(\Omega)$ and $\Omega = \Omega_x$ along the axis x with the maximum moment of inertia, was analyzed in Ref. [23]. The new features here are two extra quasi-Goldstone modes $\phi_\pm = [j_y \pm i j_z, \rho]$ with the frequencies $\pm\Omega$, apart from the Goldstone mode

$\phi_x = [j_x, \rho]$ with the zero frequency, and the wobbling mode [11,24]. The interesting possibility (still not accomplished) is to study the large amplitude rotational motion corresponding to the periodic solutions of the Euler equations (29) which are expressed in classical mechanics in terms of elliptic functions.

4.7. Large amplitude collective motion

The collective hamiltonian for an adiabatic large amplitude motion was derived in Ref. [15]. The collective space under study is described here by collective coordinates Q_i and conjugate momenta P_i. The main equation (15) is solved with the use of regular expansion of the CDM in powers of P_i. The collective operators should saturate the dynamics within the band, therefore in higher orders they are renormalized that Eq.(14) be fulfilled.

In some respects this method is similar to the ATDHF, see for example [25]. The essential difference is that in the GDM method all quantities keep their character as quantal operators. Therefore the resulting collective hamiltonian is also quantal and one does not need any procedure of requantization. The price one has to pay for such an advantage, is that the algorithm of actual solution is rather complicated. It includes, for example, the operator consistency of collective coordinates and the local RPA modes at each point of the collective path. Therefore the method was tested on rather simple models only [15,26].

4.8. Mean field out of chaos

This problem was discussed in Subsect.3.4. The extension of the CDM to the GDM [9] allows one to describe the full chaotic dynamics in the coarse-grained form. The MF generated from chaos is governed by the dynamics slightly different from the conventional thermal HF approximation. The difference comes from taking into account the average fluctuations of the MF which are neglected in the HF description of the single MF created by the equilibrium distribution function of particles. This point may be of significant importance for phenomena induced by the particles near the Fermi surface in hot nuclei, especially if the gradients of the nuclear density are involved.

The whole problem still presents a long list of open questions which hopefully will be addressed in the nearest future.

4.9. Fluctuations and dissipation of collective motion

The previous Subsection was devoted to the pure chaotic dynamics and the MF as its average component. The collective modes have to be included explicitly in the same way as it was done earlier. But now they interact with the chaotic background so that the collective motion becomes fluctuating and damped.

The consistent description of such effects should distinguish clearly the distortion of the CDM due to the dependence on the collective coordinates along the collective path from the irreversible dissipation (friction determined by the relative velocity of the collective mode with respect to the background of incoherent excitations which is some sort of medium or heat bath). This discrimination is very distinct in the GDM approach, see Refs. [8,13,27] where applications to deep inelastic heavy ion reactions were considered..

As a result, Langevin-type description of the collective motion arises. It gives a reliable base for the consideration of related problems like Marcoffian or non-Marcoffian dynamics, statistical characteristics of the noise, interaction between slow and fast collective modes, limiting temperature of collective vibrations [28], kinetics of preequilibrium processes and so on.

Acknowledgements

Hospitality and support from JIHIR/ORNL and University of Tennessee are highly appreciated.

References

1. A.B.Migdal, *Theory of Finite Fermi Systems* (Wiley, New York, 1967).
2. S.T.Belyaev and V.G.Zelevinsky, *Yad. Phys.* **2** (1965) 615 [*Sov. J. Nucl. Phys.* **2** (1966) 442].
3. A.Klein and A.Kerman, *Phys. Rev.* **132** (1963) 1326; **138B** (1965) 1323.
4. S.T.Belyaev and V.G.Zelevinsky, *Yad. Phys.* **11** (1970) 741 [*Sov. J. Nucl. Phys.* **11** (1970) 416].
5. S.T.Belyaev and V.G.Zelevinsky, *Yad. Phys.* **16** (1972) 1195 [*Sov. J. Nucl. Phys.* **16** (1973) 657]; *Yad. Phys.* **17** (1973) 525 [*Sov. J. Nucl. Phys.* **17** (1973) 269].
6. V.G.Zelevinsky and M.I.Shtokman, *Sov. Phys. Izvestia, ser. fiz.* **36** (1972) 2577.
7. M.I.Shtokman, *Yad. Phys.* **22** (1975) 479 [*Sov. J. Nucl. Phys.* **22** (1976) 247].
8. V.G.Zelevinsky, *Prog. Theor. Phys. Suppl.* **74-75** (1983) 251.
9. V.G.Zelevinsky, *Preprint* NBI-92-11; *Nucl. Phys.*, in press.
10. V.G.Zelevinsky, in *New Trends in Nuclear Collective Dynamics*, Springer Lecture Notes, in press.

11. A.Bohr and B.Mottelson, *Nuclear Structure*, vol.II (Benjamin, Reading, 1975).

12. J.P.Elliott, in *Selected Topics in Nuclear Theory* (IAEA, Vienna, 1963).

13. V.G.Zelevinsky, in *Materials of VII Winter School LINP* (Leningrad, 1972), part 2, p.147.

14. S.T.Belyaev, *Math.-Fys. Medd. Kgl. Dan. Vid. Selsk.* **31** *No.*11 (1959).

15. V.G.Zelevinsky, *Nucl. Phys.* **A337** (1980) 109.

16. D.J.Thouless and J.G.Valatin, *Nucl. Phys.* **31** (1962) 211.

17. D.Janssen, *Nucl. Phys.* **A331** (1979) 311,329.

18. S.T.Belyaev and V.G.Zelevinsky, *Nucl. Phys.* **39** (1962) 582.

19. O.K.Vorov and V.G.Zelevinsky, *Nucl. Phys.* **A439** (1985) 207.

20. S.G.Rogozinski, *Acta Phys. Polonica* **35** (1969) 783.

21. A.Kuriyama, *Prog. Theor. Phys.* **56** (1976) 1786; R.Okamoto and M.Nakano, *Nucl. Phys.* **A**, in press.

22. D.Thouless and J.G.Valatin, *Phys. Rev. Lett.* **5** (1960) 509.

23. V.G.Zelevinsky, *Nucl. Phys.* **A344** (1980) 109.

24. E.R.Marshalek, *Nucl. Phys.* **A331** (1979) 429.

25. P.Ring and P.Schuck, *Nucl. Phys.* **A292** (1977) 20.

26. V.V.Mazepus, *Theor. Math. Phys.* **22** (1975) 406.

27. P.N.Isaev, *Yad. Fiz.* **32** (1980) 978; **34** (1981) 717, 1207 [*Sov. J. Nucl. Phys.* **32** (1980) 505; **34** (1981) 399, 671].

28. P.F.Bortignon, A.Bracco, D.Brink and R.A.Broglia. *Phys. Rev. Lett.* **67** (1991) 3360.

GROUP 7

Particle–Core Models

PARTICLE-CORE COUPLING MODELS

P. B. SEMMES
Physics Department, Tennessee Technological University
Cookeville, TN 38501 USA

ABSTRACT

The microscopic motivation for particle-core coupling models is reviewed within the equation of motion method. Many phenomenological models are obtained if self-consistency is abandoned and the Pauli principle is relaxed. The Core-Quasiparticle Coupling Model and the Particle-Rotor Model are described in detail, and critically evaluated.

1. Introduction

Phenomenological particle-core coupling models for odd-mass nuclei have been used for about forty years in nuclear physics [1, 2], and since then many versions have been developed. In general, these describe the low energy structure of a nucleus in terms of a small set of valence particles that are treated explicitly, and a collective core that describes the rest of the nucleus. The total Hamiltonian for the system is written

$$H_{tot} = H_{core} + H_{part} + H_{int} \tag{1}$$

where H_{core} describes the collective core and H_{part} describes the set of particles (e.g., a shell model Hamiltonian, perhaps including interactions between particles). The two systems, core plus particles, are coupled by the interaction H_{int}. Similarly, the total angular momentum is written

$$\vec{I} = \vec{R} + \vec{j} \tag{2}$$

and a basis for the combined system is constructed from the direct product space $|(j \otimes R)I>$. Here j and R denote the angular momenta of the particles and the collective core, respectively, and I is the total angular momentum. Other operators are likewise divided into "core" and "particle" contributions. Thus these models attempt to correlate and interpret the properties of nuclei in terms of a few variables related to the collective core (e.g., "the nuclear shape"), the system of particles (e.g., residual interactions between them) and their coupling.

At this stage the number of valence particles has not yet been specified, and the usual shell model can even be recovered if the core is taken to be an inert doubly magic system, and all protons and neutrons outside the closed shells are treated explicitly. This article concentrates on the other extreme, in which only a single particle is treated explicitly, and the effective core corresponds to the collective structure of an even-even system. A wide variety of collective cores have been used, including boson models and geometric (e.g., vibrational and rotational) descriptions. Furthermore, "weak", "strong" or "intermediate" coupling schemes for the odd-mass eigenstates may result depending on the strength of H_{int}. As a result, one natural division is between models that are formulated in the lab frame (using a weak coupling basis) and those formulated in the intrinsic frame (using a strong coupling basis). Clearly it is hopeless to try to provide a detailed overview of all the varieties of particle-core coupling (even with the restriction to a single particle noted above), and so drastic limits must be set on the scope of this article.

This contribution presents one perspective of particle-core coupling models and will emphasize the main assumptions and restrictions common to most. However, not every version can be commented upon separately, and alternative theoretical justifications are certainly known for specific models. (See also the contribution to these proceedings by Hamamoto.) A very broad view of particle-core coupling is possible through the microscopic method of Klein and coworkers [3, 4]. This approach will be adopted here, with particular attention to the steps required in the passage from microscopic theory to phenomenological model. This formalism contains many of the odd-A collective models in use today, including particle-rotor models and some particle-vibrator models. The Core-Quasiparticle Coupling Model of Dönau and Frauendorf [5] is a direct descendant of this approach, and will be considered in some detail as an example of particle-core coupling in the lab frame. In addition, the "Lund version" of the particle + triaxial rotor model [6] is described to illustrate a formulation in the intrinsic frame.

2. A General Approach to Particle-Core Coupling

A derivation for a general particle-core coupling scheme has been presented in ref. [3, 4], using the equation of motion method. The method is also reviewed elsewhere in these proceedings by Klein. Briefly stated, the formalism is based on a fermionic shell model and seeks expressions for the matrix elements of direct interest for the low-lying collective states of an odd-mass nucleus and its even-even neighbors. These matrix elements include the energy eigenvalues, the single particle coefficients of fractional parentage and diagonal and off-diagonal matrix elements of electromagnetic operators. The key assumptions are that the most important parts of the shell model interaction can be identified (e.g., spherical mean field + pairing + quadrupole-quadrupole) and that only a few low-lying collective states in the even-even cores are important for the odd-A structure. In principle this formalism can be used to obtain a selfconsistent theory, but usually selfconsistency is abandoned and the core properties are simply accepted as input quantities. As described below, this formalism contains most of the usual particle-core coupling models but still allows for a number of possible generalizations. The derivation given below mainly follows ref. [7].

We start with an identity for a state $|I>$ of an odd-mass nucleus

$$|I>=\frac{1}{\Omega}\sum_{j}(a_j^{\dagger}a_j+\overline{a}_j\overline{a}_j^{\dagger})|I> \tag{3}$$

which follows from the anticommutation relations for the fermion creation and destruction operators $a_j^\dagger$ and a_j. Here j denotes the spherical single particle quantum numbers $(nljm)$, $\Omega = \sum(2j+1)$ and $\bar{a}_{jm} = (-1)^{j+m} a_{j-m}$. Next, completeness relations are inserted for the two cores, labeled by $|R, A-1>$ and $|R', A+1>$

$$|I> = \frac{1}{\Omega} \sum_{j,R,R'} (a_j^\dagger |R, A-1>< R, A-1|a_j|I> + \bar{a}_j |R', A+1>< R', A+1|\bar{a}_j^\dagger)|I> \tag{4}$$

This decomposition indicates that the states $|I>$ in the odd-A nucleus are quasiparticle states obtained by a Bogolyubov transformation from particle-like states $(a_j^\dagger|R, A-1>)$ and hole-like states $(\bar{a}_j|R', A+1>)$. The amplitudes of this decomposition are to be determined, and are denoted u_j and v_j. Thus

$$|I> = \frac{1}{\Omega} \sum_{j,R,R'} u_j a_j^\dagger |R, A-1> + v_j \bar{a}_j |R', A+1> \tag{5}$$

and these states have good angular momentum and particle number. The equations of motion are obtained by forming commutators between the fermion operators and the Hamiltonian, and then taking appropriate matrix elements

$$< I|[h, a_j^\dagger]|R, A-1> = (E_I - E_R^{A-1}) < I|a_j^\dagger|R, A-1> \tag{6}$$

$$< I|[h, \bar{a}_j]|R', A+1> = (E_I - E_R^{A+1}) < I|\bar{a}_j|R', A+1> \tag{7}$$

On the right side of these equations the energy eigenvalues E are obtained because the states $|I>$, $|R, A-1>$ and $|R', A+1>$ are chosen to be eigenstates of the Hamiltonian h.

To proceed further, the Hamiltonian must be specified. We assume a general form in which the effective interaction can be written as a multipole expansion

$$h = h_{sp} - \frac{1}{4}\sum_{\lambda\mu} \kappa_\lambda Q_\mu^\lambda Q_{\bar{\mu}}^\lambda - \frac{1}{4}\sum_{\lambda\mu} g_\lambda P_\mu^{\lambda\dagger} P_\mu^\lambda \tag{8}$$

This contains the spherical single particle Hamiltonian

$$h_{sp} = \sum_j \varepsilon_j a_j^\dagger a_j \tag{9}$$

and the particle-hole and particle-particle multipole operators

$$Q_\mu^\lambda = \sum_{j,j'} < j|r^\lambda Y_\mu^\lambda|j'> a_j^\dagger a_{j'} \tag{10}$$

$$P_\mu^{\lambda\dagger} = \sum_{j,j'} < j|r^\lambda Y_\mu^\lambda|j'> a_j^\dagger \bar{a}_{j'} \tag{11}$$

If this Hamiltonian is used in the left hand side of Eqns. (6) and (7) and the completeness relations are inserted inside the product terms like $a_j^\dagger Q_\mu^\lambda$, then the resulting equation can be written in the form of a 2 x 2 matrix

$$\begin{pmatrix} \varepsilon + E_R^{A-1} + \Gamma^{A-1} & -\Delta^\dagger \\ -\Delta & -\varepsilon' + E_{R'}^{A+1} - \Gamma^{A-1} \end{pmatrix} \begin{pmatrix} u \\ v \end{pmatrix}_I = E_I \begin{pmatrix} u \\ v \end{pmatrix}_I \tag{12}$$

The block matrices ϵ and E contain the spherical single particle energies and the core energies, respectively, and are diagonal in this basis

$$\epsilon = \epsilon_j \delta_{j,j'} \delta_{R,R'} \tag{13}$$

$$E_R^{A-1} = E^{A-1} \delta_{j,j'} \delta_{R,R'} \tag{14}$$

but Γ and Δ are more complicated

$$\Gamma^{A\pm1} = -\frac{1}{2} \sum_\lambda \kappa_\lambda < j||r^\lambda Y^\lambda||j' >< R, A \pm 1||Q^\lambda||R, A \pm 1 > W(j'\lambda IR; jR') \tag{15}$$

$$\Delta = \frac{1}{2} \sum_\lambda g_\lambda < j||r^\lambda Y^\lambda||j' >< R, A-1||P^\lambda||R, A+1 > W(j'\lambda IR; jR') \tag{16}$$

The Fermi level parameter λ naturally enters if the core energies are expressed relative to the ground state values. Thus

$$\lambda = \frac{1}{2}(E_{gs}^{A+1} - E_{gs}^{A-1}) \tag{17}$$

and the single particle energies ϵ are replaced by $(\epsilon - \lambda)$. Constraints imposed by the Pauli principle are also written as matrix elements, taken for example with respect to the core states in the even-even neighbors

$$< R_1|a_\alpha^\dagger a_\beta + a_\beta a_\alpha^\dagger|R_2 >= \delta_{\alpha,\beta} \delta_{R_1,R_2} \tag{18}$$

This becomes a sum rule in terms of the particle and hole amplitudes if a complete set of odd-A states $|I>$ is inserted between the fermion creation and destruction operators.

At this stage the equations are still formally exact (with this choice of the Hamiltonian) and the core matrix elements refer to the physical even-even neighboring nuclei. In principle, this formalism could lead to a selfconsistent theory, or with suitable approximations many phenomenological particle-core models can be obtained [8, 4], which we now consider.

The critical assumptions entering at this point are the following:

- The essential core properties (energies, core matrix elements of Q^λ and P^λ) are regarded as phenomenological input, rather than quantities to be determined selfconsistently.
- A collective subspace is assumed in the even-even cores, defined by the core states connected by large matrix elements of Q^λ or P^λ, thus truncating the sums over complete sets of core states (e.g., in Eqn. 4).
- Pauli principle constraints on the odd-A wave function expressed in sum rules (e.g., derived from Eqn. 18) are replaced by a simple normalization condition on the particle and hole amplitudes of the wave function (Eqn. 5).

Although the severity of these approximations may be difficult to judge in general, in a few simple limiting cases it is possible to recognize some characteristic difficulties. For example, if only static pairing and the single particle energies are retained (i.e., coupling to the ground states of the cores only), then the odd-A solutions include both the quasiparticle and quasihole states with $E = \pm\sqrt{(\epsilon - \lambda)^2 + \Delta^2}$. The u and v amplitudes of the positive energy solutions are identical to the usual BCS results (Eqns. 23 and 24), while the negative energy solutions are related to these by

$$\begin{pmatrix} u_- \\ v_- \end{pmatrix} = \begin{pmatrix} -v_+ \\ u_+ \end{pmatrix} \tag{19}$$

In the usual BCS theory, the positive energy solutions are recognized as the physical 1-quasiparticle excitations, and the negative energy solutions are simply dropped. More generally, the set of states $|I>$ in the combined particle and hole basis is clearly overcomplete by a factor of two, and some method for identifying the physical solutions is necessary. A method based on a density matrix constructed from the overlaps of the basis states is suggested in ref. [7], but this remains largely untested in practice. As a second example, consider a pairing + quadrupole-quadrupole interaction for a vibrational core (quadrupole vibrations about a spherical shape), giving a quasiparticle + phonon model. Then with the assumptions listed above, the microscopic structure of the quadrupole phonon is neglected, and the Pauli correction term [9] which is known to be important for the $I = j - 1$ "anomalous coupling state" is not included. This Pauli correction is due to a process in which the odd quasiparticle exchanges with one of the quasiparticles in the quadrupole phonon. For a general collective core it is very difficult to estimate how seriously the Pauli principle is violated, but at least in this case it is rather clear. Finally, note that if selfconsistency between the odd-A eigenstates and the core matrix elements is abandoned, then all odd-A states must have the same effective core (or $A \pm 1$ cores), and clearly this effective core is not necessarily the physical even-even neighbor. Thus, it is doubtful that configuration dependent deformations (or shape coexistence) can be accomodated without reinstating some part of the selfconsistency.

These examples suggest some of the difficulties that must follow these approximations, but the flexibility of the remaining formalism is very attractive. Note that the multipole orders of the interaction (in both the p-h and p-p channels) are still unspecified, so this could be a starting point to investigate more general pair fields [7, 10] or perhaps octupole shape vibrations or static deformations.

3. The Core-Quasiparticle Coupling Model

The CQCM of Dönau and Frauendorf [5] is one realization of this general approach in which the interaction is limited to quadrupole-quadrupole coupling and static monopole pairing. Thus, the Γ matrix in Eqn. 15 contains only $\lambda = 2$ terms and Δ in Eqn. 16 is just a constant diagonal matrix. In practice no attempt is made to achieve selfconsistency, and projection onto the physical subspace is achieved with an approximate density matrix constructed from the "adiabatic" part of the Hamiltonian. For the special case of a well deformed axially symmetric rotor core, this projection is equivalent to discarding the negative energy deformed quasiparticle states. In order to illustrate the general procedure, this example is considered in some detail. (Additional information can be found in refs. [5, 7, 11]).

In Eqn. (12), the upper left block describes a particle coupled to the A-1 core, while the lower right block is for a hole coupled to the A+1 core. In order to simplify the following discussion, the two $A \pm 1$ cores are assumed to be identical, although the distinction can be retained. It is advantageous to split the Hamiltonian of the coupled system into parts that are odd and even under this particle-hole conjugation, thus

$$H_{tot} = H_0 + H_c \tag{20}$$

with

$$H_0 = \begin{pmatrix} \epsilon + \lambda + \Gamma & -\Delta \\ -\Delta & -\epsilon - \lambda - \Gamma \end{pmatrix}, \; H_c = \begin{pmatrix} E_R & 0 \\ 0 & E_R \end{pmatrix} \tag{21}$$

The "adiabatic" part of the Hamiltonian, H_0, has the symmetry of an HFB matrix and can be diagonalized in two steps. First, the block matrix $(\epsilon - \lambda + \Gamma)$ is diagonalized leading to eigenvalues $e(k,\beta) - \lambda$ and eigenvectors described by the transformation coefficients $< k,\beta|j,R >$. For the axially symmetric rotor core, these eigenstates are just the deformed Nilsson orbitals, and the labels β and k specify the core quadrupole field and the adiabatic single particle state, respectively. The transformation coefficients $< k,\beta|j,R >$ are the amplitudes of these Nilsson states expressed in the weak coupling basis. The diagonal pairing matrix (Δ) is not affected by this transformation, and the H_0 matrix contains many 2 x 2 matrices in this representation. In the second step, the particle and hole states are coupled, leading to quasiparticle states with energies

$$E^{\pm}(k,\beta) = \pm\sqrt{(e(k,\beta) - \lambda)^2 + \Delta^2} \tag{22}$$

The u and v amplitudes of the positive energy solutions are identical to the usual BCS expressions,

$$u^2_{k,\beta} = \frac{1}{2}(1 + \frac{e_{k,\beta} - \lambda}{E_{k,\beta}}) \tag{23}$$

$$v^2_{k,\beta} = \frac{1}{2}(1 - \frac{e_{k,\beta} - \lambda}{E_{k,\beta}}) \tag{24}$$

The low energy states in odd-A nuclei are associated with these positive energy quasiparticle excitations; the negative energy solutions are the quasihole states, and are discarded.

The remaining portion of the Hamiltonian is H_c, and for the rigid symmetric rotor the core spectrum is given by the familiar $R(R+1)$ energy law. The final step in the calculations is the diagonalization of H_c in the adiabatic quasiparticle basis. When written in this basis, pairing factors of the form $(uu' + vv')$ appear on all H_c terms, and the off-diagonal matrix elements correspond to the usual Coriolis terms (which mix the adiabatic states). Note that the Coriolis and recoil terms never explicitly appear because the model is formulated in the lab frame; nevertheless, for the rigid rotor core, this method is in principle identical to the usual particle-rotor model (with the $(uu' + vv')$ pairing factors). Clearly, this discussion can be immediately generalized to include other rotor cores with a fixed nuclear shape. For example, a rigid triaxial rotor core can also be directly accomodated, as can higher multipoles in the deformed field (e.g. hexadecapole) if appropriate terms are added to the particle-core interaction. Similarly, a particle + octupole rotor model can be obtained that would be formally identical to the intrinsic frame description given by [12]. The formulation of particle-rotor models in the lab frame is also discussed in [13].

Finally, the procedure outlined above is generalized in the CQCM and is applied for any quadrupole collective core: First, the core properties are defined simply by a set of numbers, the core energy eigenvalues and the reduced matrix elements of the quadrupole operator between all core states. These numbers are accepted as phenomenological input, as are the energies of the spherical j-shells included in the basis and the Fermi level λ and the gap paramter Δ (which appears as the diagonal pairing matrix elements connecting the $A \pm 1$ cores). The adiabatic part of the coupled Hamiltonian is diagonalized, and quasiparticle states are constructed from these. The core Hamiltonian is projected onto the space of positive energy quasiparticle states, and diagonalized. The desired properties in the odd-A nucleus (e.g. electromagnetic matrix elements) are computed from these final eigenvectors. In general, any desired operator is divided into a contribution from the unpaired particle and a core part, which is defined simply in terms of (input) core matrix elements.

Thus the CQCM retains great flexibility and can be applied to a wide variety of collective cores, which may be based on boson models (e.g., the Interacting Boson Model) or to geometric rotors. In a somewhat different example of a collective core, high spin calculations for 3-quasiparticle states have been made in which an odd proton is coupled to a 0 + 2-quasineutron rotor core [14]. As long as the form of the interaction is sufficient and an appropriate core description can be obtained, then the CQCM can provide a convenient framework for calculating the properties of collective states in odd-A nuclei. Additional details and some further explorations can be found in refs. [7, 10].

4. The Particle-Rotor Model

Probably the most widely used particle-core coupling model is one in which the core is a quadrupole deformed rotor with a fixed shape. It is very commonly used to interpret electromagnetic data (primarily M1 and E2 transition rates, mixing ratios and static moments) at low and moderate spins. Since the model considered here treats only a single unpaired nucleon explicitly, it is generally not appropriate for describing high spin states with $I > 29/2$ or so. In the following sections, the formalism underlying the "Lund version" of the particle + triaxial rotor model [6, 15] is presented. In earlier descriptions of the particle + triaxial rotor model, the odd particle was restricted to a single intrinsic orbital [16], or to a single j-shell [17]. The Lund version essentially allows a more flexible treatment of the single particle space, as described below. Many general features of the model are discussed by Meyer-ter-Vehn [17], including characteristic energy spectra for unique parity states as a function of the triaxiality parameter γ and the Fermi level λ. Also, some important features of M1 and E2 properties are discussed in refs. [9, 17, 18, 19, 20].

4.1. Outline of the Formalism

From the particle-core coupling point of view, we start with the Hamiltonian for the odd-A nucleus

$$H_{tot} = H_{core} + (H_{sph} + H_{p-h} + H_{pair}) \tag{25}$$

where the first term inside the parentheses, H_{sph}, represents the spherical shell model Hamiltonian for the unpaired nucleon, and the next two are the particle-hole (H_{p-h}) and particle-particle (H_{pair}) interactions. Typically, the most important terms here are the quadrupole-quadrupole (p-h) and the monopole pairing (p-p) interactions. The even-even collective core (H_{core}) is treated as a rigid triaxial rotor. With the restriction to a rigidly deformed core, the sum ($H_{sph} + H_{p-h}$) can be transformed into the intrinsic frame where it takes the form of a one-body deformed shell model Hamiltonian. The eigenstates of this static deformed potential (Nilsson states) provide a convenient basis for further calculations; in particular, the pairing term is easily included by a BCS treatment on these adiabatic single particle states. In the strong coupling basis, all these single particle terms are diagonal, and finally the core Hamiltonian can be diagonalized within a basis constructed from some set of quasiparticle states close to the Fermi level. Finally, electromagnetic matrix elements or other desired properties are calculated for the energy eigenstates by transforming the appropriate spherical tensor operator from the lab frame to the intrinsic system [21, 9]. Some details are discussed below, and a brief comparison with some other approaches is given.

4.2. The Deformed Single-Particle Hamiltonian

Some general features of Nilsson orbitals for triaxial shapes are described here, and some specific conventions that are used in the particle-rotor model are discussed. These conventions are useful whether the mean field is parameterized as a Modified Oscillator potential (as in ref. [6]), or a Woods-Saxon potential, etc.

The mean fields under consideration here are reflection symmetric but not necesarily axially symmetric. Thus, parity is a good quantum number in the intrinsic system, but Ω need not be (i.e., j_z is not necessarily conserved). Even so, it is useful to use Ω as a label on the intrinsic single particle basis states. The "ν'th" adiabatic deformed orbital can be expanded in such a basis,

$$|\nu> = \sum_{\alpha,\Omega} c^{(\nu)}_{\alpha,\Omega} \, |\alpha\,\Omega> \tag{26}$$

where $c_{\alpha,\Omega}$ is an expansion coefficient and $|\alpha\,\Omega>$ is the corresponding basis state. The label α stands for all quantum numbers other than Ω needed to specify the basis state. Due to the time reversal symmetry, these deformed orbitals are doubly degenerate. In the axially symmetric case, the two degenerate orbitals correspond to positive and negative values of Ω and one set can be chosen as "normal orbitals" and the other as "conjugate orbitals", related to each other by time reversal or by rotation of π about the intrinsic x- or y- axis. In the axially asymmetric case, Ω is not a good quantum number, but the double degeneracy remains. Because the triaxial potential contains terms with $Y_{2,+2}$ and $Y_{2,-2}$, the mixing between different Ω basis states occurs only in steps of ± 2. In ref. [6] this is exploited by adopting the following convention for "normal" and "conjugate" orbitals: The normal orbital $|\nu>$ has components with Ω restricted to ..., -3/2, +1/2, +5/2, ... etc., and its conjugate orbital (with components $\Omega =$ ..., -5/2, -1/2, +3/2, ... etc.) is given by

$$|\tilde{\nu}> = R_1(\pi)|\nu> = -T|\nu> \tag{27}$$

where R_1 is a rotation about the intrinsic 1-axis (x-axis), and T is the time reversal operator. (This convention for the conjugate orbital differs slightly from that adopted in ref. [9], and has an extra factor of -1.) For example, for the $|N\,\ell\,j\,\Omega>$ basis (with the set of quantum numbers N, ℓ, j represented by α above), used with the Modified Oscillator potential [6] the orbital conjugate to $|\nu>$ is given by

$$|\tilde{\nu}> = \sum_{\alpha,\Omega}(-1)^{(j-\Omega)}\, c^{(\nu)}_{\alpha,\Omega}|N\,\ell\,j\,-\Omega> \tag{28}$$

Various single particle matrix elements must be calculated from the adiabatic states; details are given in ref. [6] for the $|N\,\ell\,j\,\Omega>$ basis.

4.3. The Pairing Interaction

Once the adiabatic orbitals have been constructed, the pairing interaction is approximately diagonalized by the BCS method. This requires constructing quasiparticle operators ($\alpha^\dagger$, α) from the particle creation and destruction operators ($a^\dagger$, a). The quasiparticle creation operators are defined by

$$\alpha^\dagger_\nu = u_\nu a^\dagger_\nu - v_\nu a_{\tilde{\nu}},\ \alpha^\dagger_{\tilde{\nu}} = u_\nu a^\dagger_{\tilde{\nu}} + v_\nu a_\nu \tag{29}$$

The intrinsic state of the odd-A nucleus in which the unpaired nucleon occupies the orbital $|\nu>$ is written as a 1-qp state

$$\alpha^\dagger_\nu|BCS> = \alpha^\dagger_\nu \prod_\mu (u_\mu + v_\mu a^\dagger_\mu a^\dagger_{\tilde{\mu}})|-> \tag{30}$$

where $|BCS>$ denotes the BCS vacuum. Note that the orbital $|\nu>$ is not excluded from the BCS vacuum; this last equation can also be written in terms of the particle operator $a^\dagger$

$$\alpha^\dagger_\nu|BCS> = a^\dagger_\nu \prod_{\mu\neq\nu} (u_\mu + v_\mu a^\dagger_\mu a^\dagger_{\tilde{\mu}})|-> \tag{31}$$

The quasiparticle energy is

$$E_\nu = \sqrt{(e_\nu - \lambda)^2 + \Delta^2} \tag{32}$$

and the u^2 and v^2 factors are the usual nonoccupation and occupation probabilities, and are given by Eqns. (23, 24).

The BCS equations that determine the gap parameter Δ and the Fermi level λ for the vacuum state are

$$\Delta = G \sum_{\mu} u_{\mu} v_{\mu} \tag{33}$$

$$N = 2 \sum_{\mu} v_{\mu}^2 \tag{34}$$

where N is the average number of particles. If the gap and the average particle number are calculated directly for the 1-quasiparticle state given above, then these expressions should be modified because the orbital $|\nu>$ is occupied by the unpaired particle and is thus blocked (e.g., see ref. [22]). If the blocking effect is included, then one obtains a different λ, Δ and pair distribution for each orbital $|\nu>$, and the different one-quasiparticle states are not necessarily orthogonal. Instead, in practice a single BCS vacuum state is constructed without any levels blocked, and which thus contains only pairs. The Fermi level is chosen so that N is the odd number of particles. Then the 1-qp states close to the Fermi level still have the correct average particle number, and different 1-qp states are automatically orthogonal. Thus the BCS vacuum state is an intrinsic state characterizing an effective core which is taken to be the same for all odd-A states in the model (including the range of spins considered) and cannot be strictly identified with either $A \pm 1$ neighboring even-even nucleus.

4.4. The Core Hamiltonian and the Strong Coupling Basis States

For each intrinsic 1-quasiparticle state defined above, a family of strong coupling basis states are generated. For an axially symmetric potential, Ω is a good quantum number and there is only one basis state for each total angular momentum I. Even in this case, however, and entire rotational band of states can be generated from one orbital, with $I = \Omega, \Omega+1, \Omega+2, ...$ etc. For a triaxial potential, Ω is not a good quantum number, and a number of different basis states can be defined for the same angular momentum I. These different basis states correspond to different orientations of I relative to the 3-axis, i.e. to different K values. Just as the total angular momentum is separated into core and particle parts (as in Eqn. 2) so are the components on the 3-axis separated (as an operator equation)

$$I_3 = R_3 + j_3 \tag{35}$$

Recall that for the even-even triaxial rotor, the K quantum number (denoted here K_{rot}) takes on values of 0, ±2, ±4, etc., where the core states with negative values of K_{rot} can be related to those with positive values by a rotation about the 1-axis, just like the conjugate and normal deformed orbitals. For the particle + rotor system, the total K contains contributions from the core (K_{rot}) and from the odd particle. Thus, the symmetrized strong coupling basis states become

$$|IKM\nu> = \sqrt{\frac{2I+1}{16\pi^2}}[D^I_{MK}\alpha^\dagger_\nu|\tilde{0}> + (-1)^{(I-K)} D^I_{M-K}\alpha^\dagger_{\bar{\nu}}|\tilde{0}>] \tag{36}$$

and it is possible to restrict K to the same values that Ω is restricted for the normal orbitals, i.e., K = ..., -3/2, +1/2, +5/2, ... The remaining K values (..., -5/2, -1/2, +3/2, ...) enter with the conjugate states in the second term.

Since the D function does not depend explicitly on the core angular momentum R, but on the total I, the core Hamiltonian of the triaxial rotor is written in terms of I and j,

$$H_{core} = \sum_{\kappa=1}^{3} A_\kappa R_\kappa^2 = \sum_{\kappa=1}^{3} A_\kappa (I_\kappa - j_\kappa)^2, \; A_\kappa = \frac{\hbar^2}{2\mathcal{J}_\kappa} \tag{37}$$

Taking the 3-axis as the quantization axis, it is more convenient to write this in terms of step operators $I_\pm$ and $j_\pm$ which gives

$$\begin{aligned} H_{core} &= \frac{1}{2}(A_1 + A_2)[I^2 - I_3^2 + j^2 - j_3^2 - (I_+ j_- + I_- j_+)] \\ &+ \frac{1}{4}(A_1 - A_2)[I_+^2 + I_-^2 + j_+^2 + j_-^2 - 2(I_+ j_+ + I_- j_-)] + A_3 (I_3 - j_3)^2 \end{aligned} \tag{38}$$

The moments of inertia $\mathcal{J}_\kappa$ (with $\kappa = 1, 2, 3$) are assumed to vary with γ according to the hydrodynamical relation

$$\mathcal{J}_\kappa = \frac{4}{3}\mathcal{J}_0 \sin^2(\gamma + \kappa\frac{2\pi}{3}) \tag{39}$$

where $\mathcal{J}_0$ sets the overall energy scale of the core rotor. The core Hamiltonian thus contains terms that are associated with the rotation of the system as a whole (I^2, etc.), terms that act on the intrinsic state alone (j^2, etc.), and terms that couple the intrinsic and rotational motions ($I_+ j_-$, etc.). Note that if $A_1 = A_2$ and $A_3 = 0$ (as in the axially symmetric case, with the 3-axis taken as the symmetry axis), then the core Hamiltonian reduces to the familiar form, including the Coriolis ($I_+ j_- + I_- j_+$) and recoil terms ($j^2 - j_3^2$). The rotational matrix elements are easily evaluated in the strong coupling basis given above; recall that the step operators I_+ and I_- lower and raise K by one unit, respectively, because these operators refer to the intrinsic frame. The matrix elements of the intrinsic operators (j_+, etc.) are not just the bare single particle matrix elements because they must be evaluated in the quasiparticle basis.

4.5. The Coriolis and Recoil Terms

A general one-body operator $\hat{F}$ can be expressed in second quantized form and in the quasiparticle representation as [23]

$$\begin{aligned} \hat{F} &= \sum_{i,f} < f|F|i > \{(1-c)v_i^2\delta_{i,f} + [u_i u_f + c v_i v_f](\alpha_f^\dagger \alpha_i - c\alpha_{\bar{i}}^\dagger \alpha_{\bar{f}}) \\ &+ [v_i u_f - c u_i v_f](\alpha_f^\dagger \alpha_{\bar{i}}^\dagger - c\alpha_{\bar{f}}\alpha_i)\} \\ &+ \sum_{i,f} < f|F|\tilde{i} > \{(u_i u_f + c v_i v_f)\alpha_f^\dagger \alpha_{\bar{i}} - v_i u_f(\alpha_f^\dagger \alpha_i^\dagger + c\alpha_{\bar{f}}\alpha_{\bar{i}})\} \\ &+ \sum_{i,f} < \tilde{f}|F|i > \{(u_i u_f + c v_i v_f)c\alpha_{\bar{i}}^\dagger \alpha_f - c u_i v_f(\alpha_{\bar{f}}^\dagger \alpha_{\bar{i}}^\dagger + c\alpha_f \alpha_i)\} \end{aligned} \tag{40}$$

where the sums are over the normal orbitals (conjugate states are explicitly indicated by a tilde), and $< f|F|i >$ is a "bare" single particle matrix element of $\hat{F}$. Note that the terms with two creation or two destruction operators give no contribution because the basis is restricted to 1-quasiparticle states only. With the convention adopted here for the normal and conjugate orbitals, the operators j_+ and j_- can only connect normal to conjugate orbitals (with $\Delta\Omega = \pm 1$), while operators like j_z, j_+^2, etc. (with $\Delta\Omega = 0$ and 2, respectively) connect only normal to normal or conjugate to conjugate orbitals. The phase factor c characterizes the transformation of $\hat{F}$ under particle-hole conjugation; for the components of j, $c = +1$, while for j^2, $c = -1$. Thus the Coriolis matrix elements (containing j_+, j_-) acquire a pairing factor $(u_i u_f + v_i v_f)$ onto the bare single particle matrix elements.

If this scheme is followed through for the quadratic operators like j_z^2 etc. (in the case of axial symmetry, these are called the recoil terms), then clearly these must be treated differently than the components of j because the phase factor c is different. Note that the term with the Kronecker delta contributes a sum over all occupied orbitals for diagonal matrix elements, e.g., for the quasiparticle state $|\hat{\beta} >$

$$< \hat{\beta}|j_z^2|\hat{\beta} > = \sum_f < f|j_z^2|f > 2v_f^2 + (u_\beta^2 - v_\beta^2) < \beta|j_z^2|\beta > \tag{41}$$

However, as Bennour et al. [24] point out, the contribution from core states in this sum is spurious because it should have already been included in the core angular momentum R^2, and furthermore, that its removal is not unique. If only the contribution from the orbital $|\beta >$ is retained and the rest discarded as spurious, then this leaves a $(u^2 + v^2)$ pairing factor for the diagonal matrix elements, and then it is natural to postulate a similar $(u_i u_f + v_i v_f)$ pairing factor for the off-diagonal matrix elements of these quadratic operators as well. Note that this approach removes slightly different "core contributions" from the different diagonal matrix elements. A similar result (identical $(u_i u_f + v_i v_f)$ pairing factors for all the off-diagonal terms) is obtained in the CQCM, as noted above, and was also given by Meyer-ter-Vehn [17].

Alternatively, it has been argued that these terms should be interpreted as effectively two-body in nature and that the proper way to evaluate them is by inserting a complete set of states (including 1-qp and 3-qp states) as is done in ref. [23]. A discussion of this view within a simplified model is presented in ref. [25]; the main point is that an intrinsic state in an odd-A nucleus that is described as a hole coupled to the $A+1$ core can also be considered as a 2-p 1-h excitation relative to the $A-1$ core. If the expectation value of j^2 for this state is to be computed relative to the $A-1$ core, then the contributions from the 3 "quasiparticles" (2 particles, 1-hole) must all be included, and both one-body and two-body terms appear.

However, only a single effective core appears in the usual particle-rotor model, and some ambiguity in identifying the core angular momentum R seems unavoidable. In practical calculations, though, the two-body treatment of ref. [23]

and the $(u_i u_f + v_i v_f)$ prescription seem to give very similar results.

4.6. Coriolis Attenuation

Empirically, the mixing between the intrinsic states induced by the Coriolis terms sometimes seems too large, especially when the low Ω orbitals from a high-j shell are involved. (The attenuation problem is considered in detail in the recent review of the particle-rotor model [31], although the discussion goes beyond the simple 1-quasiparticle states considered here. Likewise, some discussion within a many BCS quasiparticles + rotor model can be found in ref. [32].) For an axially symmetric rotor (with $K = \Omega$) this means that the matrix element $(u_i u_f + v_i v_f) < I, K+1|I_-|I, K >< j, \Omega+1|j_+|j, \Omega >$ is too large compared to $|E_\Omega - E_{\Omega+1}|$, the difference in quasiparticle energies. Clearly, the relative strength of the Coriolis term can be reduced by decreasing the pairing factor $(u_i u_f + v_i v_f)$ (by small adjustments in the pairing parameters), by decreasing the single particle matrix element of j_+ (by adjusting the content of the Nilsson orbital, perhaps by adjusting the deformation or other means) or by increasing the quasiparticle energy denominator (e.g., by adjusting the pairing and/or by adjusting the deformation parameters). All these adjustments may be legitimately possible within the particle-rotor model, but examples exist which clearly demonstrate that the Coriolis mixing is too large even after such reasonable adjustments have been made. One clear example is provided in the neutron $i_{13/2}$ band of ^{163}Er [26, 27]. In such cases, the attenuation has been achieved simply by introducing an attenuation factor ξ that multiplies the Coriolis matrix elements, or by raising the $(u_i u_f + v_i v_f)$ pairing factor to some power $\eta > 1$. Typically, $\xi \approx 0.6 - 0.8$ or $\eta \approx 3 - 5$ have been required. Of course, such an *ad hoc* procedure may achieve the desired numerical result but cannot address the underlying difficulty. Many physical effects have been proposed as the cause for the overestimation of the Coriolis mixing in this model, including

- The usual restriction of the pairing interaction to monopole terms is too schematic, and quadrupole terms must be included. The Y_{20} component of the pair field leads to state dependent pair gaps which would alter the quasiparticle energies and the u, v amplitudes [28, 29]. Furthermore, the rotation induced Y_{21} component leads to a Coriolis attenuation similar to the phenomenological exponent η, namely, no attenuation for the decoupling factor and greatest attenuation for states across the Fermi level [30].

- An improved treatment of pairing is needed, in which several valence nucleons are treated as active particles and the (monopole) pairing interaction is diagonalized without resorting to the BCS treatment [31]. In this approach, the recoil terms clearly include two-body operators, and this extended version of the particle-rotor model seems to avoid the necessity of any Coriolis attenuation. Calculations within a many BCS quasiparticles + rotor formulation [32] still encounter the need for attenuation.

- Pauli principle effects, as represented in the boson-fermion "exchange term" in the Interacting Boson-Fermion Model [33]. This term effectively controls the

Coriolis mixing for an SU(3) core [34].

- A finite particle number effect, proposed in the context of the Fermion Dynamical Symmetry Model [35]. Another suggestion from the FDSM is that the Coriolis attenuation may originate from the spin-spin interaction of the (effective) nucleon-nucleon force [36].
- Any neglected octupole couplings would dilute the $< j_+ >$ matrix elements from the high-j shells [12].
- The assumed constant pairing gap and moment of inertia for all core states is too rough an approximation; calculations based on energy minimization [37] suggest that small variations in the shape, pair field, etc. are responsible.

In a specific case, one of these mechanisms might be expected to to be particularly important, but as yet, no clear consensus has emerged on the source of the attenuation problem or its solution.

4.7. Electromagnetic Matrix Elements

In practice, the particle-rotor model has been an important tool for analyzing electromagnetic properties of odd-A nuclei. In particular, the E2 and M1 matrix elements are often sensitive probes of the nuclear shape and the single particle structure, respectively. Some basic results are given here.

As usual, the relevant operators are written as a sum of contributions from the core and the odd particle. Also, it is convenient to transform these operators to the intrinsic frame where the components have simple forms. Thus, for the core part of the E2 operator,

$$T(E2)_\mu = \sqrt{\frac{5}{16\pi}} e[Q_0 D^2_{\mu,0} + Q_2(D^2_{\mu,2} + D^2_{\mu,-2})] \tag{42}$$

where μ indicates the spherical component in the lab frame. The intrinsic moments are usually obtained by assuming a homogeneous charge distribution inside the shape defined by the potential parameterization, i.e.,

$$eQ_0 = \int \rho_{chg}(3z^2 - r^2)d^3r = \sqrt{\frac{16\pi}{5}} \int \rho_{chg} Y_{2,0} d^3r \tag{43}$$

and

$$eQ_2 = \sqrt{\frac{3}{2}} \int \rho_{chg}(x^2 - y^2)d^3r = \sqrt{\frac{4\pi}{5}} \int \rho_{chg}(Y_{2,2} + Y_{2,-2})d^3r \tag{44}$$

The collective contribution to the E2 operator usually dominates the single particle part, so the latter is not written here.

Similarly, the M1 operator is written

$$T(M1)_\mu = \sqrt{\frac{3}{4\pi}} \sum_\nu D^1_{\mu,\nu}(g_R R_\nu + g_\ell \ell_\nu + g_s s_\nu) \tag{45}$$

or

$$T(M1)_\mu = \sqrt{\frac{3}{4\pi}} \sum_\nu D^1_{\mu,\nu}[g_R I_\nu + (g_\ell - g_R) j_\nu + (g_s - g_\ell) s_\nu] \tag{46}$$

where the second form is more convenient since it uses I and j instead of R and ℓ. The orbital and spin g-factors for the odd particle are g_ℓ and g_s. The core g-factor, g_R, is assumed to be constant and is typically estimated as Z/A. The g_s factor of the odd particle is renormalized due to the spin polarization effect, and is usually taken at 60% - 70% of the free value [9].

The matrix elements are easily obtained with the basis states given in Eqn. (36), and those results are given in ref. [6] (although the pairing factors are omitted from the single particle matrix elements). The usual strong coupling results for an axially symmetric core are special cases in which the initial and final states correspond to the same intrinsic orbital [9]:

$$B(E2; I_i, K \rightarrow I_f, K) = \frac{5}{16\pi}(eQ_0)^2 < I_i K 20 | I_f K >^2 \tag{47}$$

and

$$B(M1; I_i, K \rightarrow I_f, K) = \frac{3}{4\pi}(g_K - g_R)^2 K^2 < I_i K 10 | I_f K >^2, K \neq \frac{1}{2} \tag{48}$$

In this last equation, g_K is the intrinsic moment of the orbital $|\nu>$,

$$g_K = \frac{1}{\Omega}(g_\ell + (g_s - g_\ell) < \nu | s_z | \nu >) \tag{49}$$

and $K = \Omega = <\nu | j_z | \nu>$ for axial symmetry. Note that the BM1 expression depends on the Nilsson orbital through the matrix element $<\nu|s_z|\nu>$ as well as the K quantum number. Thus, for strong coupled bands, the relative strength of M1 and E2 transitions is often a sensitive indicator of the single particle structure. Similarly, magnetic moments are directly sensitive to the orbital occupied by the odd particle, while an E2 moment is proportional to Q_0, and thus reflects the deformation. Numerous examples are given in ref. [9].

If $K = 1/2$ states are considered, then an additional term appears in the strong coupling BM1 expression because the j_+ and s_+ operators can connect the conjugate orbital $|\tilde{\nu}>$ (with $\Omega = -1/2$) to the normal orbital $|\nu>$. This additional contribution changes sign as I changes in the rotational band, and produces a staggering of the BM1 strength. Similarly, a staggering of the level energies also appears in this strong coupling limit due to a diagonal contribution from the Coriolis term and contains the $<\nu|j_+|\tilde{\nu}>$ matrix element (the decoupling parameter). Thus, for strong coupled $K = 1/2$ bands, there is a direct link between staggering (signature splitting) in the level energies and in the BM1 strengths.

Furthermore, for axially symmetric shapes, $K = 1/2$ components can be mixed into a rotational band by the Coriolis interaction, and thus produce signature splitting in both the energies and the BM1 strengths. Bands built on a high-j (unique parity) orbital are particularly sensitive to the rotational perturbation because the Coriolis matrix elements can be large. Also in this case, the signature splitting in the BM1 strengths can be directly related to the signature splitting in the energies

[18], which in turn can be related to a geometrical alignment of the spin of the odd particle with the total angular momentum.

In triaxial nuclei, $K = 1/2$ components can be present in a rotational band due to the mixing of Ω from the triaxial potential as well as the rotational perturbation. Calculations for unique parity bands have shown that the staggering in the BM1 strengths can be very different than that observed for the axially symmetric shapes [20], and that the $B(E2; I \to I-1)$ strengths develop a characteristic staggering pattern. A recent discussion of the observable consequences of a triaxial shape in yrast unique parity states and a comparison with available experimental data is given in ref. [38].

5. Comparisons Between Different Approaches

In this contribution, two particle-core coupling models have been considered in some detail, the CQCM and the particle + triaxial rotor model. Each is intended to serve as an example for a broader class of models, one that works in the weak coupling basis (i.e., the lab frame) and the other in the strong coupling basis (in the intrinsic frame). In principle, of course, the physical observables cannot depend on the choice of the basis used for calculations as long as a complete basis is used, but in practice some approximations must always be made. As a result, each basis may be better suited for certain approximation schemes and thus entails characteristic advantages and disadvantages. In this section these two approaches are compared, and their relative strengths and limitations discussed. Also, a brief comparison between the particle-rotor model and the one-dimensional cranking formalism is presented.

First, both the particle-rotor model and the CQCM (and their relatives) provide a framework for calculating collective states in odd-mass nuclei in which the angular momentum is conserved. Experience has shown that these models are very useful for calculating level energies, electromagnetic matrix elements and other properties as well (e.g., spectroscopic factors for single nucleon transfer) and then relating these observable properties to a few model parameters. These parameters usually describe the collective core (e.g., the deformation, softness/stiffness, etc.) and the single particle orbitals coupled to the core.

The main advantage of the strong coupling basis employed in the particle-rotor model is that it is possible to diagonalize the deformed shell model + pairing Hamiltonian first, and only then truncate the single particle space to a few adiabatic orbitals. This allows a very complete and efficient treatment of the single particle space, and well studied potentials can be directly implemented (e.g., Modified Oscillator, Woods-Saxon, etc.). Through these potentials, contact can be maintained with other mean field methods, including various cranking calculations. The rather complete treatment of the single particle space allows pairing calculations in which the Fermi level and gap parameter are not fitted quantities but are derived, and when standard potentials are used together with predicted deformations for the even-even neighbors (e.g. from Nilsson-Strutinsky total energy calculations), some predictive capabilities result. Particle-core coupling models that begin in a spherical basis (like the CQCM) may allow more than a single j-shell, but become rather cumbersome when many j-shells are necessary to descibe a particular Nilsson state.

However, the limitation of the strong coupling basis is that it is awkward to include core degrees of freedom other than rigid rotation at a fixed shape, which limits the applicability of the particle-rotor model to well deformed and perhaps transitional nuclei. The effects of core softness can be partly taken into account by using a variable moment of inertia (VMI) to describe the core energy spectrum [39], but any accompanying variation in deformation or pairing is usually neglected. Some justification for this method at low to moderate spins is presented in ref. [6]. Also, it should be noted that the coupling of a particle to the lowest γ vibration [40, 41] is very similar to the coupling to a rigid triaxial shape with small γ, and thus for such transitional nuclei a rigid triaxial description will be very similar to the γ vibrational model. As yet, no clear feature has been found in odd-A nuclei that can distinguish between γ-soft cores and rigid triaxiality [5, 6, 42]. A clear strength of the CQCM and the weak coupling basis is the great flexibility in describing the collective core properties, which can include vibrational as well as rotational states. The microscopic foundation of the CQCM through the equation of motion method is useful especially as a source for possible extensions and improvements.

Finally, it is interesting to make some comparisons with 1-dimensional cranking [43, 44], which is not a particle + core model. The cranking approach is designed to be appropriate in the high-spin limit, and treats the nuclear rotation semiclassically as an external constraint added to the single particle motion in the deformed potential. In this way, all the nucleons are treated equivalently , as particles moving in a rotating ("cranked") mean field and with no separation between a core and a set of valence particles. This allows a treatment of multi-quasiparticle states within the same framework as for 0 or 1-qp states. Considerable predictive power has been achieved for a few quantities, e.g., nuclear shapes, softness/stiffness of the shapes, rotation alignment effects, etc., and how these evolve with Z, N, and rotational frequency. However, the very approximations that make these calculations feasible preclude accurate calculation of other observables. For example, cranking wave functions contain a large spread of angular momenta, and electromagnetic properties (especially transitions) usually can only be estimated at best. Also, since the angular momentum vector is constrained to point in a particular direction (along a principal axis), low spin and non-yrast states are poorly described if at all. In contrast, the particle-rotor model does conserve angular momentum and does not constrain its direction but allows the dynamics of the system (through the model Hamiltonian) to determine the relative orientations of $\vec{R}$, $\vec{j}$ and $\vec{I}$. This freedom is crucial for a quantitative understanding of some properties such as the relationship between signature splitting (staggering of energy levels within a rotational band) and the shape asymmetry, where a 1-dimensional treatment of the rotation is too restrictive especially for small triaxiality [45]. Furthermore, the particle-rotor model can provide accurate calculations of M1/E2 matrix elements, including static moments, transitions rates and mixing ratios. The M1 and E2 properties give direct information on the single particle configuration and the nuclear shape, and sometimes allow unambiguous assignments of Nilsson orbitals even when spins are not firmly established experimentally. However, it must be noted that the present versions of the particle-rotor model are restricted to a single fixed core shape, and that shape is not predicted by the model but must be input. Coexisting nuclear shapes cannot be addressed directly. Also, the extension to multiquasiparticle states is much more complicated and only a few such calculations have been made.

In summary, a microscopic derivation for particle-core coupling based on the equation of motion method has been reviewed. Many phenomenological models can be obtained if selfconsistency is abandoned and the Pauli principle is partially relaxed. The severity of these approximations remains somewhat unclear. The Core-Quasiparticle Coupling Model and the particle-rotor model were described in detail as examples of models that are formulated in the lab and intrinsic systems, respectively. These were compared to each other and to the cranking model, and the specific strengths and limitations of each were discussed. These particle-core coupling models provide a valuable tool for interpreting experimental observables, and form a crucial link to other theoretical approaches.

6. Acknowledgements

This work is supported in part by the U.S. Department of Energy under the grant DE-FG05-92ER40694.

7. References

1. A. Bohr, *Mat. Fys. Medd. Dan. Vid. Selsk.* **26** (1952) No. 14.
2. A. Bohr and B. Mottelson, *Mat. Fys. Medd. Dan. Vid. Selsk.* **27** (1953) No. 16.
3. A. K. Kerman and A. Klein, *Phys. Rev.* **132** (1963) 1326.
4. A. Klein, *Prog. Part. and Nucl. Phys.* **10** (1983) 39.
5. F. Dönau and S. Frauendorf, *Phys. Lett.* **71B** (1977) 263.
6. S. E. Larsson et al., *Nucl. Phys.* **A307** (1978) 189.
7. F. Dönau, *Proc. XV Summer School on Nuclear Physics,* Mikolajki, Poland, 1983.
8. G. Do Dang et al., *Nucl. Phys.* **A114** (1968) 481; 501.
9. A. Bohr and B. Mottelson, *Nuclear Structure, Vol. 2* (W. A. Benjamin, 1975)
10. F. Dönau, *Proc. Winter School on Nuclear Structure,* Zakopane, Poland, 1984.
11. F. Dönau and S. Frauendorf, *J. Phys. Soc. Japan* **44** (1977) Suppl. 526.
12. G. A. Leander and Y. S. Chen, *Phys. Rev.* **C37** (1988) 2744.
13. D. Brink et al., *J. Phys.* **G13** (1987) 629.
14. Y. S. Chen et al., *Phys. Rev.* **C34** (1986) 1935.

15. I. Ragnarsson and P. B. Semmes, *Hyperfine Interactions* **43** (1988) 425.
16. K. T. Hecht and G. R. Satchler, *Nucl. Phys.* **32** (1962) 286.
17. J. Meyer-ter-Vehn, *Nucl. Phys.* **A249** (1975) 111,141.
18. I. Hamamoto, *Phys. Lett.* **106B** (1981) 281.
19. I. Hamamoto and B. R. Mottelson, *Phys. Lett.* **132B** (1983) 7.
20. G. B. Hagemann and I. Hamamoto, *Phys. Rev.* **C40** (1989) 2862.
21. A. Bohr and B. Mottelson, *Nuclear Structure, Vol. 1* (W. A. Benjamin, 1969)
22. P. Ring and P. Schuck, *The Nuclear Many Body Problem* (Springer Verlag, 1980)
23. E. Osnes et al., *Nucl. Phys.* **A253** (1975) 45.
24. L. Bennour et al., *Nucl. Phys.* **A465** (1987) 35.
25. J. Rekstad, T. Engeland and E. Osnes, *Nucl. Phys.* **A330** (1979) 367.
26. S. A. Hjorth et al., *Nucl. Phys.* **A144** (1970) 513.
27. G. Lövhöiden et al., *Nucl. Phys.* **A194** (1972) 463.
28. D. R. Bes, et al., *Phys. Lett.* **40B** (1972) 338.
29. R. Casten, et al., *Phys. Lett.* **40B** (1972) 333.
30. I. Hamamoto, *Nucl. Phys.* **A232** (1974) 445.
31. T. Engeland, in *International Review of Nuclear Physics, Vol. 2,* ed. T. Engeland et al. (World Scientific, 1984)
32. J. Almberger et al., *Physica Scripta* **22** (1980) 331.
33. F. Iachello and O. Scholten, *Phys. Rev. Lett.* **43** (1979) 679.
34. P. B. Semmes et al., *Phys. Rev.* **C33** (1986) 1476.
35. H. Wu et al., *Phys. Rev.* **C37** (1988) 1739.
36. H. Wu et al., *Phys. Lett.* **198B** (1987) 119.
37. R. Bengtsson et al., to be published.
38. I. Hamamoto, *Nucl. Phys.* **A520** (1990) 297c.
39. H. Toki and A. Faessler, *Nucl. Phys.* **A253** (1975) 231.
40. A. Ikeda and T. Shimano, *Phys. Rev.* **C42** (1990) 149.
41. A. Ikeda, *Nucl. Phys.* **A439** (1985) 317.
42. G. Leander, *Nucl. Phys.* **A273** (1976) 286.
43. R. Bengtsson and S. Frauendorf, *Nucl. Phys.* **A314**(1979) 27; **A327** (1979) 139.
44. M. J. A. deVoigt et al., *Rev. Mod. Phys.* **55**(1983) 949.
45. I. Hamamoto, *Phys. Lett.* **211B** (1988) 1.

CRITIQUE OF THE PARTICLE–CORE MODEL

R. BENGTSSON
UNISOR
Oak Ridge Institute for Science and Education
P.O.B. 2008, Oak Ridge, TN 37831, USA

and

P. O. LIPAS
Department of Physics, University of Jyväskylä
P.O.B. 35, SF-40351, Jyväskylä, Finland

Report of working group:

Pertti Lipas, chairman, University of Jyväskylä
Ragnar Bengtsson, UNISOR
Jutta Escher, Louisiana State University
Niels Walet, University of Pennsylvania

ABSTRACT

The particle–core model, particularly in its particle–rotor version, is evaluated. The main points discussed are the scope of the model, the determination of its parameters and its predictive power. Consistency of deformation and the moment of inertia, and the Coriolis attenuation factor are recognized as crucial problems. Comparison is made with the cranking model.

1. Introduction

The particle–core model dates back to the definitive works of Bohr, Mottelson and Nilsson[1]. It is then understood mainly as a model for odd-A nuclei such that a single nucleon is coupled to the doubly even core. The particle and core degrees of freedom are assumed separable although connected by an interaction*. By way of introduction we first sketch the basic tenets of the particle–core model. Then, in the subsequent sections, we discuss in more detail properties and problems of the model.

The particle–core Hamiltonian has the structure

$$H = H_{\rm core} + H_{\rm ple} + H_{\rm int} \equiv H_0 + H_{\rm int} \,. \tag{1}$$

The core can be taken as spherical or deformed. The corresponding model versions are referred to as *weak* and *strong* coupling. In both cases the total angular momentum is the sum of the contributions of the core and the particle: $\mathbf{I} = \mathbf{R} + \mathbf{j}$ with $\mathbf{R}$ and $\mathbf{j}$ commuting.

---

* In a wider context the shell model could also be viewed as a particle–core model. Likewise the interacting-boson–fermion model/approximation (IBFM/IBFA) singles out the "last" particle.

1.1. Weak Coupling

The eigenstates ψ_0 of H_0 have the angular-momentum structure $\psi_0 = |RjIM\rangle$. For very weak coupling the energies are simply

$$E = E_{\text{core}} + E_{\text{ple}} + \langle RjIM|H_{\text{int}}|RjIM\rangle \,. \tag{2}$$

Otherwise the states $|RjIM\rangle$ serve as a basis for diagonalizing the complete Hamiltonian (1).

The core can, but need not necessarily, be described in terms of a spherical quadrupole vibrator, of classical shape

$$R(\theta,\phi;t) = R_0\Big[1+\sum_m \alpha_m(t)\, Y^*_{2m}(\theta,\phi)\Big] \equiv R_0[1+\alpha(t)\cdot Y_2(\theta,\phi)] \,. \tag{3}$$

Upon quantization the vibrational amplitudes α_m are replaced by phonon creation and annihilation operators: $\alpha_m \propto b^\dagger_m + (-)^m b_{-m}$. The particle–core interaction is taken to be

$$H_{\text{int}} = -\kappa(r)\alpha\cdot Y_2(\theta,\phi) \,, \tag{4}$$

where at the very simplest κ is a constant. The effect of this interaction, seen at its simplest in Eq. (2), is to split the one-phonon 2^+ state of a neighbouring even–even nucleus into a multiplet with $I = 2+j, \ldots |2-j|$. The nucleus ^{209}Bi offers a good example of this (see I. Hamamoto's contribution to this workshop).

1.2. Strong Coupling

In the strong-coupling version of the particle–core model the core is described as a deformed rotor. The model is then commonly referred to as the *particle–rotor model.* Although it is not required, we shall here assume axial symmetry, whereupon there can be no rotation of the core about its symmetry axis, $R_3 = 0$. We then have

$$\begin{aligned} H_{\text{core}} &= \frac{\hbar^2}{2\mathcal{J}}\mathbf{R}^2 = \frac{\hbar^2}{2\mathcal{J}}(\mathbf{I}-\mathbf{j})^2 = \frac{\hbar^2}{2\mathcal{J}}(\mathbf{I}^2 - 2\mathbf{I}\cdot\mathbf{j} + \mathbf{j}^2) \\ &= \frac{\hbar^2}{2\mathcal{J}}\left[(\mathbf{I}^2 - I_3^2) - (I_+ j_- + I_- j_+) + (\mathbf{j}^2 - j_3^2)\right] \\ &\equiv H_{\text{rotor}} + H_{\text{Coriolis}} + H_{\text{recoil}} \,. \end{aligned} \tag{5}$$

This conventional nomenclature assigns the *total* angular momentum* to "rotation". The "recoil" term is appended to H_{ple} as it contains only particle operators. The "Coriolis" term is the particle–core interaction H_{int} in the simple strong-coupling model.

The single-particle Hamiltonian H_{ple} of the particle–rotor model is the deformed Nilsson Hamiltonian or one of its later modifications. The complete Hamiltonian (1) in the strong-coupling case now takes on the form

$$H = (H_{\text{rotor}} + H'_{\text{ple}}) + H_{\text{Coriolis}} \equiv H_0 + H_{\text{int}} \,, \tag{6}$$

* It is well to note that the body-fixed components of the total angular momentum obey the "backward" commutation relations $[I_1, I_2] = -iI_3$ etc., while $[j_1, j_2] = +ij_3$ etc.

where H'_{ple} incorporates the recoil term. The noninteracting part is called H_0 and the interaction H_{int} although they are not the same as in Eq. (1). The unperturbed eigenstates are of the form

$$\psi_0 = D^{I*}_{MK}(\alpha\beta\gamma)|\Omega\rangle + \text{symm.}, \tag{7}$$

where the j_3 quantum number Ω must be equal to the I_3 quantum number K and where the second term is due to $\pm K$ symmetrization. It is important to note that the *total* angular momentum $\mathbf{I}$ is the conserved quantity whose quantum numbers I, M, K label the rotational wave function D^{I*}_{MK}. The core angular momentum $\mathbf{R}$ and the particle angular momentum $\mathbf{j}$ are *not* conserved except that their 3-components are good: $R_3 = 0$, $j_3 = \Omega$.

The unperturbed energy is

$$E_0 = \frac{\hbar^2}{2\mathcal{J}}\left[I(I+1) - K^2\right] + E_\Omega\,. \tag{8}$$

The interaction H_{int} is treated in perturbation theory or, more accurately, the complete Hamiltonian is diagonalized in the basis of the states (7).

The resultant description of odd-A nuclei constituted the early success of the Bohr–Mottelson–Nilsson unified model; for examples see Ref. 1.

1.3. Pairing

An essential improvement of the strong-coupling description is obtained through the inclusion of the pairing interaction among the nucleons involved, i.e. those in the core and the odd one. That way the core is recovered as a many-fermion system rather than a structureless collective element. We then have the replacement

$$H_{\text{int}} \to H_{\text{int}} + H_{\text{pairing}}\,. \tag{9}$$

This entails the transformation from particles (creation operator $a^\dagger_\nu$) to quasiparticles:

$$a^\dagger_\nu \to \alpha^\dagger_\nu = u_\nu a^\dagger_\nu - v_\nu a_{\bar\nu}\,. \tag{10}$$

Pairing brings about the gap parameter Δ and the Fermi level λ. The particle energies go over into quasiparticle energies,

$$E_{\text{ple}} \to E_{\text{qp}} = \sqrt{(E_{\text{ple}} - \lambda)^2 + \Delta^2}\,. \tag{11}$$

Consistent with experiment, the quasiparticle levels are more densely spaced than the corresponding particle levels.

2. Separation into Core and Particle(s)

Up to a certain point one can formally make a distinction between a core and the "outside" particle(s) and then expand the states of a nucleus as sketched in the introduction. Let us now address the definition of the core in more precise terms.

2.1. Core of Filled Shells

The most conservative definition of a core is that it is constituted by the filled major shells. This view underlies the (spherical) shell model. This definition, however, leaves the number of valence nucleons normally too large for practical handling. We do not pursue this line any further in the present context.

2.2. Few Particles Outside

The working definition in the particle–core context is that the core is formed by the bulk of the nucleus with only one or a few particles excluded. Ambiguities occur in selecting those particles to be treated explicitly, particularly in deformed nuclei, where the spherical subshells are heavily mixed.

One possibility is to treat the particles in an intruder subshell explicitly and the rest of the nucleus as a core, as has been done e.g. by Faessler *et al.*[2]. With the pairing interaction, the number of particles in the intruder subshell is not constant: pairs of particles are scattered between the intruder subshell and the core. Properly treated, this requires that separate cores with different particle numbers be introduced. Thus the wave functions for a nucleus with A particles will have the general form

$$|A\rangle = \sum_n a_n |A-n\rangle \otimes |n\rangle\,, \tag{12}$$

where n is the number of particles in the intruder subshell.

A particle–core model of this kind has great flexibility. In particular, it allows particles in the intruder subshell to align their angular momenta so that band crossings can be described in a satisfactory way. Although this approach is very appealing, it has been very little used in practical calculations, since already a modest average number of particles in the intruder subshell (say 4) requires a large configuration space and makes calculations time consuming.

2.3. Maximal Core

In the simplest view, as sketched in the introduction, the core for an odd-A (odd–odd) nucleus is defined as the whole nucleus minus one (two) particles. At this level computation times are short, and the model is a very useful tool for practical applications. In contrast to the cranking approach[3] to deformed nuclei, angular-momentum conservation is automatically included.

This simplest version of the particle–core model, however, is beset by severe limitations. Band crossings or alignment effects cannot be described because they would require more than one (two) active particles. Moreover, the core properties cannot be easily calculated. In practice they are simply prescribed, taken as input parameters or fitted to the data at hand. The problem of core description is compounded in the approach of Subsect. 2.2 where the properties of several different cores, for different core-particle numbers, have to be specified.

3. Description of the Core

We now limit our discussion mainly to the simple particle–core model in the strong coupling option, i.e. the particle–rotor model, with axial symmetry assumed. This model has been applied not only to hard, well-deformed nuclei but also to soft, transitional nuclei. The latter area of application, however, does present problems which we discuss below.

We see the description of the core as the most central problem with the particle–rotor model. Either there is a theoretical concept for determining the core properties (primarily its moment of inertia) or there is not. In the first case, the model can make predictions of the properties of unknown nuclei. In the second case, the parameters describing the core become fitting parameters and the model becomes merely descriptive. Indeed, the model is often used in this latter way, with both the core moment of inertia and the deformation of the single-particle potential as fitting parameters. Even further fitting parameters may be used, such as the Coriolis attenuation factor discussed below.

The particle–rotor model involves a number of parameters: deformation, pairing gap, Fermi energy, moment of inertia, attenuation parameter and possibly a few more, but they are not independent. Thus the moment of inertia depends on the shape of the nucleus and on the pairing gap, and so does the Fermi energy. And the pairing gap, when determined self-consistently, depends on the deformation of the nucleus.

The moment of inertia of the core is in principle determined once the deformation and pairing gap are chosen. Irrespective of exactly *how* the moment of inertia depends on deformation and pairing gap, the choice of feasible values will be highly restricted. We find it essential that known (and assumed) relationships between the model parameters be built into the particle–rotor model. In particular, the properties of the core are to be derived in a consistent way from the parameters used for the single-particle potential and the pairing field.

A potent possibility is to use the cranking model[3] to calculate the remaining, independent parameters (deformation, pairing gap, ...) of the particle–rotor model. The particle–rotor model can then give results which the cranking model by itself cannot produce. This complementary use of the two models was indeed suggested by the spokesman for the particle–core model.

4. Core-Polarization and Deformation Changes

A critical assumption underlying the particle–rotor model is that the deformation and the pairing field are the same for all bands and for all individual states of a given nucleus. On the other hand, both experimental data and cranking calculations point to sizeable variations in shape and pairing field in many nuclei, caused by the phenomena of core polarization, centrifugal stretching, Coriolis anti-pairing and pairing blocking. This causes ambiguities in the calculation of e.g. transition probabilities.

We can make simple estimates of the effects of centrifugal stretching and Coriolis anti-pairing as follows. Consider only the core and neglect the influence of the odd particle. Assuming that the energy can be expanded in terms of a deformation parameter ε and the pairing gap Δ about the ground-state minimum, we have

$$E = E_0 + a(\varepsilon - \varepsilon_0)^2 + b(\Delta - \Delta_0)^2 + \dots \tag{13}$$

Similarly the moment of inertia can be expanded about its ground-state value as

$$\mathcal{J} = \mathcal{J}_0 + c(\varepsilon - \varepsilon_0) + d(\Delta - \Delta_0) + \dots \tag{14}$$

Finally we can write the Routhian, i.e. the energy in the rotating frame, as

$$E_{\mathrm{R}} = E - \tfrac{1}{2}\mathcal{J}\omega^2 . \tag{15}$$

Substituting Eqs. (13) and (14) in the Routhian and minimizing it for a given frequency, we find the equilibrium values

$$\varepsilon = \varepsilon_0 + \frac{c}{4a}\omega^2 , \qquad \Delta = \Delta_0 + \frac{d}{4b}\omega^2 . \tag{16}$$

These equations, respectively, describe centrifugal stretching and Coriolis anti-pairing. The coefficients a, b and c are positive while d is negative.

Substituting the equilibrium values (16) into Eq. (14), we have for the moment of inertia

$$\mathcal{J} = \mathcal{J}_0 + \left(\frac{c^2}{4a} + \frac{d^2}{4b}\right)\omega^2 \equiv \mathcal{J}_0 + \mathcal{J}_1\omega^2 , \tag{17}$$

in agreement with the empirical Harris expression.

The effects of deformation changes and Coriolis anti-pairing, contained in the $\mathcal{J}_1$ term of Eq. (17), are experimentally well documented. While small in well deformed nuclei, they can hardly be neglected in soft transitional nuclei. The same general statement is true about core-polarization effects caused by the odd particle. Consequently the assumption of a constant deformation made in practical particle–rotor calculations becomes risky, and it should be carefully considered from case to case. The consequences of the frequency dependence of ε and Δ are investigated in the literature[4]. Figure 1 shows an example[5] of the importance of optimizing the model parameters for each state.

The use of a variable moment of inertia, instead of a constant one, was suggested in the presentation of the particle–rotor model. A basically equivalent alternative is to use experimental ground-band energies of neighbouring even–even nuclei for describing the core. Such an approach can be justified only insofar as it is accompanied by consistent changes in deformation. We see no basic obstacle to using a variable deformation; only transitions between states of different deformation would present a problem.

On occasion, the basic particle–rotor model has been used indiscriminately to fit experimental data. By adjusting the model parameters freely, in particular

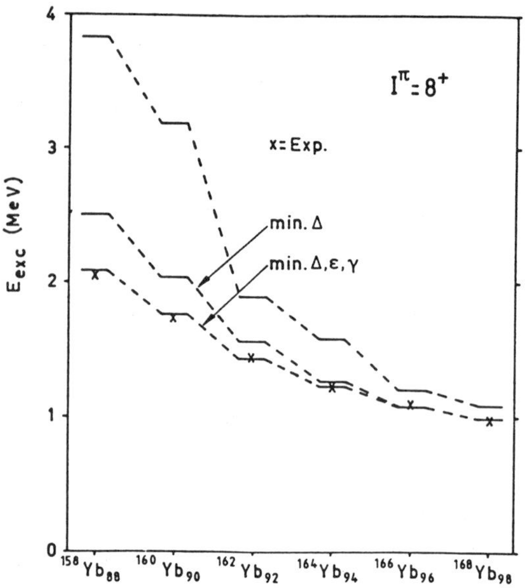

Fig. 1. Excitation energy of the 8_1^+ state in Yb isotopes. The highest levels are calculated with constant model parameters (ground-state values). The next ones represent minimization of the 8_1^+ energy with respect to the pairing-gap parameter Δ, and the lowest ones minimization with respect to Δ *and* the deformation parameters ϵ and γ. The experimental values are marked with crosses.

by determining the moment of inertia independently of the deformation and the pairing gap, and allowing the use of the physically unjustified (see Sect. 5) Coriolis attenuation factor, it is often possible, even in very soft nuclei, to get a reasonably good *reproduction* of experimental data. However, *predictions* for soft nuclei are out of the question.

5. Coriolis and Recoil Terms

The Coriolis and recoil terms of the Hamiltonian (5) present irksome problems in the practical application of the particle–rotor model. We now address each of them in turn.

5.1. The Coriolis Attenuation Problem

The Coriolis attenuation problem is most obvious in bands where the odd particle is in a low-K orbital that can be traced back to a large j. Experience shows that there is no way a good fit of data can be achieved in such cases, if a set of constant (i.e. spin-independent) parameters are used. To make the energies of a low-K, high-j band agree with experiment, one needs a correction which is proportional to the total angular momentum I.

Changing any one of the basic model parameters will not have the desired effect on the spectrum. This can be easily seen for the moment of inertia, which appears as $1/\mathcal{J}$ in front of the $\mathbf{R}^2$ term in Eq. (5). But R is not proportional to I. The average value of R is lowest when $I = j$ and increases for both lower and higher values of I. Therefore a change of the moment of inertia cannot remedy the energy spectrum. The other model parameters (deformation, pairing gap etc.) work no better. However, in bands with a high j and a low K, the particle angular momentum $\mathbf{j}$ and the total angular momentum $\mathbf{I}$ are more or less parallel for all values of I. As a result, the $\mathbf{I} \cdot \mathbf{j}$ term increases in a nearly linear fashion when I increases. Therefore this term can be conveniently used for adjusting the particle–rotor energy spectrum to the experimental data.

In practice the adjustment is done by introducing a factor, the so-called Coriolis attenuation factor, in front of the Coriolis term. Typical values of this factor vary between 0.6 and 0.8. To our knowledge, however, there is no physical justification for manipulating the strength of the Coriolis term. This practice is all the more dubious since the particle–rotor model ignores changes in deformation and pairing gap, as discussed in Sect. 4 above.

5.2. The Recoil Term

In the presentation of the particle–rotor model no definite conclusion was reached about the $\mathbf{j}^2$ recoil term of Eq. (5). In particular, the question was whether it should be treated as a one-body or a two-body operator. The inclusion of pairing complicates the appearance of the issue, but is not essential.

Without pairing, i.e. with plain particles rather than quasiparticles, it seems clear to us that the operator $\mathbf{j}^2$ is a one-body operator. Allowing for any number of particles i and single-particle states ν, we can write the operator equivalently in coordinate representation and occupation-number representation:

$$\sum_i \mathbf{j}_i^2 = \sum_{\nu\nu'} \langle \nu | \mathbf{j}^2 | \nu' \rangle a_\nu^\dagger a_{\nu'} \,. \tag{18}$$

The operator $\sum_i \mathbf{j}_i^2$ describes a single-particle property; the fact that $\sum_i \mathbf{j}_i \equiv \mathbf{J}$ alone is a one-body operator does not make $\sum_i \mathbf{j}_i^2$ a two-body operator. Similarly the total momentum $\mathbf{P} = \sum_i \mathbf{p}_i$ and the kinetic energy $T = \sum_i \mathbf{p}_i^2/2m$ are both one-body operators.

Another question touched upon in the presentation is whether the operator $\mathbf{j}^2$ can be properly associated with the odd particle alone, without duplicating contributions assigned to the core. No clear conclusion was reached. Altogether the problem with the recoil term appears relatively minor compared with the problem of Coriolis attenuation.

6. Negative-Energy Solutions

For this last item of critique of the particle–core model we depart from the simple view of a single core. We now examine the model with specific reference to pairing.

The quasiparticle formalism leads to a set of (basis) states that is overcomplete by a factor of two. The reason for this is that a given (basis) state can be viewed either as a particle coupled to an $A-1$ core or as a hole coupled to an $A+1$ core. A physical state will in general correspond to a mixture of the two types. In the limit of zero rotational energy, i.e. $\mathcal{J} \to \infty$, each positive-energy state possesses a conjugate state with the opposite particle–hole composition and a negative energy of equal magnitude[6].

The practical way of dealing with the overcompleteness problem is to disregard all intrinsic states with negative energy (usually referred to as "unphysical") and keep only those with positive energy (the "physical" states). We question this practice and propose that all states be coupled to the rotor and the separation between physical and unphysical solutions be made only at the end of the calculation. Such a procedure is in fact followed in cranking Hartree–Fock–Bogoliubov calculations. Due to a coupling existing between the physical and unphysical states, the particle–core results would be affected by the inclusion of the negative-energy states.

As a concession to the prevalent practice, we can conjecture that the mixing of the two types of state is small. This appears reasonable because in the limit of no rotation they are separated by at least twice the pairing gap. As long as the angular momentum remains moderate, one may expect a clear separation to persist.

7. Conclusions

One of the goals of the workshop was to find answers to questions like "What can the model describe?", "How are the model parameters determined?" and "What is the predictive power of the model?". For the particle–core model, mainly in the particle–rotor version, our evaluation has resulted in the following answers.

7.1. Scope of the Model

The use of the particle–core model is restricted mainly to odd-A nuclei with only the odd quasiparticle treated individually. In odd–odd nuclei two quasiparticles are similarly singled out. The weak-coupling version is particularly suitable for nuclei without a strong permanent deformation, whereas the strong-coupling version, i.e. the particle–rotor model, indeed requires a nucleus with rotational behaviour.

Compared with its nearest rival, the cranking model, the particle–rotor model has the advantage of a well-defined total angular momentum. Another strong point of the particle–rotor model is its capability of yielding reliable electromagnetic matrix elements, whereas the cranking model only provides rough estimates.

A severe limitation of the particle–rotor model is its inability to describe band crossings involving the alignment of quasiparticle pairs. Likewise the model does not extend to configurations of several quasiparticles. Generalizations are known for handling such many-particle phenomena, but they have not been developed to a well-established, practically useful level. The cranking model, on the other hand, embodies band crossings etc. in a natural way.

7.2. Determination of Parameters

The crucial model parameters are deformation(s), moment(s) of inertia and the Coriolis attenuation factor. Other parameters are those of the mean field and the pairing gaps. The mean-field parameters are routinely adopted from standard single-particle potentials. The pairing gaps are estimated from odd–even mass differences or they are calculated with BCS theory, once the deformation parameters are selected.

The deformation and moment-of-inertia parameters are determined by fitting a particular set of data, or their values are taken from independent model calculations. In the first case one is liable to make incompatible choices of the parameters since the model has no built-in self-consistency.

The Coriolis attenuation factor is merely a phenomenological fitting parameter as long as there is no theory justifying its use, let alone providing its value through a calculation.

7.3. Predictive Power

Strictly speaking, the particle–rotor model taken by itself has no predictive power since it does not provide for the model parameters, the most important of which are the deformation and the moment of inertia. The parameter values must be introduced from outside the model. Thus *another model* can infuse the particle–rotor model with actual predictive power.

Nearest at hand is the cranking model. In fact, a symbiosis of the two models appears fruitful. The cranking model can provide the particle–rotor model with the necessary parameter values; and the particle–rotor model can be used for calculating additional properties (e.g. transition probabilities) outside the reach of the cranking model.

UNISOR is a consortium of universities, the State of Tennessee, Oak Ridge Associated Universities and Oak Ridge National Laboratory and is partially supported by them and by the U.S. Department of Energy under Contract No. DE-AC05-OR00033 with Oak Ridge Associated Universities.

References

1. A. Bohr and B. R. Mottelson, *Nuclear Structure*, Vol. 2 (Benjamin, New York, 1975).
2. F. Grümmer, K.W. Schmid and A. Faessler, Nucl. Phys. **A326** (1979) 1
3. M.J.A. de Voigt, J. Dudek and Z. Szymański, Rev. Mod. Phys. **55** (1983) 949
4. R. Bengtsson, Proc. Int. Symp. on Contemporary Topics in Nuclear Structure Physics, Cocoyoc, Mexico (World Scientific, Singapore, 1988), p. 317
5. R. Bengtsson, Proc. 18th Summer School on Nuclear Physics, Mikolajki, Poland, 1986, Harwood Academic Publishers, Nuclear Science Research Conference Series, vol. 12, p. 1
6. R. Bengtsson, T. Berggren and P. Olanders, to be published

PARTICLE-CORE COUPLING

IKUKO HAMAMOTO
Department of Mathematical Physics
Lund Institute of Technology, University of Lund
Lund, Sweden

Abstract

Some aspects of a version of "particle-core coupling model", which is different from the one described in the lecture given by the "spokesman of particle-core coupling model" in the present workshop, P. Semmes, is presented. A few observed quantities are chosen to illustrate how unambiguously one may extract some physics in a relatively simple manner by using the model.

1. Introduction

What I present in the following will be quite a different view of "particle-core coupling model" from the lectures given by the "spokesman of particle-core coupling model in the present workshop", P. Semmes. It seems to me that the difference lies not only in the "younger" and "older" work (namely, Semmes' talk started from a formalism of particle-core coupling in 1977, while the contents of my talk is based on the works published before 1977), but also in the intention of constructing and using the model so as to extract certain physics.

When one says the particle-core coupling model it does not necessarily mean that the system described is an odd-A nucleus, the particle is an odd nucleon and the core corresponds to an even-even nucleus. For example, in the particle-rotor model the "particle" may be taken as some intrinsic degree of freedom, such as a vibration, which is not included in the degree of collective rotation. (See, for example, ref. 1.) Another example is the particle-vibration coupling model, which is described somewhat in detail in the present talk and in which one can calculate also the structure of the collective vibration in even-even nuclei itself.

The particle-core coupling model is constructed from the consideration that the dynamics of a nuclear system results from the superposition of approximately independent elementary modes of excitations, single-particle excitations and various kinds of collective excitations. (See ref. 2.) The one-particle field generated by a collective distortion in the average nuclear density plays a central role in exploring the effects of the collective modes on the motion of the individual nucleons and the significance of the shell structure for collective dynamics.

Having a limited time of my present talk I would like to concentrate on a particular version of the particle-vibration coupling model (based on ref. 1) and to pick up a few selected observed quantities to illustrate how unambiguously one may extract the physics in a relatively simple manner by using the model. A model in nuclear physics is, after all, an approximation to the complicated nuclear many-body problem so as to explain (or understand) particular aspects of the physics. Therefore, somewhere one should naturally expect the presence of the limit of its applicability.

2. Particle-Vibration Coupling

To leading order in the vibrational amplitude, α, the variations in the one-body potential can be written in the form

$$\delta V = \kappa \alpha F(x) \tag{1}$$

where $F(x)$ is a one-particle operator describing the dependence of the potential on the nucleonic variables x (space, spin, isospin etc.), while the coupling constant κ characterizes the relationship between potential and density. The dependence of F on variables x comes from the structure of the collective mode and is determined based ultimately on the conditions of self-consistency. The basic matrix element of the coupling (1), which is illustrated in fig. 1, involves a single particle transition accompanied by the creation (or annihilation) of a single quantum of the collective vibration.

For example, for shape oscillations the coupling constant can be obtained from our knowledge of observed static potentials. If an oscillator potential

$$V = \frac{1}{2} m \omega_0^2 r^2 \tag{2}$$

is used, the following result[1] is well known and has been widely used for the last years:

Isoscalar Mode	Isovector Mode
$F_{\lambda\mu} = r^{\lambda} Y_{\lambda\mu}$	$F_{\lambda\mu,\tau=1,\mu_\tau=0} = \tau_z r^{\lambda} Y_{\lambda\mu}$
$\kappa_\lambda = -\frac{4\pi}{2\lambda+1} \frac{m\omega_0^2}{A\langle r^{2\lambda-2}\rangle}$	$\kappa_{\lambda,\tau=1} = \frac{\pi V_1}{A\langle r^{2\lambda}\rangle}$

In the literature[1,3] it is shown that a consistent treatment of the coupling δV in (1) gives a systematic method for treating; (i) overcompleteness of degrees of freedom; (ii) anti-symmetrization of particles, including those that are involved in collective motions; (iii) orthogonality of different excitation quanta; and so on.

As successful attempts to establish connections between different phenomena (see, for example, ref. 1,4,5), among others I may mention: (a) Particle-phonon interaction (i.e. estimate of anharmonic effects); (b) Renormalization of single-particle moments (i.e. core-polarization effect); (c) Structure of the vibration itself; (d) One-particle transfer with excitation of a phonon; (e) Quadrupole moment of a phonon; (f) Renormalization in the effective interaction between two particles; (g) Coupling between

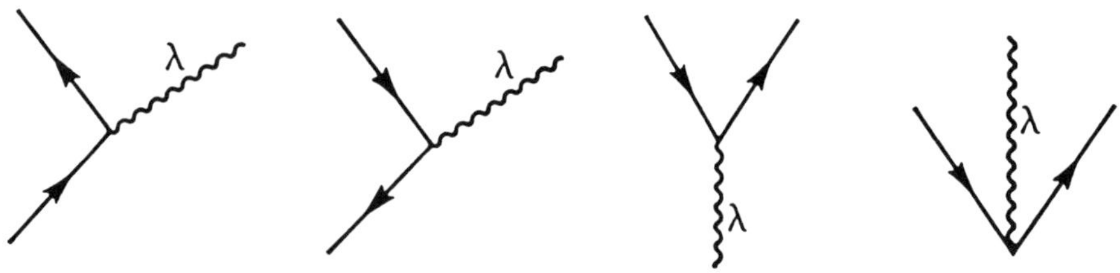

Fig. 1. Basic matrix elements of the particle-vibration coupling (1). The straight line with an arrow expresses a particle or a hole, while the wavy line represents a vibration.

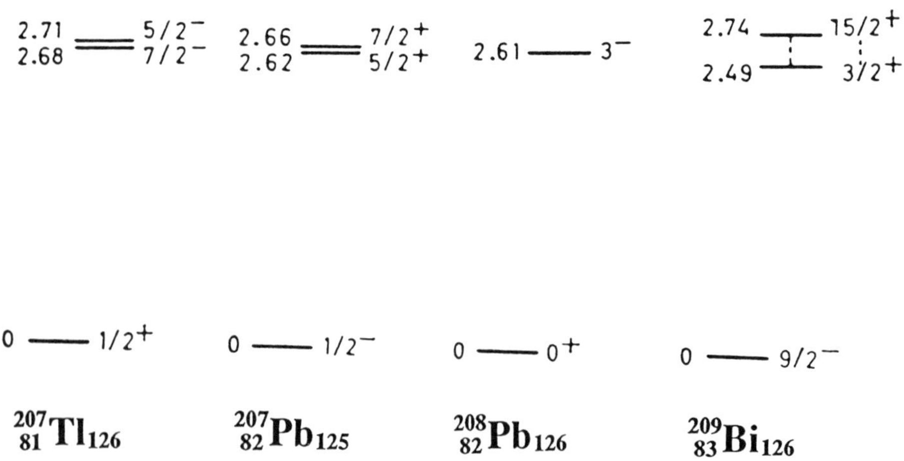

Fig. 2. Observed spectra of octupole vibrational states in nuclei around ^{208}Pb.

the states with one and two phonons; (h) Interaction between different elementary modes; (i) Anharmonicity in the two-phonon spectra.

It is impossible to go through in the present talk the whole basic idea and the results of the particle-vibration coupling model. Since they are already well-documented and published (for example, see ref. 1,3,4,5,6), in the following I take an example of the isoscalar octupole shape oscillation in ^{208}Pb and explain the above phenomena (a) in sect. 3, (b) and (c) in sect. 4, and (d) in sect. 5. The octupole vibration in ^{208}Pb is sufficiently collective and, moreover, the related particle-vibration coupling is well-defined and is weak so that the perturbation treatment can well be justified in most cases. Therefore, I would think that this is a classical example in which one can show how the particle-vibration coupling model can be used in a relatively simple manner to understand the physics of a variety of observed phenomena.

The total Hamiltonian consists of the particle part H_p, the core part H_c and the coupling term H_{coupl},

$$H = H_p + H_c + H_{coupl} \tag{3}$$

The Hamiltonian for particles is written as

$$H_p = T + V(r) \tag{4}$$

where T expresses the kinetic energy and

$$V(r) = V_{Coulomb} + V_c f(r) - V_c v \left(\frac{\lambda}{2}\right)^2 \frac{1}{r}\frac{df}{dr} \left(\vec{\sigma} \cdot \vec{\ell}\right) \tag{5}$$

$$f(r) = \left[1 + \exp\left(\frac{r-R}{a}\right)\right]^{-1} \tag{6}$$

The parameters appeared in the potential (5) are determined so that the shell-model single-particle energies in the ^{208}Pb region are reproduced. The particle-vibration coupling term can be obtained from the expansion

$$f\left(r, \alpha_{\lambda\mu}\right) = f(r) - R\frac{\partial f}{\partial r} \sum_{\mu} Y^*_{\lambda\mu} \alpha_{\lambda\mu} + \ldots \tag{7}$$

Namely,

$$H_{coupl} = -V_c R \frac{\partial f}{\partial r} \sum_{\mu} Y^*_{\lambda\mu} \alpha_{\lambda\mu} \tag{8}$$

In order to proceed calculations using the present model, we need only two properties of the octupole vibration, namely the vibrational frequency and the vibrational amplitude. To begin with, they are taken from two measured quantities in the lowest-lying 3^- state of ^{208}Pb,

$$\hbar\omega = 2.61 \text{ MeV} \tag{9}$$

$$B(E3) = 32 B_w(E3)$$

however, later we can check the consistency of those two input numbers by estimating the properties of the vibration itself (see, for example, sect. 4) within the model.

3. Particle-Phonon Interaction

In fig. 2 we show the observed spectra of octupole vibrational states both in ^{208}Pb and in its neighbouring nuclei. In the limit of vanishing particle-vibration coupling the multiplet of octupole vibrational states in those neighbouring nuclei should be degenerate and should be found at 2.61 MeV. The observed small energy splitting of the multiplet members shows that the coupling is weak in those examples.

The energy splitting of the multiplet $(j_1\lambda)I$, where the particle angular momentum j_1 couples to the phonon angular momentum λ to produce the total angular momentum I, can be obtained from the estimate of the four diagrams shown in fig. 3. There is a way of estimating the contributions from fig.3a together with fig. 3b (and fig. 3c together with fig. 3d) summing over all particle and all hole states, which have angular momentum j and which could contribute to the splitting (refs. 4,7). The obtained result is known to be in a very good agreement with measurements. (See, foe example, ref. 4.) The agreement is certainly very nice, however, it is known that the energy is the easiest quantity to reproduce and may not be used as a stringent test of the model. Therefore, in sect. 4 and sect. 5 we choose the quantities to be calculated in the model, which are much more sensitive to the structure of calculated wave functions.

It may be worth noting that the diagram in figs. 3b, 3c and 3d are exchange diagrams shown in figs. 4b, 4c and 4d, respectively. Since we have used observed properties of the vibration as well as single-particle states, we should not take into account the contributions from fig. 4. However, the diagrams in figs. 3b, 3c and 3d express the contributions, which exist first in the presence of the extra odd particle, and, thus, have to be included.

The comparison of the diagrams in figs. 3 and 4 elucidates the fact that in the present perturbation treatment of the particle-vibration coupling the anti-symmetrization of particles (and the symmetrization of phonons) as well as the overcompleteness of degrees of freedom is systematically taken care of, though in the zeroth order both the particle and the vibration are treated as independent building blocks in the model.

4. Core-Polarization Effect and Structure of Vibration Itself

An enhanced E3 transition between two shell model single-particle states can be used to check the structure of the particle-octupole-vibration coupling. The strong enhancement can be obtained from the polarization of the core, which is the first-order effect of the coupling and is illustrated in fig. 5. In the estimate of the effective charge in fig. 5 one has to include all possible isoscalar as well as all possible isovector shape oscillations. The effective charge expressed by fig. 5 is just a number, though it may in principle depend on involved particle orbits. Compared with B(E3) values

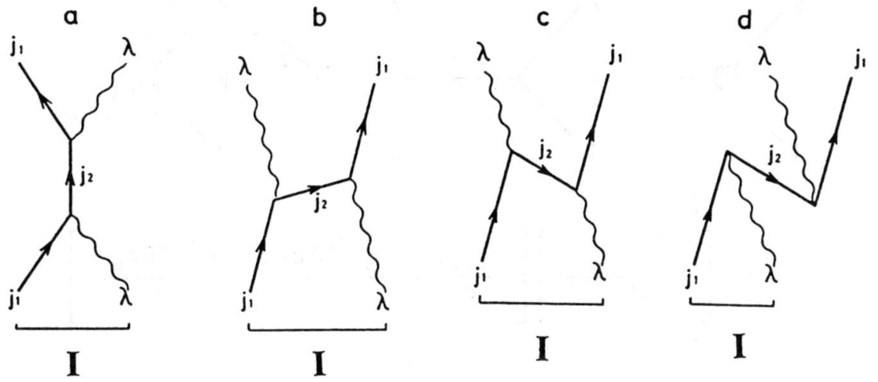

Fig. 3. Leading order (i.e. second order) terms in effective particle-phonon interactions.

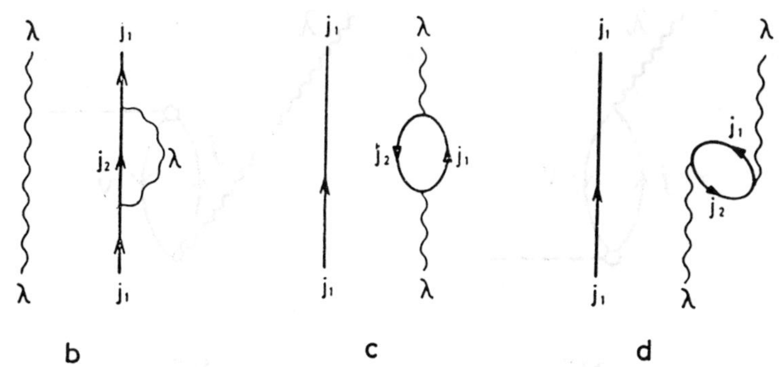

Fig. 4. Exchange parts of the diagrams in figs. 3b, 3c and 3d.

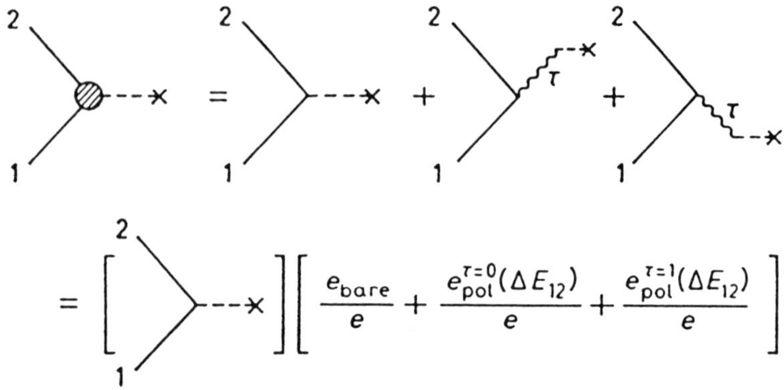

Fig. 5. Renormalization of single-particle moments.

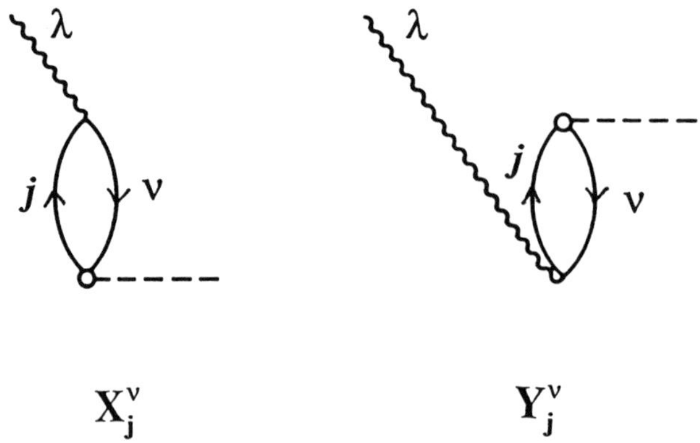

Fig. 6. Amplitudes of particle configurations in vibrational modes λ.

which are proportional to the square of the effective charge, a form factor in electron scatterings obtained as a function of momentum transfer provides a much richer information to test the structure of the coupling. Therefore, in the following I try to explain the structure of a polarization taking the example of the form factor of the octupole vibration itself.

The detailed form factor populating the octupole vibrational state in ^{208}Pb is obtained from inelastic electron scattering experiments. All the nuclear structure information is contained in the radial transition charge density, which may be written as

$$\begin{aligned}\rho_\lambda^{tr}(r) &\equiv (2\lambda+1)^{-1/2}\left\langle \lambda \left\| \sum_k e_k \frac{\delta(r-r_k)}{r^2} Y_\lambda(\Omega_k) \right\| 0 \right\rangle \\ &= e(2\lambda+1)^{-1/2} \sum_{j\nu} \langle j \| Y_\lambda \| \nu \rangle \left[X_j^\nu(r) + Y_j^\nu(r) \right] R_\nu(r) \end{aligned} \tag{10}$$

where $R_\nu(r)$ is the radial wave function of the nucleon in the orbit ν and the summation extends over the proton orbits. In terms of the diagram the estimate of $X_j^\nu(r)$ and $Y_j^\nu(r)$ corresponds to the calculation of fig. 6, in which the hole orbit ν is fixed while the contributions from all particle orbits with angular momentum j should be summed over. For example, the equation to determine $X_j^\nu(r)$ is written as

$$\begin{aligned}\frac{d^2\varphi_j}{dr^2} &- \frac{\ell(\ell+1)}{r^2}\varphi_j + \frac{2m}{\hbar^2}\left[\hbar\omega + \varepsilon_\nu - V(r)\right]\varphi_j \\ &= \frac{2m}{\hbar^2}(-1)^{\nu-j}\langle \nu \| Y_\lambda \| j \rangle \langle n_\lambda = 1 | \alpha_{\lambda\mu} | n_\lambda = 0 \rangle k(r)\phi_\nu \\ &- \frac{2m}{\hbar^2}\sum_{e_j<e_F}\left\langle \left(j\nu^{-1}\right)_\lambda, n_\lambda = 0; \lambda \left| H_{coupl} \right| n_\lambda = 1 \right\rangle \phi_j \end{aligned} \tag{11}$$

where

$$rR_\nu(r) \equiv \phi_\nu(r) \quad , \quad rX_j^\nu(r) \equiv \varphi_j(r) \quad \text{and} \quad k(r) = -V_c R \frac{\partial f}{\partial r} \tag{12}$$

Since the number of occupied orbits (i.e. hole orbits) is finite and the number of the possible particle angular momentum j for a given set of ν and $\lambda = 3$ is also finite, the summation in (10) can be easily carried out.

In fig. 7 we illustrate an example of the polarization effect which appeared in the contributions to the radial transition density. Namely, keeping the proton hole orbit fixed as the $2d_{3/2}^{-1}$ orbit, the dashed line shows the contribution from the proton particle orbit fixed as the $1h_{9/2}$ orbit which is an eigenstate of the potential in (5), while the solid line exhibits the contribution obtained from summing over all particle orbits with $\ell = 5$ and $j = 9/2$. If I use the terminology of oscillator potential, the dashed line expresses the $\Delta N = 1$ contribution, while the solid line includes all contributions (i.e. $\Delta N = 3, 5, \ldots$, in addition to $\Delta N = 1$). The comparison between the solid and the dashed line in fig. 7 shows that the effect of the inclusion of the $\Delta N = 3, 5, \ldots$, excitations cannot be expressed just by a constant multiplication

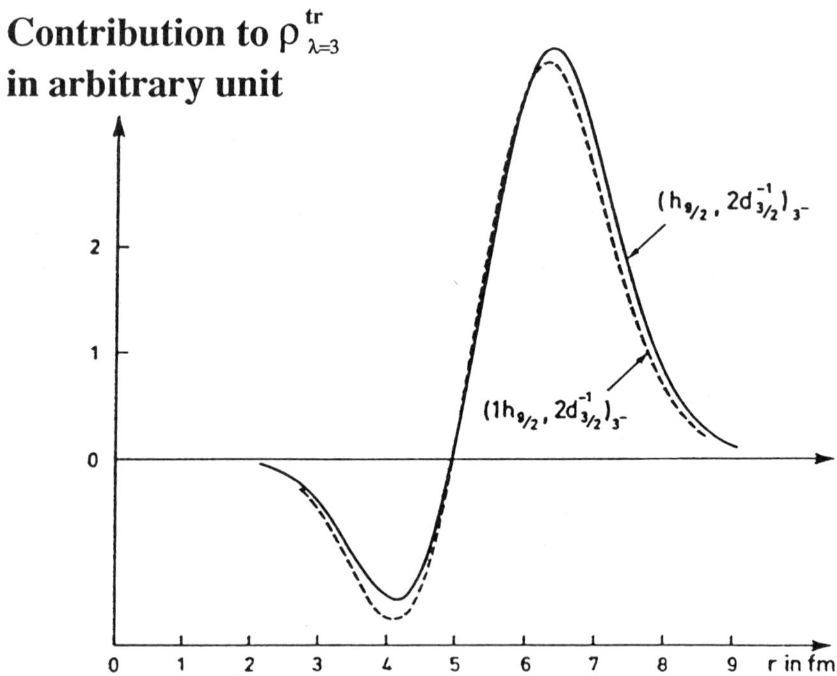

Fig. 7. Example of polarization effect, which appeared in the contribution to the radial transition density populating the octupole vibrational state in ^{208}Pb. See text. The figure is taken from ref. 8.

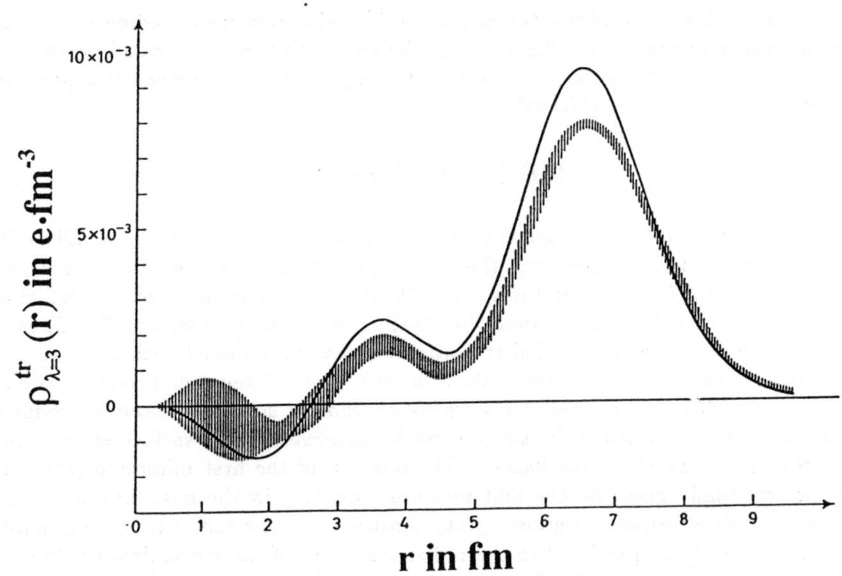

Fig. 8. Radial transition charge density of the octupole vibrational state at 2.61 MeV in ^{208}Pb. The solid line shows the calculated value, while experimental values are shown by the dashed band. The figure is taken from ref. 4.

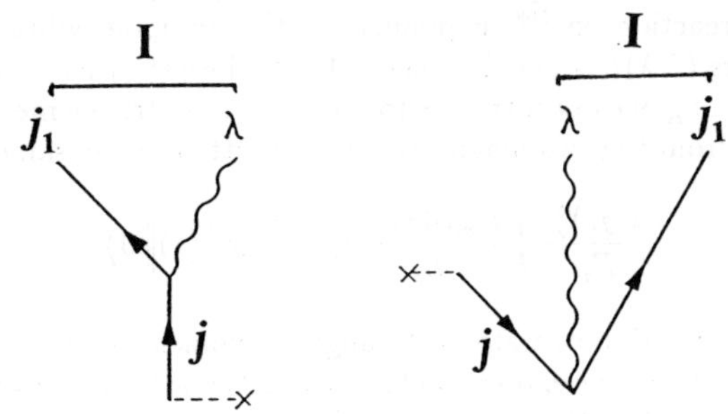

Fig. 9. Diagrams for one-particle transfer with excitation of a phonon.

factor, since the shape of the transition density calculated by including all possible excitations is appreciably different from the shape obtained by including only the $\Delta N = 1$ excitations. In fact, it is very interesting to examine the radial dependence of this difference. The B(E3) value,

$$B(E\lambda) \propto \left[\int_0^\infty \rho_\lambda^{tr}(r) r^{\lambda+2} dr\right]^2 \tag{13}$$

which is calculated by using the transition density estimated by including all excitations, is almost twice larger than the one estimated by using the transition density including only $\Delta N = 1$ excitations. We note that the information on the radial dependence of the polarization cannot be obtained from the discussion of B(E3) value.

In fig. 8 the calculated radial transition density is compared with the measured transition density. The presence of the observed two relatively small peaks inside the nucleus in addition to the well-known peak around the surface is nicely reproduced. The large peak just inside the surface can be understood as a surface effect, which is a feature of the shape oscillation. The position of the first minimum (around 5 fm) is essentially given by the first maximum existing in the quantum-mechanical oscillatory structure of the ground-state density, since the radial transition density is approximately proportional to the radial derivative of the static density. In order to obtain a reliable estimate of this minimum in the transition density, it is essential to use the realistic potential with a surface such as the Woods-Saxon potential. The two small peaks inside the nucleus are certainly a volume effect. The structure of these small peaks depends on the microscopic shell structure of the octupole phonon in addition to the general quantum-mechanical oscillatory structure of the static density.

5. One-Particle Transfer with Excitation of Phonon

In the limit of vanishing particle-vibration coupling the cross section of one-nucleon transfer reaction on ^{208}Pb populating the octupole vibrational state with multiplet members $(j_1\lambda)I$ is equal to zero. In the lowest order of the coupling two terms illustrated in fig. 9 can contribute to the nuclear matrix element of the reaction. The contribution from the two diagrams can be written in the same form

$$\frac{\langle (j_1\lambda)\, I \,| H_{coupl} |\, j\rangle}{E_{j_1} + \hbar\omega - E_j} \delta(j, I) \left\langle j \left\| a^+(j) \right\| 0 \right\rangle \tag{14}$$

where the nucleon creation operator with angular momentum j is denoted by $a^+(j)$. Thus, one should sum up the expression (14) over all single-particle orbits with j. The quantity which is equal to the sum can be directly calculated by using the solution of the following equation,

$$\left(E_{j_1} + \hbar\omega - H_0\right) Z_j(r) = \left\langle (\mathcal{Y}_{\ell j} \otimes \Phi_0)_{jm} \left| H_{coupl} \right| (\mathcal{Y}_{\ell_1 j_1} \otimes \Phi_\lambda)_{jm} \right\rangle R_{j_1}(r) \tag{15}$$

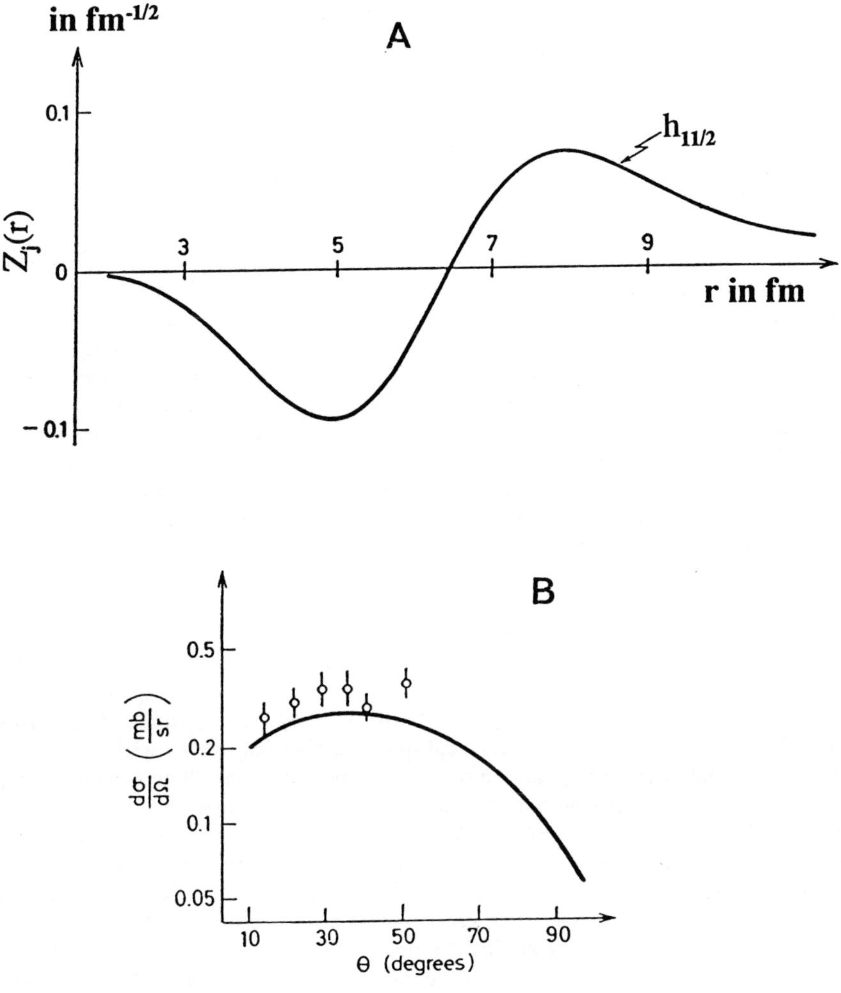

Fig. 10. A: The $h_{11/2}$ single-particle radial wave function admixed into the septuplet member $(g_{9/2}3^-)11/2^-$ of ^{209}Pb. B: Cross section of the ^{208}Pb(t,d)^{209}Pb reaction ($E(t)_{lab} = 20.0$ MeV) to the septuplet member $(g_{9/2}3^-)11/2^-$. The figures are taken from ref. 4.

where the single-particle orbital j_1 is assumed to be the eigenstate of the Hamiltonian (4) as

$$(E_{j_1} - H_0) R_{j_1}(r) = 0 \tag{16}$$

and

$$\begin{aligned} H_0 &= -\frac{\hbar^2}{2m}\frac{d^2}{dr^2} + \frac{\hbar^2}{2m}\frac{\ell(\ell+1)}{r^2} + V(r) \\ \hbar\omega &= 2.61 \text{ MeV} \end{aligned}$$

The solution $Z_j(r)$ of eq. (15) expresses the result of the coherent interference of the single-particle components admixed in the target and the final state. The radial wave function $Z_j(r)$ is calculated at the energy $E = (E_{j_1} + \hbar\omega)$ which is the unperturbed energy of the multiplet members. Therefore, it has the correct dependence on the radial variable at this energy. Since the calculated cross section of one-nucleon transfer reactions is very sensitive to the radial shape of the single-particle wave function involved, one expects that the radial form factor $Z_j(r)$ evaluated in the present way is suitable for being used in calculating the cross section.

In the case that the above solution $Z_j(r)$ is not so different from the eigenfunction of the Hamiltonian (4), the one-nucleon transfer cross section would be relatively large and the calculated cross sections by using the present method are not so different from those obtained by other methods available in the market. An example of this kind is the ^{208}Pb (^{3}He,d) ^{209}Bi reaction populating a septuplet member in ^{209}Bi, $(h_{9/2}3^-)I^\pi$ with $I^\pi = 13/2^+$.

In order to show a special advantage of the present method, as a numerical example I choose the small peak observed at 2.59 MeV in the ^{208}Pb(t,d)^{209}Pb reaction. Since the nucleus ^{209}Pb is not stable and, thus, it cannot be used as a target in inelastic scatterings or Coulomb excitations, it is not easy to find the octupole vibrational state built on the ground state, $(g_{9/2}3^-)I^\pi$ with $I^\pi = 3/2^-, \ldots, 15/2^-$. In fig. 10 we show the solution of eq. (15) for the septuplet member $(g_{9/2}3^-)I^\pi$ with $I^\pi = 11/2^-$ and the calculated cross section compared with measured one. The calculated $h_{11/2}$ single-particle wave function has the unique radial dependence, which is something between that of the $1h_{11/2}$ and the $2h_{11/2}$ single-particle wave functions, and has a relatively large amplitude outside the nuclear surface in spite of the small amplitude inside the nucleus. Therefore, the calculated angular distribution obtains a peculiar shape and the calculated stripping cross section becomes appreciable even though the total admixed probability of this component into the octupole vibrational state is very small. Since there is no other candidate state, which has the observed magnitude of the cross section of the one-nucleon transfer reaction with the unique shape of angular distribution and of which the energy is expected around 2.59 MeV, the observed state is identified as the septuplet member $(g_{9/2}3^-)I^\pi$ with $I^\pi = 11/2^-$.

References

1. A. Bohr and B.R. Mottelson, *Nuclear Structure*, Vol.II, W.A.Benjamin Inc., Reading, Massachusetts (1975).

2. The physics resulting from the superposition of approximately independent excitation modes were considered already at an early stage in the study of collective nuclear motion (L.L. Foldy and F.J. Milford, *Phys.Rev.* **80** (1950) 751; A. Bohr and B.R. Mottelson, *Mat. Fys. Medd. Dan. Vid. Selsk.* **27** (1953) no.16). The generality of such a coupling scheme was emphasized by de-Shalit, *Phys.Rev.* **122** (1961) 1530.

3. D.R. Bes, G.G. Dussel, R.A. Broglia, R. Liotta and B.R. Mottelson, *Phys. Lett.* **52B** (1974) 253.

4. I. Hamamoto, Proc. Int. School of Physics Enrico Fermi, course 69, 1976, p. 252-270; *Physics Reports*, **10C** (1974) 63.

5. B.R. Mottelson, Proc. Int. Conf. on Nuclear Structure, Tokyo, 1967; Nobel Lectures 1975.

6. D.R. Bes and R.A. Broglia, *Phys. Rev.* **3C** (1971) 2349.

7. I. Hamamoto, *Nucl. Phys.* **A126** (1969) 545.

8. I. Hamamoto, *Phys. Lett.* **66B** (1977) 410.

GROUP 8

Shell Model Approaches

MONSTER AND VAMPIR : VARIATIONAL APPROACHES TO THE NUCLEAR SHELL MODEL

K. W. SCHMID
Institut für Theoretische Physik, Universität Tübingen
Auf der Morgenstelle 14, D-7400 Tübingen, FRG

ABSTRACT

Many nuclear structure problems require the use of model spaces which are far too large to allow for a complete diagonalisation of some suitably chosen effective many-body Hamiltonian as it is done in the conventional shell-model approach. Being interested in such problems therefore suitable truncation methods have to be developed. The methods discussed here use general symmetry-projected Hartree-Fock-Bogoliubov configurations as basic building blocks and determine the underlying mean fields as well as the configuration mixing by chains of variational calculations. In this way the selection of the relevant degrees of freedom is entirely left to the dynamics of the considered system and the complete shell-model expansion of the nuclear wave functions can be approximated with, at least in principle, arbitrary accuracy. The present status of these methods is illustrated with various results. Advantages as well as shortcomings and possible improvements are discussed.

1. Definition of the Problem

"Being a complex composite system of many constituents, the nucleus displays a multitude of various excitations, which may be classified by their constants of motion : energy, angular momentum, parity and at least in light systems also the total isospin. The energetically lowest state of each angular momentum is usually called the yrast state, the higher ones are referred to as non-yrast excitations. Both yrast as well as non-yrast states may be either of single particle or collective nature according to the number of constituents taking part in the particular excitation. Furthermore, we usually distinguish between the nuclear continuum, being high enough in excitation energy so that a decay via nucleon (or even nucleus) emission is possible, and the discrete levels lying below the emission threshold for hadronic particles and hence being able to decay only by photon emission (or electron conversion) into the energetically lower states of the same or, for example, by β decay of neighbouring nuclear systems.

To explore these discrete excitations and their measurable properties all over the nuclear mass table is the classical field of experimental nuclear spectroscopy ; to explain the complex interplay of collective and single particle degrees of freedom causing these properties is the central aim of theoretical nuclear structure physics." (from Ref. [1])

Today we believe that the elementary constituents of the nucleus are quarks and gluons. Thus a really "microscopic" description of the nucleus should be based on the relativistic quantum field theory for quarks and gluons, Quantum Chromo Dynamics (QCD). However, not even for a single nucleon QCD could be solved up to now. Trying to do it for the nucleus is preposterous. Thus our aim is more modest : We "only" want to construct quantum mechanical many-nucleon wave functions for the ground and low excited discrete states of all nuclei with mass numbers in between $8 \leq A \leq 240$ from one and the same model. However, even the solution of this more modest problem is still far beyond our capacities. Thus we have to make some approximations :

(i) We shall work with non-relativistic, structureless nucleons as elementary constituents.

(ii) We shall assume that for the desription of the low lying discrete states of a nucleus the use of a finite (if necessary very large) model space is sufficient.

(iii) We shall assume that for this model space an appropriate, effective many-nucleon Hamiltonian does exist, does contain only one- and two-body terms, and that we know its matrix elements.

NOTHING ELSE WILL BE ASSUMED

The remaining problem is still extremely hard. However, now there is at least some *hope* that we can solve it.

2. General Considerations

How did we arrive at the above approximations? Let us have a short look on

2.1 The Background

We know that nucleons are not elementary. They have excited states like the Δ and the Roper resonance. Lepton scattering tells us that nucleons are composite : there are "partons" inside. Today we believe that the elementary constituents of the nucleons (and hence the nucleus) are quarks and gluons. Thus the really "microscopic" theory of the nucleus is QCD. However, on the energy scale of the low excited discrete levels we encounter "infra-red slavery" : perturbative QCD doesn't work. Should we all start doing lattice QCD calculations?

Fortunately we need not : in fact we know how for low energy nucleon-nucleon scattering the solution should look like. More than 40 years of nuclear physics have taught us that the concept of elementary nucleons interacting with each other via meson exchange forces is a rather good one. One Boson Exchange Potentials (OBEPs) as investigated in great detail by the Bonn group [2] do reproduce the nucleon-nucleon scattering data at not too high energies as well as even rather sensitive data in the three nucleon systems rather nicely. So the hope is that at low energies it should be possible to eliminate the quark and gluon (and even the meson) degrees of freedom from the problem and to obtain some kind of "effective" meson exchange theory like the OBEP model. Nucleon-nucleon scattering will then be described by a "T-matrix" being the solution of a Lippmann-Schwinger equation (or its relativistic equivalent) with an OBEP as input. Obviously this can only work at low energy (not too small distances). At higher energies the "cluster structure" of the nucleons would be seen, which leads to all kinds of complicated many-body effects like the occurence of many-body forces, energy dependencies, strong non-localities, and so on. Thus here the whole concept of a nucleon-nucleon potential would loose its meaning.

Unfortunately the assumption of an effective nucleon-nucleon potential is not the only approximation we have to make. In nuclear structure physics we are not much interested in the interaction between two "free" nucleons but in the force which acts between the "bound" nucleons inside the nucleus. This force is considerably different from the "free" interaction discussed above. Unlike the latter case, where for the final states of the two scattered nucleons the whole phase space is available, inside the nucleus all the states occupied by the other nucleons do not participate in the scattering. Because of this so-called "Pauli blocking" the interaction inside the nuclear medium is not given by the T-matrix, but by the "G-matrix" being the solution of a modified Lippmann-Schwinger equation, in which the occupied states are excluded by the so-called "Pauli operator" and furthermore the energy denominator takes care of the binding energies which the nucleons aquire due to their interactions in the medium. This equation is called the Bethe-Goldstone, its relativistic equivalent the Bethe-Salpeter equation. Now it is obvious that the Pauli operator as well as the binding energies are determined by the structure of the considered system. The structure, however, depends on the interaction, and the interaction in turn depends on the Pauli operator and the binding energies. Thus we encounter a kind of "what is first the egg or the hen" problem which in principle should be solved selfconsistently. An additional complication results from the fact that also the meson propagators entering the OBEP should be modified inside the nuclear medium. Thus the determination of the effective interaction inside the nucleus is in itself a complicated many-body problem which at least at present we are not able to solve. Therefore we shall introduce an additional approximation and leave this "circulus vitiosus" by assuming that some kind of "average" Pauli operator is good enough, and that we can obtain a reasonable interaction by furthermore solving the Bethe-Goldstone equation for an OBEP (with, if necessary, "medium corrected" meson parameters) only approximatively, either in infinite nuclear matter, or better within the Brueckner-Hartree-Fock approximation for a finite nuclear system. This obviously yields a non-relativistic two-body G-matrix. Is that sufficient? Well, a deeply bound nucleon has only a separation energy of about 50 MeV as compared to its mass of 1 GeV, so why bother? Well, Dirac phenomenology[3] tells us that these 50 MeV may be the difference of two large numbers, an attractive scalar (−400 MeV) and a repulsive vector potential (+350 MeV). These numbers, however, are of the same order of magnitude as 1 GeV. Unfortunately, the Bethe-Salpeter equation is tremendously more complicated than its non-relativistic version. Therefore we shall ignore all relativistic effects in the following and simply assume that something like an effective, non-relativistic nucleon-nucleon interaction in the nucleus does exist and can be represented as a non-relativistic Brueckner two-body G-matrix.

But now we are there, aren't we? Not yet. Our G-matrix still acts in the full Hilbert space. In practical calculations, however, we can only include a finite model space. It is by no means obvious that such an assumption is justified, even if the chosen model space is large. Furthermore, due to the restriction to a finite model space yet another renormalisation of the effective interaction should be performed, being once more a many-body problem in itself. This again does introduce all types of additional complications like, e.g., the occurence of many-body forces, non-localities, etc.. We shall ignore this problem and make our final approximation : We shall assume that for the description of low lying discrete nuclear excitations a finite model space is sufficient, and that even for this finite model space an "average", energy independent, effective interaction of two-body nature can be found. In fact we shall even assume that we "know" such an interaction for any particular model space, which we shall use later on, at least approximately (e.g., a G-matrix or some more phenomenological interaction). Solving the remaining nuclear structure problem for such an approximate Hamiltonian, we may then hope to learn something about its deficiencies and to use this information to obtain an improved version of the effective interaction.

Finally, we have reached our goal : we have thrown away more than 99 percent of the original problem and are now left with something which we shall call

2.2 The Restricted Nuclear Many-Body Problem

This, in fact, can be solved exactly, at least in principle. Let us define our model space by a finite, M_b dimensional set $\{|i\rangle, |k\rangle, ...\}_{M_b}$ of orthonormal single nucleon states, e.g., harmonic oscillator or some suitably discretized Woods-Saxon wave functions. The corresponding creation and annihilation operators will be denoted by $\{c_i^\dagger, c_k^\dagger, ...\}_{M_b}$ and $\{c_i, c_k, ...\}_{M_b}$, respectively. They fulfill the usual anticommutation relations for Fermion field operators. The particle vacuum $|0\rangle$ is defined by $c_i|0\rangle \equiv 0$ for all $i = 1, ..., M_b$. Using this basis representation the chosen effective many-nucleon Hamiltonian can be written as

$$\hat{H} = \sum_{i,r=1}^{M_b} t(ir) c_i^\dagger c_r + \frac{1}{4} \sum_{i,k,r,s=1}^{M_b} v(ikrs) c_i^\dagger c_k^\dagger c_s c_r \tag{1}$$

where $t(ir) \equiv \langle i|\hat{t}|r\rangle$ are the matrix elements of the one-body operator of the kinetic energy (or, if an inert core is used, some suitably chosen single particle energies) and $v(ikrs) \equiv \langle ik|\hat{v}|rs - sr\rangle$ are the antisymmetrized matrix elements of the effective two-body interaction. The "only" thing left to do now is to distribute the A nucleons of our problem in all the possible (Pauli allowed) ways over the M_b orbits of our single particle basis and to diagonalize the Hamiltonian (1) in the complete space of the resulting Slater determinants

$$\mathbf{C}(A) \equiv \Big\{|D_i\rangle \equiv \Big(\prod_{k\in D_i}^{(A)} c_k^\dagger\Big)|0\rangle \quad ; \quad i = 1, ..., n = \binom{M_b}{A}\Big\} \tag{2}$$

This yields then the "exact" solutions of our "restricted nuclear many-body problem". The task can be simplified even more : Except for the conservation of the total nucleon number A, in (2) no symmetry of the Hamiltonian (1) has been used. Now $\hat{H}$ does conserve besides A also the charge number Z (or, equivalent, the 3-component of the total isospin $T_z = \frac{1}{2}(A - 2Z)$), the total parity π (at least, if as here weak interactions are neglected), and last but not least the total angular momentum I and its 3-component M (For charge symmetric forces furthermore also the total isospin is a "good" quantum number, however, since this symmetry is already broken by the Coulomb interaction, we shall not consider the isospin as a good symmetry in the following). Thus, instead of using the configuration space (2) we may split the problem into various diagonalisations in the subspaces of (2) corresponding to a definite symmetry $s \equiv AT_z I^\pi$

$$\mathbf{C}(s, M) \equiv \{|\sigma; sM\rangle \equiv \sum_{i=1}^{n} |D_i\rangle\langle D_i|\sigma; AT_z I^\pi M\rangle \; ; \; \sigma = 1, ..., n(s) \leq n\} \tag{3}$$

Obviously, in general $n(s) \ll n$. The resulting wave functions for the states with symmetry s have now the form

$$|\rho; sM\rangle = \sum_{\sigma=1}^{n(s)} |\sigma; sM\rangle g^s_{\sigma\rho} \quad ; \quad \rho = 1, ..., n(s) \tag{4}$$

This is the so-called Shell-model Configuration-Mixing (SCM) approach[4] to our problem. Its solutions are "exact", so why going on with these lectures? Well, the problem is the dimension $n(s)$. If we take for example as single particle basis the 0s, 0p, 1s0d, 1p0f - oscillator shells plus the $0g_{9/2}$- orbit (these are 100 single nucleon states) and calculate the number of $I^\pi = 0^+$ configurations for the nucleus ^{28}Si we obtain

$$n(A = 28, T_z = 0, I^\pi = 0^+) = 136\,124\,437\,576\,139\,270\,616 \tag{5}$$

i.e. 1.36×10^{20}. It is worthwhile to contemplate this number a little bit : If you have a super computer with a cycle time of 1 nanosecond (the best today give you about 4-6 ns), it will take more than 4000 years to run once linearly over these configurations without calculating anything. Furthermore, if you assume that you must store about 100 Bit's of information for each configuration and you want to do this on CD laser discs, better contact the nearest factory right away. If this factory is able to sell you 1 million CD's a month, you'll still have to wait for more than 80000 years before you have collected all you need. It is obvious that such proposals are ridiculous. It is simply impossible to do "exact" calculations in large model spaces at all.

In addition, as some of you may have noticed already, calling the above SCM approach "exact" I've been cheating. It is obvious that measuring the spectrum of a nucleus in Tübingen and in Oak Ridge should yield the same result. Thus all the interaction terms of the effective many-body Hamiltonian, however complicated they may be, have to display

2.3 Translational Invariance

i.e., they cannot depend on the center of mass (COM) coordinate

$$\vec{R} \equiv \frac{1}{A}\sum_{j=1}^{A} \vec{r}_j \tag{6}$$

of the considered system but only on some suitably chosen relative coordinates with respect to the latter. Furthermore, the internal motion of the nucleons should not depend on the motion of the nucleus as a whole and thus Galilean invariance is required, too. As a consequence of these two invariances the total Hamiltonian can always be splitted into two parts

$$\hat{H} = \hat{H}_{int} + \frac{\hat{P}^2}{2Am} \tag{7}$$

where $\hat{H}_{int}$ describes the internal motion of the nucleons and depends besides on the spin and isospin quantum numbers only on the relative coordinates and the operators of the corresponding conjugate momenta while the second term with m being the nucleon mass and

$$\hat{P} \equiv \sum_{j=1}^{A} \hat{p}_j \tag{8}$$

the center of mass momentum operator is the kinetic energy operator for the center of mass motion. Due to this decomposition the eigenfunctions of (7) must be of the form

$$|\Psi\rangle = |\Psi_{int}\rangle|\Psi_{CM}\rangle \tag{9}$$

with

$$\langle\vec{R}|\Psi_{CM}\rangle \equiv (2\pi)^{-\frac{3}{2}}e^{i\vec{P}\cdot\vec{R}} \tag{10}$$

being a plane wave with momentum $\vec{P}$ while $|\Psi_{int}\rangle$ depends besides on spin and isospin only on the relative coordinates and is thus translationally invariant. Obviously it is only this latter part of the wave function (9) we are interested in.

Now the wave function (4) obviously does not have this invariance property. It depends on all A coordinate vectors and hence in general does contain components which correspond to the motion of the nucleus as a whole and thus have to be considered as "spurious".

As has been shown already more than 30 years ago by Elliot and Skyrme[5] these spurious components can be eliminated with the help of group theoretical methods provided one uses a harmonic oscillator single particle basis as well as a particular truncation scheme for the shell model configuration spaces (3). For "complete" SCM calculations of the above described type as well as for more general basis functions, however, these methods can not be applied.

Nevertheless, the center of mass motion can still be treated correctly even in the general case if one uses projection techniques[6] for the restoration of the translational invariance. As has been discussed at length in a recent paper[7], the operator

$$\hat{C}(0) \equiv (2\pi)^{-3} \int d^3\vec{a} \exp\left\{i\,\vec{a}\cdot\hat{P}\right\} \tag{11}$$

with the integration to be performed over the whole 3-dimensional space extracts from any arbitrary A-nucleon configuration (3) the component with total center of mass momentum $\vec{P}=0$, i.e., it projects into the center of mass rest frame* . This operator is a true mathematical projector. It carries neither linear nor angular momentum and thus it does not matter whether it is applied before or after the Slater determinants (2) are coupled to SCM configurations of the type (3) having the required symmetry s.

Acting with the operator (11) on all the configurations (3) of a particular symmetry $s \equiv AT_zI^\pi$ we thus obtain a configuration space of the form

$$\mathbf{C}^0(\vec{P}=0,s,M) \equiv \{\,\hat{C}(0)|\sigma;sM\rangle \quad ; \quad \sigma = 1,...,n(s)\} \tag{12}$$

consisting out of translationally invariant but now non-orthogonal configurations. The complete set of linear independent (i.e., non-spurious) wave functions which can be constructed in this model space

$$|\rho^0;sM\rangle = \sum_{\sigma=1}^{n(s)} \hat{C}(0)|\sigma;sM\rangle g^{s0}_{\sigma\rho} \qquad ; \qquad \rho = 1,...,n'(s) \le n(s) \tag{13}$$

then results from the solution of the generalized eigenvalue problem

$$[\,H^{s0} - E^{s0}\,N^{s0}\,]\;g^{s0} = 0 \tag{14}$$

being complemented by the constraint

$$g^{s0\dagger}\,N^{s0}\,g^{s0} = \mathbf{1}_{n'(s)} \tag{15}$$

in order to ensure the orthonormality of the states (13). Here $n'(s)$ is the number of non-zero eigenvalues of the hermitean and positive definite $(n(s)\times n(s))$ dimensional overlap matrix

$$N^{s0}_{\sigma\sigma'} \equiv \langle\sigma;sM|\hat{C}^\dagger(0)\hat{C}(0)|\sigma';sM\rangle \equiv \langle\sigma;sM|\hat{C}(0)|\sigma';sM\rangle \tag{16}$$

---

* Note, that the configurations (3) are all bound states which are confined to a small region of space. Thus the diagonal norm integrals for the operator (11) are always finite and do never vanish. Note furthermore, that we are not interested to reconstruct a plane wave for the COM motion explicitely but only in extracting the internal motion part of the configurations (3).

The number of vanishing eigenvalues $n(s) - n'(s)$ of this matrix is identical to the number of spurious states being contained in the configuration space (12).

The Hamiltonian matrix in (14) is defined as

$$H^{s0}_{\sigma\sigma'} \equiv \langle\sigma; sM|\hat{C}^{\dagger}(0)\hat{H}\hat{C}(0)|\sigma'; sM\rangle = \langle\sigma; sM|\hat{H}\hat{C}(0)|\sigma'; sM\rangle \tag{17}$$

where use has been made of the fact that the projector (11) commutes with $\hat{H}$. It should be stressed here that in the matrix elements (17) besides the configurations (3) also configurations with the same symmetry s lying outside this model space occur as intermediate states in between $\hat{H}$ and $\hat{C}(0)$. Thus the Hamiltonian (1) can in fact not be restricted to the finite basis used for the construction of (3) but is needed in the representation of a considerably larger basis system.

Besides this additional complication, however, the translationally invariant SCM approach presented here looks rather similar to the conventional one given in the last section. Especially the dimensionality of the matrices is not changed and thus both approaches are equally impossible in large model spaces like those discussed above.

Fig.1 : Skyrme Hartree Fock charge form factors for ^{16}O. From Ref.[9].

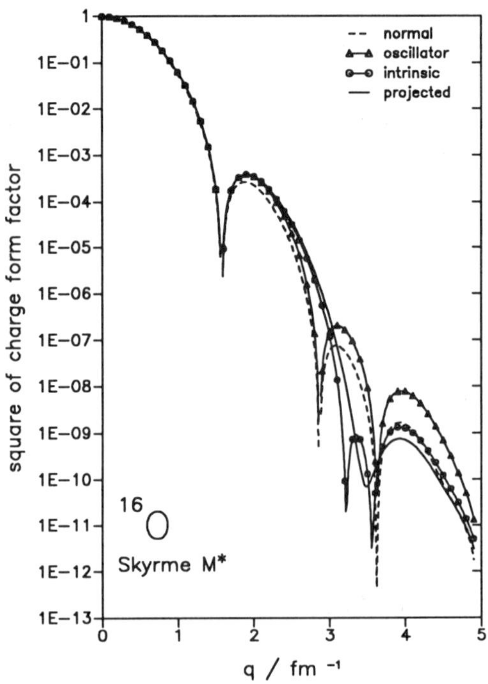

Now you may ask, is it really necessary to restore translational (and Galilean) invariance? After all it should be an $1/A$ effect. Well, it depends on what you look for[8,9]. Fig.1 shows the charge form factor as obtained for a spherical Skyrme-Hartree-Fock solution for the nucleus ^{16}O in four different ways (a) not worrying about the center of mass at all ("normal"), (b) folding out a harmonic oscillation ("oscillator"; this is the standard correction method usually applied in electron scattering analyses), (c) neglecting the projection into the rest frame but at least taking the correct relative coordinates in the charge operator ("intrinsic"), and, finally (d) the full translational invariant projected description ("projected"). As can be seen the effects of the COM motion are indeed small for small momentum transfers. For transferred momenta larger than about 2 fm^{-1}, however, the proper treatment becomes rather important* . Preliminary results for quasielastic electron scattering[10] show that there the effects are even more drastic, and, last but not least, the proper treatment of the COM motion is very essential to avoid spurious contaminations of the wave functions.

Furthermore you may ask, is it really necessary to consider such large model spaces? Isn't it sufficient to take something like the 1s0d-oscillator shell as basis and then calculate ^{28}Si? Then, because only a single major shell is used, no spurious center of mass components have to be eliminated[5]. Furthermore in this basis there are only 3372 different 0^+ configurations, and the calculation becomes "easy". However, what if you are interested in the negative parity states? Then you should include in addition at least the 0p and the 1p0f shells, and you are already in trouble. Furthermore, even the "one-shell" description of the low lying positive parity states in light 1s0d shell nuclei is not quite sufficient : while the electromagnetic transition probabilities can still be in most cases reasonably well reproduced, if an effective charge is assumed, the dependence of an inelastic formfactor on the momentum transfer, e.g., can for many states only be described assuming a larger single particle basis [11]. As much for light nuclei. In medium heavy and heavy nuclei the situation becomes even worse. Here usually even more single particle orbits than discussed above have to be included in the model space.

So, what to do? It is obvious that we have to truncate the complete SCM expansion of the nuclear wave functions to a numerically feasible number of configurations. The question is only, how this can be done without loosing those degrees of freedom, which are essential for any particular state under consideration. Truncate according to the unperturbed energy of the configurations (3) ? You'll run into terrible convergence problems[12]. Better let the Hamiltonian itself do the truncation for you. This is where

2.4 Mean Fields

enter the picture. The idea is to expand the nuclear wave function not with respect to some ad hoc chosen (and hence somewhat arbitrary) basis creating one-body potential as it is done in the SCM approach, but with respect to the optimal "mean

---

* The restoration of full Galilean invariance requires either the projection before the variation as presented in the present article or a double projection method. In Ref. [9] the projection was done after the HF-variation. Thus there only translational invariance had been restored.

field" each of the nucleons feels due to its interactions with all the others and to extract this average potential directly from the Hamiltonian (1) via variational principles (this is what "optimal" means). The best known approaches along this lines are the Hartree-Fock (HF) [13] and its more general extension, the Hartree-Fock-Bogoliubov (HFB) theory [14]. We shall base our considerations on the latter in the following.

Mathematically, a mean field is introduced by a linear "quasi particle" transformation over the creators and annihilators of the spherical basis states. Thus we introduce quasi particle creators and annihilators via

$$a_\alpha^\dagger \equiv \sum_{i=1}^{M_b} (A_{i\alpha} c_i^\dagger + B_{i\alpha} c_i) \tag{18}$$

and hence

$$a_\alpha = \sum_{i=1}^{M_b} (B_{i\alpha}^* c_i^\dagger + A_{i\alpha}^* c_i) \tag{19}$$

respectively. Eqs. (18) and (19) can be combined to a single matrix equation

$$\begin{pmatrix} a^\dagger \\ a \end{pmatrix} = \begin{pmatrix} A^T & B^T \\ B^\dagger & A^\dagger \end{pmatrix} \begin{pmatrix} c^\dagger \\ c \end{pmatrix} \equiv F \begin{pmatrix} c^\dagger \\ c \end{pmatrix} \tag{20}$$

with F being a $(2M_b \times 2M_b)$ dimensional matrix. In order to ensure anticommutation relations for the quasi particle operators (18) and (19) this matrix has to be unitary

$$FF^\dagger = F^\dagger F = \mathbf{1}_{2M_b} \tag{21}$$

Eqs. (20) and (21) define the famous HFB transformation [11] being the most general linear transformation conserving the anticommutation relations, which can be constructed within the chosen finite single particle basis.

The vacuum $|F\rangle$ for the quasi particle annihilators (19) is defined by

$$a_\alpha |F\rangle \equiv 0 \quad \text{for all} \quad \alpha = 1, ..., M_b \tag{22}$$

and may be represented as [12]

$$|F\rangle = \left(\prod_{\alpha=1}^{M_b'} a_\alpha\right)|0\rangle \quad \text{with} \quad M_b' \leq M_b \tag{23}$$

where the product runs only over those α's with $a_\alpha|0\rangle$ being different from zero.

However, being a generalized Slater determinant (23) is obviously not translationally invariant. In addition, since the transformation (20) sums over all the quantum numbers characterizing the single particle basis states (isospin-3-projection, orbital angular momentum, total angular momentum, the 3-projection of the latter, and the node-number), $|F\rangle$ is neither an eigenstate of the square of the total angular momentum operator $\hat{I}^2$ nor of its 3-component $\hat{I}_z$. Furthermore particle number and charge conservation are violated and, in general, the vacuum (23) has no definite parity either. The only symmetry required by the Hamiltonian which is still conserved is the so-called "number parity" [15], i.e., $|F\rangle$ contains either only components with even or with odd total nucleon numbers A. Expanding (23) in the complete SCM configuration spaces (3) we therefore obtain

$$|F\rangle = \sum_{AT_zI\pi K} \sum_{\sigma=1}^{n(s)} |\sigma; sK\rangle\langle\sigma; sK|F\rangle \tag{24}$$

This, however, is not quite what we wanted, since the physical states (13) are not mixtures of different symmetries s. The sum over σ is fine : in this way we want to account for many of the configurations (3) via a single quasi particle determinant (23) ; the sum over s and K, however, i.e. the symmetry breaking, is the price we have to pay for this attempt. So, how to obtain physical configurations? For this purpose we shall now consider the

2.5 Restoration Of The Broken Symmetries

From the vacuum (23) one can construct configurations with the desired symmetry quantum numbers $s \equiv AT_zI^\pi$ using the operator

$$\hat{\Theta}^s_{MK} \equiv \sum_{\sigma=1}^{\infty} |\sigma; sM\rangle\langle\sigma; sK| \tag{25}$$

with the sum running over *all* and not only the $n(s)$ model space configurations with the same symmetry s. Since, however, the vacuum (23) has by construction no components outside the model space, in

$$\hat{\Theta}^s_{MK}|F\rangle = \sum_{\sigma=1}^{n(s)} |\sigma; sM\rangle\langle\sigma; sK|F\rangle \tag{26}$$

the sum runs then only over the model space configurations.

It is easy to show [16] that the operator (25) can be represented as

$$\hat{\Theta}^s_{MK} \equiv \hat{P}(IM;K)\hat{Q}(2T_z)\hat{Q}(A)\hat{p}(\pi) \tag{27}$$

Here

$$\hat{p}(\pi) \equiv \frac{1}{2}(1 + \pi\hat{\Pi}) \tag{28}$$

with $\hat{\Pi}$ being the parity operator, projects onto definite parity π.

$$\hat{Q}(A) \equiv \frac{1}{2\pi}\int_0^{2\pi} d\varphi \exp\{i\varphi(A - \hat{A})\} \tag{29}$$

with $\hat{A}$ being the nucleon number operator, restores the desired total nucleon number A [17], and

$$\hat{Q}(2T_z) \equiv \frac{1}{2\pi}\int_0^{2\pi} d\chi \exp\{i\chi(N - Z - \hat{N} + \hat{Z})\} \tag{30}$$

with $\hat{N}$ and $\hat{Z}$ being the neutron and proton number operators, respectively, selects the components with a definite total isospin-3-projection $2T_z = N - Z$. Finally

$$\hat{P}(IM;K) \equiv \frac{2I+1}{8\pi^2} \int^{(4\pi)} d\Omega D^{I*}_{MK}(\Omega)\hat{R}(\Omega) \tag{31}$$

with $\hat{R}(\Omega)$ being the usual rotation operator [18] and $D^I_{MK}(\Omega)$ denoting its representation in angular momentum eigenstates constructs from the $I_3 = K$ component in the symmetry breaking "intrinsic" frame of reference a configuration in the laboratory frame with total angular momentum I and 3-component $I_z = M$ [19].

As discussed in section 2.3, furthermore translational invariance should be ensured. This can be achieved by using instead of (25) the operator

$$\hat{\Theta}^{s0}_{MK} \equiv \hat{C}(0)\,\hat{\Theta}^{s}_{MK} \equiv \hat{\Theta}^{s}_{MK}\,\hat{C}(0) \tag{32}$$

with $\hat{C}(0)$ being the projector into the center of mass rest frame as given by eq.(11). Since it conserves total angular momentum and parity as well as the proton and neutron neutron number, this operator obviously commutes with the symmetry projector (25).

Via the K quantum number the configuration obtained by acting with the operator (32) on the HFB vacuum (23) does still depend on the orientation of the intrinsic quantisation axis. Now the wave function in the laboratory frame of reference should obviously not depend on our particular choice for this axis. This can be achieved by taking the linear combinations

$$\begin{aligned} |\phi^0_\rho; sM\rangle &\equiv \sum_{K=-I}^{+I} \hat{\Theta}^{s0}_{MK}|F\rangle f_{K;\rho} \\ &= \sum_{\sigma=1}^{n(s)} \hat{C}(0)|\sigma; sM\rangle \sum_{K=-I}^{+I} \langle\sigma; sK|F\rangle f_{K;\rho} \\ &= \sum_{\sigma=1}^{n(s)} \hat{C}(0)|\sigma; sM\rangle y_{\sigma;\rho}(F,f) \end{aligned} \tag{33}$$

as physical configurations and by determining the f's by variation (diagonalisation of the chosen Hamiltonian). Even if only a single determinant is considered, the restoration of the rotational symmetry thus introduces additional configuration mixing coefficients f, which together with the intrinsic degrees of freedom of the underlying HFB transformation will have to be determined by variational principles.

The symmetry projected configuration (33) looks formally identical to the complete translationally invariant SCM expansion (13). The matrix elements $\langle\sigma; sK|F\rangle$, however, project onto the subspace of linear independent SCM configurations being contained in the HFB vacuum (23). These matrix elements do obviously depend on the underlying HFB transformation F. Using (33) as a test wave function and minimizing the corresponding expectation value of the Hamiltonian with respect to arbitrary variations of the underlying transformation F as well as of the configuration mixing degrees of freedom f thus yields the energetically optimal linear combination of the SCM configurations (12) for the considered symmetry s, which can be reached from a single determinant of the form (23). This obviously is only true if the variation is performed after the restoration of the broken symmetries via the operator (32).

Now we have the basic ingrediences. What kind of meal we can cook from them will be presented in the following.

3. Variational Strategies

In the following we shall discuss four approaches using translationally invariant symmetry projected HFB type quasi particle determinants of the type (33) as basic building blocks : the VAMPIR (from **V**ariation **A**fter **M**ean field **P**rojection **I**n **R**ealistic model spaces)[1,16,20,21] and its extension the EXCITED VAMPIR[1,16,21,22] approach work with essentially only one single configuration of the type (33) for each state being considered. They can therefore still be considered as "mean field" approaches, though very sophisticated ones. Additional correlations are then taken into account in the FED (from **FE**w **D**eterminant) VAMPIR and its extension the EXCITED FED VAMPIR approach[23]. The latter methods use instead of only one always a few projected HFB determinants for each states and thus go beyond the mean field approximation. On the other hand "small amplitude vibrations" around the VAMPIR solutions can always be described within the MONSTER (**MO**del handling many **N**umber and **S**pin projected **T**wo quasi-particle **E**xcitations in **R**ealistic model spaces)[1] or within RPA-type[24] expansions.

3.1 VAMPIR and EXCITED VAMPIR

In the VAMPIR approach we take a single center of mass momentum and symmetry projected HFB vacuum of the form (33) as approximation to the energetically lowest ("yrast") state of a particular symmetry s. The corresponding energy-functional has the form

$$E_1^s[\,F_1^s, f^{s1}\,] = \frac{\langle \phi_1; sM|\hat{H}|\phi_1; sM\rangle}{\langle \phi_1; sM|\phi_1; sM\rangle} \tag{34}$$

Minimizing this functional with respect to arbitrary variations of the underlying HFB transformation F_1^s as well as of the configuration mixing degrees of freedom f^{s1} yields then the "optimal" representation of the yrast-state of a given symmetry by a single configuration of the type (33). Here the superscript s indicates the dependence of the HFB transformation and the configuration mixing on the considered symmetry, and the index "1" makes clear that the lowest state of this symmetry is meant. For simplicity the additional superscript "0" indicating the projection into the center of mass rest frame will be left out in the following.

At the solution we have

$$\sum_{K,K'=-I}^{+I} f_{K;1}^{s1\,*} \langle F_1^s|[\hat{H} - E_1^s]\hat{\Theta}_{KK'}^{s0} a_\alpha^\dagger(F_1^s) a_\beta^\dagger(F_1^s)|F_1^s\rangle f_{K';1}^{s1} \equiv 0 \tag{35}$$

for all $\alpha, \beta = 1, ..., M_b$. This is a kind of generalized "Brillouin theorem" which expresses the stability of the VAMPIR solution $|\phi_1; sM\rangle$ with respect to energetical admixtures of arbitrary symmetry projected two quasi particle excitations being based on the same HFB vacuum. Thus in many cases the VAMPIR approach will already yield a rather good approximation to the yrast-states of a nucleus. Note that each of these states is based on a different mean field $\{F_1^s, f^{s1}\}$, which is determined by an independent variational calculation. Thus, e.g., if the nucleus wants to be prolate at angular momentum 2^+, then prefers an oblate deformation at spin 4^+, and finally gets a cubic shape at angular momentum 6^+, the variational procedure will select the appropriate configurations automatically. In other words : the method is able to describe arbitrary drastic changes of the structure of the yrast-states with angular momentum.

However, what about the excited non-yrast states ? Assume for the moment that we would be interested only in such excited states of a particular symmetry s, whose structure is similar to that of the corresponding yrast-state (e.g., small amplitude vibrations). Then, obviously, it is a good idea to expand the wave functions for these states around the VAMPIR solution for the latter. If one takes for example besides the symmetry projected vacuum all the symmetry projected two quasi particle excitations with respect to the latter into account and diagonalizes the Hamiltonian in the resulting non-orthogonal configuration space, one obtains the MONSTER(VAMPIR) approach[1]. In a similar way also a kind of VAMPIR-RPA can be derived[24]. These approaches, however, will not be discussed in detail in the present lectures.

Instead we shall turn our attention to the description of "general" excited states with no restriction being made on their particular structure. Here now the VAMPIR has to become EXCITED. Let's assume that we are interested in the m energetically lowest states with a particular symmetry s. How we are going to describe them ? By the following procedure :

In a first step we shall just do a VAMPIR calculation. This yields the optimal one-determinant approximation

$$|\psi_1; sM\rangle \equiv |\phi_1; sM\rangle\beta_1^1 \equiv |\phi_1; sM\rangle \equiv \sum_{K=-I}^{+I} \hat{\Theta}_{MK}^{s0}|F_1^s\rangle f_{K;1}^{s1} \tag{36}$$

to the yrast-state with this symmetry. In the next step we want now to obtain the optimal one-determinant approximation for the first excited state with the same symmetry. Obviously we have to ensure that the resulting state is orthogonal to the VAMPIR solution (36). Therefore we use as test wave function

$$|\psi_2; sM\rangle \equiv |\phi_1; sM\rangle\beta_1^2 + |\phi_2; sM\rangle\beta_2^2 \tag{37}$$

with $|\phi_1; sM\rangle$ being kept fixed and

$$|\phi_2; sM\rangle \equiv \sum_{K=-I}^{+I} \hat{\Theta}_{MK}^{s0}|F_2^s\rangle f_{K;1}^{s2} \tag{38}$$

being based on a new mean field $\{F_2^s, f^{s2}\}$, which is then obtained by requiring a minimum for the expectation value of the Hamiltonian within the test wave function (37) with respect to arbitrary variations of the HFB-transformation F_2^s as well as of the configuration mixing f^{s2}. Note that the mixing-coefficients β_1^2 and β_2^2 are here not additional variational parameters but result from the requirement that the wave function (37) should be normalized and orthogonal with respect to the yrast-solution (36) (Gram-Schmidt orthogonalisation). After the solution (37) for the first excited state with the symmetry s has been calculated, in the third step now both the solutions (36) and (37) are eliminated from the variational space (again by Gram-Schmidt orthogonalisation), and the procedure is repeated for the second excited state, and so on up to finally the m-th solution

$$|\psi_m; sM\rangle \equiv \sum_{i=1}^{m} |\phi_i; sM\rangle \beta_i^m \tag{39}$$

has been constructed. These m solutions $|\psi_i; sM\rangle$ $(i = 1, ..., m)$ form a basis of orthonormal A-nucleon configurations with the symmetry s, in which finally the residual interaction

$$H_{ik} \equiv \langle\psi_i; sM|\hat{H}|\psi_k; sM\rangle \quad \text{for} \quad i, k = 1, ..., m \tag{40}$$

is diagonalized. This diagonalisation yields then the final EXCITED VAMPIR wave functions

$$|\varphi_\alpha; sM\rangle \equiv \sum_{i=1}^{m} |\psi_i; sM\rangle g_{i;\alpha}^s \tag{41}$$

Note, that each of the m configurations $|\psi_i; sM\rangle$ results from an independent variational calculation and thus, if the nucleus wants to be prolate in, e.g., the 2^+ yrast-state, prefers an oblate deformation in the first excited 2^+, and then goes cubic in the second excited 2^+, no problem : the variational procedure will reproduce this behaviour. In other words : now also arbitrary changes of the structure of the states having the same symmetry with increasing excitation energy become accessible. The method automatically selects those degrees of freedom, which are relevant for a particular excitation, no matter whether these are of single particle or of a more collective nature.

Furthermore, each of these m configurations is stable against energetical admixtures of the symmetry projected two quasi particle excitations being based on its particular mean field. Thus the residual interaction (40) in between these configurations is automatically minimized : the procedure creates successively an optimal basis of A-nucleon configurations with a particular symmetry.

Last but not least, for m sufficiently large the method converges to the exact SCM expansion and can hence be considered as a rather powerful truncation scheme.

However, as usual "*Grau, teurer Freund, ist alle Theorie...*" (Mephistopheles, in : Faust, part I, by J.W. von Goethe) : Unfortunately we have not been able up to now to apply the above procedure in "full beauty". Instead, up to the about three years ago, we had to introduce some rather severe in order to simplify the numerics. This was done by imposing certain symmetries for the underlying HFB transformations (20) :

(i) **Time-reversal invariance was required.**
(ii) **Axial symmetry was assumed.**
(iii) **Proton-neutron mixing was neglected.**
(iv) **Parity-mixing was not allowed for.**
(v) **Only real transformation matrices were admitted.**

Now, according to the Bloch-Messiah theorem [25], any HFB transformation of the form (20) can be decomposed into three successive transformations. Using the above symmetry restrictions the first of these gets here the form

$$b_\alpha^\dagger \equiv \sum_{i=1}^{M_b} \delta(\tau_i,\tau_\alpha)\delta((-)^{l_i},\pi_\alpha)\delta(m_i,m_\alpha)D_{i\alpha}c_i^\dagger \quad ; \quad m_\alpha > 0 \tag{42}$$

and

$$b_{\bar\alpha}^\dagger \equiv \sum_{i=1}^{M_b} \delta(\tau_i,\tau_\alpha)\delta((-)^{l_i},\pi_\alpha)\delta(m_i,m_\alpha)D_{i\alpha}c_{\bar i}^\dagger \quad ; \quad m_\alpha > 0 \tag{43}$$

Here $\bar\alpha$ and $\bar\imath$ denote the time-reversed partners of the orbits α and i, respectively, and D is a **real** (orthogonal) transformation matrix. The second Bloch-Messiah transformation is now of "BCS" type

$$\eta_\alpha^\dagger \equiv u_\alpha b_\alpha^\dagger - v_\alpha b_{\bar\alpha} \tag{44}$$

and

$$\eta_{\bar\alpha}^\dagger \equiv u_\alpha b_{\bar\alpha}^\dagger + v_\alpha b_\alpha \tag{45}$$

respectively, where the u_α and v_α are all real numbers with $u_\alpha \geq 0$ and $u_\alpha^2 + v_\alpha^2 = 1$ for all α. In terms of these quasi particle operators the HFB vacuum gets then the "canonical" form

$$|F\rangle = \prod_{\tau=p,n}\Big\{\prod_{\pi=+,-}\Big[\prod_{m=\frac{1}{2}}^{m_{max}}\Big(\prod_{\alpha}^{(\tau,\pi,m)}\big[u_\alpha + v_\alpha b_\alpha^\dagger b_{\bar\alpha}^\dagger\big]\Big)\Big]\Big\}|0\rangle \tag{46}$$

and thus it becomes obvious that the symmetry restricted vacuum (46) does contain only components with even proton and neutron numbers and positive parity. Furthermore (46) is time-reversal invariant and an eigenstate to the 3-component of the total angular momentum operator with eigenvalue $K = 0$. Its basic "building block" can be represented as

$$\begin{aligned} b_\alpha^\dagger b_{\bar\alpha}^\dagger = & \sum_{\mathbf{i}\leq\mathbf{k}}^{(m_\alpha\tau_\alpha\pi_\alpha)} [\,1+\delta(\mathbf{i},\mathbf{k})\,]^{-1}\sum_I (-)^{j_k+l_k-m_\alpha}(j_i j_k I|m_\alpha \; -m_\alpha \; 0) \\ & \cdot\Big\{ D_{i\alpha}D_{k\alpha}\,[\,1+(-)^I\,]\Big[\,c_{\mathbf{i}}^\dagger c_{\mathbf{k}}^\dagger\,\Big]_{12\tau_\alpha}^{I0}\Big\} \end{aligned} \tag{47}$$

where i denotes the triple of quantum numbers n_i, l_i, j_i of the basis state i and the creators of the coupled two-nucleon states are given by

$$\left[c_{\mathbf{i}}^{\dagger} c_{\mathbf{k}}^{\dagger}\right]_{TT_z}^{IM} \equiv \sum_{m_i m_k \tau_i \tau_k} (j_i j_k I | m_i m_k M)(1/2\, 1/2\, T | \tau_i \tau_k T_z)\, c_i^{\dagger} c_k^{\dagger} \tag{48}$$

The basic building block, and hence the total vacuum, too, thus only contains "like"-nucleon pairs coupled to even angular momenta and positive parity.

Using the above symmetry restrictions therefore only states with even angular momentum and positive parity in doubly even nuclei can be calculated. This is obviously a rather severe restriction (which we shall release in section 3.4). Furthermore, for simplicity we have up to now not projected into the center of mass rest frame. Thus the resulting wave functions are not translationally invariant. For the low-lying positive parity states being considered here this may not be very serious (since the first spurious excitation lies about $2\hbar\omega$ in excitation energy away), but it is obviously an additional approximation. We shall see, however, that we can go rather far even with this translationally non-invariant, symmetry restricted version of the above approaches. We shall demonstrate this by

3.3 Some examples

The first of these examples concerns VAMPIR calculations for the yrast-bands of the two nuclei ^{130}Ce and ^{128}Ba [26]. For these calculations the $2s_{1/2}$, $1d_{3/2}$, $1d_{5/2}$, $0g_{7/2}$, $1f_{7/2}$, $0h_{9/2}$, and $0h_{11/2}$ oscillator orbits for both protons and neutrons had been taken as model space. Slightly modified standard single particle energies had been used. Starting point for the effective interaction was a nuclear matter G-matrix derived from an early OBEP parametrisation of the free nucleon-nucleon interaction [27]. The pairing component of this force had been increased by adding two short-range (0.7 fm) Gaussians with strengths of -120 MeV and -50 MeV to the isospin T=1 relative matrix elements for the protons and neutrons, respectively. This renormalisation tends to make the nuclei more spherical. In order to compensate this effect all the T=0 matrixelements of the G-matrix were scaled by a factor of 1.20 . Hamiltonian and model space were exactly the same as in a rather successful systematic analysis of more than 50 nuclei in this mass region performed with the MONSTER(HFB) approach [28,1].

Fig.2 presents the excitation energies of the yrast-states in ^{130}Ce as function of the total angular momentum. Compared are the results of a HFB calculation with symmetry projection **after** the variation (i.e. projecting the whole band from a fixed intrinsic HFB solution of the form (46)), a MONSTER(HFB) calculation in which the Hamiltonian was diagonalized in the space of the projected HFB vacuum plus about 130 different symmetry projected two quasi particle excitations with respect to it, a VAMPIR calculation with symmetry projection **before** the variation as discussed above, and the experiment [29]. The insert displays a so-called "backbending plot", i.e. the dependence of the γ transition energies on the angular momentum. It is easily seen that the VAMPIR approach using only a single symmetry projected determinant for each member of the yrast-bands does equally well as the MONSTER(HFB) approach in which about 130 two quasi particle configurations were needed in order to reproduce the experimental energies. This is not only true for the energies but also for other observables like, e.g., the B(E2) values [26]. Thus here the VAMPIR does indeed provide a very good approximation to the yrast states. The agreement with the experimental data for ^{128}Ba is of similar quality and shall hence not be presented here.

Fig.2 : Yrast band in ^{130}Ce. From Ref.[26].

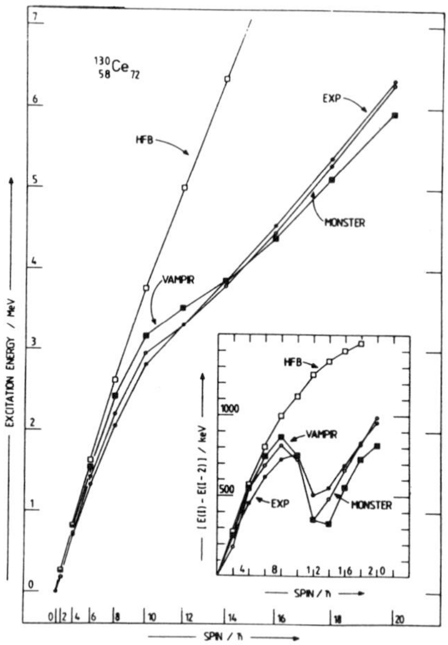

As much to the first example. As second example I would like to present some results of EXCITED VAMPIR calculations [30] performed for nuclei in the $A \sim 70$ mass region. These nuclei show a couple of interesting features like the occurence of many low-lying excited 0^+ states as well as a kind of "shape transition" from oblate to prolate deformation with increasing neutron number. Qualitatively this behaviour can be understood from the Nilsson level scheme : there are many gaps corresponding to stable configurations at various deformations for the nucleon numbers being of interest here. Thus one expects a kind of "shape coexistence", i.e. the occurence of, e.g., prolate and oblate deformed states with the same angular momentum at comparable excitation energies. This mass region should therefore be an ideal testing ground for the EXCITED VAMPIR approach, which, as we have discussed above, does not care whether the excited states with a particular symmetry have a similar or a totally different structure as the corresponding yrast-state.

Fig.3 : Even spin, positive parity bands in the nucleus ^{68}Ge. From Ref.[30].

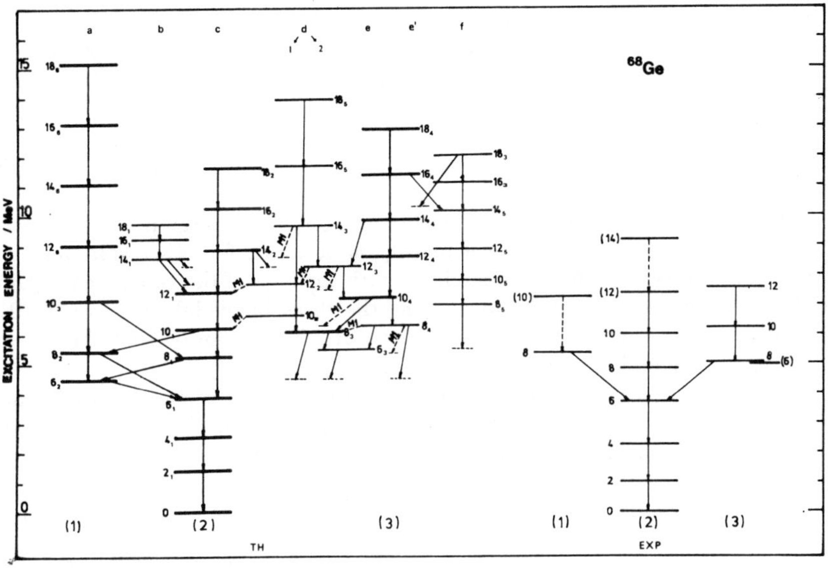

In the calculations [30] an even larger model space than in the $A \sim 130$ mass region had been used : the full N=3 and N=4 shells for both protons and neutrons and for the latter in addition the $0h_{11/2}$ orbital. Thus 112 single particle states above a ^{40}Ca core were available for the about 30 valence nucleons of the $A \sim 70$ nuclei. As effective interaction again a renormalized G-matrix derived from the same OBEP [27] as before had been taken. Again two short-range (0.7 fm) Gaussians here with strengths of -50 MeV and -40 MeV were added to the T=1 proton-proton and neutron-neutron relative matrix elements, respectively. The readjustment of the deformation, however, was now not made by scaling, but by a constant "monopole" shift of -115 KeV to all the T=0 matrix elements of the form $\langle 0g_{9/2}0f|\hat{G}|0g_{9/2}0f\rangle^{I}_{T=0}$. Furthermore a two-body spin-orbit force with Gaussian shape (range : 0.5 fm ; strenth : -1500 MeV) was added. This last renormalisation was motivated by our experiences in light nuclei, however, plays almost no role in the mass region considered here. The single particle energies had been adjusted in MONSTER(HFB)

calculations for "well deformed" odd mass nuclei in the $A \sim 50$ and $A \sim 80$ mass regions. The such defined Hamiltonian was then kept fixed and used for EXCITED VAMPIR calculations in several doubly even Ge and Se isotopes.

Indeed the approach is able to describe the shape coexistence phenomena very nicely. However, the situation seems to be even more complicated than described above : for low angular momenta configurations with different deformations are not only "coexisting" but do mix strongly with each other. This holds especially for the 0^+ states, however, though with decreasing importance, considerable mixing is obtained also in the 2^+, 4^+ and higher spin states. The shape coexistence persists even at rather high angular momenta, and extremely complicated decay patterns are obtained for the nuclei in this mass region. As an example I shall show you the spectrum of the nucleus ^{68}Ge as obtained with the EXCITED VAMPIR approach in Fig.3. Though in this figure for the sake of transparence most of the low spin states have been omitted, the remaining theoretical spectrum looks still rather complex. So, e.g., five 8^+ states are obtained within an excitation energy interval of less than 1.5 MeV and a similar high level density is obtained for other spin values, too. Analyzing the various decay possibilities with the help of the calculated B(E2) and B(M1) values one ends up by identifying the three bands labelled as (1), (2) and (3) with the experimental bands taken from Ref.[31]. As can be seen the agreement between theory and experiment is surprisingly good.

However, what about the negative parity states? What about odd angular momenta? What about odd and odd-odd systems? What if other nucleon-nucleon couplings, e.g., pairing correlations between protons and neutrons are important? It is clear that with the symmetry restricted version of the EXCITED VAMPIR approach these questions cannot be answered. Thus, here only some intermediate conclusions can be drawn :

(i) Even in its symmetry restricted version the EXCITED VAMPIR approach seems to be an excellent tool for nuclear structure studies. So, e.g., the very complicated spectra as they are encountered in the mass $A \sim 70$ can be described on a completely microscopic basis. To our knowledge no other approach can do this.

(ii) There are still a lot of open questions. Other correlations than those taken into account may be important. Thus the quantitative results discussed above may change. However, any further improvement will make the situation only more complex : additional states and more mixing are to be expected. The qualitative features of the results for the mass 70 nuclei like the strong mixing of configurations with different deformations and the other "shape coexistence" phenomena obtained with the present version of the approach will persist in any case.

Nevertheless, the strong limitations of the approach due to the symmetry restrictions on the underlying HFB-transformations cause an "uneasy feeling". The EXCITED VAMPIR approach in the above version is still a rather badly educated beast. Can we tame it? Can we educate it to behave more "predictable"? We shall try this by introducing

3.4 Complex Mean Fields

Do you remember our preliminary approximations out of section 3.2 ? Now we are going to skip all of them except the first two, as it has been proposed in Ref. [16]. Only

(i) **Time-reversal invariance** and
(ii) **Axial symmetry**

will still be required for the underlying HFB transformations. However, now parity- as well as proton-neutron mixing is taken into account and essentially complex mean fields will be admitted. Consequently the first Bloch-Messiah transformation gets then the form

$$b^\dagger_\alpha \equiv \sum_{i=1}^{M_b} \delta(m_i, m_\alpha) D^*_{i\alpha} c^\dagger_i \quad ; \quad m_\alpha > 0 \tag{49}$$

and

$$b^\dagger_{\bar\alpha} \equiv \sum_{i=1}^{M_b} \delta(m_i, m_\alpha) D_{i\alpha} c^\dagger_{\bar i} \quad ; \quad m_\alpha > 0 \tag{50}$$

instead of (42) and (43), respectively, and D is now an **essentially complex** (unitary) transformation matrix. The second Bloch-Messiah transformation keeps the same form as before, the corresponding "canonical" HFB vacuum, however, becomes more general than (46). We obtain

$$|F\rangle = \left\{ \prod_{m=\frac{1}{2}}^{m_{max}} \left(\prod_{\alpha}^{(m)} \left[u_\alpha + v_\alpha b^\dagger_\alpha b^\dagger_{\bar\alpha} \right] \right) \right\} |0\rangle \tag{51}$$

Furthermore, instead of the rather "poor" basic building block (47) we get now

$$\begin{aligned} b^\dagger_\alpha b^\dagger_{\bar\alpha} = & \sum_{\tau=p,n} \sum_{\mathbf{i}\le\mathbf{k}}^{(m_\alpha \tau)} [\,1+\delta(\mathbf{i},\mathbf{k})\,]^{-1} \sum_I (-)^{j_k+l_k-m_\alpha} (j_i j_k I | m_\alpha \; - m_\alpha \; 0) \\ & \cdot \Big\{ \Big[\, Re\,(D^*_{i_\tau\alpha} D_{k_\tau\alpha}) \, [\,1+(-)^{l_i+l_k+I}\,] \\ & + \; i\, Im\,(D^*_{i_\tau\alpha} D_{k_\tau\alpha}) \, [\,1-(-)^{l_i+l_k+I}\,] \Big] \Big[c^\dagger_{\mathbf{i}} c^\dagger_{\mathbf{k}} \Big]^{I0}_{12\tau} \Big\} \\ & + \sum_{\mathbf{i}}^{(m_\alpha p)} \sum_{\mathbf{k}}^{(m_\alpha n)} \sum_{IT} (1/2\;1/2\;T| -1/2\;1/2\;0)(-)^{j_k+l_k-m_\alpha} (j_i j_k I | m_\alpha \; - m_\alpha \; 0) \\ & \cdot \Big\{ \Big[\, Re\,(D^*_{i_p\alpha} D_{k_n\alpha}) \, [\,1+(-)^{l_i+l_k+I}\,] \\ & + \; i\, Im\,(D^*_{i_p\alpha} D_{k_n\alpha}) \, [\,1-(-)^{l_i+l_k+I}\,] \Big] \Big[c^\dagger_{\mathbf{i}} c^\dagger_{\mathbf{k}} \Big]^{I0}_{T0} \Big\} \end{aligned} \tag{52}$$

It is obvious that this basic building block, and hence the total vacuum (51), too, does now contain all possible two-nucleon-couplings. Thus for example pairing correlations between like nucleons are included as well as both the T=0 and the T=1 pairing correlations between protons and neutrons. It is furthermore evident that the use of essentially complex transformations is very important. As can be seen from (52) natural parity couplings come always with real expansion coefficients while the unnatural parity ones have purely imaginary amplitudes. Thus only if the latter are non-zero time-odd correlations are accounted for by the vacuum (51).

Using generalized vacua of the form (51) instead of the strongly symmetry restricted ones (46), now states with arbitrary parity and angular momentum in both doubly even and doubly odd systems can be described.

It should however be stressed that the vacuum (51) is still "symmetry restricted". So the assumption of time-reversal invariance introduces a two-fold degeneracy in the quasi particle spectrum and thus eliminates from the original vacuum (23) all components with an odd total number of nucleons. Thus odd systems cannot be described even with the extended version of the EXCITED VAMPIR approach discussed here. Furthermore, the assumption of axial symmetry, though having no effects on the two-nucleon couplings, eliminates particular four- and more-nucleon couplings from (23). This is due to the fact that two natural or unnatural parity nucleon pairs always yield a natural parity four-nucleon state while for unnatural parity four-nucleon states always a natural and an unnatural parity pair have to be coupled. On the other hand the couplings of two natural or unnatural parity pairs to an unnatural parity four-nucleon wave function are not accounted for via (51) and the same holds for the coupling of an unnatural and a natural parity pair to natural parity. Nuclear states which are dominated by such "missing couplings" can hence not be described even with the generalized version of the theory presented in Ref. [16].

Unfortunately no simple "geometrical interpretation" for the missing configurations is possible. In fact one can show that out of all possible shell-model configurations of, e.g., the $(0d_{5/2})^4$ - system just one $I^\pi = 3^+$ and one 5^+ state both with T=0 are missing, while in the $(0d_{5/2})^6$ - system one 4^+ and one 6^+ configuration (again both with T=0) are not accounted for. In more general basis systems the precise knowledge which states are in and which out of the range of the wave function (51) requires in each case the information about the "parentage" of such missing configurations and is hence not easy to achieve.

Nevertheless, it is obvious that the "complex" versions of the VAMPIR and EXCITED VAMPIR approaches (hereafter referred to as CVAMPIR and CEV, respectively) are a remarkable improvement with respect to their "real" brothers (RVAMPIR and REV). This will be demonstrated by

3.5 Some Applications

to 1s0d shell nuclei [21]. In order to be able to compare the results with those of exact SCM calculations in these applications only the 1s0d shell had been taken as model space. This has furthermore the advantage that the projection (11) into the center of mass rest frame is not necessary. The single particle energies have been taken here from experiment [32]. As effective interaction the mass-dependent version [33] of the Chung and Wildenthal force $\hat{V}(A) = \hat{V}(18) * (\frac{18}{A})^\alpha$, however, with $\alpha = 1/3$ instead of 0.3 has been used. This force had been adjusted by its authors to a few hundred experimental data in various 1s0d shell nuclei and is generally believed to be "the standard" interaction for this model space.

Fig.4 : Spectrum of ^{18}F. *From Ref.* [21].

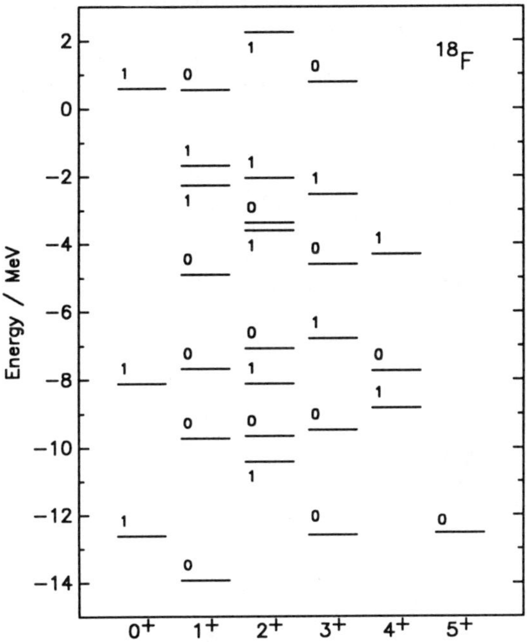

As a first example the spectrum of the nucleus ^{18}F is considered. Here the complete SCM-diagonalisation yields 28 different states the energies of which are displayed in Fig.4. The CEV approach reproduces all these states exactly. Each of them is represented by a single symmetry projected configuration of the type (39) with no residual interaction remaining. This is in fact not surprising, since, as it has been been discussed above, any HFB vacuum of the form (51) does account for all possible two-nucleon couplings. Because now the symmetry projection is performed before the variation, it becomes obvious that each state of the two-valence-nucleon system must be representable by a single projected quasi particle determinant, provided there are enough linear independent variables available. A simple counting of their number reveals that this is indeed the case, not only for small model spaces like the 1s0d - shell but for arbitrary finite basis systems independent from their actual size. Thus for two valence nucleons the CEV approach is always exact and furthermore yields "pure" configurations of the type (39) for each state.

Obviously, to apply the complex variational procedures discussed above to a simple case like two nucleons in the 1s0d - shell is a rather expensive method to diagonalize a couple of at most (5×5) - dimensional SCM matrices. The perfect reproduction of the exact results, however, is a highly non-trivial test of the computer program. Note, that being an odd-odd system the nucleus ^{18}F had not been describable at all in the older, more restricted "real" RVAMPIR- and REV-approaches.

As the next example we consider the ground state band (0^+, 2^+, 4^+, 6^+, and 8^+) of the nucleus ^{20}Ne. Fig.5 displays the absolute energy deviations of the different members of this band as obtained with various approximations from the exact SCM results. Forget about the other curves and compare only the results of the "real" RVAMPIR approach with those of the "complex" CVAMPIR calculation. While in the former approach deviations in between 158 KeV for the 8^+ and 570 KeV for the 0^+ state do remain, the CVAMPIR calculation reproduces the 0^+, 6^+, and 8^+ members of the SCM ground state band exactly, each of them by a single symmetry projected determinant of the form (36). For the 4+, and 6^+ states deviations of only 66 KeV and 98 KeV, respectively, are obtained being extremely small as compared to the total SCM ground state energy of -41.412 MeV.

Fig.5 : Ground band of ^{20}Ne. From Ref.[21].

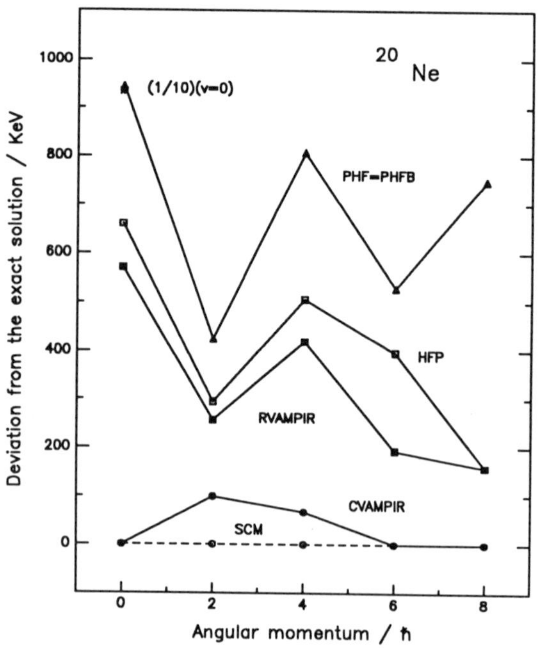

This result shows clearly the importance of the unnatural parity (mainly proton-neutron) pairing correlations in this nucleus, which had not been included in the "real" version of the approach.

Obviously, four nucleons in the 1s0d-shell have still "small" SCM configuration spaces (3) with maximum dimension of only the order of 10^2. Nevertheless, it is non-trivial that, at least for the ground state band, the whole SCM expansion of the wave function can be represented almost entirely by a single symmetry projected HFB vacuum for each state.

However, even the CEV description of 1s0d shell nuclei is far from being perfect. This holds especially if heavier nuclei in the middle of the shell are considered, where the shell-model configuration spaces have dimensions of the order of 10^4. So, e.g., if the 0^+, 2^+, and 4^+ members of the ^{28}Si ground state band are considered, the inclusion of unnatural parity pairing correlations in the CVAMPIR approach causes again remarkable improvements with respect to the more restricted RVAMPIR approximation, however, it leaves still discrepancies of 1116 KeV, 939 KeV, and 989 KeV for the three states, respectively. Though being again small with respect to the total SCM ground state energy of -137.495 MeV and though the resulting relative excitation energies are not bad at all, these discrepancies are obviously too large to be acceptable.

So how can we improve? Well, the problem is that in the EXCITED VAMPIR approach for each new state only one new symmetry projected configuration is added. Thus except for the diagonalisation of the residual interaction (40) this approach is still essentially a "mean field" approximation. One cannot expect that it is always possible to account for all important correlations via a single configuration. Thus if we want to improve our description we have to go

3.6 Beyond Mean Fields

as it has been proposed in a recent article[23]. The idea is quite simple. Instead of using only one additional configuration for each new state of a given symmetry s now a linear combination of several configurations is constructed.

Let's assume for the moment that we are interested in the yrast-state with a given symmetry s. Then as first step again a VAMPIR calculation is performed. This yields a solution

$$|\psi_1^{(1)}; sM\rangle \equiv \sum_{K=-I}^{+I} \hat{\Theta}_{MK}^{s0} |F_{11}^s\rangle f_{1K;1}^{s(1)} \tag{53}$$

In the second step then F_{11}^s is kept fixed and the expectation value of the Hamiltonian is minimized for the test wave function

$$|\psi_1^{(2)}; sM\rangle \equiv \sum_{K=-I}^{+I} \hat{\Theta}_{MK}^{s0} \left\{ |F_{11}^s\rangle f_{1K;1}^{s(2)} + |F_{21}^s\rangle f_{2K;1}^{s(2)} \right\} \tag{54}$$

with respect to arbitrary variations of the additional mean field F_{21}^s as well as of the configuration mixing $f_{iK;1}^{s(2)}$ $(i = 1, 2)$. Note that again the yrast-energy is minimized. The procedure selects therefore that particular determinant $|F_{21}^s\rangle$ which yields a maximal contribution to the VAMPIR solution (53). In the next step then the first two HFB transformations are kept fixed and a third determinant is added and so on up to finally a linear combination

$$|\psi_1^{(n)}; sM\rangle \equiv \sum_{K=-I}^{+I} \hat{\Theta}_{MK}^{s0} \sum_{i=1}^{n} |F_{i1}^{s}\rangle f_{iK;1}^{s(n)} \tag{55}$$

of n such configurations has been constructed. In each step of this procedure always that additional configuration is selected which has a maximal residual interaction with the previous approximation for the considered yrast state, no matter at which particular excitation energy this configuration may be. Furthermore, any additional configuration cannot perturb the current approximation more than the last one added previously. In other words : the method selects always the most important additional correlations. This approach we shall call the FED (from **FEw Determinant**) VAMPIR.

The extension of this approach to the description of the excited non-yrast states is straightforward : the yrast solution (53) is eliminated from the variational space by Schmidt-orthogonalisation and the above procedure is repeated for the first excited state, and so on, in complete analogy to the EXCITED VAMPIR procedure discussed in section 3.1. The only difference of this EXCITED FED VAMPIR approach with respect to the former procedure is that for each new state with the considered symmetry s instead of only one now always several configurations are constructed by successive variational calculations in the above described way. Finally, then again the (now even smaller) residual interaction between the lowest m states of the considered symmetry is diagonalized.

We shall demonstrate the power of the complex FED VAMPIR approach (shorthand notation CFV(n), with n being the number of configurations taken into account) as well as of its extension the complex EXCITED FED VAMPIR (CEFV) approach again first by

3.7 Some Examples

out of the 1s0d shell. These calculations have been done using exactly the same basis and Hamiltonian as those discussed in the last section.

As first example we consider the lowest three $I^\pi = 1^+$ states of the nucleus ^{20}Ne. The corresponding 0^+ ground state resulting from the complex CVAMPIR calculation [21] has already been discussed above. It was identical to the exact SCM result so that here a comparison of the excitation energies alone is sufficient. This is done in Fig.6. The extreme-left column (CEV) of this figure displays the excitation energies of the lowest three 1^+ states (having isospins T=1, T=0, and T=1, respectively) resulting from the complex EXCITED VAMPIR calculation with just three different configurations. In the next column (CEFV) then the EXCITED FED VAMPIR results are shown, which have been obtained here by creating successively three different configurations for the lowest and two more for both of the two higher excited states. The second-but-last column (SCM) presents the results of the complete SCM diagonalization. Finally, in the extreme-right column (EXP) the excitation energy of the experimentally known [34] strong isovector M1 state is displayed.

Fig.6 : The lowest three 1^+ *states in* ^{20}Ne*. From Ref.*[23].

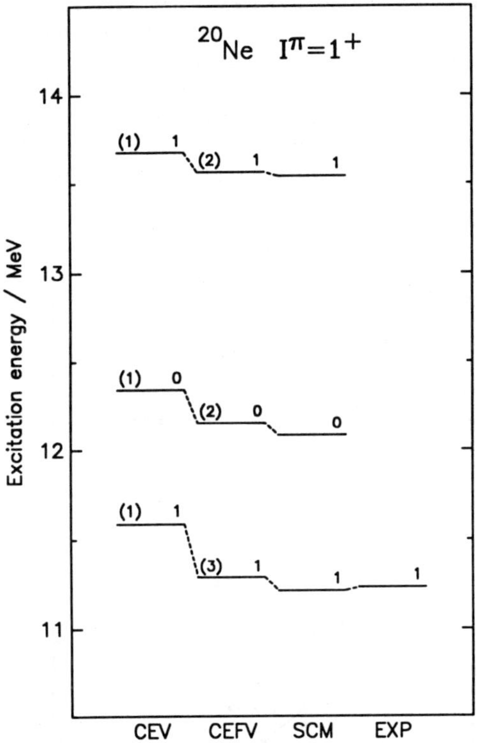

As can be seen here already the CEV description is rather good, although, if compared to the exact reproduction of the ground state, obviously not perfect. The energy differences with respect to the exact spectrum are 373, 259 and 132 KeV for the three states, respectively. The CEFV approach improves these results once more considerably. Here only energy differences of 75, 70 and 23 KeV do remain. Thus in this particular case the exact results can be reproduced with in fact very few optimized configurations. It is interesting to note here that the ratio of orbital to spin contributions to the total M1 transition amplitude describing the M1 transition from the ground to the first excited 1^+ state is 1.37 thus indicating the similarity of this excitation with the so-called "collective M1 states" observed in heavier nuclei [35]. The total CEFV B(M1) value is here $1.89\ \mu_n^2$ as compared to the experimental value of $1.99 \pm 0.38\ \mu_n^2$ [36]. As second example we want to consider again the ground band of the nucleus ^{28}Si. As discussed already above here the complex CVAMPIR calculations [21] left a discrepancies of 1116, 939, and 989 KeV with respect to the exact SCM results for the lowest three states of this band, respectively.

Fig.7 shows how the addition of further configurations in the complex FED VAMPIR approach does improve the agreement with the exact result for the ground state considerably. For $n = 7$ now only 337 KeV deviation from the latter do remain and it is obvious that even this discrepancy can still be decreased by increasing the number of configurations. Furthermore, the importance of the unnatural parity pairing correlations even in the ground state of this nucleus is obvious : the complex CVAMPIR needs only a single determinant to obtain the same absolute binding energy being reached by six configurations in the real FED VAMPIR approach and the final CFV solution is 737 KeV more bound than the corresponding RFV result.

Fig.7 : The ground state of ^{28}Si in various approximations. From Ref. [23].

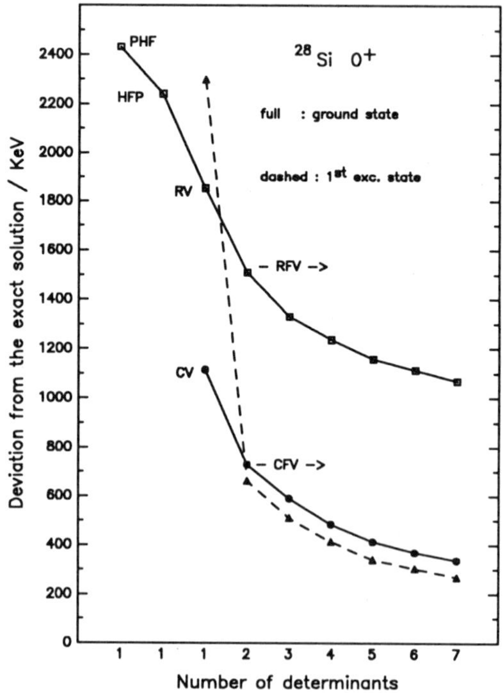

Similar improvements are obtained for the 2^+ and 4^+ members of the ground band, too. The results can be looked up in Ref. [23] and shall hence not be presented here.

Instead, we want to come once more back to the A~70 mass region. How do the results change there, if we go beyond the mean field approach? Fig. 8 shows the results of EXCITED FED VAMPIR calculations [37] for the nucleus ^{68}Ge using exactly the same force and single particle energies as for the EXCITED VAMPIR results presented in Fig.3. Note that here only *real* HFB transformations had been admitted. The experimental data are from Ref.[38]. Fortunately, as can be seen by comparison with Fig.3 the qualitative structure of this complicated spectrum does not change due to the additional correlations. Quantitatively, however, considerable differences can be seen. So, for example, the ground state turns out to be more correlated than the higher spin states and thus the 0^+ to 2^+ energy difference looks now much worse than in the more restricted calculations. This is a clear indication that the renormalizing components of the effective interaction, which had been adjusted only in the EXCITED VAMPIR approach, will have to be modified, in fact they should be smaller.

Fig.8 : Even spin positive parity bands in ^{68}Ge. From Ref.[37,38].

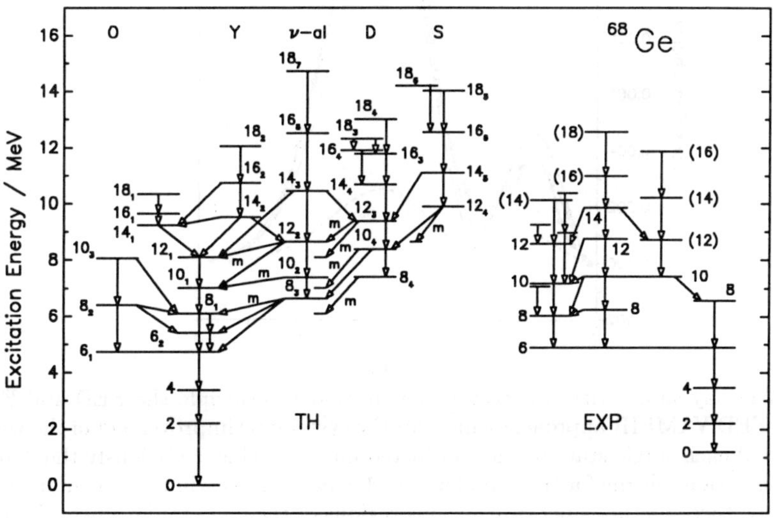

Such a slight modification has been made in Ref.[39], where charge transition densities for the lowest two 2^+ excitations in various Ge isotopes have been calculated using *real* EXCITED FED VAMPIR wave functions and the results have been compared to electron scattering data[40]. A typical example of these calculations is presented in Fig.9.

Calculations within this mass region using *essentially complex* mean fields are in progress and shall soon be published.

Fig.9 : Transition charge density to the yrast 2^+ in ^{70}Ge. From Ref.[39,40].

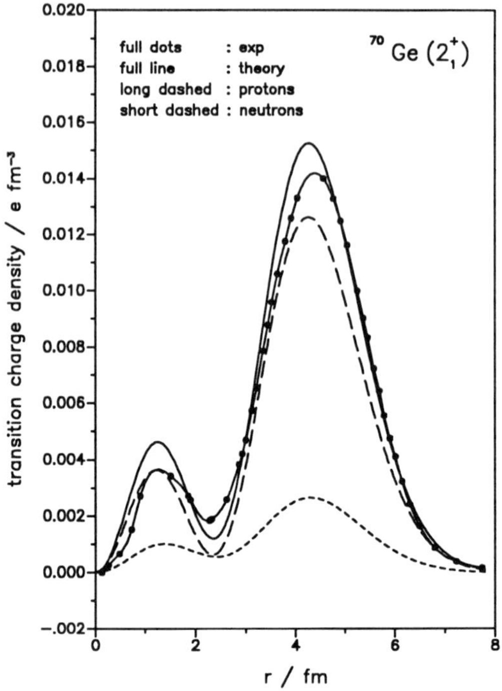

We may summerize this section saying that in principle the FED and EXCITED FED VAMPIR approaches allow for the systematic improvement of the wave function of each single state in the considered nucleus. This is obviously the "optimal" truncation scheme for the complete SCM expansion. It will be very interesting to see how these methods perform various regions of the mass table, where the latter are inaccessible. There also the effects of parity-mixing (octupole-deformations etc.) could be studied. This will be done in the very near future.

Obviously there remains still a lot to be done. Beasts like the members of the "VAMPIR family" have many faces and not only friendly ones. It's time for some

4. Conclusions

It is obvious that the methods discussed in the present lectures are rather powerful tools for microscopic nuclear structure physics. They leave the selection of the relevant configurations entirely to the dynamics of the considered system and determine them by a chain of variational calculations. In this way a unique possibility is provided to approximate the complete SCM-expansion of the nuclear wave functions with almost any desired accuracy. This holds especially for the most sophisticated of the above approaches, the EXCITED FED VAMPIR. Unlike the SCM model, however, the methods discussed here can still be applied in rather large model spaces. Thus for the first time really "large-scale" nuclear structure studies with one and the same model all over the nuclear mass table have become possible.

Should we thus be content ? Obviously not ! I have already mentioned several times that even the "complex" versions of the above approaches are still "symmetry restricted". Time-reversal invariance makes an application to odd nuclei impossible and axial symmetry causes particular "missing" couplings in the wave functions. Thus even in doubly even or doubly odd nuclei not all states are accessible. Furthermore, in larger basis systems, obviously the center of mass motion cannot be neglected any more as it has been done in the calculations discussed here.

These, however, are not "principal" shortcomings of the above approaches but only "technical" limitations. If the remaining two symmetry restrictions on the underlying HFB transformations would be released and in addition the projection into the center of mass rest frame would be performed none of these drawbacks would remain. Then indeed "really optimal" mean fields and correlations would be obtained.

Unfortunately, the numerical effort of such completely "unrestricted" calculations would be still orders of magnitude larger than that of the already rather sophisticated "complex" approaches presented here. At least at the moment we see no chance for such an extension. This, however, may change if the powerfull new computers with enormous parallel processing facilities will become available. Such computers should be particularly suited to handle multi-dimensional integrations as they are required by the methods presented here. Thus there is some hope that even the remaining problems can be overcome "some day".

5. Acknowledgements

I would like to thank at this place my collaborators. Frank Grümmer from Bochum first place, but also the others who contributed to various aspects presented in these lectures : Esko Hammarén from Helsinki, Alexandra Petrovici from Bucharest, Zheng Ren-Rong from Chonqing, Mauro Kyotoku from Joâo Pessoa, Paul-Gerhard Reinhard from Erlangen, Luis Egido from Madrid, and last but not least Burkhard Fladt and Amand Faessler from Tübigen.

6. References

1. K. W. Schmid and F. Grümmer, *Rep. Progr. Phys.* **50** (1987) 731.
2. see, e.g., H. Machleidt, *Adv. Nucl. Phys.* **19** (1989) 189.
3. see, e.g., B. D. Serot and J. D. Walecka, *Adv. Nucl. Phys.* **16** (1986) 1.

4. see, e.g., J. B. McGrory and B. H. Wildenthal, *Annu. Rev. Nucl. Part. Sci.* **30** (1980) 383.
5. J. P. Elliot and T. H. R. Skyrme, *Proc. Roy. Soc.* **24** (1955) 561.
6. see, e.g., P. Ring and P. Schuck, *The Nuclear Many-Body Problem*, Springer, New York/Heidelberg/Berlin, 1980.
7. K. W. Schmid and F. Grümmer, *Z. Phys.* **A336** (1990) 5.
8. K. W. Schmid and F. Grümmer, *Z. Phys.* **A337** (1990) 267.
9. K. W. Schmid and P.-G. Reinhard, *Nucl. Phys.* **A530** (1991) 283.
10. K. W. Schmid and L. Egido, work in progress.
11. K. W. Schmid and R. Smith, *Nucl. Phys.* **A381** (1982) 195.
12. P. W. M. Glaudemans, in *Nuclear Structure at High Spin, Excitation and Momentum transfer*, (H. Nann, Ed.) AIP, New York 1985, p.316.
13. D. R. Hartree, *Proc. Cambr. Philos. Soc.* **24** (1928) 89; A. Fock, *Z. Phys.* **61** (1930) 126.
14. N. N. Bogoliubov, *Zk. Eksp. Teor. Fiz.* **34** (1958) 58; N. N. Bogoliubov, *Usp. Fiz. Nauk.* **67** (1959) 541; N. N. Bogoliubov and V. G. Soloviev, *Dokl. Akad. Nauk.* **124** (1959) 1011; J. G. Valatin, *Phys. Rev.* **122** (1961) 1012.
15. H. J. Mang, *Phys. Rep.* **18** (1975) 325.
16. K. W. Schmid, F. Grümmer and Amand Faessler, *Ann. Phys.(NY)* **180** (1987) 1.
17. B. F. Bayman, *Nucl.Phys.* **15** (1960) 33.
18. A. R. Edmonds, *Angular Momentum in Quantum Mechanics*, Princeton Univ. Press, Princeton, 1957.
19. R. E. Peierls, and I. Yoccoz, *Proc. Phys. Soc.*, **A70** (1957) 381; F. Villars, in *Varenna Lectures*, (C. Bloch, Ed.), Academic Press, New York, 1966, p. 1 .
20. K. W. Schmid, F. Grümmer and Amand Faessler, *Nucl. Phys.* **A431** (1984) 205.
21. Zheng Ren-Rong, K. W. Schmid, F. Grümmer, and Amand Faessler, *Nucl. Phys.* **A494** (1989) 214.
22. K. W. Schmid, F. Grümmer, M. Kyotoku and Amand Faessler, *Nucl. Phys.* **A452** (1986) 493.
23. K. W. Schmid, Zheng Ren-Rong, F. Grümmer, and Amand Faessler, *Nucl. Phys.* **A499** (1989) 63.
24. K. W. Schmid, M. Kyotoku, F. Grümmer and Amand Faessler, *Ann. Phys.(NY)* **190** (1989) 182; K. W. Schmid, M. Kyotoku, F. Grümmer and Amand Faessler, *Phys. Rev.* **C41** (1990) 284.
25. C. Bloch and A. Messiah, *Nucl.Phys.* **39** (1962) 95; B. Zumino, *J. Math. Phys.* **3** (1962) 1055.
26. K. W. Schmid, F. Grümmer, E. Hammarén and Amand Faessler, *Nucl. Phys.* **A436** (1985) 417;

K. W. Schmid, F. Grümmer, E. Hammarén, M. Kyotoku and Amand Faessler, *Nucl. Phys.* **A456** (1986) 437.
27. K. Holinde, K. Erkelenz and R. Alzetta, *Nucl. Phys.* **A194** (1972) 161.
28. E. Hammarén, K. W. Schmid, F. Grümmer, Amand Faessler and B. Fladt, *Nucl. Phys.* **A437** (1985) 1; *ibid* **A454** (1986) 301.
29. D. M. Todd, R. Aryaeinejad, D. J. G. Love, A. H. Nelson, P. J. Nolan, P. J. Smith and P. J. Twin, *J.Phys. G : Nucl. Phys.* **10** (1984) 1407.
30. A. Petrovici, K. W. Schmid, F. Grümmer, Amand Faessler and T. Horibata, *Nucl. Phys.* **A483** (1988) 317;
A. Petrovici, K. W. Schmid, F. Grümmer and Amand Faessler, *Nucl. Phys.* **A504** (1989) 277.
31. A. P. de Lima, A. V. Ramayya, J. H. Hamilton, B. van Nooijen, R. M. Ronningen, H. Kawakami, R. B. Piercey, E. de Lima, R. L. Robinson, H. J. Kim, L. K. Peker, F. A. Rickey, R. Popli, A. J. Caffrey and J. C. Wells, *Phys. Rev.* **C23** (1981) 213.
32. F. Ajzenberg-Selove, *Nucl. Phys.* **A460** (1986) 1.
33. B. H. Wildenthal, *Progr. Part. Nucl. Phys.* **11** (1983) 5.
34. F. Ajzenberg-Selove, *Nucl. Phys.* **A392** (1983) 1.
35. A. Richter, in *Proc. Int. Conf. Nucl. Phys.*, Florence, Aug. 29 - Sep. 3, 1983, Edizione Compositori, Florence, Italy .
36. W. L. Bendel, L. W. Fagg, S. K. Numrich, E. C. Jones Jr. and H. F. Kaiser, *Phys. Rev.* **C3** (1971) 1821.
37. A. Petrovici, K. W. Schmid, F. Grümmer and A.Faessler *Nucl. Phys.* **A517** (1990) 108.
38. L. Chaturvedi, X. Zhao,, A. V. Ramayya, J. H. Hamilton, J. Kormicki, S. Zhu, C. Girit, H. Xie, W.-B. Gao, J.-R. Jiang, A. Petrovici, K. W. Schmid, A. Faessler, N. R. Johnson, C. Bactash, I. Y. Lee, F. K. McGowan, M. L. Halbert, M. A. Riley, J. H. McNeill, M. O. Kortelahti, J. D. Cole, R. B. Piercey and H. Q. Jin *Phys. Rev* **C43** (1991) 2541.
39. A. Petrovici, K. W. Schmid, F. Grümmer and A.Faessler *Z. Phys.* **A339** (1991) 71.
40. D. Goutte, PhD.-thesis, Orsay 1984, Nr. d'ordre : ORSAY 2803.

VARIATIONAL APPROACH TO THE NUCLEAR SHELL MODEL

Victor A. Madsen
Physics Department, Oregon State University, Physics, Weniger 301 Corvallis OR 97331, USA

Zdzisław Szymański
Institute for Theoretical Physics, University of Warsaw, Hoża 69 00-681 Warsaw, Poland

and

E. C. Halbert
Oak Ridge National Laboratory, Oak Ridge, Tennessee 37831, USA

ABSTRACT

This is the report of working group 8 on the Hartree-Fock-Bogoliubov (HFB) nuclear structure technique used at Tübingen and presented by Professor K.W. Schmid. This version of HFB is characterized by projections onto spaces of conserved quantities, such as angular momentum, before variation. This report presents a summary of the method as understood by the group and a list of questions from both the group and the audience, along with Schmid's answers.

1. The Tübingen Approach

Professor Schmid began his lecture with a convincing demonstration that, beyond the sd shell, it is impossible to do shell-model calculations by simply diagonalizing the assumed one- plus two-body Hamiltonian operator in a "full" shell-model space when there are several active particles, even if only one major shell is considered. His demonstration gives strong motivation to find ways of limiting the otherwise astronomically large space of the multiparticle basis states. There are several different procedures for doing this, some of which, the boson expansion method, the fermion dynamic symmetry model, and the symplectic model, were presented during this workshop.

The Tübingen approach is based on the Hartree-Fock-Bogoliubov (HFB) theory, which is the most general mean field theory involving only nucleons. It takes into account both Hartree-Fock (HF) and pairing correlations. Thus the Tübingen method is based on the same theory as the self-consistent HF plus BCS and the HFB methods presented in this workshop by P.H.-Heenen and by J.F. Berger. In all these approaches one applies the variational principle to let the dynamics of the problem determine the single-fermion basis. The differences between the procedure presented by Schmid and those presented by Heenen and by Berger lie in the calculational techniques.

In the Tübingen approach, the single-fermion HFB states are expanded in terms of spherical shell-model basis states, and the generalized many-fermion Slater determinants are subjected to as many kinds of projections onto good-quantum-number spaces as feasible before variation to minimize the expectation value of the assumed effective one- plus two-body Hamiltonian. Projections are included to give good neutron and proton numbers, angular momentum and, in a few cases in light nuclei, even the center-of-mass momentum. Different versions of this variational approach are labeled under the family names VAMPIR and MONSTER. All of the versions involve sophisticated numerical computations.

At the basis of all the variational strategies is VAMPIR, which does minimization after projection onto good particle number, parity, and angular momentum. For each such set of good quantum numbers the VAMPIR method yields a ground-state nuclear energy. EXCITED VAMPIR includes successive calculations of the first, second, etc., excited states, each obtained by imposing the constraint of orthogonality to previous (lower) solutions of the same quantum numbers before redoing the variation. Both methods, VAMPIR and EXCITED VAMPIR, start by restricting the ground state and each excited state to the projection of a single generalized Slater determinant. In EXCITED VAMPIR the resulting ground and excited state vectors are then used together as a basis for diagonalizing the residual interaction. FED VAMPIR and FED EXCITED VAMPIR are analogous to VAMPIR and EXCITED VAMPIR but allow linear combinations of a few Slater determinants (subjected to projection before variation) rather than a single determinant. In the MONSTER technique, one diagonalizes the Hamiltonian including in the basis all symmetry-projected two quasiparticle states based on the mean field of a given VAMPIR solution.

2. Advantages

(1) For a given basis, projection before variation has to yield a lower ground state energy than projection after variation. This must necessarily be more accurate in the sense of giving better agreement with results of exact diagonalization in the full shell-model space.

(2) Projection before variation has numerical advantages in the search procedures.

(3) The Tübingen approach can handle large shell-model bases that are not possible to treat by straightforward diagonalization in the shell model space.

(4) The calculations are always done in the laboratory frame (thus avoiding the introduction of a cranking-model approximation).

(5) These methods use an effective two-nucleon force (for example the Bonn OPEP plus Gaussian terms) determined in part by fitting data. The method takes

advantage of the best theoretical information we have about the force between two free nucleons.

(6) Comparison with full sd calculations in ^{20}Ne show good accuracy.

3. Disadvantages

(1) By always using a spherical basis, the Tübingen approach may sacrifice a level of accuracy which is possible with a deformed basis, for nuclei which turn out to be deformed.

(2) The single-particle bases that can be handled are still so limited that effective charges are needed.

(3) The one-boson interaction used probably cannot give as good agreement with experimentally observed nuclear structure data as the Skyrme or Gogny forces used by Heenen and by Berger, which are fitted to properties of nuclei.

4. Questions and Answers

This section reports on Professor Schmid's answers to questions put to him about the Tübingen approach.

(1) The first question was about the importance of prior projection. This was answered very emphatically by Schmid: he cited calculations in ^{28}Si in which the energies of 0^+ states from the prior-projected calculations were quite different from those deduced by constrained Hartree-Fock without projection. Projection before variation is absolutely crucial to the technique.

(2) The second question had to do with the use of complex coefficients in the basis. These resulted in large improvements in accuracy in the comparisons against shell-model results. They are needed if the variational procedure includes pairing of neutrons and protons.

(3) Schmid was asked about the use of the symplectic model or some other procedure for limiting the shell-model basis before variation. His response gives insight into the philosophy of the Tübingen approach. He does not wish to introduce into the variational calculation any biases other than restriction to selected major shells. His goal is to find the low-lying eigenstates of the assumed Hamiltonian in the chosen spherical shell-model space. He is willing to sacrifice close agreement with the results of experimental observations, if necessary, in order to work with

such a well-defined nuclear model and calculate its consequences accurately.

5. Concluding Remarks

The Tübingen approach entails sophisticated numerical calculations. It requires large computer programs and a great deal of attention to the efficiency of computational techniques, because of the enormousness of the calculations needed. It is an earnest approach to the nuclear many-body problem, an approach in which one takes seriously the model of a realistic one- plus two-body Hamiltonian acting in a chosen spherical shell-model space, and then tries to push that model to its practical limit in the sense of solving accurately for properties of its low-lying nuclear states.

6. Acknowledgments

This report is based on comments and suggestions from all members of working group 8, consisting of Chairul Bahri, Eugene Marshalek, and Luis Robledo, in addition to the three authors. We are grateful to Professor Schmid for his help in answering questions put to him by individual members of the group.

SHELL-MODEL CALCULATIONS WITH REALISTIC EFFECTIVE INTERACTIONS

E. OSNES, M. HJORTH-JENSEN, T. ENGELAND and A. HOLT
Department of Physics, University of Oslo
P.O. Box 1048 Blindern, N-0316 Oslo, Norway

ABSTRACT

We review recent attempts to calculate the shell-model effective interaction, starting from the basic nucleon-nucleon interaction. The resulting interaction is applied to explorative shell-model calculations of nuclei with two and several valence nucleons in the oxygen and calcium regions. Perspectives for large-scale shell-model calculations with realistic effective interactions in heavier nuclei are discussed.

1. Introduction

One of the fundamental, yet unresolved problems of nuclear theory is to describe the properties of complex nuclei in terms of the constituent nucleons and the interactions among them. Firstly, this represents a quantal many-body problem which can only be approximately solved. Secondly, the basic nucleon-nucleon interaction is not well known. Thus, one does not know whether an eventual failure to solve this problem should be ascribed to the many-body method or the interaction model used.

Ideally, one should derive the free nucleon-nucelon interaction from the interaction among quarks. Although attempted, this program has not been quantatively successful. Thus, one would be content to start from a nucleon-nucleon interaction derived from meson-exchange models and reproducing the two-nucleon data. Examples of such potential models are the Paris, Bonn-Jülich and Bonn potentials. Once the basic nucleon-nucleon interaction has been established, it should be employed in a quantal many-body approximation to the actual nuclear structure problem of interest. Here, we shall be concerned with the nuclear shell model which has provided a successful microscopic approach to nuclear structure.

Thus, in view of the approximative nature of the shell model, we will be faced with the problem of deriving an effective Hamiltonian for the shell model. In Section 2 we will briefly review the basic elements of effective interaction theory and apply it to prototype nuclei in the oxygen and calcium regions with two valence nucleons. Then, in Section 3 we will present results of shell-model calculations for nuclei in the oxygen and calcium regions with several valence nucleons using the most recent effective interactions. Conclusions are given in Section 4 together with a discussion of the perspectives for applying realistic effective interactions to extensive shell-model calculations in heavier nuclei.

2. Calculation of the Effective Interaction

2.1. Theoretical Framework

Consider the nuclear structure problem for ^{18}O. This nucleus may be regarded as the prototype nucleus for effective interaction calculations, since its shell-model description consists of eight protons and eight neutrons in closed $0s$ and $0p$ shells and two valence neutrons in the $1s0d$ shell. Thus, instead of solving the eigenvalue problem for 18 nucleons

$$H\Psi_i(1,2,...,A) = E_i\Psi_i(1,2,...,A) \tag{1}$$

with $A = 18$, we have to solve the two-nucleon problem

$$H_{eff}\Phi_m = E_m\Phi_m, \tag{2}$$

where H_{eff} is the effective Hamiltonian and Φ_m the shell-model wave function for two valence neutrons in the $1s0d$ shell. More formally, one defines a projection operator

$$P = \sum_{i=1}^{d} |\psi_i\rangle \langle\psi_i| \tag{3}$$

which projects onto the chosen shell-model space (or P-space), ψ_i being the basis functions spanning this space and d being the size of the space. We also define the complement of P, $Q = 1 - P$, which then projects out of the shell-model space. The effective Hamiltonian acting in the P-space is then defined by the requirement

$$PH_{eff}P\Psi_m = E_mP\Psi_m, \tag{4}$$

where E_m is the m-th eigenvalue and Ψ_m the m-th eigenstate of the original Hamiltonian H of Eq. (1). Thus, the eigenfunctions of H_{eff} are projections of the true eigenfunctions of H onto the P-space.

For ^{18}O, the basis functions ψ_i of the P-space will generally have eight protons and eight neutrons in closed $0s$ and $0p$ shells and two neutrons in the valence $1s0d$ shell. It is possible to eliminate the closed-shell core from Eq. (4), and thus we obtain Eq. (2) for the valence nucleons only. The resulting H_{eff} can be written as

$$H_{eff} = H_0 + V_{eff}, \tag{5}$$

where H_0 is the unperturbed, one-body Hamiltonian and V_{eff} is the so-called effective interaction which for two valence nucleons consists of a one-body and a two-body part

$$V_{eff} = V_{eff}(1) + V_{eff}(2). \tag{6}$$

It is customary to add the one-body parts of the effective Hamiltonian, writing

$$H_{eff} = \tilde{H}_0 + V_{eff}(2), \tag{7}$$

with $\tilde{H}_0 = H_0 + V_{eff}(1)$. This allows us, as in the shell model, to replace the eigenvalues of $\tilde{H}_0$ by the empirical single-particle energies for the valence nucleons, e.g. for the

$0d_{5/2}$, $1s_{1/2}$ and $0d_{3/2}$ valence-neutron orbitals in the case of ^{18}O. Thus, the remaining quantity to calculate is the two-body effective interaction $V_{eff}(2)$. The procedure for this is well established and will not be repeated here [1]. The resulting $V_{eff}(2)$ can be expressed in terms of all valence-linked, irreducible and connected two-body Feynman-Goldstone diagrams.

We shall express the effective interaction in the $\hat{Q}$-box formulation of Kuo and coworkers [2]:

$$V_{eff} = \hat{Q} - \hat{Q}' \int \hat{Q} + \hat{Q}' \int \hat{Q} \int \hat{Q} - \dots \tag{8}$$

Here, $\hat{Q}$ is the so-called $\hat{Q}$-box, which represents the (infinite) sum of all the non-folded, valence-linked and irreducible two-body diagrams. Further, $\int$ represents the folding operator (note that factorization is obtained by virtue of generalized time ordering). Note that the lowest order diagram to the left of the first folding operation must be at least second order in the interaction, hence $\hat{Q}'$ contains all the diagrams of $\hat{Q}$ which are at least second order in the interaction. We shall refer to Eq. (8) as the folded-diagram expansion of the effective interaction.

By using a degenerate model-space, we can express V_{eff} of Eq. (8) in terms of energy derivatives of $\hat{Q}$. Defining

$$\hat{Q}_m = \frac{1}{m!} \frac{d^m \hat{Q}}{d\omega^m}\bigg|_{\omega=E_0}, \tag{9}$$

where the derivatives are taken at the (unperturbed) starting energy $\omega = E_0$, we can write V_{eff} as

$$V_{eff} = \sum_{m=0}^{\infty} \hat{Q}_m V_{eff}^m = \sum_{k=0}^{\infty} F_k. \tag{10}$$

Here, F_k denotes the term containing k number of foldings. We have

$$\begin{aligned} F_0 &= \hat{Q} \\ F_1 &= \hat{Q}_1\hat{Q} \\ F_2 &= \hat{Q}_2\hat{Q}\hat{Q} + \hat{Q}_1\hat{Q}_1\hat{Q} \\ F_3 &= \hat{Q}_3\hat{Q}\hat{Q}\hat{Q} + \hat{Q}_2\hat{Q}_1\hat{Q}\hat{Q} + \hat{Q}_2\hat{Q}\hat{Q}_1\hat{Q} + \hat{Q}_1\hat{Q}_2\hat{Q}\hat{Q} + \hat{Q}_1\hat{Q}_1\hat{Q}_1\hat{Q} \end{aligned} \tag{11}$$

and so forth.

Examples of diagrams contained in the $\hat{Q}$-box are shown in Fig. 1. Here, the dashed vertices represent the free nucleon-nucleon interaction which contains a strong short-range repulsion. Thus, each of the diagrams $A1, A2, A3, \dots$ and $B2, \dots, B6, \dots$ is very large and we cannot evaluate them individually. However, by summing the infinite sequence of ladder diagrams $A1 + A2 + A3 + \dots$, we obtain the Brueckner G-matrix shown by diagram A and defined by the integral equation

$$G(\omega) = V + VQ_2 \frac{1}{\omega - Q_2TQ_2} Q_2 G(\omega), \tag{12}$$

where V is the free nucleon-nucleon interaction, Q_2 a projection operator projecting out of the two-nucleon subspace for which the G-matrix is calculated, ω the starting energy for the two nucleons and T the kinetic energy operator. Although the individual diagrams $A1, A2, A3, ...$ are all very large, diagram A for the G-matrix is finite, typically of the order of 5 MeV. Similarly, we can sum the infinite series of diagrams $B2 + ... + B6 + ...$, obtaining diagram B, which corresponds to replacing the V-vertices of diagram $B2$ by G-vertices. By continuing this procedure, we can rewrite the entire diagram expansion for the $\hat{Q}$-box in terms of G-matrix vertices. This is easily done by replacing the V-vertices of the original diagrams by G-vertices, leaving out diagrams having particle-particle intermediate states to avoid double counting. We remark in passing that modern G-matrix calculations tend to leave out the low-lying particle-particle intermediate states (the so-called double-partition scheme). These then have to be included in the $\hat{Q}$-box expansion in terms of G.

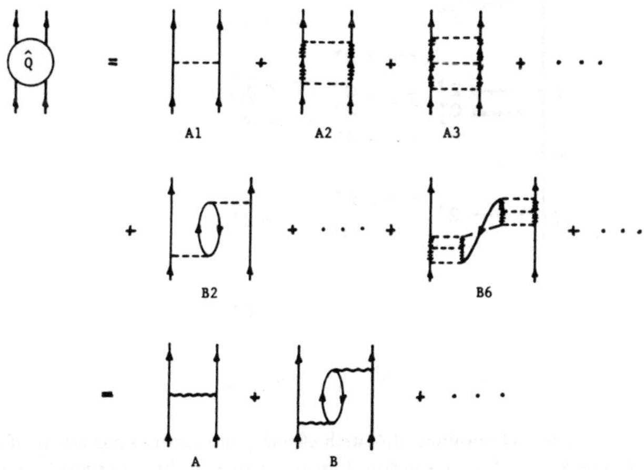

Figure 1. The $\hat{Q}$-box expressed in terms of diagrams with V-vertices (dashed lines) and G-vertices (wavy lines). The railed lines denote particle states excluded from the two-nucleon subspace for which the G-matrix is calculated. See the text for further discussion of the expansion.

2.2. Early Calculations

In their pioneering calculation [3,4] of the effective interaction Kuo and Brown found that an effective interaction including diagrams A and B of Fig. 1 gave a reasonable description of the energy spectrum of ^{18}O. This is shown in Fig. 2. Whereas diagram A representing the G-matrix gave too little binding energy and too compressed spectrum, the correction to second order in G represented by diagram B served to increase the ground-state binding energy and open up the spectrum. The latter diagram (often denoted by G_{3p-1h} since it has three particles and one hole in its intermediate states) accounts for the polarization of the closed-shell core by the valence nucleons. This is decribed by a particle-hole excitation (represented in the diagram by a "bubble") being exchanged between the valence-nucleon lines. Detailed calculations show that the main contribution to this diagram comes from isoscalar quadrupole particle-hole intermediate states, thus giving rise to an effective quadrupole interaction. Such a force is of fairly long range and is, as known from phenomenological shell-model calculations, needed in addition to a short-range, δ-function type of force in order to obtain a reasonable description of nuclear spectra. In fact, the G-matrix interaction is of rather short range and thus the Kuo-Brown effective interaction was roughly similar to a δ-function plus quadrupole interaction.

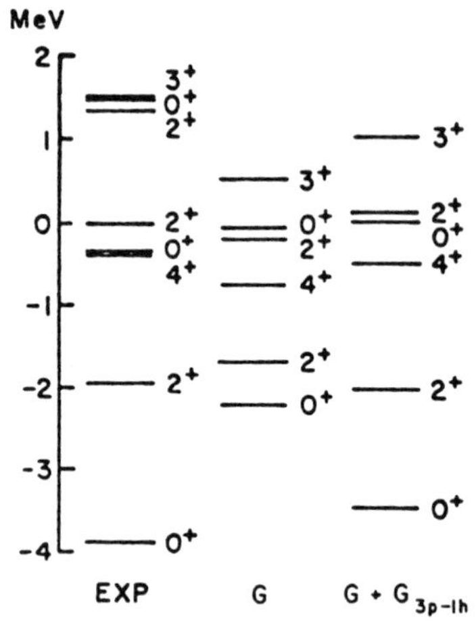

Figure 2. Energy spectra of ^{18}O obtained [3,4] with effective interactions consisting of diagram A (G) and of diagrams A plus B ($G+G_{3p-1h}$) of Fig. 1, as compared to the experimental spectrum. (Note that the observed second 0^+-state is a core-deformed intruder state and should thus be excluded from the comparison.)

2.3. Subsequent Developments

The pioneering work of Kuo and Brown was followed by several extensive studies of the effective interaction. In view of the importance of the second-order corrections, it was of interest to examine the higher order contributions. One line of development was to go to the next higher order, thus including all third-order diagrams in $\hat{Q}$ (non-folded) and in $\hat{Q}_1\hat{Q}$ (folded). In fact, Barrett and Kirson [5] found that many of the third-order contributions were as large as and of opposite sign to the second-order contributions, indicating that the effective interaction expansion did not converge smoothly when summed order by order in G. Similar conclusions were reached by Goode and Koltun [6] who studied the average contributions from selcted fourth-order topologies.

A different line of study focussed on the evaluation of the core-polarization process to arbitrary order. Summing the particle-hole bubbles to infinite order, as shown in Fig. 3A, corresponds to evaluating the core excitations in the Tamm-Dancoff approximation. Including in addition backward-going bubbles, as in Fig. 3B, one obtains the core excitations in the random-phase approximation. The latter gives rather collective core vibrations, thus enhancing the core-polarization contribution to the effective interaction. However, by including corrections to the particle-hole vertices and to the vertices coupling the particle-hole excitations to the valence-nucleons, as shown in Fig. 3C, one recovered essentially the second-order core-polarization result, thus providing justification for the Kuo-Brown approximation. It is worth noting that the diagrams of Fig. 3 can be included self-consistently to all orders by a multiple scattering approach [7]. In the effective interaction thus obtained, the short-range correlations induced by the strong nucleon-nucleon interaction are incorporated via the reaction matrix G, while the important long-range correlations are accounted for by the diagrams of Fig. 3.

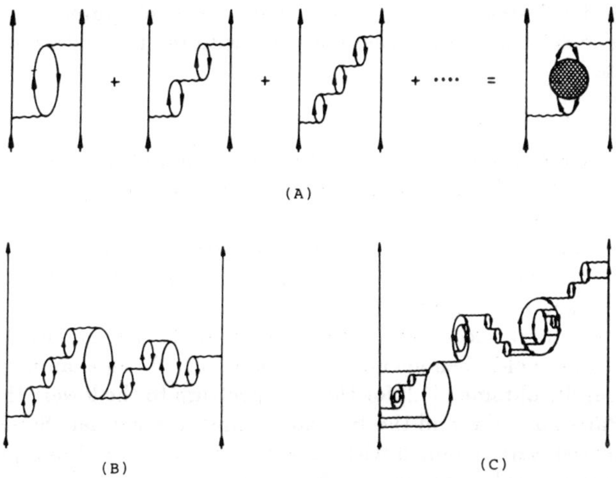

Figure 3. Diagrams contributing to the core-polarization process in the Tamm-Dancoff (A), random-phase ($A + B$) and multiple scattering ($A + B + C$) approximations.

On the other hand, a serious problem was discovered by examining in more detail the second-order core-polarization contribution. While Kuo and Brown truncated the summation over intermediate particle-hole states at $2\hbar\omega$ harmonic oscillator excitation energy, Vary, Sauer and Wong [8] included intermediate states up to $22\hbar\omega$ excitation energy. Thus, they found that the contribution from $2\hbar\omega$ intermediate states was cancelled by the contribution from the higher lying intermediate states, thus essentially wiping out the entire core-polarization correction. The large contribution from high intermediate states was attributed to the strong tensor-force component contained in the G-matrix interaction. This presumably reflected a similarly strong tensor-force component of the free nucleon-nucleon interaction

V. In a one-boson exchange model of the nucleon-nucleon potential the important tensor-force contributions come from exchanges of π- and ρ-mesons. These contributions have opposite signs, with the π-meson contribution dominating. Thus, the strength of the tensor force depends on the degree of cancellation between the π- and ρ-meson contributions, which again depends on the relative strengths of the πN (N for nucleon) and ρN coupling constants. The early calculations were based on phenomenological NN potentials with small effective ρN coupling constants [9]. In modern meson-exchange potentials, such as the Paris, Bonn-Jülich and Bonn potentials, a larger ρN coupling, close to the empirical value proposed by Höhler and Pietarinen [10], has been employed, with a resulting weaker tensor-force component. Indeed, using the Bonn-Jülich potential, Sommermann et al. [11] found that while the $2\hbar\omega$ contribution to the second-order core-polarization diagram was typically -1 MeV, the contribution from intermediate states above $2\hbar\omega$ was $+0.1$ MeV, thus justifying truncation of the intermediate-state summation at $2\hbar\omega$ excitation. This is gratifying, as it facilitates the calculation of higher order diagrams. (It would indeed be prohibitive to calculate third-order diagrams with intermediate-state excitations up to, say, $22\hbar\omega$.)

2.4. Recent Developments

The more recent developments in the calculation of the effective interaction may be said to follow the arrival of modern meson-exchange potentials with relatively weak tensor-force components, thus allowing truncation of intermediate-state summations at low excitation energy. This has made it possible to embark on reliable higher order calculations which, on the other hand, have been considerably facilitated by improved computers.

To low (i.e. first, second and third) order in G the new potentials gave rather similar results to the old phenomenological potentials. As an example, we compare in Fig. 4 the results obtained [12] for the ^{18}O spectrum to the lowest three orders in G using the Paris potential and the Hamada-Johnston potential. Similar results to the Paris potential were obtained with various versions of the Bonn potential.

Then, we proceed to incorporate folded diagrams by the $\hat{Q}$-box method. This can be done either by the folded-diagram expansion shown in Eqs. (10) and (11) or by using the iterative method proposed by Lee and Suzuki [13]. In the latter, the effective interaction is given by

$$H_{eff} = H_0 + \lim_{n\to\infty} R_n, \tag{13}$$

with

$$R_n = \left[1 - \hat{Q}_1 - \sum_{m=2}^{n} \hat{Q}_m \prod_{k=n-m+1}^{n-1} R_k\right]^{-1} \hat{Q}. \tag{14}$$

The structural difference between the folded-diagram expansion and the Lee-Suzuki method can be seen already after the first iteration, namely

$$F_1 = \hat{Q}_1 \hat{Q} \tag{15}$$

and

$$R_1 = \frac{1}{1 - \hat{Q}_1}\hat{Q}. \tag{16}$$

By expanding the denominator in Eq. (16), we see that R_1 contains $\hat{Q}_1$ to arbitrary order, whereas F_1 of Eq. (15) is only first order in $\hat{Q}_1$. In fact, recent calculations [14] have shown that for a given $\hat{Q}$-box the Lee-Suzuki method converges much more quickly than the folded-diagram expansion. The Lee-Suzuki expansion has been found to converge even when intruder states are present. In this case the Lee-Suzuki procedure yields the energy of those states which have their energy closest to the unperturbed energy. In contrast, the folded-diagram expansion will give those states which have the largest overlap with the unperturbed states.

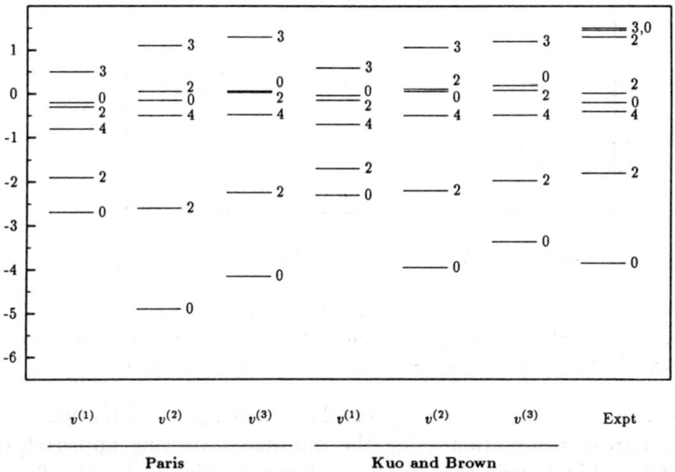

Figure 4. The low-lying $T = 1$ mass-18 spectrum for effective interactions calculated to the lowest three orders in the G-matrix using the Paris and Hamada-Johnston potentials [12]. All energies are in MeV.

When evaluating the effective interaction by the $\hat{Q}$-box method, caution must be exercised to retain only the two-body connected diagrams, as the one-body diagrams are already included by using empirical single-particle energies. The procedure for doing this is discussed in some detail in refs. [14,15].

The remaining problem in a calculation of the effective interaction using either the folded-diagram or the Lee-Suzuki method, is what to include in the $\hat{Q}$-box. In Fig. 5 we show the results obtained for the $T = 1$ mass-18 spectrum using $\hat{Q}$-boxes calculated to second and third order in G. The folded-diagram summation is performed using the Lee-Suzuki method (LS) and the results are compared with those obtained in perturbation theory (PT) to second and third order. It is seen that LS(2) is much closer to LS(3) than PT(2) is to PT(3). This may indicate that the results are less sensitive to the approximation used to evaluate the $\hat{Q}$-box (as long as it is reasonable) than to a proper inclusion of the folded diagrams. Similar results were obtained for the mass-42 system, and we refer to the literature for details [16].

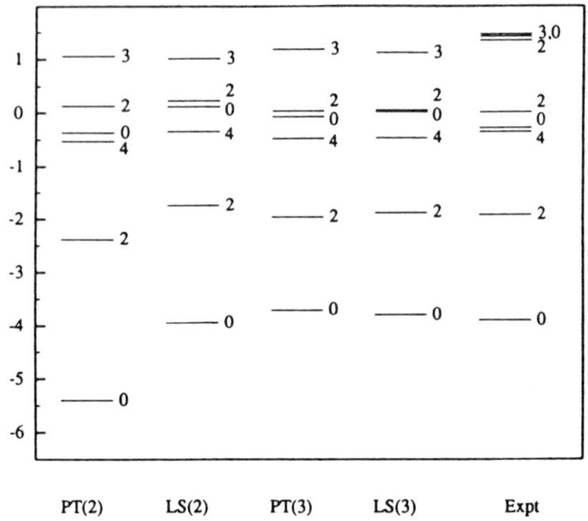

Figure 5. Comparison of $T = 1$ mass-18 spectra evaluated in perturbation theory (PT) to second and third order and in the Lee-Suzuki (LS) approach using $\hat{Q}$-boxes evaluated to second and third order [14]. The Bonn A potential was used, and the energies are given in MeV.

We conclude this section by briefly reviewing a calculation of the $\hat{Q}$-box by infinite partial summation using the multiple-scattering approach disussed in Subsect. 2.3 and including the diagrams shown in Fig. 3. In Fig. 6 we show the spectra obtained by Chakravarti et al. [17] for the $T = 1$ mass-18 system in various approximations for the $\hat{Q}$-box using the Paris and Bonn-Jülich potentials. Consider the Bonn-Jülich case. As usual the bare G-matrix gives too little ground-state binding energy and too compressed a spectrum. The ground-state binding energy is strongly increased and the spectrum opened up when including the second-order core-polarization term [3p-1h]. Then, by summing all diagrams of the type shown in Fig. 3C self-consistently to all orders, one essentially recovers the second-order result, provided $2\hbar\omega$ particle-particle ladders be included and spurious center-of-mass motion corrected for. In the next step an approximate Hartree-Fock basis is used, leading to considerable loss of binding energy for the lowest states, especially for the ground state. Finally, by explicitly including the low-lying core-deformed 4p-2h states known from the phenomenological shell-model calculation of Ellis and Engeland [18], one gets the low-lying states back in place. The agreement with experiment is indeed very impressive, but one has to keep in mind that folded diagrams are not included. (Note that it is justified to compare with experiment only when the 4p-2h intruder states are included in the calculation; thus in the previous figures where only 2p states are considered, the experimental levels are given only for orientation.)

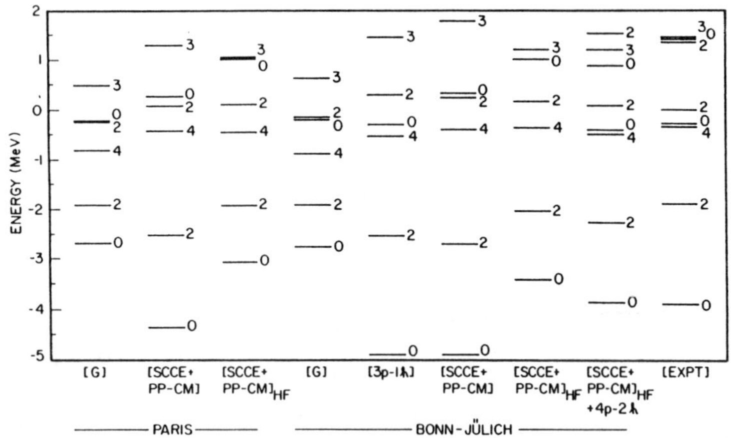

Figure 6. The $T = 1$ mass-18 spectra for various approximations of the effective interaction [17]. See the text for further explanation.

3. Shell-Model Calculations for Several Valence Nucleons

Having examined various approximations for the effective interaction in the previous section by applying them to prototype two-valence-nucleon systems, we are now prepared to expose them to further tests by carrying out shell-model calculations for systems of several valence nucleons. Ideally, many-body effective forces ought to be included for such systems. Here, we shall leave out such forces, under the assumption that they be negligible. This may be a reasonable approximation as far as the $\hat{Q}$-box is concerned. However, calculations seem to indicate that the effect of folded many-body diagrams can be sizeable.

We are going to consider the effective interaction obtained by applying the Lee-Suzuki method to a $\hat{Q}$-box evaluated to third order, using the Bonn A potential. This interaction will be applied to several nuclei in the *sd* and *pf* shells. In the *sd* shell we will also apply the $\hat{Q}$-box interaction obtained by Chakravarti et al. [17] using the self-consistent multiple-scattering approach discussed above.

3.1. The 1s0d Shell

In Fig. 7 are shown the ^{20}Ne and ^{22}Ne spectra obtained [14] using a third-order $\hat{Q}$-box and the Lee-Suzuki folding procedure. The agreement with experiment is astounding, in particular when folded diagrams are included. With non-folded diagrams only, the spectra are reasonable, although the ground-state binding energy is strongly overestimated. The contribution from folded diagrams is repulsive, bringing the ground-state binding energies in agreement with experiment. The excited levels are less affected, and thus the folding also results in a more compressed spectrum, thus improving the agreement with experiment.

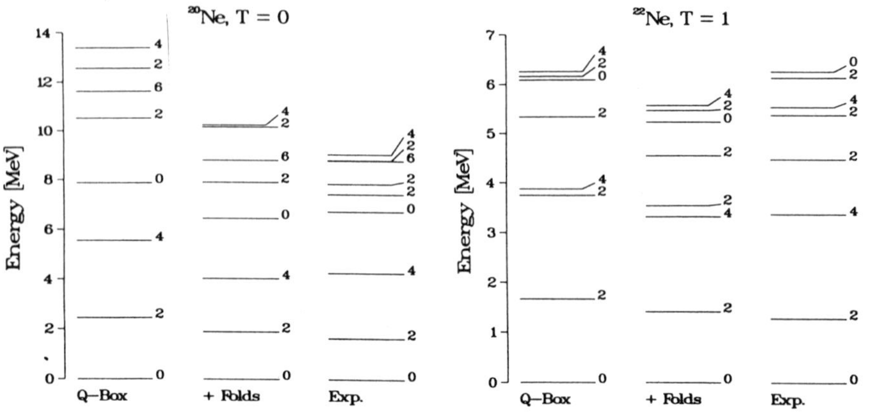

Figure 7. Energy spectra for ^{20}Ne and ^{22}Ne calculated with a $\hat{Q}$-box interaction including all terms up to third order in G and with a fully folded interaction obtained from the $\hat{Q}$-box using the Lee-Suzuki folding procedure [14].

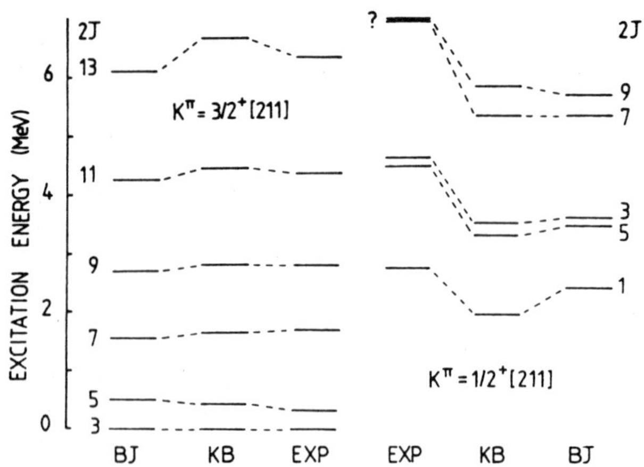

Figure 8. The positive-parity spectrum of ^{21}Ne obtained [19,20] with the interaction (BJ) calculated by Chakravarti et al. [17] including all the diagrams of Fig. 3 to arbitrary order and with the interaction (KB) of Kuo and Brown [3,4].

In Fig. 8 we show the ^{21}Ne spectrum obtained [19,20] with the $\hat{Q}$-box interaction of Chakravarti et al. [17]. It is gratifying that a realistic shell-model calculation is capable of reproducing the rotational level structure observed for this nucleus. The $3/2^+[211]$ ground band is in good agreement with experiment. The excited $1/2^+[211]$ band comes, however, too high, indicating insufficient Coriolis mixing as compared to experiment. This may also be seen from the electromagnetic interband transitions.

3.2. The 1p0f Shell

In the pf shell we have evaluated [16] the spectra for nuclei with four valence nucleons. The $T = 2$, $T = 1$ and $T = 0$ spectra of the four-valence-nucleon system correspond to the low-lying spectra of ^{44}Ca, ^{44}Sc and ^{44}Ti, respectively. We use an interaction similar to the one used in Fig. 7, i.e. being calculated from the Bonn A potential employing a $\hat{Q}$-box to third order with the folded diagrams included by the Lee-Suzuki method. The spectra are shown in Figs. 9-11. Here, "Deg" and "Non-deg" indicate use of degenerate and non-degenerate single-particle unperturbed energies. In the latter case approximate, although not self-consistent Hartree-Fock energies were used. The results are seen to be reasonable for both ^{44}Ca and ^{44}Ti, especially when keeping in mind that the second 0^+, 2^+ and 4^+ states presumably are intruder states. The ^{44}Sc spectrum is not so well reproduced. In particular, the lowest 1^+ state is too strongly bound, reflecting the problem of obtaining appropriate $JT = 10$ two-particle matrix elements. This problem is more apparent still in ^{42}Sc.

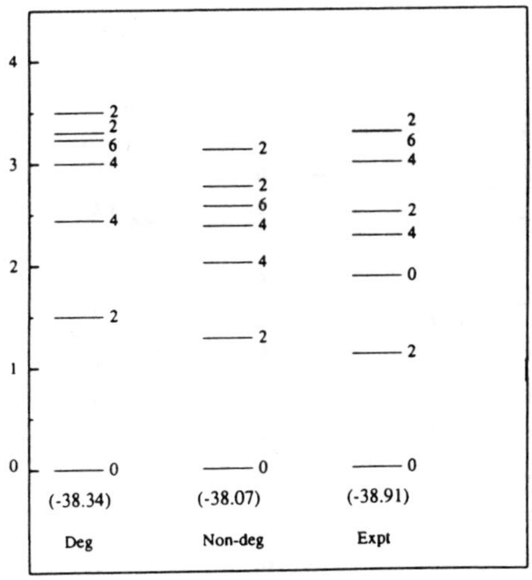

Figure 9. The low-lying spectrum of ^{44}Ca calculated with the folded interaction of Fig. 7 [16]. The ground-state binding energies with respect to ^{40}Ca are shown in brackets. "Deg" and "Non-deg" refer to use of degenerate and non-degenerate unperturbed single-particle energies. All energies are in MeV.

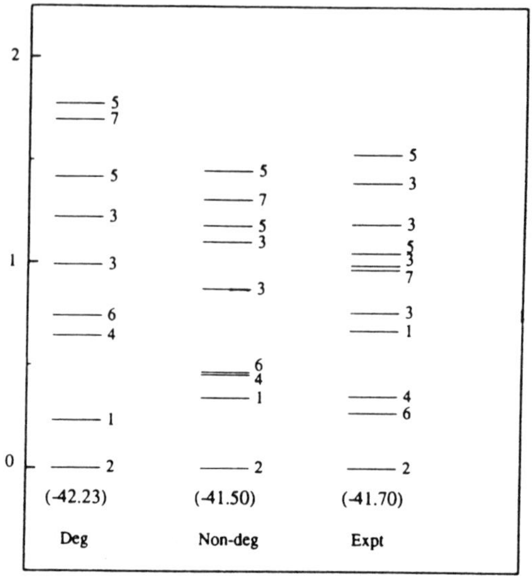

Figure 10. The low-lying spectrum of ^{44}Sc [16]. See legend of Fig. 9.

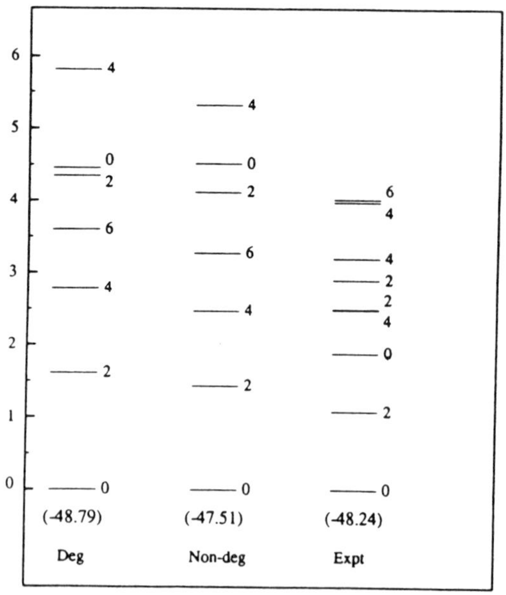

Figure 11. The low-lying spectrum of ^{44}Ti [16]. See legend of Fig. 9.

4. Summary and Perspectives

We have argued that in calculating the shell-model effective interaction from the free nucleon-nucleon interaction, it is essential to start from meson-exchange potentials containing a strong ρN coupling. This will give rise to a weak tensor-force component in the nucleon-nucleon interaction, thus allowing truncation at low-energy intermediate states in evaluating the higher order contributions to the effective interaction. This in turn facilitates the calculation of the effective interaction to second and higher order in the reaction matrix G.

For the calculation of the effective interaction we have used the $\hat{Q}$-box formalism of Kuo and coworkers and have included in $\hat{Q}$ all diagrams up to third order in G. The folded diagrams have been generated from this $\hat{Q}$ using the iterative folding procedure of Lee and Suzuki. Further, we have seen that it is possible to include in $\hat{Q}$ all important long-range correlations self-consistently to all orders. For the latter, however, no folding has yet been done.

The effective interactions were evaluated from the Paris, Bonn-Jülich and Bonn meson-exchange potentials. The resulting interactions reproduce nicely the $T = 1$ two-valence-nucleon spectra in both the sd and pf shells. The $T = 0$ effective interaction is, however, less successful and has to be further studied.

It is worth noting that although the old phenomenological nucleon-nucleon potentials contained strong tensor-force components invalidating truncation at $2\hbar\omega$ excitation energy in the intermediate-state summations, they yielded results very similar to the modern meson-exchange potentials up to $2\hbar\omega$ intermediate-state summation. Thus, the old Kuo-Brown effective interaction based on the Hamada-Johnston potential may still be used in shell-model calulations.

We have also performed exploratory shell-model calculations on systems with several valence nucleons and obtained good results for those spectra which are not strongly sensitive to the $T = 0$ two-body interaction.

In view of the results of these exploratory calculations, it is appropriate to ask what the perspectives are for more extensive shell-model calculations using modern realistic effective interactions. It is probably futile to expect that such calculations one day will supersede calculations based on fits to the observed levels. In fact, a slight discrepancy in the two-body interaction may be magnified in a system of many valence nucleons. Further, the role of effective three-body forces should be clarified before embarking on extensive shell-model calculations of nuclei with a large number of valence nucleons. It is good reason to suspect that effective many-body forces be responsible for the A-dependence of empirical sd-shell matrix elements obtained by Brown and Wildenthal and coworkers [21].

Even if one were able to calculate the shell-model effective interaction accurately starting from "first principles", it is doubtful if it would be economical to perform actual large-scale shell-model calculations using such interactions. One would probably still prefer the more simple phenomenological approach. However, a successful calculation of the effective interaction from "first principles" would provide a justification of the phenomenological approach.

In the future, there will certainly be need for extensive shell-model calculations in order to obtain e.g. reliable beta-decay rates in nuclei which are present in pre-supernova matter and double beta-decay rates as a test of the standard model. It is important that such calculations be firmly based on a proper understanding both of how two nucleons interact and how this interaction is modified by the presence of other nucleons.

References

1. P. J. Ellis and E. Osnes, *Rev. Mod. Phys.* **49** (1977) 777 and references therein.
2. T. T. S. Kuo and E. Osnes, *Folded-Diagram Theory of the Effective Interaction in Nuclei, Atoms and Molecules,* Springer Lecture Notes in Physics, Vol. **364** (Springer, Berlin, 1990) and references therein.
3. T. T. S. Kuo and G. E. Brown, *Nucl. Phys.* **85** (1966) 40.
4. T. T. S. Kuo, *Nucl. Phys.* **A103** (1967) 71.
5. B. R. Barrett and M. W. Kirson, *Nucl. Phys.* **A148** (1970) 145.
6. P. Goode and D. S. Koltun, *Nucl. Phys.* **A243** (1975) 44.
7. F. L. Goodin and P. J. Ellis, *Nucl. Phys.* **A334** (1980) 229.
8. J. P. Vary, P. U. Sauer and C. W. Wong, *Phys. Rev.* **C7** (1973) 1776.
9. M. R. Anastasio and G. E. Brown, *Nucl. Phys.* **A285** (1977) 516.
10. G. Höhler and E. Pietarinen, *Nucl. Phys.* **B139** (1975) 210.
11. H. M. Sommermann, H. Müther, K. C. Tam, T. T. S. Kuo and A. Faessler, *Phys. Rev.* **C23** (1981) 1765.
12. M. Hjorth-Jensen and E. Osnes, *Phys. Lett.* **228B** (1989) 281.
13. S. Y. Lee and K. Suzuki, *Prog. Theor. Phys.* **64** (1980) 2091.
14. M. Hjorth-Jensen, E. Osnes and H. Müther, *Ann. of Phys.* **213** (1992) 102.
15. J. Shurpin, T. T. S. Kuo and D. Strottman, *Nucl. Phys.* **A408** (1983) 310.
16. M. Hjorth-Jensen, T. Engeland, A. Holt and E. Osnes, *Nucl. Phys.* **A541** (1992) 105.
17. S. Chakravarti, P. J. Ellis, T. T. S. Kuo and E. Osnes, *Phys. Lett.* **109B** (1982) 141.
18. P. J. Ellis and T. Engeland, *Nucl. Phys.* **A144** (1970) 161; **A181** (1972) 368.
19. F. Brut, E. Osnes and D. Strottman, in *Proc. Int. Nucl. Phys. Conf., Harrogate, 1986,* Vol. **1** (Institute of Physics, Bristol, 1986) p. 77.
20. E. Osnes, in *Windsurfing the Fermi Sea, Proc. Int. Conf. and Symp. on Unified Concepts of Many-Body Problems, Stony Brook, 1986,* eds. T. T. S. Kuo and J. Speth, Vol. **2** (North-Holland, Amsterdam, 1987) p. 226.
21. B. A. Brown, W. A. Richter, R. E. Julies and B. H. Wildenthal, *Ann. of Phys.* **182** (1988) 192.

GROUP 9

Boson Expansion Methods

RECONSTRUCTING BROKEN SYMMETRIES FROM DEFORMED BOSON EXPANSIONS

E. R. MARSHALEK
Department of Physics, University of Notre Dame
Notre Dame, IN 46556

ABSTRACT

After a brief introduction to the boson mapping approach and an overview of various types of applications, the remainder of the presentation concentrates on the perturbative application of boson expansions to systems with broken symmetries. The discussion includes the case of two-dimensional rotation in a toy model and the extraction of low-spin rotational band structure in three dimensions.

1. Introduction

Many-body theorists have long known that under suitable circumstances pairs of fermions can behave like bosons, especially when the density of such pairs is low enough that they do not get in each other's way, thereby minimizing the effects of the Pauli principle. Under these circumstances, the low-lying excitations of the system may exhibit properties resembling those of a low-dimensional system of harmonic oscillators with weak anharmonic couplings. However, even in situations in which the interactions between the collective oscillators is strong and the spectrum exhibits rotational properties, the introduction of a boson description can be technically and conceptually advantageous as will be shown later.

The boson methods are intended to exploit this kind of situation by replacing the degrees of freedom of fermion pairs *directly* with *exact* boson degrees of freedom, an idea first introduced into nuclear physics by Belyaev and Zelevinsky.[1] Contact with the actual fermion problem is maintained by performing a mapping from the original many-fermion Hilbert space into another Hilbert space called the *ideal space*.[2] In the case of a system with an even number of fermions, the ideal space is generated by boson creation and destruction operators, with one boson degree of freedom corresponding to each fermion pair of interest. For the case of an odd number of fermions, there are two similar approaches. In one, the ideal space is taken as the tensor product of the boson space for the even system with the usual fermion Fock space, subject, however, to the stipulation that states with more than one fermion are unphysical, since pairs of fermions are prempted by bosons.[2] In the second approach, the odd fermion is replaced by a *quasifermion* whose algebraic properties guarantee that no more than one can be created.[3] This built-in "birth control" on the number of quasifermions serves as a kind of Ockham's razor to preclude redundant degrees of freedom. In a variant of the second approach, applicable to both even and odd systems, only a subset of the fermion-pair degrees of freedom, corresponding to collective superpositions, are mapped into bosons, while the remainder are mapped into kinematically independent quasifermions whose algebraic properties guarantee the fulfillment of nonredundancy constraints.[4] In all cases, the replica of the fermion space that resides in the ideal space (called the *physical subspace*) is a small (possibly finite) subspace of the infinite-dimensional ideal space. Therefore, the mapping is basically an injective one.

The reader is referred to the extensive review paper of Klein and Marshalek[5] for the early history, derivations and details of implementations of the boson mapping technique in nuclear physics. The first part of this talk will provide an overview of the main types of boson mappings and a brief survey of their applications. The second part will concentrate on the details of a particular type of application, namely, a solution of the problem of broken symmetries when dealing with deformed nuclei.

2. Introduction to Boson Mappings

The most common types of boson mappings are all closely related, either through what are essentially canonical transformations, or through rearrangments of infinite expansions. They can be derived in many equivalent ways, involving, for example, direct mappings between Hilbert spaces, use of generator coordinates, path-integral methods, and, most recently, with the aid of the theory of Lie algebras, which is the most natural. Here, the interest will be focused on the structure of the expansions, rather than the derivations, which can be found in the review paper (Ref. 5).

2.1. The SU(2) Paradigm

The notion of boson mappings becomes most transparent in the context of the rotation group SU(2), for which two versions of such mappings have been known for approximately a half century. One is the Dyson map[6] and the other the Holstein-Primakoff (HP) map.[7] Proceeding logically rather than chronologically, consider first the Dyson map, which is given by

$$(J_0)_D = -j + N \,,\, N \equiv B^\dagger B \,,\, [B, B^\dagger] = 1, \tag{2.1a}$$

$$(J_-)_D = B \,,\, (J_+)_D = B^\dagger(2j - N), \tag{2.1b}$$

where the symbol $(\,)_D$ denotes the Dyson image of a generator and j labels an irrep of SU(2). It is easily checked that this realization in terms of the single boson degree of freedom represented by the operators B, $B^\dagger$ indeed fulfills the SU(2) commutation rules $[J_+, J_-] = 2J_0$, $[J_0, J_\pm] = \pm J_\pm$. What is distinctive about the Dyson realization is that one of the ladder operators, namely, J_-, is represented by a single boson annihilation operator. Thus, although the commutation rules are satisfied, no attempt is made to preserve Hermitian conjugation since $(J_-)_D{}^\dagger \neq (J_+)_D$. However, this feature only means that the Dyson realization is a nonunitary one that is equivalent to a unitary realization, as will be seen. It is also obvious that there exists a Hermitian-conjugate realization in which $(J_+)_D = B^\dagger$, but the choice (2.1) is the conventional one.

The mapping (2.1) preserves the matrix elements of an SU(2) irrep lableled by j only within a subspace of the infinite-dimensional oscillator space. This subspace, the physical subspace, is spanned by the basis vectors

$$|jm)_D = \left[\frac{(j-m)!}{(2j)!\,(j+m)!}\right]^{1/2} (J_+)_D{}^{j+m}|0) = \mathcal{N}_{jm}\,\frac{(B^\dagger)^{j+m}}{[(j+m)!\,]^{1/2}}|0), \quad -j \le m \le j, \tag{2.2}$$

where $B|0) = 0$, and $\mathcal{N}_{jm} = [(2j)!/(j-m)!]^{1/2}$. It is straightforward to verify that the operators (2.1) generate the standard SU(2) matrix elements when acting on these basis vectors. Corresponding to (2.2) is the dual basis:

$$ {}_{\mathrm{D}}\overline{(jm}| = (0|(J_-)_{\mathrm{D}}{}^{j+m}\left[\frac{(j-m)!}{(2j)!\,(j+m)!}\right]^{1/2} = {}_{\mathrm{D}}(jm|(\mathcal{N}_{jm})^{-2}, \tag{2.3} $$

which signifies that the vectors (2.2) are not normalized, though mutually orthogonal. In general, the physical basis vectors in realizations of the Dyson type are not normalized and, in fact, may not even be orthonormal. This property precisely reflects the nonunitarity of the representation. When there is a physical subspace there must also be an unphysical subspace, which is the orthogonal complement, in this case spanned by the vectors $(B^\dagger)^n|0)$ with $n > 2j$.

Let us now consider the HP map of SU(2). This is obtained by means of a similarity transformation of the Dyson map:

$$ (J_\mu)_{\mathrm{HP}} = K(J_\mu)_{\mathrm{D}}K^{-1}, \tag{2.4} $$

where K is chosen to make the HP realization unitary, or, equivalently, to normalize the physical basis. The result is:

$$ (J_0)_{\mathrm{HP}} = -j + N\,,\ (J_+)_{\mathrm{HP}} = (J_-)_{\mathrm{HP}}{}^{\dagger} = B^\dagger(2j - N)^{1/2}. \tag{2.5} $$

It is now easy to see that the HP mapping correctly reproduces the matrix elements of the generators in the basis of *normalized* oscillator states

$$ |j\,m) = [(j+m)!]^{-1/2}(B^\dagger)^{j+m}|0)\,,\ \ -j \le m \le j. \tag{2.6} $$

Although the Dyson and HP realizations are mathematically equivalent, their structural differences make them suitable for different types of approximations. The Dyson map of any finite function of the generators is a finite function of the bosons. This makes it suitable for nonperturbative applications. The nonunitary nature of the Dyson map implies that the Hamiltonian may not be a Hermitian operator, but that is only a minor inconvenience.[8] The HP map is characterized by a *smallness parameter*, $(2j)^{-1/2}$. The images of the Hamiltonian and other Hermitian operators depending on $J_\pm$, though preserving Hermitian conjugation, become infinite expansions in this parameter. Thus, the Taylor expansion of $(J_+)_{\mathrm{HP}}$ is given by

$$ (J_+)_{\mathrm{HP}} = (2j)^{1/2}B^\dagger - \tfrac{1}{2}B^\dagger N(2j)^{-1/2} - \tfrac{1}{8}B^\dagger N^2(2j)^{-3/2} - \dots . \tag{2.7} $$

Therefore, the HP mapping is most naturally suited for perturbative expansions in the small parameter, although nonperturbative applications are by no means ruled out.

Finally, we come to a related mapping of more recent vintage, called the Marumori map[2], whose original incarnation was not the elementary SU(2) algebra, but rather the more general SO($2n$) fermion-pair algebra that will be discussed later. Historically, however, the understanding of the connection between the SU(2) Marumori and HP realizations set the stage for a deeper insight into their SO($2n$) counterparts.[9]

In order to convey the general flavor of the Marumori approach, denote by $\{|p\rangle\}$ an orthonormal basis for the original, e.g., fermion space, and let Ω be an arbitrary operator in this space. In the ideal space, i.e., the boson space, denote an orthonormal basis for the physical subspace by $\{|p)\}$ and for its orthogonal complement, the unphysical subspace, by $\{|u)\}$. This implies the condition $(u|p) = 0$. The operator Ω can be written in the form

$$\Omega = \sum_p \langle p' | \Omega | p \rangle | p' \rangle \langle p |. \tag{2.8}$$

The mapping $|p\rangle \to |p)$ defines the Marumori image Ω_M of Ω, given by:

$$\Omega_M = \sum_p \langle p' | \Omega | p \rangle | p')(p |. \tag{2.9}$$

By construction, Ω_M duplicates all physical matrix elements, but for any unphysical vector $|u)$

$$\Omega_M | u) = 0. \tag{2.10}$$

It then follows that the relation between the Marumori operator and the corresponding HP operator defined on the same physical subspace is just

$$\Omega_M = P\Omega_{HP} P , \tag{2.11}$$

where P is the projector to the physical subspace (in many cases, P can be dropped from the left or right side, depending on whether Ω_{HP} leaves the physical or unphysical subspace invariant).

In contrast to the HP expansion (2.7), the Marumori operator converts automatically to a normal-ordered boson form with the aid of the substitutions

$$|p) = (B^\dagger)^p |0)/\sqrt{p!}, \; (p' | = (0 | B^{p'} /\sqrt{p'!}, \tag{2.12}$$

and the normal-ordered expansion of the vacuum projector

$$|0)(0| = :\exp(-B^\dagger B): . \tag{2.13}$$

For the SU(2) generators, one then obtains

$$(J_0)_M = \sum_{n=0}^{\infty} c_n (B^\dagger)^n (B)^n, \; (J_+)_M = (J_-)_M^\dagger = B^\dagger \sum_{n=0}^{\infty} d_n (B^\dagger)^n (B)^n, \tag{2.14}$$

where

$$c_n = \sum_{p=0}^{p_n} (-1)^{n-p} \frac{p-j}{p!(n-p)!} \; , \; d_n = \sum_{p=0}^{p_n} (-1)^{n-p} \frac{(2j-p)^{1/2}}{p!(n-p)!} \; , \tag{2.15}$$

with $p_n = n$ if $n \le 2j$ and $p_n = 2j$ if $n > 2j$. We see that all Marumori operators have infinite expansions as well as being normal-ordered.

2.2. SO(2n+1) Mappings for Fermions

The analogs of the SU(2) mappings can readily be derived for the problem of interacting fermions, when the Lie algebra corresponds to the group SO(2n+1) in the most general case. Now, the fermion Fock space is generated by n single-particle creation and destruction operators $c_i^\dagger$, c_i, respectively, obeying the familiar anticommutation rules

$$\{c_i, c_j^\dagger\} = \delta_{ij}, \{c_i^\dagger, c_j^\dagger\} = \{c_i, c_j\} = 0. \tag{2.16}$$

From this, it is easily demonstrated that the fermion pair operators $c_i^\dagger c_j^\dagger$, $c_j c_i$ and $c_j^\dagger c_i + \frac{1}{2}\delta_{ij}$ generate the group SO(2n). Adding the single-particle operators $c_i^\dagger$, c_i themselves to the set then gives the group SO(2n+1), which describes systems with both even and odd particle numbers.

In the ideal space, the building blocks include, first, the antisymmetric boson creation and destruction operators $B_{ij}^\dagger$, B_{ij}, which prempt the degrees of freedom of fermion pairs $c_i^\dagger c_j^\dagger$ and obey the commutation rules

$$[B_{ij}, B_{kl}^\dagger] = \delta_{ik}\delta_{jl} - \delta_{il}\delta_{jk},$$

$$[B_{ij}^\dagger, B_{kl}^\dagger] = 0, [B_{ij}, B_{kl}] = 0. \tag{2.17}$$

An odd nucleon, if present, is described by the *quasifermion* creation and destruction operators $a_i^\dagger$, a_i. These may be chosen to satisfy the following algebraic rules:

$$a_i^\dagger a_j^\dagger = a_j a_i = 0, a_i a_j^\dagger = \delta_{ij} P_0,$$

$$P_0 a_i^\dagger = a_i P_0 = 0, P_0 = P_0^2 = 1 - \sum_i a_i^\dagger a_i, \tag{2.18}$$

where P_0 is the projector to the purely bosonic subspace. These algebraic rules for the quasifermions are designed to ensure that the ideal space accommodates at most one quasifermion (two for odd-odd nuclei).[10–11]

The generalized Dyson (GD) and generalized Holstein-Primakoff (GHP) maps of the SO(2n+1) generators correspond to the spinor irreps of this group. Detailed derivations are provided in Refs. 10-11. The GD map is the nonunitary realization and, as is true of all Dyson-type maps, the images are all finite.

The images of the pair operators are:

$$(c_i^\dagger c_j)_D = (\mathbf{B}^\dagger \mathbf{B})_{ji}^T + a_i^\dagger a_j, \tag{2.19a}$$

$$(c_j c_i)_D = B_{ij}, \tag{2.19b}$$

$$(c_i^\dagger c_j^\dagger)_{\rm D} = B_{ij}^\dagger - [\mathbf{B}^\dagger(\mathbf{B}^\dagger\mathbf{B})^{\rm T}]_{ji} + \sum_k B_{jk}^\dagger a_i^\dagger a_k - \sum_k B_{ik}^\dagger a_j^\dagger a_k\,, \tag{2.19c}$$

and the images of one-nucleon operators are:

$$(c_i)_{\rm D} = a_i + \sum_j B_{ij} a_j^\dagger\,,\quad (c_i^\dagger)_{\rm D} = a_i^\dagger + \sum_j \left[B_{ij}^\dagger a_j - (\mathbf{B}^\dagger\mathbf{B})_{ji}^{\rm T} a_j^\dagger\right]. \tag{2.20}$$

Here, the matrix notation $(\mathbf{B}^\dagger)_{ji} = B_{ij}^\dagger = -B_{ji}^\dagger$ was used for convenience. Note that $(\mathbf{B}^\dagger\mathbf{B})_{ji}^{\rm T} = \sum_k B_{ik}^\dagger B_{jk}$, where "T" stands for the matrix transpose. Also,

$$[\mathbf{B}^\dagger(\mathbf{B}^\dagger\mathbf{B})^{\rm T}]_{ji} = -[\mathbf{B}^\dagger(\mathbf{B}^\dagger\mathbf{B})^{\rm T}]_{ij} = \sum_{kl} B_{ik}^\dagger B_{lj}^\dagger B_{lk}, \tag{2.21}$$

which guarantees the antisymmetry of the image (2.19c). The GD mapping clearly does not preserve Hermitian conjugation, as expected. Apart from the quasifermion degree of freedom, the structural similarities between Eqs. (2.1) and (2.19) are obvious. In particular, the operators (2.19a), which generate a unitary subalgebra, play a role analogous to $(J_0)_{\rm D}$, while the ladder operators (2.19b) and (2.19c) are analogous to $(J_-)_{\rm D}$ and $(J_+)_{\rm D}$, respectively.

Just as in the SU(2) case, the GHP map can be obtained by unitarizing the GD map, or, equivalently, normalizing the basis vectors. The images of the pair operators are then given by

$$(c_i^\dagger c_j)_{\rm HP} = (c_i^\dagger c_j)_{\rm D} = (\mathbf{B}^\dagger\mathbf{B})_{ji}^{\rm T} + a_i^\dagger a_j\,, \tag{2.22a}$$

$$(c_i^\dagger c_j^\dagger)_{\rm HP} = (c_j\, c_i)_{\rm HP}^\dagger = \left\{\mathbf{B}^\dagger[\mathbf{I} - (\mathbf{B}^\dagger\mathbf{B})^{\rm T}]^{1/2}\right\}_{ji} + \sum_{kl}\left[([\mathbf{I} - (\mathbf{B}^\dagger\mathbf{B})^{\rm T}]^{1/2})_{kl}\,,\, B_{ij}^\dagger\right] a_k^\dagger a_l\,, \tag{2.22b}$$

while those of the one-nucleon operators are given by

$$(c_i^\dagger)_{\rm HP} = (c_i)_{\rm HP}^\dagger = \sum_j a_j^\dagger \left\{([\mathbf{I} - (\mathbf{B}^\dagger\mathbf{B})^{\rm T}]^{1/2})_{ji} + B_{ij}^\dagger\, a_j\right\}. \tag{2.23}$$

Eq. (2.23) has the interesting property that if the bosons are replaced by c-numbers and the quasifermions by fermions, then it becomes a general Bogoliubov-Valatin transformation. For this reason, it is called a *quantized Bogoliubov transformation*, a term first applied to an SU(2) analog.[12]

It is worthwhile mentioning that the derivation of Eqs. (2.22) and (2.23) is facilitated by the following Pauli-principle identities,[10–11] which are valid only inside the physical subspace for arbitrary integer powers p:

$$B^\dagger_{ij}(2\hat{N}_B)^p = \left\{\mathbf{B}^\dagger[-(\mathbf{B}^\dagger\mathbf{B})^T]^p\right\}_{ji}, \quad (2\hat{N}_B)^p\,\hat{n} = \sum_{ij}\left\{[-(\mathbf{B}^\dagger\mathbf{B})^T]^p\right\}_{ij} a^\dagger_i a_j,$$

$$a^\dagger_i(2\hat{N}_B)^p = \sum_j a^\dagger_j\left\{[-(\mathbf{B}^\dagger\mathbf{B})^T]^p\right\}_{ji}, \tag{2.24}$$

where

$$2\hat{N}_B \equiv \sum_{ij} B^\dagger_{ij}B_{ij}, \quad \hat{n} \equiv \sum_i a^\dagger_i a_i. \tag{2.25}$$

The GD and GHP images are defined only inside the physical subspace of antisymmetric states first introduced by Marumori *et. al.*[2] This subspace carries the *spinor representations* of SO(2n+1). Its orthogonal complement, of course, is the unphysical subspace, which has no physical relation to the Pauli principle. The normalized Marumori basis is given by:

$$|i_1, j_1, i_2, j_2, \ldots, i_m, j_m) = (c^\dagger_{i_1})_{HP}(c^\dagger_{j_1})_{HP}(c^\dagger_{i_2})_{HP}(c^\dagger_{j_2})_{HP}\ldots(c^\dagger_{i_m})_{HP}(c^\dagger_{j_m})_{HP}|0)$$

$$= [(2m-1)!!]^{-1/2}\sum_{\mathcal{P}}(-1)^{\mathcal{P}}\mathcal{P}B^\dagger_{i_1j_1}B^\dagger_{i_2j_2}\ldots B^\dagger_{i_mj_m}|0) \text{ (even subspace)},$$

$$|i, i_1, j_1, i_2, j_2, \ldots, i_m, j_m) = (c^\dagger_i)_{HP}|i_1, j_1, i_2, j_2, \ldots, i_m, j_m)$$

$$= [(2m+1)!!]^{-1/2}\sum_{\mathcal{P}}(-1)^{\mathcal{P}}\mathcal{P}a^\dagger_i B^\dagger_{i_1j_1}B^\dagger_{i_2j_2}\ldots B^\dagger_{i_mj_m}|0) \text{ (odd subspace)}. \tag{2.26}$$

where the sums run over all permutations $\mathcal{P}$ of the indices.

While the GHP mapping is formally similar to its SU(2) counterpart, the former lacks an explicit expansion parameter. In fact, such a parameter is usually introduced into the theory *aposteriori*, for example, by transforming to RPA bosons. The parameter is then provided by the RPA zero-point amplitudes for the various normal modes. A parameter can also be provided by introducing bosons coupled to good angular momentum, but this is most appropriate for spherical nuclei. In lieu of an explicit parameter, it is sometimes convenient to replace the identity matrix I by z**I** in (2.22) and (2.23) and use the parameter $z^{-1/2}$ to keep track of orders. At the end of the calculation, of course, one must set $z = 1$. The boson expansion of the Hamiltonian and other operators can then be obtained in a straightforward way.

3. A Survey of Applications of Boson Mappings

In this section, a brief survey is presented of the various kinds of applications of boson mapping methods that have been attempted so far; for additional details, see Ref. 5, or the original papers. The applications fall mainly into two broad categories: perturbative and nonperturbative. The latter includes boson shell-model calculations and boson mean-field approximations. The order of presentation does not necessarily reflect the importance of the application.

3.1. Perturbative Corrections Beyond the Mean-Field and the RPA

Boson expansions in a small parameter are used to derive the RPA and higher-order corrections as well. If the zeroth approximation violates continuous symmetries, then the Marshalek-Weneser (MW) method can be used to reconstruct operators with the correct symmetry properties.[13–17] This topic will be the main focus of the later discussion.

3.2. Boson Shell-Model Calculations

In this approach, one diagonalizes a boson-mapped Hamiltonian in a truncated basis of states generated by correlated bosons, such as those corresponding to Tamm-Dancoff or RPA normal-modes. Work along these lines is best exemplified by the efforts of Takada et. al.[18–24] on the one hand and those of Tamura, Kishimoto, Sakamoto and coworkers[25–32] on the other. The first group favors the use of the generalized Dyson mapping, which leads to a finite boson Hamiltonian. The second group has used truncated infinite boson expansions that are intermediate between the generalized Holstein-Primakoff and Marumori expansions, but closer to the latter. A longstanding and difficult problem in the boson shell model is the identification and elimination in the calculated spectrum of spurious states associated with the unphysical subspace. The nature of the problem can already be appreciated by considering an independent $2N$-fermion Hamiltonian of the form

$$H = \sum_{i=1}^{n} E_i c_i^\dagger c_i \,, \tag{3.1}$$

($2N < n$) which maps into [Eq. (2.22a)] the following independent-boson Hamiltonian:

$$H_{\mathrm{HP}} = H_{\mathrm{D}} = \frac{1}{2}\sum_{ij} (E_i + E_j) B_{ij}^\dagger B_{ij} \,. \tag{3.2}$$

The Hamiltonian (3.2) certainly includes all the physical states among its eigenvectors, but it also has an infinite number of unphysical eigenvectors as well. In this case, one knows apriori which are the physical eigenvectors, but not in the general case with interactions. Although the Hamiltonian map does not mix physical and unphysical vectors, truncations of the boson expansions and also of the boson basis usually introduce such a mixing. The problem of distinguishing which eigenvectors are mainly physical and which are mainly spurious then becomes especially acute. Various ingenious solutions have been proposed,[33–41] but so far, most of the applications of these ideas have been limited to simple models.

3.3. Boson Mean-Field Approximations

Among the novel approximations that emerge from boson mappings are boson mean-field theories. Depending on the formulation, these may differ from fermion mean-field approximations, possibly including more correlation effects. An example of a boson mean-field theory is the boson factorization approximation (BFA),[42] suggested a number of years ago. This method is a prescription involving the self-consistent factorization of the boson-mapped Hamiltonian in such a way that an RPA-like equation is obtained, but with a matrix that depends nonlinearly on its eigenvectors. These equations preserve conservation laws in the same sense as the usual linear RPA. In the application to the Lipkin model, the BFA gave good results well beyond the phase transition in the Hartree-Fock approximation.

In recent years, there has been a revival of interest in bosonic mean-field approximations, spearheaded by the Argentinian group.[43–45] Their method, dubbed the

Hartree-Bose approximation (HBA), which is distinct from the BFA, is applicable only to Hamiltonians that conserve the number of bosons. A boson analog of the Hartree-Fock method, the HBA employs a trial variational state vector that is an N-boson condensate in which the boson is of the Tamm-Dancoff type. Like the usual Hartree-Fock method and unlike the BFA, the HBA does violate conservation laws, which, however, can be restored by a subsequent bosonic RPA derived from the equations of motion. If the bosons correspond to fermion pairs then this RPA actually introduces four-body correlations, unlike the usual RPA or the BFA, which explicitly bring in only two-body correlations.

Recently, Bijker, Pittel and Dukelsky[46–48] introduced an ingenious improvement on the mean-field approach, which is an iterative scheme involving boson mappings and the Hartree-Bose approximation (ITHBA), that works as follows. The solution of the first HBA determines a complete set of correlated Tamm-Dancoff bosons. A second boson mapping is introduced in which the new boson represents a pair of the first HBA bosons. Since the latter boson corresponds to a fermion pair then the new boson corresponds to a quartet. A second HBA can then be performed in this quartet boson representation, which then determines a correlated Tamm-Dancoff quartet boson. This procedure can then in principle be iterated indefinitely, with repeated mappings to octet, etc. bosons, each mapping being followed by the HBA. The procedure can be applied to systems that are initially bosonic or fermionic. For interacting boson systems, the convergence is very rapid. This is attributed to the fact that the unphysical excitations lie high and thus have little weight in the wave functions. Dukelsky and Pittel have also studied the iteration scheme for the two-level pairing model of fermions. Here the results are not as good owing to the fact that unphysical states tend to lie low in fermion systems. Very recently, the procedure has been improved so that the mapping to multi-index bosons can be done in one fell swoop.[48] This makes it possible to study n-body correlations when n is not a power of 2. So far, the applications have been limited to simple bosonic systems.

3.4. Large-Amplitude Collective Motion

When the bosons are properly replaced by c-numbers in a GHP-mapped Hamiltonian, one recovers time-dependent mean-field theories (TDMF) such as the Hartree or Hartree-Bogoliubov approximations.[49–51] What one finds is the TDMF equations expressed in the form of Hamilton's equations with the c-number bosons and their complex conjugates acting as canonically conjugate coordinates. As a result, the c-number form of the generalized Holstein-Primakoff mapping has proven very useful in theories of large-amplitude collective motion formulated within the framework of TDMF approximations, such as the self-consistent coordinate (SCC) theory developed by Marumori *et. al.*[52–55] and the large-amplitude theory of Klein *et. al.*.[56–58]

3.5. Applications to Group Theory

We have seen that a boson mapping is, after all, nothing more than a realization of the generators of a Lie algebra and one or more irreps in terms of boson operators. Recent insights into the relation between boson mappings and Lie algebras has not only led to a simplification of the derivation of boson mappings, but, conversely, boson mappings are proving to be an invaluable tool in the mathematical theory of group representations. The latter developments have been embodied recently in the theory of the vector-coherent state (VCS) and the so-called K matrix where the bosons usually appear in the garb of Bargmann space.[59–68] Very recently, a close connection has been found between the quantized Bogoliubov transformation, alluded to in Sec. 2.2, and the VCS realization of operators lying outside of the group algebra.[69–71]

3.6. Solution of Toy Models

Boson mappings and toy models have long enjoyed a symbiotic relation. Boson mappings have been constructed to solve toy models, thereby providing insights into nuclear structure, and conversely, toy models have been constructed to take advantage of available or easily formulated boson mappings. One of the simplest example is the application of the SU(2) mappings to the Lipkin model.[9] Another example involving SU(2) × SU(2) will be presented later. A rather exhaustive catalog of toy models can be found in the review of Ref. 5.

3.7. Nuclei at Nonzero Temperature

Although the usual boson mappings were intended for use when the temperature $T = 0$, it is possible to extend the domain of application to situations when $T \neq 0$, which is a "hot" topic these days. In that case, it is obvious that a blithe application of Bose statistics would be incorrect since the unphysical states would completely swamp the partition function. The problem of developing thermal boson expansions was first addressed by da Providência and Fiolhais[72] and applied to the Lipkin model. More general approaches were subsequently suggested by Hatsuda,[73] who reexamined the problem from the viewpoint of *thermo-field dynamics* (TFD),[74–75] which seems ideally suited for boson expansions. An advantage of TFD is that it allows the computation of a thermal average to be replaced by a vacuum expectation value with the help of the trick of doubling the dimension of the Hilbert space. As a consequence, a thermal Bogoliubov-Valatin transformation plays a key role just as in the Hartree-Fock Bogoliubov method. The boson method of Hatsuda, however, can be used to extend the statistical mechanics beyond the mean-field approximation, along lines similar to the $T = 0$ treatment to be discussed in this lecture. This work, in turn, stimulated an elegant study of TFD and dynamical symmetries by Walet and Klein within the framework of a simple model.[76] Closely related to this work are the recent studies of Kuriyama, da Providência and Yamamura within a quantized mean-field approach.[77] The subject of TFD and boson mappings is still in the nascent stage, but appears to be a topic of increasing interest and importance that fits in nicely with the program of perturbative boson expansions.

3.8. Baryon Mappings

Very recently, the mapping techniques have been applied to constituent quark models.[78–81] For example, the quark-diquark model of a nucleon is subjected to a boson-fermion mapping. Alternatively, the nucleons are described by a mapping from single-index fermions (quarks) to ideal three-index fermions (nucleons). It is much too early to judge the success of this type of application, but the possibilities are intriguing.

4. Perturbative Boson Expansions With Broken Symmetries

Because of the diversity of applications just outlined, only one of the topics will be covered in depth, namely, *perturbative applications to systems with broken symmetries*. This approach, the MW method alluded to in Sec. 3.1, is applicable whenever the mean-field (MF) approximation (Hartree-Fock or Hartree-Bogoliubov) is a good zero-order approximation and the RPA is a good first-order approximation. By using a boson mapping of the generalized Holstein-Primakoff (GHP) type as a Taylor expansion in a small parameter, not only can the MF and RPA be recovered, but also higher-order corrections, which appear as anharmonic perturbations to the harmonic RPA. This does not necessarily require that the perturbation series in the small parameter converge; it may be sufficient to have an asymptotic series. Indeed, the latter possibility is more likely in realistic models. Moreover, in ideal cases, it may be possible to partially resum the series using Padé approximation.

The approach described here is not expected to apply to all nuclei and may be limited to only part of the spectrum. For example, soft transitional nuclei exhibiting large-amplitude collective motion would not be good candidates. However, once the soft modes are separated off by other methods, the remaining orthogonal modes may be amenable to the perturbative approach. As another example, the method may not be applicable without significant modification in the neighborhood of a rotational band crossing, especially when there is a backbend, because the cranked mean-field approximation breaks down,[82–84] a point often nonchalantly ignored by cranking aficionados. In such a situation, special treatments like the use of diabatic orbitals provide a partial solution.[85–86] In certain nuclei, perturbation theory may run into difficulties in some order because of small resonance denominators,[87] a problem well known in classical mechanics. While this problem deserves more study, it is likely to be soluble, either by methods analogous to those used in classical mechanics[88] or by means of higher-order degenerate perturbation theory.

The chief message to be given here is that when the perturbative approach is appropriate, then broken symmetries arising in the MF approximation and the attendant Goldstone modes that manifest themselves in the RPA pose no serious obstacles. Indeed, the MW procedure is able to reconstruct from the local symmetry-violating expansions the Hamiltonian and also transition operators having the correct global dependence on the constants of motion and the conjugate angle variables.

It should also be mentioned that there is a close relationship between the MW approach and two other systematic approximations. The first is *nuclear field theory* (NFT),[89] whose aim is to generate the same perturbation expansion, but which turned out by itself to be ill-suited for treating the problem of broken symmetries. The second is the Birkhoff-Gustavson method (BG),[90] which is a basically classical method that reduces the full Hamiltonian to an expansion in powers of oscillator Hamiltonians by means of a series of canonical transformations. Quantization can later be effected by a simple Bohr-Sommerfeld ansatz. This procedure is paralleled in the MW approach by a series of unitary transformations that diagonalize the Hamiltonian as a power series in the boson number operators. However, while the BG method may miss quantal fluctuation terms, the MW approach, which is quantal from the outset, appears to include them all. Moreover, the BG method has not, to my knowledge, been formulated yet to repair broken symmetries, as does the MW method. Since the accuracy of the BG expansion is believed to be correlated with the presence of chaos, an interesting topic for the future is the corresponding relation for the MW expansion.

The MW method has so far been applied to spherical nuclei, high-spin states,[91] two-dimensional models of deformed nuclei,[13–15] and, most recently, three-dimensional rotational band structure at low-spins.[16–17] Because of time limitations, only the last two topics will be discussed here.

4.1. The Villars Expansion for Two-Dimensional Rotation

A convenient conceptual starting point is provided by the Villars expansion[92] for two-dimensional rotation. Let $H(\mathbf{p},\mathbf{q})$ be the Hamiltonian for a two-dimensional nuclear model, where **p**, **q** denotes a generic set of nucleon canonical variables. It is irrelevant whether these are defined directly on configuration space, result from a boson mapping, or even represent fermion creation and annihilation operators. It is assumed that the Hamiltonian is rotationally invariant so that

$$[H(\mathbf{p},\mathbf{q})\;,\; J(\mathbf{p},\mathbf{q})] = 0, \tag{4.1}$$

where $J(\mathbf{p},\mathbf{q})$ is the angular-momentum perpendicular to the plane of motion. In fact, J could represent any continuous constant of motion such as the particle-number operator. Given a canonically conjugate angle variable $\varphi(\mathbf{p},\mathbf{q})$,

$$[\varphi(\mathbf{p},\mathbf{q}), J(\mathbf{p},\mathbf{q})] = i, \tag{4.2}$$

one may expand H as follows:

$$H = \sum_{n=0}^{\infty} h_n J^n / n!, \tag{4.3}$$

in which the h_n are *intrinsic* operators, i.e.,

$$[h_n, J] = [h_n, \varphi] = 0, \tag{4.4}$$

given by

$$h_n = \sum_{k=0}^{\infty} (-1)^k J^k C_k / k!, \tag{4.5}$$

Here, the operators C_k are defined iteratively by

$$C_0 = H\,,\; C_{k+1} = [C_k\,, i\varphi],\;\; k = 0, 1, 2,\ldots\;. \tag{4.6}$$

The following observations can now be made. First of all, it is clear that the J-expansion of H depends on the choice of $\varphi(\mathbf{p},\mathbf{q})$, which is not unique. This choice determines how much additional diagonalization one must do and how fast the results converge. If $\varphi(\mathbf{p},\mathbf{q})$ is defined by an expansion in a small parameter, then the intrinsic operators h_n are also given by an expansion in this parameter. The strategy used here is not to guess φ beforehand, but instead to use a boson expansion in a small parameter. The expansion of φ is developed concomittantly with the expansions of H and J. Once the the leading term in the expansion of φ is fixed by the mean-field and RPA, higher-orders are obtained systematically. The expansions of the h_n in terms of intrinsic degrees of freedom can then be reconstructed to a given order.

4.2. The Problem of Goldstone Modes and Two-Dimensional Rotation

Let H be an HP or GHP-mapped Hamiltonian, and J a constant of motion (continuous symmetry). The expansion of H about a mean-field extremum is given by

$$H = E^{(0)} + H^{(2)} + H^{(3)} + H^{(4)} + \ldots, \tag{4.7}$$

where $E^{(0)}$ is the MF ground-state energy, $H^{(2)} = H_{\text{RPA}}$, representing a boson quadratic form, and, in general, the anharmonic term $H^{(n)}$ is an nth-order boson polynomial. The linear boson term is absent in (4.7) since the expansion is about a mean-field extremum. Since J is the generator of a *broken symmetry* then it has an expansion

$$J = J^{(1)} + J^{(2)} + J^{(3)} + \ldots, \tag{4.8}$$

assuming that J is defined so that the mean-field average $J^{(0)} \equiv \langle J \rangle_0 = 0$. It is important to realize that *the presence of a linear boson term* $J^{(1)}$ *is the hallmark of a broken symmetry* in the mean-field approximation. Otherwise, J would be diagonal and thus would not depend on the ladder operators of the underlying Lie algebra.

Given any three (boson-mapped) operators Ω, Λ, Γ, such that $[\Omega, \Lambda] = \Gamma$, with expansions

$$\Omega = \Omega^{(0)} + \Omega^{(1)} + \Omega^{(2)} + \ldots, \tag{4.9}$$

etc., the commutator relation is preserved in the form

$$\sum_{k=1}^{n} [\Omega^{(n-k+1)}, \Lambda^{(k)}] = \Gamma^{(n-1)}, \quad n=1, 2, \ldots . \tag{4.10}$$

Therefore, the conservation law

$$[H, J] = 0 \tag{4.11}$$

implies

$$\sum_{k=1}^{n} [H^{(n-k+1)}, J^{(k)}] = 0, \quad n = 2, 3, \ldots \tag{4.12}$$

(note $H^{(1)} = 0$). *This order-by-order preservation of the conservation law is characteristic of HP-type expansions*, but not, for example, for the Marumori expansions.

Let us first consider the RPA order. The Hamiltonian in diagonal form looks like

$$H^{(2)} = E^{(2)} + \sum_{\mu} \hbar\omega_{\mu} \mathscr{B}_{\mu}^{(1)\dagger} \mathscr{B}_{\mu}^{(1)} + \frac{(\hbar J^{(1)})^2}{2\mathscr{J}}, \tag{4.13}$$

where $E^{(2)}$ is the RPA zero-point energy, and the boson operator $\mathscr{B}_{\mu}^{(1)\dagger}$ creates an RPA normal mode with energy $\hbar\omega_{\mu}$. The last term corresponds to the *Goldstone mode* arising from the broken symmetry and $\mathscr{J}$ is the corresponding inertial parameter. From (4.12), the conservation law in the RPA order becomes:

$$[H^{(2)}, J^{(1)}] = 0, \tag{4.14}$$

which corresponds to the zero-frequency Goldstone mode. In order to obtain a complete set of RPA normal modes, it is customary to define a canonically conjugate RPA angle $\varphi^{(1)}$,

$$[\varphi^{(1)}, J^{(1)}] = i\,, \tag{4.15}$$

which is also required to commute with the bosons $\mathscr{B}_{\mu}^{(1)\dagger}$, $\mathscr{B}_{\mu}^{(1)}$. Then $\varphi^{(1)}$ must be the solution of the Thouless-Valatin equation

$$[H^{(2)}, \varphi^{(1)}] = -i\hbar^2 J^{(1)}/\mathcal{J}, \tag{4.16}$$

and the inertial parameter is given by

$$\hbar^2/\mathcal{J} = [\varphi^{(1)}, [H^{(2)}, \varphi^{(1)}]]. \tag{4.17}$$

While the Hamiltonian (4.13) exhibits a rotational term that correctly identifies the inertial parameter $\mathcal{J}$ to leading order (equal to the self-consistent cranking value) it also has a severe pathology, which has long been known.[13] While J may have a *discrete* spectrum, $J^{(1)}$ and $\varphi^{(1)}$ have a *continuous* spectrum. As a consequence, the RPA ground state has an *infinite norm*, although this corresponds to an infinite norm factor associated with the Goldstone degree of freedom only.[13] Nevertheless, one cannot use the RPA eigenvectors in a literal sense as basis vectors for a perturbation treatment of the higher-orders, which is what is meant by the "infrared catastrophe".

The situation, however, is not hopeless. It is not necessary to embark on a totally different path. It should be kept in mind that what we have is a *local* Taylor expansion. Here, the angle variable φ is approximated by a small angle $\varphi^{(1)}$ that does not range over 2π. This is what causes the loss of the quantization of J. Having diagnosed the basis of the problem, one can solve it by reconstructing the *global* Hamiltonian (and transition operators too) whose Taylor series we have found. Villars' construction assures one that the global form exists. The MW method provides a convenient recipe for the reconstruction.

4.3. The MW Recipe for Two-Dimensional Rotation

Starting with the RPA approximation $\varphi^{(1)}$ to the angle variable and the known expansion (4.8) of J, one can find an expansion of φ that satisfies the canonical commutator order by order:

$$[\varphi^{(1)}, J^{(1)}] = i\,,$$

$$[\varphi^{(2)}, J^{(1)}] + [\varphi^{(1)}, J^{(2)}] = 0\,,$$

$$[\varphi^{(3)}, J^{(1)}] + [\varphi^{(2)}, J^{(2)}] + [\varphi^{(1)}, J^{(3)}] = 0,$$

$$\bullet\bullet\bullet\bullet\bullet\bullet$$

$$\sum_{k=1}^{n} [\varphi^{(n-k+1)}, J^{(k)}] = 0\,. \tag{4.18}$$

Solutions of this hierarchy can be found successively for $\varphi^{(1)}$, $\varphi^{(2)}$, ..., etc. to the desired order. From Villars' construction, this expansion of φ and that of H is sufficient to determine an expansion of the intrinsic coefficients h_n.

The intrinsic coefficients h_n should be expressed in terms of intrinsic boson operators $\mathcal{B}_\mu^\dagger$, $\mathcal{B}_\mu$ that commute with both J and φ. The local expansion of the intrinsic operators,

$$\mathcal{B}_\mu^\dagger = \mathcal{B}_\mu^{(1)\dagger} + \mathcal{B}_\mu^{(2)\dagger} + \mathcal{B}_\mu^{(3)\dagger} + ..., \tag{4.19}$$

can be found from the requisite commutation rules. However, there is an equivalent but easier method. In this method, one seeks a *formal unitary transformation* with generator S,

$$S = S^{(3)} + S^{(4)} + \ldots, \tag{4.20}$$

such that

$$\exp(iS)J^{(1)}\exp(-iS) = J^{(1)} + J^{(2)} + J^{(3)} + \ldots . \tag{4.21}$$

This requires sequential solution of only one hierarchy to the desired order, namely,

$$\begin{aligned} &[iS^{(3)}, J^{(1)}] = J^{(2)}, \\ &[iS^{(4)}, J^{(1)}] = J^{(3)} + \tfrac{1}{2}[J^{(2)}, iS^{(3)}], \\ &\bullet\bullet\bullet\bullet\bullet\bullet. \end{aligned} \tag{4.22}$$

Given a solution for S to some order, the expansions of φ, $\mathscr{B}_\mu^\dagger$ and $\mathscr{B}_\mu$ are then generated from the RPA values as follows:

$$\begin{aligned} \varphi = \exp(iS)\varphi^{(1)}\exp(-iS) &= \varphi^{(1)} - [\varphi^{(1)}, iS^{(3)}] + \tfrac{1}{2}[[\varphi^{(1)}, iS^{(3)}], iS^{(3)}] - [\varphi^{(1)}, iS^{(4)}] + \ldots \\ &= \varphi^{(1)} + \varphi^{(2)} + \varphi^{(3)} + \ldots, \end{aligned}$$

$$\begin{aligned} \mathscr{B}_\mu = \exp(iS)\,\mathscr{B}_\mu^{(1)}\exp(-iS) &= \mathscr{B}_\mu^{(1)} - [\mathscr{B}_\mu^{(1)}, iS^{(3)}] + \tfrac{1}{2}[[\mathscr{B}_\mu^{(1)}, iS^{(3)}], iS^{(3)}] - [\mathscr{B}_\mu^{(1)}, iS^{(4)}] + \ldots \\ &= \mathscr{B}_\mu^{(1)} + \mathscr{B}_\mu^{(2)} + \mathscr{B}_\mu^{(3)} + \ldots . \end{aligned} \tag{4.23}$$

Since the transformation is unitary, the correct commutation rules, namely,

$$\begin{aligned} &[\varphi\,, J] = i\,,\ [\mathscr{B}_\mu\,, \mathscr{B}_\nu^\dagger] = \delta_{\mu\nu}\,, [\mathscr{B}_\mu\,, \mathscr{B}_\nu] = 0\,,\ [\mathscr{B}_\mu^\dagger, \mathscr{B}_\nu^\dagger] = 0\,, \\ &[\varphi\,, \mathscr{B}_\mu] = 0\,,\ [J\,, \mathscr{B}_\mu] = 0\ \text{(and H.c. eqs.)}, \end{aligned} \tag{4.24}$$

are guaranteed to be satisfied order by order.

At this point, Villars' construction is dispensable. Instead, one may proceed as follows. For any function $\Omega(J^{(1)}, \varphi^{(1)}, \mathscr{B}_\mu^{(1)\dagger}, \mathscr{B}_\mu^{(1)})$ of the original RPA modes, define the operator

$$\tilde{\Omega} \equiv \exp(iS)\Omega\exp(-iS) = \Omega(J, \varphi, \mathscr{B}_\mu^\dagger, \mathscr{B}_\mu). \tag{4.25}$$

Then Ω may be expanded in terms of the set J, φ, $\mathscr{B}_\mu^\dagger$, $\mathscr{B}_\mu$ using the inverse transformation:

$$\Omega = \exp(-i\tilde{S})\tilde{\Omega}\exp(i\tilde{S}) = \Omega(J, \varphi, \mathscr{B}_\mu^\dagger, \mathscr{B}_\mu) + [\Omega(J, \varphi, \mathscr{B}_\mu^\dagger, \mathscr{B}_\mu), i\tilde{S}^{(3)}] + \ldots . \tag{4.26}$$

When applied to H, this procedure generates the new expansion

$$H = E^{(0)} + H'^{(2)} + H'^{(3)} + H'^{(4)} + \dots, \tag{4.27}$$

where $H'^{(n)} = H'^{(n)}(J, \mathcal{B}_\mu^\dagger, \mathcal{B}_\mu)$, i.e., H is independent of φ as guaranteed by the rotational invariance relations (4.11) - (4.12). For example, the reconstructed RPA Hamiltonian is given by

$$H'^{(2)} = E^{(2)} + \sum_\mu \hbar\omega_\mu \mathcal{B}_\mu^\dagger \mathcal{B}_\mu + \frac{J^2}{2\mathcal{J}}. \tag{4.28}$$

In general, $H'^{(n)}$ is a homogeneous n^{th}-order polynomial in J, $\mathcal{B}_\mu^\dagger$, $\mathcal{B}_\mu$.

At this point, let us pause to consider a couple of technicalities. First, from Eq. (4.22) one must solve in each order an equation of the form

$$[iS^{(n)}, J^{(1)}] = \Lambda^{(n-1)}. \tag{4.29}$$

However, the solution for $iS^{(n)}$ is not unique; to any solution may be added an arbitrary function $F^{(n)}(J^{(1)}, \mathcal{B}_\mu^{(1)\dagger}, \mathcal{B}_\mu^{(1)})$ to give another solution. But this is not a problem. On the contrary, this arbitrariness may be used to shape the final form of H, for example, to *diagonalize* H to the given order. The solution for iS with all arbitrary $F^{(n)} = 0$ will be called the *kinematic* solution. It can be shown that the kinematic solution is given by the following series of nested commutators and anticommutators:[15]

$$\begin{aligned} iS^{(n)} = & -\frac{1}{2}\left\{i\varphi^{(1)}, A^{(n-1)}\right\} - \frac{1}{8}\left\{i\varphi^{(1)},\left\{i\varphi^{(1)},\left[\Lambda^{(n-1)}, J^{(1)}\right]\right\}\right\} \\ & - \frac{1}{48}\left\{i\varphi^{(1)},\left\{i\varphi^{(1)},\left\{i\varphi^{(1)},\left[\left[\Lambda^{(n-1)}, J^{(1)}\right], J^{(1)}\right]\right\}\right\}\right\} - \dots \\ & - \frac{1}{2^k k!}\Big\{\underbrace{i\varphi^{(1)},\Big\{i\varphi^{(1)}, \dots,\Big\{i\varphi^{(1)}}_{k\ i\varphi^{(1)}\text{'s}},\Big[\Big[\dots\Big[\Lambda^{(n-1)}, \underbrace{J^{(1)}\Big], J^{(1)}\Big] \dots, J^{(1)}}_{k-1\ J^{(1)}\text{'s}}\Big]\Big\}\Big\}\dots\Big\} \\ & - \dots . \end{aligned} \tag{4.30}$$

The series terminates when a multiple commutator vanishes. The solution can be generated via computer algebra (eg., REDUCE) using a simple algorithm.

The second technicality concerns the existence of solutions of Eq. (4.21). On the one hand, one can certainly, generate the expansion $S = S^{(3)} + S^{(4)} + \dots$ by solving Eqs. (4.22). On the other hand, it is also true that *no unitary transformation exp(iS) exists* satisfying the equation $\exp(iS)J^{(1)}\exp(-iS) = J$ if J has a discrete spectrum, since $J^{(1)}$ has a continuous spectrum. The implication is clear: *the series for S cannot converge.* This situation is not as serious as might seem at first sight, since one can show that the nonconvergent parts are just those associated with the Goldstone-mode operators $\varphi^{(1)}$ and $J^{(1)}$. Moreover, their contribution cancels out in all the final reconstructed operators so no problem occurs. One way of viewing this situtation is that since errors must occur in cancelling pairs to produce correct final results, the nonconvergence of S can be regarded as the "error" that compensates the initial "error" of working with a small-angle approximation.

The theory would not be complete without a brief discussion of transition operators. Rotational tensors T_M in two-dimensions satisfy

$$[J, T_M] = MT_M \, , \tag{4.31}$$

where M is an integer. This can be written in the global form:

$$T_M = \frac{1}{2}\left\{e^{iM\varphi}, \hat{T}_M\right\}, \tag{4.32}$$

where $\hat{T}_M$ is a *scalar* and therefore like H has a Villars expansion in J. The local expansion of T_M can be transformed to the final variables J, φ, $\mathscr{B}_\mu^\dagger$, $\mathscr{B}_\mu$. This contains powers of φ arising from the expansion of $e^{iM\varphi}$. The scalar operator $\hat{T}_M$ can then be gotten from

$$\hat{T}_M = \lim_{\varphi \to 0} T_M \, . \tag{4.33}$$

This can be inserted in the global form (4.32) to calculate transition matrix elements.

4.4. The Moszkowski Model

The Moszkowski model[93] is one of several that provide a concrete test of the MW method for two-dimensional rotation.[15] This model is a system of n_0 fermions confined to a two-dimensional oscillator major shell with N quanta and a spin-orbit splitting, which interact via the analog of the quadrupole-quadrupole (Q-Q) interaction. In other words, this is a two-dimensional Elliott model[94] with the added feature of the spin-orbit coupling.

To define the model, one first introduces the quasispin operators t_x, t_y, and t_z by

$$t_x = \frac{1}{4}\left[m\omega_0(x^2 - y^2) + (p_x^{\,2} - p_y^{\,2})/(m\omega_0)\right] , \; t_y = \frac{1}{2}\left[m\omega_0 xy + p_x p_y/(m\omega_0)\right] ,$$

$$t_z = \frac{1}{2}\, l_z = \frac{1}{2}(xp_y - yp_x). \tag{4.34}$$

The rotational tensors $t_\pm \equiv t_x \pm it_y$, t_z obey an SU(2) algebra (we set $\hbar = 1$ in this subsection). The orbitals in the oscillator potential are labeled by $|N\, m\pm\rangle$, where m is the eigenvalue of l_z and $\pm$ distinguishes the two spin states. One can now define the components of the one-body rotational tensor T_μ by:

$$T_\mu(\pm) = \sum_{mm'} \langle Nm|t_\mu|Nm'\rangle c^\dagger_{m\pm} c_{m'\pm} \, , \tag{4.35}$$

where $c^{\dagger}_{m\pm}$ creates a fermion in the orbital $|N\,m\pm\rangle$. The sets of operators for spin-up and spin-down, $T_{\mu}(+)$ and $T_{\mu}(-)$, form two commuting SU(2) algebras. The total quasispin is the vector sum

$$\mathbf{T} = \mathbf{T}(+) + \mathbf{T}(-). \tag{4.36}$$

The total angular momentum of the system L_z is given by

$$L_z = 2T_z\,. \tag{4.37}$$

Since

$$[L_z\,, T_{\pm}] = \pm\, 2T_{\pm}\,, \tag{4.38}$$

the operators $T_{\pm} \equiv T_x \pm iT_y$ are two-dimensional "quadrupole" operators.

The Hamiltonian of the Moszkowski model is given by

$$H = -\tfrac{1}{4}G^2(T_+T_- + T_-T_+) - c[T_z(+) - T_z(-)], \tag{4.39}$$

where G^2 is the strength of the Q-Q interaction and c that of the spin-orbit splitting. When $c = 0$, H reduces to the SU(2) Elliott model, whose energy levels are given by

$$E(T,M) = -\tfrac{1}{2}G^2T(T+1) + \tfrac{1}{8}G^2M^2, \tag{4.40}$$

where $M = 0, \pm 2, ..., \pm 2T$ is the eigenvalue of L_z, and $T(T+1)$ that of $\mathbf{T}\bullet\mathbf{T}$. In this limit, one has an ideal but finite rotational band structure, with the moment of inertia $\mathcal{J} = 4/G^2$. The different bands are labeled by $T = T_0 - n$, $n = 0, 1, ...T_0$, where $T_0 \equiv \frac{1}{4}n_0(2N-n_0+2)$ for even particle numbers n_0. In the other extreme when $G = 0$, the spectrum is perfectly harmonic but truncated. However, this spectrum is not collective but merely reflects the spin-orbit splitting of the degenerate oscillator shell.

The first step in the analysis consists of the mean-field approximation, in this case, the Hartree approximation. The independent-particle Hartree Hamiltonian corresponding to (4.39) is

$$H_{\text{MF}} = E^{(0)} - :\Big\{c[T_z(+) - T_z(-)] + G^2T_0\alpha(\cos 2\varphi\, T_x + \sin 2\varphi\, T_y)\Big\}: \tag{4.41}$$

This is subject to the self-consistency condition:

$$\langle\Psi_0|T_x|\Psi_0\rangle = T_0\alpha\cos 2\varphi,\;\; \langle\Psi_0|T_y|\Psi_0\rangle = T_0\alpha\sin 2\varphi, \tag{4.42}$$

where $|\Psi_0\rangle$ is the Hartree ground state, α the *intrinsic shape parameter* and φ the orientation angle of the elliptically deformed field. One may choose $\varphi = 0$, since $E^{(0)}$ is independent of φ.

The solutions of Eqs. (4.42) are given by

$$\alpha = 0 \quad \text{if } \chi < 1 \text{ (circular shape)}, \tag{4.43a}$$

$$\alpha \equiv \alpha_0 = (1 - 1/\chi^2)^{1/2} \quad \text{if } \chi > 1 \text{ (elliptic shape)}, \tag{4.43b}$$

where the parameter χ is given by

$$\chi \equiv G^2 T_0 / c. \tag{4.44}$$

In the case of the deformed solution (4.43b), H_{MF} is diagonalized by the quasispin rotation:

$$U(\gamma) \equiv \exp[-i\gamma T_y(+)] \bullet \exp[i\gamma T_y(-)], \quad \gamma \equiv \sin^{-1}\alpha_0. \tag{4.45}$$

From now on, the discussion is restricted to the deformed regime given by (4.43b). The next step is to transform the full Hamiltonian (4.39) using (4.45) so that it is expressed in terms of the rotated quasispins $\mathbf{T}'(\pm)$ given by

$$\mathbf{T}'(\pm) \equiv U(\gamma)\mathbf{T}(\pm)U(\gamma)^\dagger. \tag{4.46}$$

The operators $\mathbf{T}'(\pm)$ are then subjected to simple HP mappings as follows:

$$T'_{\mp}(\pm) \to B^\dagger_\pm (T_0 - N_\pm)^{1/2}, \; T_\pm{}'(\pm) \to (T_0 - N_\pm)^{1/2} B_\pm,$$

$$T'_z(\pm) \to \pm(\tfrac{1}{2}T_0 - N_\pm), \; N_\pm \equiv B^\dagger_\pm B_\pm. \tag{4.47}$$

This involves two kinds of bosons B_+ and B_-, which are *deformed bosons*, i.e., they do not carry good angular momentum. The asymmetry in the mapping of spin-up and spin-down operators is due, of course, to the asymmetry introduced in the Hartree ground state by the spin-orbit force. The total angular momentum operator, which is given by

$$L_z = 2\left\{\cos\gamma T'_z - \sin\gamma\,[T'_x(+) - T'_x(-)]\right\}, \tag{4.48}$$

is also mapped with the aid of (4.47).

One can now proceed to expand the boson-mapped operators H and L_z in powers of the small parameter $T_0^{-1/2}$ as follows:

$$H = E^{(0)} + H^{(2)} + H^{(3)} + H^{(4)} + \ldots, \tag{4.49}$$

$$L_z = L_z^{(1)} + L_z^{(2)} + L_z^{(3)} + \ldots, \tag{4.50}$$

where $E^{(0)}$ is the Hartree ground-state energy. The RPA Hamiltonian is given by

$$H^{(2)} = E^{(2)} + \omega \mathscr{B}^{(1)\dagger} \mathscr{B}^{(1)} + \frac{(L_z^{(1)})^2}{2\mathscr{I}}, \tag{4.51}$$

where

$$\omega = G^2 T_0 \alpha_0 , \quad \mathscr{I} = 4\alpha_0^{\ 2}/G^2 , \tag{4.52}$$

and

$$\mathscr{B}^{(1)\dagger} = \tfrac{1}{2}(2\alpha_0)^{-1/2}[(1+\alpha_0)(B_+^\dagger + B_-^\dagger) - (1-\alpha_0)(B_+ + B_-)], \tag{4.53}$$

$$L_z^{(1)} = - T_0^{\ 1/2}\alpha_0(B_+^\dagger - B_-^\dagger + B_+ - B_-). \tag{4.54}$$

The canonically conjugate RPA angle variable is

$$\varphi^{(1)} = \tfrac{1}{4}i(\alpha_0^{\ 2}T_0)^{-1/2}(B_+^\dagger - B_-^\dagger - B_+ + B_-). \tag{4.55}$$

The rest of H and L_z can also be expressed in terms of the RPA modes $L_z^{(1)}$, $\varphi^{(1)}$, $\mathscr{B}^{(1)\dagger}$, $\mathscr{B}^{(1)}$, but these terms will not be explicitly exhibited here (see Ref. 15). The MW procedure can be carried out now just as outlined in Sec. 4.2.

One finds that the *kinematic solution* for S gives rise to off-diagonal terms in the reconstructed H. In order to compare with exact results, S should be chosen instead to *diagonalize H*. This can also be done by a second unitary transformation after the kinematic one. The final reconstructed Hamiltonian to the order of interest is found to be

$$H = E^{(0)} + H'^{(2)} + H'^{(4)} + ..., \tag{4.56}$$

(Note that $H'^{(3)} = 0$ is required for diagonality) where

$$E^{(0)} = - G^2 T_0^{\ 2}(1 - \tfrac{1}{2}\alpha_0^{\ 2}), \tag{4.57a}$$

$$H'^{(2)} = E^{(2)} + G^2 T_0 \alpha_0 \mathscr{B}^\dagger \mathscr{B} + \frac{G^2}{8\alpha_0^{\ 2}} L_z^{\ 2}, \tag{4.57b}$$

$$\begin{aligned} H'^{(4)} = E^{(4)} &- \left[\frac{3-\alpha_0^{\ 2}}{4\alpha_0^{\ 2}}\right] \mathscr{B}^{\dagger 2} \mathscr{B}^2 - \frac{(1-\alpha_0)(3+\alpha_0)}{2\alpha_0^{\ 2}} \mathscr{B}^\dagger \mathscr{B} - \frac{(1-\alpha_0^{\ 2})^2}{32 T_0^{\ 2}\alpha_0^{\ 8}} L_z^{\ 4} \\ &+ \frac{(1-\alpha_0^{\ 2})(6-4\alpha_0-\alpha_0^{\ 2})}{16 T_0 \alpha_0^{\ 5}} L_z^{\ 2} + \frac{(1-\alpha_0^{\ 2})(6-\alpha_0^{\ 2})}{8 T_0 \alpha_0^{\ 5}} L_z^{\ 2} \mathscr{B}^\dagger \mathscr{B}. \end{aligned} \tag{4.57c}$$

Here, $E^{(2)}$ and $E^{(4)}$ are correlation corrections to the ground-state energy given by

$$E^{(2)} = -G^2T_0(1-\tfrac{1}{2}\alpha_0),\ E^{(4)} = -\frac{(1-\alpha_0)(3-\alpha_0)}{8\alpha_0^{\ 2}}. \tag{4.58}$$

These results exhibit clearly the expansion in the small parameter T_0^{-1}. It can easily be shown that the cranking model gives the same leading-order coefficients of the $L_z^{\ 2}$ and $L_z^{\ 4}$ terms. However, the cranking model cannot provide the $1/T_0$ renormalizations of these coefficients that one finds using the MW method, since these are quantal corrections absent in the cranking model.

In the diagonal boson representation, the eigenvectors have a very simple form, namely,

$$|n,M) = \frac{e^{iM\varphi}}{\sqrt{2\pi}} \otimes \frac{(\mathcal{B}^\dagger)^n}{\sqrt{n!}}|0), \tag{4.59}$$

where $\mathcal{B}|0) = 0$. From Eqs.(4.57)-(4.58) the eigenvalues describing the rotation-vibration band structure can then be immediately written down in terms of the angular-momentum quantum number M and the number of vibrational quanta n as follows:

$$E(n,M)/G^2 = -T_0^{\ 2}(1-\tfrac{1}{2}\alpha_0^{\ 2}) - T_0 - \frac{3(1+\alpha_0)^2}{16\alpha_0^{\ 2}} + \left[T_0\alpha_0 + \frac{1}{\alpha_0}\right](n+\tfrac{1}{2}) - \left[\frac{3-\alpha_0^{\ 2}}{4\alpha_0^{\ 2}}\right](n+\tfrac{1}{2})^2$$

$$+ \left[\frac{1}{8\alpha_0^{\ 2}} - \frac{1-\alpha_0^{\ 2}}{4T_0\ \alpha_0^{\ 4}}\right]M^2 - \frac{(1-\alpha_0^{\ 2})^2}{32T_0^{\ 2}\alpha_0^{\ 8}}M^4 + \frac{(1-\alpha_0^{\ 2})(6-\alpha_0^{\ 2})}{8T_0\alpha_0^{\ 5}}(n+\tfrac{1}{2})M^2 + \dots . \tag{4.60}$$

To the given order, one picks up higher-order renormalizations of the ground-state moment of inertia, and the vibrational frequency. In addition, the term proportional to $(n+\frac{1}{2})M^2$ accounts for differences between the moments of inertia of ground-state and excited rotational bands, a very interesting topic that has never been investigated in realistic models.

As an example of the MW treatment of transition operators, we consider the "quadrupole" operators $T_\pm$. After performing the unitary transformation (4.45) to the Hartree basis, introducing the deformed bosons $B_\pm^\dagger$, $B_\pm$ and finally the unitary transformation to the diagonal boson representation, the operators $T_\pm$ are expanded in terms of the operators $\mathcal{B}^\dagger$, $\mathcal{B}$, L_z, φ. As a variant of Eq. (4.32), it proves convenient to write $T_\pm$ in the global form

$$T_+ = T_-^{\ \dagger} = e^{i\varphi}\,\hat{T}_+ e^{i\varphi}, \tag{4.61}$$

where $\hat{T}_+$ is a scalar given by

$$\hat{T}_+ = \lim_{\varphi\to 0} T_+. \tag{4.62}$$

In this way, one finds the expansion

$$\hat{T}_+ = T_0\alpha_0 + \hat{T}'^{(1)}_+ + \hat{T}'^{(2)}_+ + \dots, \tag{4.63}$$

where

$$\hat{T}'^{(1)}_+ = \left[\frac{T_0(1-\alpha_0^2)}{2\alpha_0}\right]^{1/2} (\mathscr{B}^\dagger + \mathscr{B}), \tag{4.64a}$$

$$\hat{T}'^{(2)}_+ = -[3-4\alpha_0-\alpha_0^2 + 2(3-\alpha_0^2)\,\mathscr{B}^\dagger\mathscr{B}]/(4\alpha_0^2) + \left[\frac{2-3\alpha_0^2}{8T_0\alpha_0^5}\right] L_z^2 + \left[\frac{1-\alpha_0^2}{4\alpha_0^2}\right](\mathscr{B}^{\dagger 2} + \mathscr{B}^2)$$
$$- \left[\frac{1-\alpha_0^2}{2T_0\alpha_0}\right]^{1/2} \frac{(2-\alpha_0^2)}{2\alpha_0^3} L_z(\mathscr{B}^\dagger - \mathscr{B}). \tag{4.64b}$$

The exercise just presented would not be so interesting if there were no possibility of comparing with an exact solution. Fortunately, this possibility does exist in the following sense. The SU(2) × SU(2) mapping just implemented provided a boson expansion about a *broken-symmetry* Hartree solution. To test the MW reconstruction, one should compare it with an expansion that preserves the rotational symmetry from the outset. This can be accomplished with the aid of the following SO(4) ⊃ SU(2) ⊃ U(1) mapping to a rotor-vibrator space, which strictly preserves angular-momentum conservation.[15] If one defines

$$L_z \equiv -i\frac{d}{d\varphi}, \ \hat{N} \equiv B^\dagger B, \ \hat{T} = T_0 - \hat{N}, \tag{4.65}$$

which already includes the mapping of L_z, the mapping of the quasispin operators is given by

$$T_z(\pm) = \tfrac{1}{4}L_z \pm \tfrac{1}{4}\left[\frac{(\hat{T}+T_0+1)(4\hat{T}^2-L_z^2)}{2\hat{T}+1}\right]^{1/2} B\frac{1}{(2\hat{T}+1)^{1/2}} \pm \frac{1}{4}\frac{1}{(2\hat{T}+1)^{1/2}}B^\dagger\left[\frac{(\hat{T}+T_0+1)(4\hat{T}^2-L_z^2)}{2\hat{T}+1}\right]^{1/2},$$

$$T_+(\pm) = T_-(\pm)^\dagger = \tfrac{1}{4}e^{i\varphi}[(2\hat{T}+1)^2-L_z^2]^{1/2}e^{i\varphi}$$

$$\mp \tfrac{1}{4}e^{i\varphi}\left\{\frac{(\hat{T}+T_0+1)[(2\hat{T}+1)^2-L_z^2]}{2\hat{T}+1}\right\}^{1/2} B\frac{e^{i\varphi}}{(2\hat{T}+1)^{1/2}} \pm \tfrac{1}{4}e^{i\varphi}\left[\frac{2\hat{T}+1-L_z}{2\hat{T}+1}\right]^{1/2} B^\dagger\left[\frac{(2\hat{T}+1-L_z)(\hat{T}+T_0+1)}{2\hat{T}+1}\right]^{\frac{1}{2}}e^{i\varphi}. \tag{4.66}$$

These operators are defined on the collective physical subspace:

$$|(T_0/2\ T_0/2)T\ M/2) = \frac{e^{iM\varphi}}{\sqrt{2\pi}}\otimes\frac{(B^\dagger)^{T_0-T}}{[(T_0-T)!]^{1/2}}|0), \ \ (T = T_0, T_0-1,\dots,0;\ M = 0, \pm 2, \pm 4,\dots, \pm 2T). \tag{4.67}$$

After mapping H and transition operators, expanding in $1/T_0$, and performing perturbative unitary transformations to diagonalize H to the order of interest, one finds *all results identical to those of the MW procedure.* This is only one of several examples verifying the MW prescription. For realistic models, of course, such symmetry-preserving mappings are not known. If they were known, the MW method would not be necessary. However, its success for simple but sufficiently general models, increases one's confidence in applications to realistic models.

4.5. The Bohr-Mottelson and Villars Representations for Three-Dimensional Rotation

In order to prepare the groundwork for extending the MW treatment to the rotational band structure of even-even axially symmetric nuclei, we must first consider the relation between the Bohr-Mottelson (BM) and Villars representations. In the general BM phenomenological theory of axially symmetric systems,[95] the Hamiltonian is given by an angular-momentum expansion

$$H_{\mathrm{BM}} = \sum_{mn} \mathcal{H}_{mn} \frac{1}{2}\left\{I'^{m}_{-}, I'^{n}_{+}\right\}, \tag{4.68}$$

where the angular-momentum components along the principal axes (PA) are denoted by $I'_{\pm} \equiv I_1 \pm iI_2$, I_3, with the 3-axis designated as the symmetry axes. The correponding components of **I** along the laboratory axis will be denoted by $I_{\pm} \equiv I_x \pm iI_y$, I_z. The coefficients $\mathcal{H}_{mn}$ are *intrinsic operators*, i.e.,

$$[\mathcal{H}_{mn}, \mathbf{I}] = 0, \quad [\mathcal{H}_{mn}, f(\varphi,\theta,\psi)] = 0, \tag{4.69}$$

where $f(\varphi,\theta,\psi)$ is an arbitrary function of the Euler angles. A unique property of the BM representation is the existence of an *intrinsic angular momentum* J_3 along the symmetry axis such that

$$[\mathbf{I}\, , J_3] = 0, \tag{4.70a}$$

$$[J_3\, , \mathcal{H}_{mn}] = (m-n)\, \mathcal{H}_{mn}. \tag{4.70b}$$

The property (4.70b) implies that the coefficients $\mathcal{H}_{mn}$ carry m-n units of intrinsic angular momentum along the symmetry axis. On the other hand, the only eigenvectors $|\ \rangle$ that are considered physical are those that satisfy the condition

$$(I_3 - J_3)|\ \rangle = 0. \tag{4.71}$$

Now, Villars[96] and, independently, V. M. Mikhailov,[97] proposed an angular-momentum expansion of the nuclear Hamiltonian given by

$$H = \sum_{mn} \mathcal{H}_{mn} \frac{1}{2}\left\{J'^{m}_{-}, J'^{n}_{+}\right\}, \tag{4.72}$$

where **J** denotes the total angular-momentum vector, with PA components $J'_{\pm} \equiv J_1 \pm iJ_2$, J_3. The correponding components of **J** along the laboratory axis will be denoted by $J_{\pm} \equiv J_x \pm iJ_y$, J_z. In addition, the coefficients $\mathscr{H}^{\circ}_{mn}$ are intrinsic operators, i.e.,

$$[\mathscr{H}^{\circ}_{mn}, \mathbf{J}] = 0, \; [\mathscr{H}^{\circ}_{mn}, f(\varphi,\theta,\psi)] = 0, \tag{4.73}$$

which can be expressed as multiple commutators of H with the Wigner D-functions, analogous to Eqs. (4.5)-(4.6), though this property will not be used.

So far, the Villars representation appears indistinguishable from the BM one, apart from the different symbol used for the angular momentum. However, there is one significant difference - *there is no intrinsic angular momentum in the Villars representation.* The coefficients $\mathscr{H}^{\circ}_{mn}$ *carry zero angular momentum.* In order to develop a microscopic theory having the form of the BM representation, which is the most convenient in practice, it is crucial to understand its relation to the Villars representation.

In the Villars representation, the Euler angle ψ for rotations about the symmetry axis, which, like the other Euler angles, is a function of nucleon coordinates, satisfies the equation

$$[J_3, e^{iK\psi}] = e^{iK\psi}. \tag{4.74}$$

Then, the operator $\mathscr{H}_{mn}$ defined by

$$\mathscr{H}_{mn} \equiv \mathscr{H}^{\circ}_{mn} e^{i(m-n)\psi} \tag{4.75}$$

satisfies the condition

$$[J_3, \mathscr{H}_{mn}] = (m-n)\, \mathscr{H}_{mn}, \tag{4.76}$$

which resembles Eq.(4.70b) in the BM representation. However, when expressed in terms of the operators (4.75) the Villars Hamiltonian becomes

$$H = \sum_{mn} \mathscr{H}_{mn} e^{i(n-m)\psi} \frac{1}{2}\left\{J'^{m}_{-}, J'^{n}_{+}\right\}. \tag{4.77}$$

While the coefficients $\mathscr{H}_{mn}$ do carry angular momentum along the symmetry axis, they are not intrinsic operators since they do not commute with any component of **J**. Moreover, the explicit ψ-dependence is an unesthetic feature that is absent in the BM representation.

The question now arises: is there a simple relation between the two representations? The answer is affirmative and the clue is provided by the condition (4.71) which suggests the presence of a *redundant variable* in the BM representation. Let us regard the Villars Hamiltonian (4.72) or (4.77) as defined on an *extended Hilbert space* - the direct product of the original space with the space of periodic functions $e^{iK\bar{\psi}}$ (K integral or half-integral). It should be noted that $\bar{\psi}$ *is a dummy variable,* whereas ψ, a true Euler angle, is a function of nucleon coordinates. But $\bar{\psi}$ is destined to assume the role of ψ.

In the extended Hilbert space, introduce the operator I_3 defined by

$$I_3 \equiv -i\frac{\partial}{\partial\bar{\psi}}, \tag{4.78}$$

satisfying

$$[I_3\,, e^{iK\bar{\psi}}] = Ke^{iK\bar{\psi}}. \tag{4.79}$$

Of course. I_3 and $e^{iK\bar{\psi}}$ commute with everything else.

It will be demonstrated that the Villars and BM representations are related by a unitary transformation on the extended Hilbert space. For this we will need the Euler-angle representation of the angular-momentum operators. While the conventional *zyz* representation is perfectly adequate for this purpose, for later purposes, it is essential to use the *xyz* definition of the Euler angles, in which the rotation operator is given by

$$R(\varphi,\theta,\psi) \equiv \exp(-i\varphi J_x)\exp(-i\theta J_y)\exp(-i\psi J_z). \tag{4.80}$$

Then the laboratory components of **J** are represented by

$$J_x = -i\frac{\partial}{\partial\varphi},$$

$$J_z \pm iJ_y = -ie^{\mp i\varphi}\left[-\tan\theta\frac{\partial}{\partial\varphi} \pm i\frac{\partial}{\partial\theta} + \frac{i}{\cos\theta}J_3\right], \tag{4.81}$$

and the PA components by

$$J'_{\pm} \equiv J_1 \pm iJ_2 = -ie^{\mp i\psi}\left[\frac{1}{\cos\theta}\frac{\partial}{\partial\varphi} \pm i\frac{\partial}{\partial\theta} - i\tan\theta\, J_3\right]$$

$$J_3 = -i\frac{\partial}{\partial\psi} \tag{4.82}$$

On the extended Hilbert space, one may define the unitary transformation

$$\mathcal{U} \equiv \exp(i\bar{\psi}J_3)\exp(-i\psi I_3), \tag{4.83}$$

which has the following effect:

$$\mathcal{U}\,J_3\,\mathcal{U}^\dagger = I_3, \qquad \mathcal{U}\,f(\psi)\,\mathcal{U}^\dagger = f(\psi+\bar{\psi}). \tag{4.84}$$

for arbitrary functions $f(\psi)$. This transformation has no effect on other degrees of freedom. The net effect on the laboratory components of **J** is to replace J_3 by I_3; denoting the transformed laboratory components by I_K, one obtains from (4.81)

$$I_k \equiv \mathcal{U}\,J_k\,\mathcal{U}^\dagger,\; k = x, y, z, \tag{4.85}$$

where

$$I_x \equiv J_x = -i\,\frac{\partial}{\partial\varphi}\,,$$

$$I_z \pm iI_y = -ie^{\mp i\varphi}\left[-\tan\theta\,\frac{\partial}{\partial\varphi} \pm i\,\frac{\partial}{\partial\theta} + \frac{i}{\cos\theta}I_3\right]. \tag{4.86}$$

If I_3 is defined as a PA component of I, its other PA components may be defined by

$$\mathcal{U}\,J'_{\pm}\,\mathcal{U}^{\dagger} = e^{\mp i\psi}I'_{\pm} = I'_{\pm}e^{\mp i\psi}, \tag{4.87}$$

where

$$I'_{\pm} \equiv I_1 \pm iI_2 = -ie^{\mp i\bar{\psi}}\left[\frac{1}{\cos\theta}\frac{\partial}{\partial\varphi} \pm i\,\frac{\partial}{\partial\theta} - i\tan\theta\,I_3\right],$$

$$I_3 \equiv -i\,\frac{\partial}{\partial\bar{\psi}}\,. \tag{4.88}$$

The commutation rules for the operators (4.86) and (4.88) are easily seen to fit the requirements for the corresponding components of an angular-momentum vector **I**.

The motivation for introducing **I** becomes clear upon transformation of the Villars Hamiltonian (4.72) or (4.77):

$$\mathcal{U}\,H\,\mathcal{U}^{\dagger} = \mathcal{U}\sum_{mn}\mathcal{H}_{mn}\,\frac{1}{2}\left\{J'^{m}_{-},\,J'^{n}_{+}\right\}\mathcal{U}^{\dagger} = \sum_{mn}\mathcal{H}_{mn}\,\frac{1}{2}\left\{I'^{m}_{-},\,I'^{n}_{+}\right\} = H_{\rm BM}\,. \tag{4.89}$$

One recovers the BM Hamiltonian, where $\mathcal{H}_{mn}$ is given by Eq. (4.75). Of course, the basis for this assertion is not the mere form of H, but also the fact that Eqs. (4.69) and (4.70) of the BM representation are now fulfilled. In particular, Eq. (4.69) is guaranteed since **I** is independent of J_3.

To complete the story, let us consider the transformation of state vectors. In the extended space, the physical Villars states $|\;\rangle$ satisfy

$$I_3|\;\rangle = 0. \tag{4.90}$$

As is easily checked,

$$\mathcal{U}\,I_3\,\mathcal{U}^{\dagger} = (I_3 - J_3). \tag{4.91}$$

Therefore, in the transformed BM representation, the physical states are defined by

$$(I_3 - J_3)|\;\rangle' = 0, \tag{4.92}$$

where $|\;\rangle' \equiv \mathcal{U}|\;\rangle$, which just corresponds to Eq. (4.71). One hopes this discussion will also dispell the mystery connecting this familiar but not thoroughly understood result.

As a final topic in this section, let us briefly consider transition operators. A spherical tensor T^L_M in Villars representation can be written in the form:

$$T^L_M = \sum_{K=-L}^{L} \hat{T}^L_K \, D^L_{MK}(\varphi,\theta,\psi), \tag{4.93}$$

where $D^L_{MK}(\varphi,\theta,\psi)$ is the Wigner rotation matrix, and $\hat{T}^L_K$ is a *scalar* operator, which, like H, has the Villars expansion:

$$\hat{T}^L_K = \sum_{mn} (\mathring{t}^L_K)_{mn} \, \frac{1}{2}\left\{J'^m_-, J'^n_+\right\}. \tag{4.94}$$

Here, $(\mathring{t}^L_K)_{mn}$ is an *intrinsic operator* with respect to **J** and φ,θ,ψ. Noting that

$$\mathcal{U}\, D^L_{MK}(\varphi,\theta,\psi)\, \mathcal{U}^\dagger = e^{iK\psi} D^L_{MK}(\varphi,\theta,\bar{\psi}), \tag{4.95}$$

one finds

$$(T^L_M)_{\rm BM} = \mathcal{U}\, T^L_M \, \mathcal{U}^\dagger = \sum_{K=-L}^{L} T'^L_K \, D^L_{MK}(\varphi,\theta,\bar{\psi}), \tag{4.96}$$

where

$$T'^L_K \equiv \sum_{mn} (t^L_K)_{mn} \, \frac{1}{2}\left\{I'^m_-, I'^n_+\right\}, \tag{4.97}$$

and

$$(t^L_K)_{mn} \equiv (\mathring{t}^L_K)_{mn} \, e^{i(K+m-n)\psi}, \tag{4.98}$$

having the properties:

$$[J_3 \,, (t^L_K)_{mn}] = (K+m-n)(t^L_K)_{mn} \,, \tag{4.99}$$

$$[\mathbf{I} \,, (t^L_K)_{mn}] = 0. \tag{4.100}$$

Of course, the operators $(t^L_K)_{mn}$ commute with all Euler angles and are therefore intrinsic operators with respect to **I**. These are just the properties associated with the phenomenological BM representation.

4.6. The Generic Model

It will be seen that the MW reconstruction leads naturally to the Villars representation, including the unesthetic $e^{iK\psi}$ factors in the Hamiltonian. But what is desired is the more appealing and convenient BM representation, and now we know how to get it with ease. Unfortunately, there seem to be no solvable microscopic three-dimensional models that is general enough for our purposes. In order to demonstrate the mechanics of the MW reconstruction for three-dimensional systems, it is sufficient to use the *generic model* - the most general *local* GHP boson expansion that can be constructed from generic RPA modes, subject to symmetry principles.[16] This model can serve as a kind of template for more specific models, microscopic or phenomenological. In fact, the method presented here was successfully tested on the Copenhagen-Frankfurt collective model, giving the same final result as a standard expansion that avoids a broken-symmetry treatment.[17]

In constructing the generic model, the following symmetry principles are assumed: the deformed mean field has axial symmetry about the z axis and inversion symmetry; H is even under time reversal, while, of course, $\mathbf{J}$ is odd; and, of utmost importance, H is rotationally invariant. The discussion will be limited to even particle numbers, although the extension to odd-particle numbers is reasonably straightforward. For simplicity, it will be assumed that the angular-momentum operators are the only broken-symmetry generators. The extension to include others, most notably, particle-number operators in superfluid nuclei, is straightforward.

Initially, one knows only the expansion of H around the axially symmetric mean-field given by

$$H = E^{(0)} + H^{(2)} + H^{(3)} + H^{(4)} + \ldots, \tag{4.101}$$

and the expansions of the broken-symmetry generators $J_\pm \equiv J_x \pm iJ_y$ given by

$$J_\pm = J_\pm^{(1)} + J_\pm^{(2)} + J_\pm^{(3)} + \ldots . \tag{4.102}$$

On the other hand, J_z (or any constant of motion) must be a *finite boson quadratic form*

$$J_z = J_z^{(2)}, \tag{4.103}$$

as required by Eq. (2.22a). The angular-momentum operators must satisfy the SU(2) commutators order by order in accord with Eqs. (4.9) - (4.10), as follows:

$$[J_+^{(1)}, J_-^{(1)}] = 2J_z^{(0)} = 0,$$

••••••

$$\sum_{k=1}^{n} [J_+^{(n-k+1)}, J_-^{(k)}] = 2J_z^{(n)}\delta_{n,2}, \tag{4.104a}$$

$$[J_z, J_\pm^{(n)}] = \pm J_\pm^{(n)}. \tag{4.104b}$$

In addition, the conservation law $[H, J_M] = 0$ is satisfied order by order as follows:

$$[H^{(2)}, J_{\pm}^{(1)}] = 0, \text{ (RPA order)},$$

••••••

$$\sum_{k=2}^{n} [H^{(k)}, J_{\pm}^{(n-k+1)}] = 0, \tag{4.105a}$$

$$[H^{(n)}, J_z] = 0. \tag{4.105b}$$

The RPA Hamiltonian has the general form[98]

$$H^{(2)} = E^{(2)} + \sum_{\mu} \hbar\omega_{\mu} B_{\mu}^{(1)\dagger} B_{\mu}^{(1)} + \hbar^2 J_{+}^{(1)} J_{-}^{(1)}/(2\mathcal{J})$$

$$= E^{(2)} + \sum_{\mu} \hbar\omega_{\mu} B_{\mu}^{(1)\dagger} B_{\mu}^{(1)} + \frac{\hbar^2}{2\mathcal{J}} (J_x^{(1)2} + J_y^{(1)2}), \tag{4.106}$$

where $E^{(2)}$ is the correlation energy, $B_{\mu}^{(1)\dagger}$ is a boson operator that creates the excitation with energy $\hbar\omega_{\mu}$ and the last term corresponds to the Goldstone modes. For a complete set of normal modes, one must include the two angle variables $\phi_{\pm}^{(1)}$ conjugate to $J_{\pm}^{(1)}$,

$$[\phi_{\pm}^{(1)}, J_{\mp}^{(1)}] = i \, , [\phi_{\pm}^{(1)}, J_{\pm}^{(1)}] = 0, \tag{4.107}$$

where both $\phi_{\pm}^{(1)}$ and $J_{\pm}^{(1)}$ commute with all the bosons $B_{\mu}^{(1)\dagger}$ and $B_{\mu}^{(1)}$. Later, it will be shown that $\phi_{\pm}^{(1)} = (\varphi^{(1)} \pm i\theta^{(1)})/2$, where $\varphi^{(1)}$ and $\theta^{(1)}$ are small-angle approximations to Euler angles φ and θ. As usual, $\phi_{\pm}^{(1)}$ are obtained by solving the Thouless-Valatin equation

$$[H^{(2)}, \phi_{\pm}^{(1)}] = -i\hbar^2 J_{\pm}^{(1)}/(2\mathcal{J}) \, , \tag{4.108}$$

which leads to the self-consistent cranking-model moment of inertia

$$\hbar^2/(2\mathcal{J}) = [\phi_{+}^{(1)}, [H^{(2)}, \phi_{-}^{(1)}]] = [\phi_{-}^{(1)}, [H^{(2)}, \phi_{+}^{(1)}]]. \tag{4.109}$$

In terms of the RPA modes, J_z can be written as

$$J_z = \sum_{\mu} K_{\mu} B_{\mu}^{(1)\dagger} B_{\mu}^{(1)} + i\phi_{+}^{(1)} J_{-}^{(1)} - i\phi_{-}^{(1)} J_{+}^{(1)}, \tag{4.110}$$

since $B_{\mu}^{(1)\dagger}$ carries K_{μ} units of angular momentum along the symmetry axis. The remaining terms are needed to satisfy the SU(2) commutator (4.104b) with $n = 1$. It should be noted that the RPA modes come in time-reversal degenerate pairs $B^{(1)\dagger}_{-\mu} \equiv \mathcal{T}\, B^{(1)\dagger}_{\mu}\, \mathcal{T}^{-1}$, where $K_{-\mu} = - K_{\mu}$ and $\mathcal{T}$ is the time-reversal operator.

The higher-order terms in the expansions of H and $J_{\pm}$ can now be constructed in terms of the RPA modes. For $H^{(3)}$ one obtains

$$
\begin{aligned}
H^{(3)} = {} & \tfrac{1}{2}\sum_{\mu\nu\lambda}\Big[h^{(3)}(00)_{\mu\nu\lambda}B^{(1)\dagger}_{\mu}B^{(1)\dagger}_{\nu}B^{(1)}_{\lambda} + \text{H.c.}\Big] + \tfrac{1}{6}\sum_{\mu\nu\lambda}\Big[h^{(3)}(00)'_{\mu\nu\lambda}B^{(1)\dagger}_{\mu}B^{(1)\dagger}_{\nu}B^{(1)\dagger}_{\lambda} + \text{H.c.}\Big] \\
& + \sum_{\mu} h^{(3)}(00)_{\mu}\Big[B^{(1)\dagger}_{\mu} + B^{(1)}_{\mu}\Big] + \Big[J^{(1)}_{-}\sum_{\mu\nu} h^{(3)}(10)_{\mu\nu}B^{(1)\dagger}_{\mu}B^{(1)}_{\nu} + \text{H.c.}\Big] \\
& + \Big\{\tfrac{1}{2}J^{(1)}_{-}\sum_{\mu\nu} h^{(3)}(10)'_{\mu\nu}\Big[B^{(1)\dagger}_{\mu}B^{(1)\dagger}_{\nu} - B^{(1)}_{-\nu}B^{(1)}_{-\mu}\Big] + \text{H.c.}\Big\} + J^{(1)}_{+}J^{(1)}_{-}\sum_{\mu} h^{(3)}(20)_{\mu}\Big[B^{(1)\dagger}_{\mu} + B^{(1)}_{\mu}\Big] \\
& + \Big\{(J^{(1)}_{-})^{2}\sum_{\mu} h^{(3)}(20)'_{\mu}\Big[B^{(1)\dagger}_{\mu} + B^{(1)}_{-\mu}\Big] + \text{H.c.}\Big\} \\
& + \tfrac{1}{2}\, i\Big[\big\{\phi^{(1)}_{+}, J^{(1)}_{-}\big\} + \big\{\phi^{(1)}_{-}, J^{(1)}_{+}\big\}\Big]\sum_{\mu} h^{(3)}(11)_{\mu}\Big[B^{(1)\dagger}_{\mu} - B^{(1)}_{\mu}\Big] \\
& + \Big\{i\phi^{(1)}_{-}J^{(1)}_{-}\sum_{\mu} h^{(3)}(11)'_{\mu}\Big[B^{(1)\dagger}_{\mu} - B^{(1)}_{-\mu}\Big] + \text{H.c.}\Big\} + \Big[i\phi^{(1)}_{-}\sum_{\mu\nu} h^{(3)}(01)_{\mu\nu}B^{(1)\dagger}_{\mu}B^{(1)}_{\nu} + \text{H.c.}\Big] \\
& \Big\{\tfrac{1}{2}i\phi^{(1)}_{-}\sum_{\mu\nu} h^{(3)}(01)'_{\mu\nu}\Big[B^{(1)\dagger}_{\mu}B^{(1)\dagger}_{\nu} + B^{(1)}_{-\nu}B^{(1)}_{-\mu}\Big] + \text{H.c.}\Big\} + \phi^{(1)}_{+}\phi^{(1)}_{-}\sum_{\mu} h^{(3)}(02)_{\mu}\Big[B^{(1)\dagger}_{\mu} + B^{(1)}_{\mu}\Big] \\
& + \Big\{\tfrac{1}{2}(\phi^{(1)}_{-})^{2}\sum_{\mu} h^{(3)}(02)'_{\mu}\Big[B^{(1)\dagger}_{\mu} + B^{(1)}_{-\mu}\Big] + \text{H.c.}\Big\},
\end{aligned}
\tag{4.111}
$$

and for $H^{(4)}$

$$
\begin{aligned}
H^{(4)} = {} & h^{(4)}(00) + \sum_{\mu\nu} h^{(4)}(00)_{\mu\nu}B^{(1)\dagger}_{\mu}B^{(1)}_{\nu} + \tfrac{1}{4}\sum_{\mu\nu\kappa\lambda} h^{(4)}(00)_{\mu\nu,\kappa\lambda}B^{(1)\dagger}_{\mu}B^{(1)\dagger}_{\nu}B^{(1)}_{\kappa}B^{(1)}_{\lambda} \\
& + J^{(1)}_{+}J^{(1)}_{-}\Big[h^{(4)}(20) + \sum_{\mu\nu} h^{(4)}(20)_{\mu\nu}B^{(1)\dagger}_{\mu}B^{(1)}_{\nu}\Big] + h^{(4)}(40)\Big[J^{(1)}_{+}J^{(1)}_{-}\Big]^{2} + \ldots\text{dispensable terms},
\end{aligned}
\tag{4.112}
$$

where the numerical coefficients $h^{(n)}(kl)$ indicate the order n and the number k of $J^{(1)}_{\pm}$ and l of

$\phi_{\pm}^{(1)}$ factors. The "dispensable terms" are those not needed to the order of interest. The required higher-order terms of $J_{\pm}$ are $J_{\pm}^{(2)}$ and $J_{\pm}^{(3)}$, where

$$J_{+}^{(2)} = J_{+}^{(1)}\sum_{\mu} j^{(2)}(10)_{\mu}\left[B_{\mu}^{(1)\dagger} + B_{\mu}^{(1)}\right] + J_{-}^{(1)}\sum_{\mu} j^{(2)}(10)'_{\mu}\left[B_{\mu}^{(1)\dagger} + B_{-\mu}^{(1)}\right]$$
$$+ i\phi_{+}^{(1)}\sum_{\mu} j^{(2)}(01)_{\mu}\left[B_{\mu}^{(1)\dagger} - B_{\mu}^{(1)}\right] + i\phi_{-}^{(1)}\sum_{\mu} j^{(2)}(01)'_{\mu}\left[B_{\mu}^{(1)\dagger} - B_{-\mu}^{(1)}\right]$$
$$+ \sum_{\mu\nu} j^{(2)}(00)_{\mu\nu} B_{\mu}^{(1)\dagger} B_{\nu}^{(1)} + \frac{1}{2}\sum_{\mu\nu} j^{(2)}(00)'_{\mu\nu}\left[B_{\mu}^{(1)\dagger} B_{\nu}^{(1)\dagger} - B_{-\nu}^{(1)} B_{-\mu}^{(1)}\right], \tag{4.113}$$

$$J_{+}^{(3)} = J_{+}^{(1)}\left[j^{(3)}(10) + \sum_{\mu\nu} j^{(3)}(10)_{\mu\nu} B_{\mu}^{(1)\dagger} B_{\nu}^{(1)}\right] + j^{(3)}(30)\, J_{+}^{(1)2} J_{-}^{(1)} + \text{dispensable terms.} \tag{4.114}$$

Of course, $J_{-}^{(n)} = J_{+}^{(n)\dagger}$. The notation for the numerical coefficients $j^{(n)}(kl)$ is similar to that for the $h^{(n)}(kl)$. There are some symmetry restrictions on these coefficients. The most important are due to the rotational invariance conditions (4.105) which imply the following relations:

$$h^{(3)}(11)_{\mu} = \hbar\omega_{\mu} j^{(2)}(10)_{\mu} + \hbar^2 j^{(2)}(01)_{\mu}/(2\mathcal{J}),$$
$$h^{(3)}(11)'_{\mu} = \hbar\omega_{\mu} j^{(2)}(10)'_{\mu} + \hbar^2 j^{(2)}(01)'_{\mu}/(2\mathcal{J}),$$
$$h^{(3)}(02)_{\mu} = -\hbar\omega_{\mu} j^{(2)}(01)_{\mu},$$
$$h^{(3)}(02)'_{\mu} = -\hbar\omega_{\mu} j^{(2)}(01)'_{\mu},$$
$$h^{(3)}(01)_{\mu\nu} = \hbar(\omega_{\mu} - \omega_{\nu}) j^{(2)}(00)_{\mu\nu},$$
$$h^{(3)}(01)'_{\mu\nu} = \hbar(\omega_{\mu} + \omega_{\nu}) j^{(2)}(00)'_{\mu\nu}. \tag{4.115}$$

4.7. Outline of the MW Scheme for the Three-Dimensional Case

The task now is to use the MW approach to make sense of the congeries of terms in the expansions of H and **J**. It is obvious that the RPA phonons should be chosen as the leading-order approximation to the intrinsic variables. But these do not commute with the angular-momentum component J_z, as is clear from Eq. (4.110). Therefore, **J** cannot be identified directly with the angular-momentum operators in the BM representation. However, this is not a serious problem since one can construct a Villars representation in which the coefficients $\mathcal{H}_{mn}$ do not, in fact, commute with **J**. The $\mathcal{H}_{mn}$ can therefore be chosen as functions of the vibrational phonons. Afterwards, one can transform to the BM representation in which $\mathbf{J} \to \mathbf{I}$ and **I** must commute with all phonons. That is the general strategy that will be pursued.

There is, however, a second problem, namely, that a unitary transformation cannot be used to directly generate $J_\pm$ from $J_\pm^{(1)}$ in analogy with the two-dimensional case, since $J_+^{(1)}$ and $J_-^{(1)}$ *commute*, while J_+ and J_- do not. An MW scheme will now be outlined to remedy this problem. Matters are greatly facilited by choosing the right Euler-angle represention - the *xyz* Euler-angle representation (4.81), which can also be written as follows:

$$J_x = p_\varphi ,$$

$$J_z \pm iJ_y = -\tan\theta \, \tfrac{1}{2}\left\{e^{\mp i\varphi}, p_\varphi\right\} \pm ie^{\mp i\varphi}p_\theta + \frac{e^{\mp i\varphi}}{\cos\theta}J_3. \tag{4.116}$$

It should be noted that the usual *zyz* representation cannot be used for a small-angle expansion since it blows up! The representation (4.116) also has the advantage that it identifies the canonical momentum p_φ with J_x, whose expansion is given. Given the RPA approximation $\varphi^{(1)}$, an expansion of φ can be then be found satisfying

$$[\varphi, p_\varphi] = [\varphi, J_x] = i, \tag{4.117}$$

as in the two-dimensional problem. By inverting (4.116) one then gets

$$p_\theta = \tfrac{1}{4}i\left\{e^{-i\varphi}, J_z - iJ_y\right\} - \tfrac{1}{4}i\left\{e^{i\varphi}, J_z + iJ_y\right\}. \tag{4.118}$$

The expansion of p_θ then follows from the known expansions of J_y, J_z and φ. The expansion of θ beginning with the RPA value can then be obtained from the relation

$$[\theta, p_\theta] = i. \tag{4.119}$$

In practice, it is convenient to define complex canonical coordinates

$$\phi_\pm \equiv (\varphi \pm i\theta)/2 , \ p_\pm \equiv p_\varphi \pm ip_\theta , \tag{4.120}$$

satisfying

$$[\phi_\pm , p_\mp] = i , [\phi_\pm , p_\pm] = 0. \tag{4.121}$$

The expansion of $p_\pm$,

$$p_\pm = p_\pm^{(1)} + p_\pm^{(2)} + p_\pm^{(3)} + ..., \tag{4.122}$$

is given by

$$p_\pm^{(1)} = J_\pm^{(1)}, p_\pm^{(2)} = J_\pm^{(2)}, p_\pm^{(3)} = J_\pm^{(3)} \pm \tfrac{1}{2}i\left\{\phi_+^{(1)} + \phi_-^{(1)}, J_z\right\} \mp \tfrac{1}{4}\left[\phi_+^{(1)} + \phi_-^{(1)}\right]^2\left[J_+^{(1)} - J_-^{(1)}\right]. \tag{4.123}$$

To solve for the expansion of $\phi_\pm$ beginning with the RPA values $\phi_\pm^{(1)}$ and that of the vibrational phonons,

$$B_\mu^\dagger = B_\mu^{(1)\dagger} + B_\mu^{(2)\dagger} + B_\mu^{(3)\dagger} + \ldots, \tag{4.124}$$

that commute with the rotational degrees of freedom, it is most convenient to use the formal unitary transformation method as in the two-dimensional case. The unitary transformation is chosen to satisfy

$$\exp(iS)\, J_\pm^{(1)} \exp(-iS) = p_\pm^{(1)} + p_\pm^{(2)} + p_\pm^{(3)} + \ldots, \tag{4.125}$$

where S has the expansion

$$S = S^{(3)} + S^{(4)} + \ldots\,. \tag{4.126}$$

This requires the sequential solution of the following hierarchy to the desired order:

$$[iS^{(3)}, J_\pm^{(1)}] = J_\pm^{(2)},$$

$$[iS^{(4)}, J^{(1)}] = J_\pm^{(3)} + \tfrac{1}{2}[J_\pm^{(2)}, iS^{(3)}] \pm \tfrac{1}{2}\, i\left\{\phi_+^{(1)} + \phi_-^{(1)}, J_z\right\} \mp \tfrac{1}{4}\left[\phi_+^{(1)} + \phi_-^{(1)}\right]^2 \left[J_+^{(1)} - J_-^{(1)}\right],$$

$$\bullet\bullet\bullet\bullet\bullet\bullet. \tag{4.127}$$

Given the solution for S, the set of RPA modes may be transformed as follows:

$$\exp(iS)\left\{J_\pm^{(1)}, \phi_\pm^{(1)}, B_\mu^{(1)\dagger}, B_\mu^{(1)}\right\}\exp(-iS) = \left\{p_\pm, \phi_\pm, B_\mu^\dagger, B_\mu\right\}. \tag{4.128}$$

Thus, it is possible to proceed much as in the two-dimensional case by carefully choosing the canonical variables.

The transformation (4.128) of H gives a function of $p_\pm, \phi_\pm, B_\mu^\dagger, B_\mu$. To eliminate the $\phi_\pm$ dependence, one needs to replace $p_\pm$ by the PA components of **J**. The PA components of **J**, denoted by J_K, are related to the given lab components J_M by

$$J_K = \sum_{M=-1}^{1} D_{MK}^{1\dagger}(\varphi,\theta,\psi)\, J_M = \sum_{M=-1}^{1} J_M\, D_{MK}^{1\dagger}(\varphi,\theta,\psi). \tag{4.129}$$

The expansion of D_{MK}^I can be found by expanding

$$D_{MK}^I(\varphi,\theta,\psi) = e^{iK\psi}\langle IK|\exp(i\theta J_y)\exp(i\varphi J_x)|IM\rangle. \tag{4.130}$$

Since $p_\pm$ is a function of the J_M and angles, it can finally be expressed as a function of the PA components J_K and angles with the help of the inverse of (4.129). To the order of interest here, one needs only the following relations:[16]

$$p_+p_- = J_1^{\,2} + J_2^{\,2} - \theta^2 p_\varphi^{\,2} + 2\theta p_\varphi J_3 + \tfrac{1}{2} + \ldots,$$

$$p_\pm = e^{\pm i\psi} J'_\pm + \text{cubic terms} = J'_\pm e^{\pm i\psi} + \text{cubic terms},$$

$$J_3 = \sum_\mu K_\mu B_\mu^\dagger B_\mu. \qquad (4.131)$$

Notice that this introduces $e^{\pm i\psi}$ factors into the Hamiltonian, as is characteristic of the Villars representation in the form of Eq. (4.77).

4.8. Transforming from the Villars to the BM Representation

If one chooses the *kinematic* solution for S, the expansion of H can be transformed into the Villars form:

$$H = E^{(0)} + H'^{(2)} + H'^{(3)} + H'^{(4)} + \ldots, \qquad (4.132)$$

where

$$H'^{(2)} = E^{(2)} + \sum_\mu \hbar\omega_\mu B_\mu^\dagger B_\mu + \frac{\hbar^2}{2\mathcal{J}} J'_+ J'_- = E^{(2)} + \sum_\mu \hbar\omega_\mu B_\mu^\dagger B_\mu + \frac{\hbar^2}{2\mathcal{J}} (J_1^{\,2} + J_2^{\,2}), \qquad (4.133)$$

$$\begin{aligned} H'^{(3)} = {} & \tfrac{1}{2}\sum_{\mu\nu\lambda}\left[h^{(3)}(00)_{\mu\nu\lambda} B_\mu^\dagger B_\nu^\dagger B_\lambda + \text{H.c.}\right] + \tfrac{1}{6}\sum_{\mu\nu\lambda}\left[h^{(3)}(00)'_{\mu\nu\lambda} B_\mu^\dagger B_\nu^\dagger B_\lambda^\dagger + \text{H.c.}\right] \\ & + \sum_\mu h^{(3)}(00)_\mu \left[B_\mu^\dagger + B_\mu\right] + \left[J'_- \mathrm{e}^{-i\psi} \sum_{\mu\nu} \Gamma^{(3)}(10)_{\mu\nu} B_\mu^\dagger B_\nu + \text{H.c.}\right] \\ & + \left[\tfrac{1}{2} J'_- \mathrm{e}^{-i\psi} \sum_{\mu\nu} \Gamma^{(3)}(10)'_{\mu\nu}\left[B_\mu^\dagger B_\nu^\dagger - B_{-\nu} B_{-\mu}\right] + \text{H.c.}\right] \\ & + (J_1^{\,2} + J_2^{\,2}) \sum_\mu \Gamma^{(3)}(20)_\mu \left[B_\mu^\dagger + B_\mu\right] \\ & + \left[J'^{\,2}_- \mathrm{e}^{-2i\psi} \sum_\mu \Gamma^{(3)}(20)'_\mu \left[B_\mu^\dagger + B_{-\mu}\right] + \text{H.c.}\right], \end{aligned} \qquad (4.134)$$

and

$$H'^{(4)} = E^{(4)} + \frac{1}{2}\sum_{\mu\nu} \Gamma^{(4)}(00)_{\mu\nu}\left[B^\dagger_\mu B_\mu + \frac{1}{2}\right]\left[B^\dagger_\nu B_\nu + \frac{1}{2}\right] + \sum_{\mu} \Gamma^{(4)}(00)_\mu\left[B^\dagger_\mu B_\mu + \frac{1}{2}\right]$$

$$+ \Gamma^{(4)}(20)(J_1{}^2 + J_2{}^2) + \Gamma^{(4)}(40)(J_1{}^2 + J_2{}^2)^2 + (J_1{}^2 + J_2{}^2)\sum_{\mu} \Gamma^{(4)}(20)_\mu\left[B^\dagger_\mu B_\mu + \frac{1}{2}\right]$$

$$+ \text{off-diagonal terms.} \tag{4.135}$$

The omitted off-diagonal terms contribute to a higher-order of perturbation theory than considered here. The numerical coefficients in Eq. (4.134) are given by

$$\Gamma^{(3)}(10)_{\mu\nu} = h^{(3)}(10)_{\mu\nu} - \frac{\hbar^2}{2\mathcal{J}} j^{(2)}(00)_{\mu\nu},$$

$$\Gamma^{(3)}(10)'_{\mu\nu} = h^{(3)}(10)'_{\mu\nu} - \frac{\hbar^2}{2\mathcal{J}} j^{(2)}(00)'_{\mu\nu},$$

$$\Gamma^{(3)}(20)_{\mu} = h^{(3)}(20)_{\mu} - \frac{\hbar^2}{2\mathcal{J}} j^{(2)}(10)_{\mu}.$$

$$\Gamma^{(3)}(20)'_{\mu} = h^{(3)}(20)'_{\mu} - \frac{\hbar^2}{2\mathcal{J}} j^{(2)}(10)'_{\mu}. \tag{4.136}$$

The expressions for the coefficients $\Gamma^{(4)}(kl)$ in (4.135), which are rather lengthy, are given in Ref. 16.

Finally, the transformation (4.83) converts the Villars to the BM representation as follows:

$$H_{\text{BM}} = \mathcal{U}\, H\, \mathcal{U}^\dagger = E^{(0)} + H^{(2)}_{\text{BM}} + H^{(3)}_{\text{BM}} + H^{(4)}_{\text{BM}} + \ldots, \tag{4.137}$$

where

$$H^{(2)}_{\text{BM}} = E^{(2)} + \sum_{\mu} \hbar\omega_\mu B^\dagger_\mu B_\mu + \frac{\hbar^2}{2\mathcal{J}}(I_1{}^2 + I_2{}^2), \tag{4.138}$$

$$H^{(3)}_{\rm BM} = \tfrac{1}{2}\sum_{\mu\nu\lambda}\left[h^{(3)}(00)_{\mu\nu\lambda}B^\dagger_\mu B^\dagger_\nu B_\lambda + \text{H.c.}\right] + \tfrac{1}{6}\sum_{\mu\nu\lambda}\left[h^{(3)}(00)'_{\mu\nu\lambda}B^\dagger_\mu B^\dagger_\nu B^\dagger_\lambda + \text{H.c.}\right]$$

$$+ \sum_\mu h^{(3)}(00)_\mu\left[B^\dagger_\mu + B_\mu\right] + \left[I'_-\sum_{\mu\nu}\Gamma^{(3)}(10)_{\mu\nu}B^\dagger_\mu B_\nu + \text{H.c.}\right]$$

$$+ \left[\tfrac{1}{2}I'_-\sum_{\mu\nu}\Gamma^{(3)}(10)'_{\mu\nu}\left[B^\dagger_\mu B^\dagger_\nu - B_{-\nu}B_{-\mu}\right] + \text{H.c.}\right]$$

$$+ (I_1{}^2 + I_2{}^2)\sum_\mu \Gamma^{(3)}(20)_\mu\left[B^\dagger_\mu + B_\mu\right]$$

$$+ \left[I'^2_-\sum_\mu \Gamma^{(3)}(20)'_\mu\left[B^\dagger_\mu + B_{-\mu}\right] + \text{H.c.}\right], \tag{4.139}$$

and

$$H'^{(4)}_{\rm BM} = E^{(4)} + \tfrac{1}{2}\sum_{\mu\nu}\Gamma^{(4)}(00)_{\mu\nu}\left[B^\dagger_\mu B_\mu + \tfrac{1}{2}\right]\left[B^\dagger_\nu B_\nu + \tfrac{1}{2}\right] + \sum_\mu \Gamma^{(4)}(00)_\mu\left[B^\dagger_\mu B_\mu + \tfrac{1}{2}\right]$$

$$+ \Gamma^{(4)}(20)(I_1{}^2 + I_2{}^2) + \Gamma^{(4)}(40)(I_1{}^2 + I_2{}^2)^2 + (I_1{}^2 + I_2{}^2)\sum_\mu \Gamma^{(4)}(20)_\mu\left[B^\dagger_\mu B_\mu + \tfrac{1}{2}\right]$$

$$+ \text{off-diagonal terms.} \tag{4.140}$$

In effect, the transition from the Villars to the BM representation involves the replacements $\mathbf{J} \to \mathbf{I}$ and $e^{iK\psi} \to 1$ in the Hamiltonian. Of course, all components of $\mathbf{I}$ commute with $B^\dagger_\mu$, B_μ.

Finally, we conclude with a few words concerning transition operators. The MW treatment leads to an expansion of the transition operator T^L_M involving the set of operators $B^\dagger_\mu$, B_μ, $\mathbf{I}$, $e^{\pm iK\psi}$, ϕ, and θ. From Eq, (4.93) and the relation

$$\lim_{\varphi\to 0\ \theta\to 0} D^L_{MK}(\varphi,\theta,\psi) = e^{iK\psi}\delta_{MK}, \tag{4.141}$$

one can find the intrinsic operators using

$$\hat{T}^L_M = e^{-iM\psi}\lim_{\varphi\to 0\ \theta\to 0} T^L_M, \tag{4.142}$$

and insert these in the global form (4.93). The BM representation of the transition operator is then given by Eq. (4.96) in which T'^L_K is effectively obtained from $\hat{T}^L_K$ by setting $\psi \to 0$ and replacing $J'_{\pm} \to I'_{\pm}$.

5. Summary and Conclusions

In this lecture, I have reviewed the basic ideas of boson mappings and have attempted to convey in the limited time available the multifarious applications to nuclear structure. The application to deformed systems was presented in considerably more detail. I hope to have convinced everyone that it is perfectly feasible to begin with a broken-symmetry mean field approximation and then to go on to calculate corrections to the energy and transition matrix elements. The calculated quantities have the correct dependence on the constants of motion involved in the broken symmetry without the use of complicated projection techniqes. The approach, of course, is perturbative, and some of its limitations have been outlined.

The MW method for three-dimensional rotational band structure can be applied in the neighborhood of superderformed minima as well as normal deformations, as long as the angular momentum is not too large. For high-spin states, an alternative formulation of the MW approach, which employs an expansion about a cranked mean field, is more appropriate.[91] The higher-order corrections to the RPA appearing in these methods, which include corrections to the moments of inertia, may play an important role in elucidating the perplexing problem of identical bands.[99]

Acknowledgments

This work was supported in part by the Department of Energy under grant DE-FG02-91ER40640.

References

1. S.T. Belyaev and V. G. Zelevinsky, *Nucl. Phys.* **39** (1962) 582.

2. T. Marumori, M. Yamamura and A. Tokunaga, *Prog. Theor. Phys.* 31 (1964) 1009; T. Marumori, M. Yamamura, A. Tokunaga and K. Takada, *Prog. Theor. Phys.* **32** (1964) 726; M. Yamamura, *Prog. Theor. Phys.* **33** (1965) 199.

3. E. R. Marshalek, *Nucl. Phys.* **A224** (1974) 221, 245.

4. E. R. Marshalek, *Nucl. Phys.* **A357** (1981) 398; A. Klein and E. R. Marshalek, *J. Math. Phys.* **30** (1989) 219.

5. A. Klein and E. R. Marshalek, *Rev. Mod. Phys.* **63** (1991) 375.

6. F. J. Dyson, *Phys. Rev.* **102** (1956) 1217.

7. T. Holstein and H. Primakoff, *Phys. Rev.* **58** (1940) 1098.

8. K. Takada, *Phys. Rev.* **C34** (1986) 750.

9. S. C. Pang, A. Klein and R. M. Dreizler, *Ann. Phys.* (N. Y.) **49** (1968) 477.

10. S. Okubo, *Phys. Rev.* **C10** (1974) 2045, 2048.

11. A. Klein and E. R. Marshalek, *Z. Phys.* **A329** (1988) 441.

12. T. Suzuki and K. Matsuyanagi, *Prog. Theor. Phys.* **56** (1976) 1156.

13. E. R. Marshalek and J. Weneser, *Ann. Phys.* (N. Y.) 53 (1969) 568.

14. E. R. Marshalek and J. Weneser, *Phys. Rev.* **C2** (1970) 1682.

15. E. R. Marshalek, *Ann. Phys.* (N. Y.) **143** (1982) 191.

16. E. R. Marshalek, *Phys. Rev.* **C35** (1987) 1900.

17. E. R. Marshalek, *Phys. Rev.* **C36** (1987) 2538.

18. K.Takada, *Nucl. Phys.* **A439** (1985) 489.

19. K.Takada and S. Tazaki, *Nucl. Phys.* **A448** (1986) 56.

20. K.Takada and Y. R. Shimizu, *Nucl. Phys.* **A468** (1987) 11.

21. K. Takada and K. Yamada, *Nucl. Phys.* **A462** (1987) 561.

22. K.Takada, *Phys. Rev.* **C38** (1988) 2450.

23. K.Takada, Y. R. Shimizu and H. Thorn, *Nucl. Phys.* **A485** (1988) 189.

24. K. Takada, K. Yamada and H. Tsukuma, *Nucl. Phys.* **A496** (1989) 224.

25. T. Kishimoto and T. Tamura, *Nucl. Phys.* **A192** (1972) 246.

26. T. Kishimoto and T. Tamura, *Nucl. Phys.* **A270** (1976) 317.

27. T. Tamura, K. J. Weeks and T. Kishimoto, *Phys. Rev.* **C20** (1979) 307.

28. T. Tamura, K. J. Weeks and T. Kishimoto, *Nucl. Phys.* **A347** (1980) 359.

29. T. Kishimoto and T. Tamura, *Phys. Rev.* **C27** (1983) 341.

30. T. Tamura, *Phys. Rev.* **C28** (1983) 2154.

31. H. Sakamoto and T. Kishimoto, *Nucl. Phys.* **A486** (1988) 1.

32. H. Sakamoto and T. Kishimoto, *Nucl. Phys.* **A501** (1989) 242.

33. J. Dobaczewski, *Nucl. Phys.* **A369** (1981) 213, 237.

34. C. T. Li, *Nucl. Phys.* **A417** (1984) 37.

35. C. T. Li, *Phys. Rev.* **C29** (1984) 2309.

36. H. B. Geyer, C. A. Engelbrecht and F. J. W. Hahne, *Phys. Rev.* **C33** (1986) 1041.

37. F. J. W. Hahne, H. B. Geyer and C. A. Engelbrecht, *Phys. Rev.* **C37** (1988) 885.

38. P. Park, *Phys. Rev.* **C35** (1987) 807.

39. C. M. Vincent, L. J. Henry and G. K. Kim, *Phys. Rev.* **C39** (1988) 1583.

40. J. Dobeš, S. P. Ivanova, R. V. Jolos and R. Pedrosa, *Phys. Rev.* **C41** (1990) 1840.

41. J. Dobaczewski, H. B. Geyer and F. J. W. Hahne, *Projection onto physical boson states in a collective subspace* (preprint, 1991).

42. E. R. Marshalek, *Phys. Lett.* **B62** (1976) 5.

43. J. Dukelsky, C. G. Dussel, R. P. J. Perazzo, S. L. Reich and H. M. Sofia, *Nucl. Phys.* **A425** (1984) 93.

44. J. Dukelsky, R. P. J. Perazzo, S. L. Reich and H. M. Sofia, *Phys. Lett.* **B158** (1985) 361.

45. M. C. Cambiaggio and J. Dukelsky, *Phys. Lett.* **B197** (1987) 479.

46. R. Bijker, S. Pittel and J. Dukelsky, *Phys. Lett.* **B219** (1988) 5.

47. J. Dukelsky and S. Pittel, *Phys. Rev.* **C39** (1989) 2001.

48. R. Bijker, S. Pittel and J. Dukelsky, *Nucl. Phys.* (to be published).

49. E. R. Marshalek and G. Holzwarth, *Nucl. Phys.* **A191** (1972) 438.

50. J. P. Blaizot and E. R. Marshalek, *Nucl. Phys.* **A309** (1978) 422.

51. J. P. Blaizot and E. R. Marshalek, *Nucl. Phys.* **A309** (1978) 453.

52. T. Marumori, T. Maskawa, F. Sakata and A. Kuriyama, *Prog. Theor. Phys.* **64** (1980) 1294.

53. F. Sakata, T. Marumori, Y. Hashimoto and T. Une, *Prog. Theor. Phys.* **70** (1983) 424.

54. F. Sakata, T. Marumori, K. Maramatsu and Y. Hashimoto, *Prog. Theor. Phys.* **74** (1985) 51.

55. Y. Hashimoto, T. Marumori and F. Sakata, *Prog. Theor. Phys.* **78** (1987) 1364.

56. A. Bulgac, A. Klein, N. R. Walet and G. Do Dang, *Phys. Rev.* **C40** (1989) 945.

57. N. R. Walet, A. Klein and G. Do Dang, *J. Chem. Phys.* **91** (1989) 2848.

58. N. R. Walet, A. Klein, G. Do Dang and A. Bulgac, *Phys. Rev.* 41 (1990) 318.

59. J. Dobaczewski, *Nucl. Phys.* A380 (1982) 1.

60. A. Klein, T. D. Cohen and C. T. Li, *Ann. Phys.* (N. Y.) 141 (1982) 382.

61. J. Deenen and C. Quesne, *J. Math. Phys.* 25 (1984) 1638, 2354.

62. D. J. Rowe, G. Rosensteel and R. Gilmore, *J. Math. Phys.* 26 (1985) 2787.

63. C. Quesne, *J. Math. Phys.* 27 (1986) 869.

64. K. T. Hecht, *The Vector Coherent-State Method and Its Application to Problems of Higher Symmetries*, Lecture Notes in Physics, Vol. 290 (Springer-Verlag, Berlin, 1987).

65. D. J. Rowe, R. Le Blanc and K. T. Hecht, *J. Math. Phys.* **29** (1988) 287.

66. K. T. Hecht, *Nucl. Phys.* A493 (1989) 29.

67. K. T. Hecht and J. Q. Chen, *Nucl. Phys.* **A512** (1990) 365.

68. R. Le Blanc, K. T. Hecht and L. C. Biedenharn, *J. Phys. A: Math. Gen.* 24 (1991) 1393.

69. F. J. W. Hahne and A. Klein, *Phys. Lett.* **B229** (1989) 1.

70. A. Klein and N. R. Walet, *Nucl. Phys.* **A515** (1990) 207.

71. K. T. Hecht, *A New Generalization of Vector Coherent State Theory and the Quantized Bogoliubov Transformation of Abraham Klein* (to be published in Proc. of the Symposium on Contemporary Physics Celebrating the Sixty-fifth Birthday of Abraham Klein).

72. J. da Providência and C. Fiolhais, *Nucl. Phys.* **A435** (1985) 190.

73. T. Hatsuda, *Nucl. Phys.* **A492** (1989) 187.

74. Y. Takahashi and H. Umezawa, *Collec. Phen.* **2** (1975) 55.

75. H. Umezawa, H. Matsumoto and M. Tachiki, *Thermo Field Dynamics and Condensed States* (North Holland, Amsterdam, 1982).

76. N. R. Walet and A. Klein, *Nucl. Phys.* **A510** (1990) 261.

77. J. da Providência, M. Yamamura and A. Kuriyama, *Prog. Theor. Phys.* **85** (1991) 939; A. Kuriyama, J. da Providência and M. Yamamura, *Prog. Theor. Phys.* **87** (1992) 135; A. Kuriyama, M. Yamamura, J. da Providência and C. Providência, *Phys. Rev.* **C45** (1992) 2196.

78. S. Pittel, J. Engel, J. Dukelsky and P. Ring, *Phys. Lett.* **B247** (1990) 185.

79. E. G. Nadjakov, *J. Phys. G: Nucl. Part. Phys.* **16** (1990) 1473.

80. J. Meyer, *J. Math. Phys.* **32** (1991) 2142.

81. J. Meyer, *Boson-Fermion and Baryon Mappings: Construction of Collective Subspaces*, I & II (to be published in *J. Math. Phys.*).

82. I. Hamamoto, *Nucl. Phys.* **A271** (1976) 15.

83. E. R. Marshalek, *Nucl. Phys.* **A294** (1978) 92.

84. E. R. Marshalek, *Nukleonika*, **23** (1978) 409.

85. Y. R. Shimizu and K. Matsuyanagi, *Prog. Theor. Phys.* **74** (1985) 1346.

86. T. Bengtsson, *Nucl. Phys.* A496 (1989) 56.

87. E. R. Marshalek and M. Sabato, *Phys. Rev.* **C5** (1972) 1130.

88. A. J. Lichtenberg and M. A. Lieberman, *Regular and Stochastic Motion* (Springer-Verlag, New York, 1983).

89. P. F. Bortignon, R. A. Broglia, D. R. Bes and R. J. Liotta, *Phys. Rep.* **30C** (1977) 305.

90. G. D. Birkhoff, *Dynamical Systems* (American Mathematical Society, New York, 1966), Vol. IX; G. F. Gustavson, *Astron. J.* **71** (1966) 670.

91. E. R. Marshalek, *Nucl. Phys.* **A275** (1977) 416.

92. F. M. H. Villars, *Nucl. Phys.* 74 (1965) 353.

93. S. A. Moszkowski, *Phys. Rev.* **110** (1958) 43.

94. J. P. Elliott, *Proc. Roy. Soc.* **A245** (1958) 128, 562.

95. Aa. Bohr and B. Mottelson, *Nuclear Structure*, Vol. II (W. A. Benjamin, Reading, 1975) Ch. 4.

96. F. M. H. Villars, in *Many-Body Description of Nuclear Structure and Reactions*, Proceedings of the International School of Physics "Enrico Fermi", Course XXXVI, ed. C. Bloch (Academic Press, New York, 1966) p. 35.

97. V. M. Mikhailov, *Izv. ANSSSR. Ser. Fiz.* (*Bull. Acad. of Sci. USSR, Phys. Ser.*) **35** (1971) 794.

98. T. Kammuri, *Prog. Theor. Phys.* 37 (1967) 1131.

99. C. Baktash, J. D. Garrett, D. F. Winchell and A. Smith, *Low-Spin Identical Bands in Neighboring Odd-A and Even-Even Nuclei: A Challenge to Mean-Field Theories* (preprint, 1992).

CRITICAL EVALUATION OF BOSON EXPANSION THEORY (GROUP 9)

FUMIHIKO SAKATA
Institute for Nuclear Study, University of Tokyo
Tanashi, Tokyo 188, Japan

Group Members: O. Castaños, M. Mukerjee, T. Otsuka and F. Sakata

1. Brief Summary of Boson Expansion Theory

Before evaluating the boson expansion theory (BET) presented by E. Marshalek, we start by briefly summarizing the history of BET to illustrate how useful it is. The BET is mainly divided into three different categories; the Schwinger type, the Dyson type and the Holstein-Primakoff type. It is also divided into two types of mapping; operator mapping type and state vector mapping type. In the former, the boson representation of the fermion pair operators is defined in such a way that the Lie algebra of fermion pair operators is satisfied by the boson representation of fermion pair operators. Since the boson space of N-boson system is larger than the fermion space of 2N-fermion system, in the latter mapping, the unitary transformation is introduced to precisely keep the one-to-one correspondence between the fermion space and "physical" boson space. The Schwinger BET belongs to the operator mapping type and the fermion Lie algebra is realized by introducing two kinds of boson operators whose physical meaning in terms of fermion operators is not clear. The Dyson BET is also classified in the operator mapping type and gives a non-unitary realization of the Lie algebra of fermion pair operators. Non-unitary property of the Dyson BET has been considered to be a major drawback. Recently, K. Takada showed how to calculate the eigenstates in the non-unitary boson representation and applied it to various realistic cases. A striking feature of these two BET rests on their finiteness of the expansion.

The Holstein-Primakoff BET is subdivided into the operator mapping type and the state vector mapping type. The former had been introduced by Belyaev-Zelevinski, followed by Kishimoto-Tamura, Marshalek-Weneser and Jannsen-Jolos-Dönau(JJD). The JJD had proposed SU(6) group theoretical structure, which gave a starting point of introducing the IBM proposed by Arima-Iachello. The latter was initiated by Marumori-Yamamura-Takada, followed by Holzwarth-Lie (Modified Marumori). This state vector mapping was further developed into the multi-phonon approach by Iwasaki-Sakata-Takada (IST), Piepenbring and Soloviev, which are widely applied in a variety of phenomena. The IST discussed how the SU(6) approximation is justified by taking account of the Pauli-principle.

2. Basic Idea of Marshalek's Treatment

Suppose there exists a constant of motion

$$\hat{J} \quad \text{satisfying} \quad [\hat{H}, \hat{J}] = 0.$$

Suppose there exists a stable mean-field $|\phi_0\rangle$ violating the symmetry

$$\delta\langle\phi_0|\hat{H}|\phi_0\rangle = 0, \quad \text{with} \quad \hat{J}|\phi_0\rangle \neq J|\phi_0\rangle.$$

Namely, the state $|\phi_0\rangle$ breaks the symmetry spontaneously. Referring to the state $|\phi_0\rangle$, the Hamiltonian is divided into

$$\hat{H} = \hat{H}_0 + \hat{H}_{\text{int}}, \quad \text{where} \quad [\hat{H}_0, \hat{J}] \neq 0.$$

As is well known, the broken symmetry is restored within the RPA, *i.e.* in the small-amplitude treatment:

$$[\hat{H}, \hat{J}_{\text{RPA}}]_{\text{RPA}} = 0.$$

which just corresponds to the Nambu-Goldstone mode. Within the RPA, one may introduce the RPA angle variable valid for in small amplitude region, as well as the inertia parameter by using the following equation;

$$[\hat{H}, \hat{\Phi}_{\text{RPA}}]_{\text{RPA}} = \frac{1}{\mathcal{J}}\hat{J}_{\text{RPA}}, \quad [\hat{J}_{\text{RPA}}, \hat{\Phi}_{\text{RPA}}]_{\text{RPA}} = -i.$$

Since the global quantities also satisfy the following relation

$$[\hat{J}, \hat{\Phi}] = -i,$$

which has the same form satisfied by the local quantities, there must exists a formal unitary transformation U which relates the local and global quantities

$$\hat{\Phi} = \hat{U}^{-1}\hat{\Phi}_{\text{RPA}}\hat{U}, \quad \hat{J} = \hat{U}^{-1}\hat{J}_{\text{RPA}}\hat{U}.$$

By exploiting a special feature of the Marshalek's Holstein-Primakoff BET where the fermion Lie algebra is satisfied order by order in the boson expansion series, Marshalek showed a general method for restoring the broken symmetry by going beyond the RPA.

3. Advantage of Marshalek's Holstein-Primakoff BET

Let us summarize the merits of Marshalek's Holstein-Primakoff BET.

a) It gives one of the general quantum theories of many-body problem

b) Since there exists a small parameter N in the Marshalek's Holstein-Primakoff BET, *i.e.*

$$\text{(fermion pair operator)} \Rightarrow B^\dagger\sqrt{1-\frac{B^\dagger B}{N}},$$

it is possible to introduce some truncation method.

c) It gives one possible way how to restore the broken symmetry. However, this method may be valid only within small amplitude region or slow angular momentum region.

d) The Marshalek's Holstein-Primakoff BET gives clear relation to the time-dependent Hartree-Fock theory. More generally, it gives a general method for the realization of the Lie algebra in the symplectic manifold

4. Disadvantage of Marshalek's Holstein-Primakoff BET

a) Since the BET intends to exactly transform the fermion system into the boson system, there arises an important question what is the merit of BET over the shell model calculation. If one solves the full asymmetric fermion system within the boson system by using the boson expansion method, the boson system has less symmetry and one is involved into much tedious calculation.

b) Since the BET is a perturbation theory, there appears a problem of small denominator. If there appears small denominator, the BET meets a difficulty of convergence. Namely, the boson operators defined at some stable mean-field is not the good boson operators any more. If one introduce some truncation by keeping only the collective boson operators, there arises question whether or not the collective boson operator has a meaning in large-amplitude region. Namely, the range of validity of the BET is not fully explored. As pointed out by Zelevinsky, it is one of the promising method to introduce some self-consistent boson operators.

c) An important aspect of the nuclear dynamics, *i.e.* the single-particle dynamics is out of the scope of the BET.

A REBUTTAL TO THE EVALUATION OF BOSON EXPANSION THEORY

E. R. Marshalek
Department of Physics, University of Notre Dame, Notre Dame, IN 46556

I have only a few minor differences with Prof. Sakata's evaluation. First of all, at the risk of appearing picayune, I would prefer to avoid the acronym of BET for the formalism I have presented because it is primarily a formalism rather than a theory, and also because the acronym is associated with a different kind of application of boson expansions, namely that of Kishimoto and Tamura. My historical perspective on the development and classification of boson mapping formalisms is also a little different from Prof. Sakata's, but that can be found in the recent review paper I did with Prof. Klein.

In the discussion of the disadvantages, Prof. Sakata comments that in the treatment of an "asymmetric" fermion system, the corresponding boson system has "less symmetry". I assume that what he means by "asymmetric" is deformed, i.e., having a broken symmetry. The main point of my talk was precisely to show that it is possible to reconstruct all the symmetries when the starting point is an expansion in deformed bosons. That the calculation of higher-order corrections is tedious is a subjective but certainly not unjustified comment. However, are there any less tedious formalisms available that achieve the same goals? In any case, once the formal manipulations are done for a sufficiently general system, they need not be repeated. In addition, the formalism can be implemented using computer algebra to relieve the tedium.

I agree that the small-denominator problem is a nuisance, but I believe it can be overcome with the aid of degenerate perturbation theory in a quantal treatment, or classically using techniques familiar to Prof. Sakata, as he himself has kindly advised me. Finally, I must emphatically disagree with the statement that single-particle dynamics is out of the scope of the formalism. Certainly, the boson-fermion expansion is capable of taking into account the degrees of freedom of odd nucleons. In addition, the formalism is not limited only to collective excitations: it is easily seen that when the two-body force is turned off, one recovers all the exact results (together with the inevitable unphysical excitations).

GROUP 10

Interacting Boson Model

INTERACTING-BOSON MODEL

P. O. LIPAS
Department of Physics, University of Jyväskylä
SF-40351 Jyväskylä, Finland

ABSTRACT

The interacting-boson model (IBM≡IBA) of nuclear structure is reviewed. Its phenomenological basic version IBA-1 is discussed as a numerical scheme and in terms of its group structure and dynamic symmetries. The proton–neutron version IBA-2 is considered both as a phenomenological model and as an approximation to the shell model. Dynamic symmetries, F spin and magnetic phenomena are examined. The microscopic basis of the IBA is viewed within the Otsuka–Arima–Iachello mapping scheme.

1. Introduction

Why do we need nuclear models other than the shell model, which is the basis of all thinking of nuclear structure? The somewhat trite answer is that the shell model can only handle light nuclei and nuclei near closed shells. Actually the modern codes and computers can handle shell-model spaces far larger than was the case only a decade ago. Dimensions as large as 62 000 in the J scheme and 320 000 in the M scheme are being quoted presently[1]. Thus the shell model reaches quite far into regions only recently thought to be inaccessible. Nevertheless, the J-scheme dimensions for single major shells in the $A \approx 150$ region are typically 10^{15}, so there is little hope for all-out shell-model calculations. Since, furthermore, only a tiny fraction of the lowest states are amenable to experiment, simplifying models have a market.

The interacting-boson model (IBM), synonymously known as the interacting-boson approximation (IBA), was proposed in an embryonic form by Iachello and Arima[2] nearly twenty years ago. The early model contained only d bosons ($J = 2$) and thus was effectively just a variation on the theme of quadrupole phonons, namely one with conserved boson number. The phonon description dates back to Aage Bohr[3], as one aspect of the Bohr–Mottelson collective model[4], and it was later generalized to a versatile form by Gneuss and Greiner[5].

The subsequent addition[6] of the s boson ($J = 0$) to the IBA gave the model its full scope which has provided much impetus to the study of nuclear structure to this day. The definitive papers by Arima and Iachello[7] were followed by a multitude of applications and extensions, a number of reviews (e.g. Refs. 8–13) and books by Iachello and Arima[14] and by Bonatsos[15]; the latter contains a comprehensive list of references to the earlier literature.

The version of the IBA referred to so far is its basic form IBA-1. Its bosons take no account of the two kinds of nucleon; it is a "one-fluid" model like the basic Bohr–Mottelson model. Moreover, the s boson of the IBA can be effectively incorporated in the quadrupole phonon model, as reviewed in e.g. Ref. 16. Thus all s operators can be replaced with d operators, e.g. $d^\dagger s \rightarrow d^\dagger(N - d^\dagger \cdot \tilde{d})^{1/2}$, where N is the constant number of IBA bosons or, equivalently, the maximum number of quadrupole phonons. In fact, such a phonon formulation by Janssen *et al.* [17] antedates the one with s bosons. Nevertheless, the s-boson formulation helps bring out the *dynamic symmetries* which have become the hallmark of the IBA. Because of the group-theoretical structure of the IBA highlighted in the dynamic symmetries, the model has received the epithet "algebraic" in contrast to the "geometric" concept of the Bohr–Mottelson model. For a fresh review of the algebraic approach see Van Isacker and Warner [18].

With due account of the two kinds of nucleon, the IBA model becomes the proton–neutron IBA, known for short as IBA-2. This version had its inception [19] rather shortly after the advent of IBA-1. The presence of two species of boson enriches the model considerably. It contains all the predictions, or descriptive capabilities, of IBA-1 essentially intact. In addition IBA-2 contains a new class of states characterized by incomplete symmetry between proton and neutron bosons. The prediction was verified experimentally in the discovery of the so-called scissors states in deformed nuclei by Bohle *et al.* [20]. The reviews and books cited above also discuss IBA-2.

A fundamentally important aspect of the model emerged with IBA-2. Already the introductory papers anchored the model to nucleon microscopy. While the bosons of IBA-1 were loosely spoken of as formed of pairs of like nucleons, the pair origin of the bosons was formulated in quantitative terms in IBA-2. In short the idea was a giant truncation of the shell-model space to S and D pairs which were then mapped onto s and d bosons [21]. In modern-day perspective the microscopic foundation of the IBA is the most interesting aspect of the model. The issue is far from closed; for an interim report see Iachello and Talmi [22] and for a new view, called democratic mapping, Skouras *et al.* [23] and Bonatsos *et al.* [24].

While IBA-1 and IBA-2 represent the main lines of development of the model, there are many extensions and offshoots, some of which even transcend the boundaries of nuclear-structure physics into molecular spectroscopy and scattering theory. True to the main line of development are versions known as IBA-3 and IBA-4, where proton–neutron pairs and isospin are tagged on to the model in the context of light nuclei; this work is due to Elliott and coworkers [25]. Furthermore, g ($J = 4$) and other bosons have been added to IBA-1 and IBA-2.

So far my cavalcade has been restricted to model versions applicable to doubly even nuclei. Odd-A nuclei and doubly odd nuclei have indeed been incorporated into the IBA scheme under the name IBFA; the book by Iachello and Van Isacker [26] contains the complete story. The idea is simply to treat the "last" unpaired nucleon (of each kind) within a single-particle shell model coupled to the IBA doubly even core.

Below I focus attention on some selected points of those mentioned in this introduction. Only IBA-1 and IBA-2 will be considered (Sects. 2 and 3). Whatever group theory comes in is interwoven in the general presentation; it is well to remark already at this point that the IBA need not be viewed as an exclusively "group-theoretic" model. Following the phenomenological presentation of these model versions, I then discuss in Sect. 4 the microscopy of IBA-2. The review is summarized in Sect. 5.

2. The Model IBA-1

2.1. Bosons and Operators

The assumptions inherent in the model IBA-1 can be listed as follows:

1. The model is intended to apply to low-lying collective states of doubly even nuclei.
2. Like valence nucleons behave as correlated pairs, i.e. linear combinations of bare pairs. Holes replace particles in the upper half-shell. There is no difference between correlated proton and neutron pairs.
3. The correlated pairs behave as identical bosons whose number is conserved.
4. The bosons may occupy two levels: $J = 0$ (s) and $J = 2$ (d). This is suggested as a first approximation by the nucleon–nucleon interaction.
5. The bosons have one-body energies and all two-body interactions in the Hamiltonian.
6. Other operators (the E2 transition operator being the most important one) are constructed so as to conserve boson number. They are taken to be one-body operators when sufficient.

Although IBA-1 is taken as a phenomenological model, points 2 and 4 above make qualitative reference to microscopy. The practical message of point 2 is that it contains an algorithm for deducing the number of bosons from shell structure. Thus, e.g., ${}^{128}_{54}\mathrm{Xe}_{74}$ has 4 valence protons and 8 valence neutron holes, whence its boson number is $N = 2 + 4 = 6$. I want to stress here that the IBA is constructed to have good boson number, in contrast to most formulations of the geometric boson model.

Let us spell out the Hamiltonian intended in assumption 5. In generic terms it is simply the boson Hamiltonian

$$H = \sum_i \epsilon_i b_i^\dagger b_i + \tfrac{1}{4} \sum_{ijkl} {}_{\mathrm{s}}\langle ij|V|kl\rangle_{\mathrm{s}}\, b_j^\dagger b_i^\dagger b_k b_l \,. \tag{1}$$

With the boson states s and d and angular-momentum coupling introduced, this becomes

$$H = \epsilon_{\mathrm{s}} s^\dagger s + \epsilon_{\mathrm{d}} d^\dagger \cdot \tilde{d} + \tfrac{1}{2} \sum_{L=0,2,4} c_L [d^\dagger d^\dagger]_L \cdot [\tilde{d}\tilde{d}]_L + \frac{\tilde{v}_2}{\sqrt{10}} ([d^\dagger d^\dagger]_2 \cdot \tilde{d} s + \mathrm{H.c.})$$
$$+ \frac{\tilde{v}_0}{2\sqrt{5}} (d^{\dagger 2} s^2 + \mathrm{H.c.}) + \frac{u_2}{\sqrt{5}}\, d^\dagger s^\dagger \cdot \tilde{d} s + \frac{u_0}{2}\, s^{\dagger 2} s^2 \,. \tag{2}$$

The two-body boson matrix elements are taken as parameters of the model and are here given in their historical notation. The tensor operator for d-boson annihilation is $\tilde{d}_m = (-)^m d_{-m}$.

The general IBA-1 Hamiltonian (2) contains 9 free parameters. Since the boson number N is fixed for a given nucleus, we can replace $s^\dagger s$ by $N - d^\dagger \cdot \tilde{d}$, so that for excitation energies the Hamiltonian only has 6 parameters. It can be put in a variety of different forms. Particularly popular in numerical applications using the code PHINT[27] is the so-called multipole form

$$H = \epsilon'' \hat{n}_\mathrm{d} + a_0 P^\dagger P + a_1 \mathbf{J}^2 + a_2 Q^2 + a_3 T_3^2 + a_4 T_4^2 \,, \tag{3}$$

$$P \equiv \tfrac{1}{2}(\tilde{d}^2 - s^2)\,; \qquad T_l \equiv [d^\dagger \tilde{d}]_l\,, \quad l = 0,\,1,\,2,\,3,\,4\,; \qquad Q \equiv s^\dagger \tilde{d} + d^\dagger s - \tfrac{1}{2}\sqrt{7}\,T_2 \tag{4}$$

and additionally $\hat{n}_\mathrm{d} = \sqrt{5}\,T_0$, $\mathbf{J} = \sqrt{10}\,T_1$.

The E2 operator of IBA-1 can be written as

$$T(\mathrm{E2}) = q(s^\dagger \tilde{d} + d^\dagger s + \chi T_2)\,, \tag{5}$$

where the parameter q is a boson quadrupole charge and χ a boson structure factor. An attractive and well-studied parametrization ("consistent Q") of the IBA-1 Hamiltonian is obtained by using the form (5) in place of Q in Eq. (3) whereupon the $P^\dagger P$ term can be eliminated[28,29].

The abundant numerical work carried out with IBA-1 is thoroughly reviewed by Casten and Warner[13]. I will not labor the point of good numerical fits. Suffice it to say that such have been obtained for the excitation energies and E2 properties of a multitude of spherical and deformed nuclei. Such fits encompass most of the collective bands for $J \lesssim 10$. To the extent that the otherwise free Hamiltonian parameters—usually fewer than 6 are used—follow smooth trends, the model has reasonable predictive power.

2.2. Basis States and Dynamic Symmetries

So far I have barely mentioned group theory; for a tutorial, directed to the needs of the IBA, see Casten *et al.*[30]. When defining the basis states of IBA-1, group theory can hardly be avoided. The six single-boson states $s^\dagger|0\rangle$ and $d_m^\dagger|0\rangle$, with $m = \pm 2, \pm 1, 0$ and $|0\rangle$ signifying the passive core, form a six-dimensional basis which we can associate with the group U(6). The single s boson has angular-momentum quantum numbers $J = 0$, $M = 0$ while the five single-d-boson states have $J = 2$, $M = m$. States containing two or more d bosons must also be eigenstates of the angular momentum. Up to three d bosons the quantum numbers $n_\mathrm{d} JM$ suffice to define the states. For $n_\mathrm{d} \geq 4$, boson seniority $v \equiv \tau$, defined analogously to fermion seniority, is a suitable choice of a necessary additional quantum number. For $n_\mathrm{d} \geq 6$ yet another label is needed to distinguish orthogonal states; in IBA-1 it is called $n_\Delta \equiv \nu_\Delta$ and its physical meaning is the number of d-boson triplets coupled to angular momentum 0.

The quantum numbers to label states with good n_d (and thence good $n_\mathrm{s} = N - n_\mathrm{d}$) can be identified as labels distinguishing completely symmetric representations of the subgroups of U(6). Thus $n_{\rm d}$ belongs to U(5) and τ to O(5), while ν_Δ is an extra label needed for the reduction of O(5) to O(3). The complete chain of subgroups with associated quantum numbers is

$$\begin{array}{ccccccccc} \mathrm{U}(6) & \supset & \mathrm{U}(5) & \supset & \mathrm{O}(5) & \supset & \mathrm{O}(3) & \supset & \mathrm{O}(2) \\ [N] & & n_{\rm d} & & \tau \quad \nu_\Delta & & J & & M \end{array} \tag{6}$$

where the notation $[N]$ refers to the complete symmetry required. I omit here a discussion of the possible values of the quantum numbers; see e.g. Ref. 30. We note that in this scheme the s bosons just "float" so as to yield good N. Accordingly the basis states are labeled as $|[N]n_{\rm d}\tau\nu_\Delta JM\rangle$.

Stripped of the s bosons, these basis states are just the same as the quadrupole-phonon states of the geometric model. The dimension of the IBA-1 Hamiltonian matrix formed in this basis is hardly larger than some tens. The matrix is formed and diagonalized routinely by the code PHINT[27], which also calculates various observables, notably E2 rates and moments.

An interesting special case arises when the Hamiltonian is immediately diagonal in the basis $|[N]n_{\rm d}\tau\nu_\Delta JM\rangle$. The simple prototype of the situation occurs for a Zeeman-type Hamiltonian $H = a\mathbf{J}^2 + bJ_z$; although its rotational symmetry is broken, its eigenstates are $|JM\rangle$ and eigenenergies $E = aJ(J+1) + bM$. Such a diagonal IBA-1 Hamiltonian is provided by Eq. (3) when its parameters a_0 and a_2 are put to zero. Then basic group theory yields for the excitation energies

$$\begin{aligned} E_{\rm I} = {} & (\epsilon'' + \tfrac{18}{35}a_4)n_{\rm d} + \tfrac{18}{35}a_4 n_{\rm d}(n_{\rm d}+4) + (\tfrac{1}{2}a_3 - \tfrac{3}{14}a_4)\tau(\tau+3) \\ & + (a_1 - \tfrac{1}{10}a_3 - \tfrac{1}{14}a_4)J(J+1)\,. \end{aligned} \tag{7}$$

In this case the Hamiltonian is said to have a *dynamic symmetry* which is designated by the first link, U(5), of the group chain (6) or simply by the ordinal I.

In addition the IBA-1 Hamiltonian has two more dynamic symmetries, SU(3) ≡II and O(6)≡III, associated with the group chains (see again e.g. Ref. 30)

$$\begin{array}{ccccccc} \mathrm{U}(6) & \supset & \mathrm{SU}(3) & \supset & \mathrm{O}(3) & \supset & \mathrm{O}(2) \\ [N] & & (\lambda\mu) \quad K & & J & & M \end{array} \tag{8}$$

and

$$\begin{array}{ccccccccc} \mathrm{U}(6) & \supset & \mathrm{O}(6) & \supset & \mathrm{O}(5) & \supset & \mathrm{O}(3) & \supset & \mathrm{O}(2) \\ [N] & & \sigma & & \tau \quad \nu_\Delta & & J & & M \end{array} \tag{9}$$

respectively. The corresponding wave functions are no longer $\psi_{\rm I} \equiv |[N]n_{\rm d}\tau\nu_\Delta JM\rangle$, but they are

$$\psi_{\rm II} = |[N](\lambda\mu)KJM\rangle, \quad \psi_{\rm III} = |[N]\sigma\tau\nu_\Delta JM\rangle\,. \tag{10}$$

Dynamic symmetry II is realized when the parameters ϵ'', a_3 and a_4 in Eq. (3) are put to zero, and then we have

$$E_{\rm II} = \tfrac{1}{2}a_2(\lambda^2 + \mu^2 + \lambda\mu + 3\lambda + 3\mu) + (a_1 - \tfrac{3}{8}a_2)J(J+1)\,. \tag{11}$$

And finally we have dynamic symmetry III when ϵ'', a_2 and a_4 vanish:

$$E_{\rm III} = -\tfrac{1}{4}a_0\sigma(\sigma+4) + \tfrac{1}{2}a_3\tau(\tau+3) + (a_1 - \tfrac{1}{10}a_3)J(J+1)\,. \qquad (12)$$

Thus IBA-1 becomes beautifully simple in three special cases. The energy formulas (7), (11) and (12) show that they correspond geometrically to a spherical anharmonic vibrator, a deformed rotor–vibrator and a γ-unstable nucleus. Technically these attractive special cases can be traced to the fact that the IBA-1 Hamiltonian can be expressed as a linear combination of the Casimir operators of the subgroups present in the chains (6), (8) and (9). However, I wish to stress that they are indeed special cases and the general case can only be solved numerically.

2.3. Summary on IBA-1

I conclude the discussion of IBA-1 by repeating some of the points made above. The model is phenomenological; its microscopic contents are restricted to (i) a qualitative notion that the bosons are made up of fermion pairs, and (ii) the algorithm of determining the boson number from the shell structure. It is rather like the geometric model when it comes to fitting data with several parameters. The special, or limiting, cases called dynamic symmetries have strong intuitive appeal and often offer a good starting point for numerical fitting. The predictive power of IBA-1 relies on parameter systematics. The model is good for an easy, semiquantitative description of quadrupole collectivity in nuclei outside the reach of microscopic theory.

3. The Model IBA-2

3.1. General Structure and Assumptions

Where IBA-1 is a one-fluid model, IBA-2 is a two-fluid model. At the level of phenomenology, IBA-2 is not fundamentally different from a geometric model consisting of a proton fluid and a neutron fluid. Only the latter kind of model, although old in concept, has received relatively little detailed attention; for a review see Ref. 29.

The idea of phenomenological IBA-2 is to have a proton IBA-1 and a neutron IBA-1 coupled together: $H = H_{\rm p} + H_{\rm n} + H_{\rm pn}$. The model thus contains two kinds of boson. They can be viewed as two separate species whose operators commute, just like proton and neutron particles. Alternatively, they can be viewed as two charge states of a general boson, analogously to isospin for nucleons. The associated "spin" is called F spin. The proton bosons of a nucleus must be symmetric under exchange of their numbering labels; likewise the neutron bosons. However, when viewed as different species, p bosons and n bosons need not have any exchange symmetry relative to one another. If, on the other hand, we opt for the F-spin alternative, the *overall* boson wave function has to be symmetric, just as a nucleon wave function has to be antisymmetric. The spatial (sd space) and F-spin parts *separately* need not be symmetric. In general all symmetries contained in two-rowed Young patterns $[N-n, n]$ are possible. There is indeed ample experimental evidence for the occurrence of $[N-1, 1]$ "mixed-symmetry" states in deformed and nondeformed nuclei, as is reviewed by Lipas *et al.*[29].

I now list the specific assumptions of IBA-2:

1. IBA-2 is a model for the collective properties of doubly even nuclei, with low-energy quadrupole properties dominating.
2. Valence (off-magic) protons and neutrons behave as correlated proton pairs and neutron pairs. Holes are equivalent to particles in counting. It is sufficient to consider S and D pairs.
3. The pairs behave as proton and neutron bosons, whose numbers $N_{\rm p}$ and $N_{\rm n}$ are conserved. Each kind of boson has its own two s and d levels.
4. The bosons have all possible one-body energies and two-body interactions.
5. Electromagnetic multipole operators are constructed as $N_{\rm p,n}$-conserving one-body operators.

The completely general IBA-2 Hamiltonian has a prohibitive number of parameters: 30 in all. They come from $9+9$ in $H_{\rm p}$ and $H_{\rm n}$ as given in Eq. (2) plus another 12 present in $H_{\rm pn}$. The constancy of $N_k = n_{sk} + n_{{\rm d}k}$, $k = {\rm p,n}$, reduces the total number of parameters to $21 = 6+6+9$ for excitation energies, still far too many for comfort.

A radical simplification based on microscopic ideas consists in the so-called Talmi Hamiltonian

$$H = \epsilon_{\rm p}\hat{n}_{\rm dp} + \epsilon_{\rm n}\hat{n}_{\rm dn} + \kappa Q_{\rm p}\cdot Q_{\rm n} + \xi\hat{M}\,, \tag{13}$$

where we have the number and quadrupole operators ($k = {\rm p,n}$)

$$\hat{n}_{{\rm d}k} = d_k^\dagger\cdot\tilde{d}_k\,,\quad Q_k = s_k^\dagger\tilde{d}_k + d_k^\dagger s_k + \chi_k[d_k^\dagger\tilde{d}_k]_2 \tag{14}$$

and the so-called Majorana operator

$$\hat{M} = (s_{\rm p}^\dagger d_{\rm n}^\dagger - s_{\rm n}^\dagger d_{\rm p}^\dagger)\cdot(s_{\rm p}\tilde{d}_{\rm n} - s_{\rm n}\tilde{d}_{\rm p}) - 2\sum_{L=1,3}[d_{\rm p}^\dagger d_{\rm n}^\dagger]_L\cdot[\tilde{d}_{\rm p}\tilde{d}_{\rm n}]_L\,. \tag{15}$$

With the usual choice $\epsilon_{\rm p} = \epsilon_{\rm n} \equiv \epsilon$ the Hamiltonian (13) has five parameters: ϵ, κ, $\chi_{\rm p}$, $\chi_{\rm n}$ and ξ. The existence of p and n bosons with s and d levels is embodied in the ϵ terms. The κ term is construed as a boson image of the long-range proton–neutron force responsible for deformation. The last term was added with the purpose of pushing up proton–neutron mixed-symmetry states which are not observed among the lowest excitations. Presumably it images the proton–neutron symmetry energy.

Apart from the qualitative microscopic background of the Hamiltonian (13), its justification lies in the success of numerical fits to wide ranges of experimental data; cf. Iachello and Arima[14]. The precise value of the Majorana strength ξ, as long as it is $\xi \gtrsim 0.2$ MeV, has little effect on the energies and E2 properties of the usual collective states, i.e. the quasi-ground, -β and -γ bands in deformed and nondeformed nuclei. A case in point is the set of global fits by Puddu *et al.*[31] to $_{54}$Xe, $_{56}$Ba and

$_{58}$Ce isotopes. As regards the Majorana force, an opposite example dealing with the same nuclei is due to Novoselsky and Talmi[32], who had $\xi = 0$ and also the other parameters rather different from those of Puddu *et al.*[31].

The most important electromagnetic transitions and moments in the realm of collective models are those of multipolarity E2. The E2 operator of IBA-2 is

$$T(\mathrm{E2}) = q_\mathrm{p} Q_\mathrm{p} + q_\mathrm{n} Q_\mathrm{n} \,, \tag{16}$$

where the quadrupole boson charges q_p and q_n are free parameters and the operators Q_p and Q_n have the form given in Eq. (14), with the χ_k two more free parameters. When the Talmi Hamiltonian (13) is used, there is no compelling *a priori* reason to have the same χ_k parameters in the Hamiltonian and in the quadrupole operator. However, in the interest of economy it is commonplace to choose them indeed to be the same. Numerical experience vindicates this choice.

3.2. Dynamic Symmetries

The group structure of IBA-2 is mathematically much richer than that of IBA-1. The largest relevant group is U(12) whose basis states are the six states of the proton boson and the six of the neutron boson. However, because p bosons and n bosons do not convert into each other, the physical starting group of IBA-2 is the subgroup $\mathrm{U_p}(6) \times \mathrm{U_n}(6)$ of U(12). Each factor group has its subgroup chains of the forms (6), (8) and (9), but at the level of O(3) the chains must be combined because the *total* angular momentum is conserved. This is expressed for the form (6) as

$$\begin{aligned}\mathrm{U_p}(6) \times \mathrm{U_n}(6) &\supset \mathrm{U_p}(5) \times \mathrm{U_n}(5) \supset \mathrm{O_p}(5) \times \mathrm{O_n}(5) \\ &\supset \mathrm{O_p}(3) \times \mathrm{O_n}(3) \supset \mathrm{O_{p+n}}(3) \supset \mathrm{O_{p+n}}(2)\,,\end{aligned} \tag{17}$$

The standard IBA-2 computer program NPBOS[33] uses basis states of this structure, i.e.

$$|[N_\mathrm{p}]n_{\mathrm{dp}}\tau_\mathrm{p}\nu_{\Delta\mathrm{p}}J_\mathrm{p}, [N_\mathrm{n}]n_{\mathrm{dn}}\tau_\mathrm{n}\nu_{\Delta\mathrm{n}}J_\mathrm{n}, JM\rangle\,. \tag{18}$$

Resultant Hamiltonian matrix dimensions of a hundred are typical.

The p and n group chains may be joined even before the O(3) stage. The earliest possibility is to do so immediately at the U(6) level:

$$\mathrm{U_p}(6) \times \mathrm{U_n}(6) \supset \mathrm{U_{p+n}}(6)\,. \tag{19}$$

Chains of the forms (6), (8) and (9) then follow, e.g.

$$\mathrm{U_{p+n}}(6) \supset \mathrm{U_{p+n}}(5) \supset \mathrm{O_{p+n}}(5) \supset \mathrm{O_{p+n}}(3) \supset \mathrm{O_{p+n}}(2)\,. \tag{20}$$

Apart from the subscripts and the—so far unspecified—quantum numbers, this is identical to the IBA-1 result (6), and we can immediately write similar counterparts to Eqs. (8) and (9). Thus it looks like we have simply recovered the three IBA-1 group chains, with associated dynamic symmetries named after U(5), SU(3) and O(6).

The IBA-1 dynamic symmetries are certainly contained in their IBA-2 counterparts. Any IBA-2 Hamiltonian possessing one of these symmetries is insensitive to the p and n labels. This means, as an important example, that for SU(3) dynamic symmetry the Hamiltonian is

$$H = a_1(J_p + J_n)^2 + a_2(Q_p + Q_n)^2\,, \quad J_k \equiv \sqrt{10}\,[d_k^\dagger \tilde{d}_k]_1\,, \tag{21}$$

where Eqs. (4) and (14) require that $\chi_p = \chi_n = -\frac{1}{2}\sqrt{7}$. Note that $Q_p \cdot Q_n$ does not have the required pn symmetry, as will become clear from the ensuing discussion of F spin. However, even for an appropriately pn-symmetric IBA-2 Hamiltonian, the wave functions are more general than those of IBA-1. In addition to the pn-symmetric wave functions that *are* essentially the same as the IBA-1 wave functions, there are wave functions that are *not* pn symmetric but of mixed pn symmetry, as already surmised above.

The quantum numbers appearing in Eq. (6), as well as in Eqs. (8) and (9), are sufficient only for the completely symmetric states within IBA-2, i.e. the IBA-1-like states. In particular the Young-pattern notation $[N]$ for the U(6) representation underlines the complete symmetry. For the two-rowed Young patterns allowed in IBA-2, the corresponding specification is $[N_1 N_2]$ with the constraint $N_1 + N_2 = N$ (N_1 and N_2 are not to be identified with N_p and N_n). Similar general labeling for the entire U(5) chain reads

$$\begin{array}{ccccccccccc} U_{p+n}(6) & \supset & U_{p+n}(5) & \supset & O_{p+n}(5) & & \supset & & O_{p+n}(3) & \supset & O_{p+n}(2) \\ {[N_1 N_2]} & & (n_1 n_2) & & (\tau_1 \tau_2) & & (\nu_{\Delta 1} \nu_{\Delta 2}) & & J & & M \end{array} \tag{22}$$

The SU(3) and O(6) chains acquire likewise more quantum numbers.

Although mathematically the separate p and n group chains may be joined at any point up to the O(3) level, only the IBA-1-like cases appear to be of major physical relevance. This can be understood as follows[18]: since the proton–neutron quadrupole force is known to be dominant for collective properties, the pn coupling should be done at the earliest level according to Eq. (19). The conclusion is that the important dynamic symmetries of IBA-2 are the same as those of IBA-1.

3.3. F Spin

Bookkeeping of pn symmetry within IBA-2 is conveniently carried out in terms of F spin, the boson analogue of nucleon isospin. Thus $F = \frac{1}{2}$ is assigned to each boson, with $M_F = +\frac{1}{2}$ $(-\frac{1}{2})$ for the p (n) boson. For a nucleus with boson numbers N_p and N_n the projection is then $M_F = \frac{1}{2}(N_p - N_n)$ and the values of F lie in the range $|M_F| \le F \le \frac{1}{2}(N_p + N_n) \equiv F_{max}$. Application of the model shows that the states with $F = F_{max}$ lie lowest; they are the symmetric IBA-1-like states. The states with $F = F_{max} - 1$ correspond to the U(6) Young pattern $[N-1, 1]$. In deformed nuclei, certain 1^+ states at about 3 MeV excitation have been identified as such[20,34]. In nondeformed nuclei, mixed-symmetry states of 2^+ character have been identified at about 2 MeV excitation[35,29]. Geometrically the former are visualised as scissors motion between the proton and neutron shapes while the latter are like counterrolling p and n tidal waves.

The F-spin operators are

$$F_+ = s_p^\dagger s_n + d_p^\dagger \cdot \tilde{d}_n \,, \quad F_- = F_+^\dagger \,, \quad F_0 = \tfrac{1}{2}(\hat{N}_p - \hat{N}_n) \tag{23}$$

and they satisfy the usual SU(2) commutation relations

$$[F_+, F_-] = 2F_0 \,, \quad [F_0, F_\pm] = \pm F_\pm \,. \tag{24}$$

An interesting and enlightning connection between the F-spin operators and the Majorana interaction (15) is[36]

$$\hat{M} = F_{\max}(F_{\max} + 1) - \mathbf{F}^2 \,. \tag{25}$$

This shows immediately the effect of the operator $\hat{M}$ on the states $|FM_F\rangle$:

$$\hat{M}|F_{\max} M_F\rangle = 0 \,, \quad \hat{M}|F_{\max} - 1, M_F\rangle = 2F_{\max} = N \,. \tag{26}$$

With the Majorana strength parameter $\xi > 0$ the mixed-symmetry states are thus raised relative to the symmetric states.

So far I have tacitly assumed F to be a good quantum number. Is it? Theoretically the goodness of F requires that the Hamiltonian be an F scalar, or at least that it commute with $\mathbf{F}^2$. A one-body operator which is an F scalar and a spatial tensor has the structure $[b_{lp}^\dagger \tilde{b}_{l'p}]_L + [b_{ln}^\dagger \tilde{b}_{l'n}]_L$, where $l = 0$ and 2 signify s and d; with a minus sign between the terms the operator would be an F vector. Here I do not pursue F-tensor algebra any further, but refer for more detail to Refs. 29 and 37. This means that the Talmi Hamiltonian (13), even with the choice $\epsilon_p = \epsilon_n$, is not an F scalar but the Hamiltonian (21) is. However, it has been established that in the usual case of a large Majorana term, which is emphatically an F scalar according to Eq. (25), F spin *is* generally a remarkably good quantum number. Experimental evidence supports the view of F spin as a valid quantum number. The matter is extensively discussed in the review by Lipas *et al.*[29].

As for E2 matrix elements, the operator (16) is generally a linear combination of an F scalar and vector. This implies selection rules $\Delta F = 0, \pm 1$. In the event $q_p = q_n$, $\chi_p = \chi_n$, which occurs for U(5)- and O(6)-like nuclei with $\chi_{p,n} = 0$, the E2 operator is a pure F scalar, and then the selection rule is $\Delta F = 0$. Such behavior is observed in some nondeformed nuclei[29].

3.4. Magnetic Phenomena

So far I have considered excitation energies and E2 properties. These two categories of nuclear data certainly constitute the main target of the IBA model. One can say that agreement with experiment is fair to good throughout these areas for IBA-1 and IBA-2 alike. Magnetic-dipole phenomena are very different. In lowest order they do not occur at all in IBA-1 whereas in IBA-2 they do. The same difference is shared by all one-fluid and two-fluid collective models, respectively.

In IBA-1 the lowest-order M1 operator is proportional to the angular momentum and therefore incapable of causing transitions. Only higher-order terms with extra parameters can bring about M1 transitions[38,39]. In IBA-2, however, M1 transitions do occur unforced. The operator is written as (SI units)

$$T(\mathrm{M1}) = \sqrt{\frac{3}{4\pi}}\frac{\mu_\mathrm{N}}{c}(g_\mathrm{p}J_\mathrm{p} + g_\mathrm{n}J_\mathrm{n}), \tag{27}$$

where the boson g factors normally have values $g_\mathrm{p} \approx 1$ and $g_\mathrm{n} \approx 0$ and the J_k are given in Eq. (21). One might think that this operator produces M1 transitions whenever allowed by the general selection rules. However, F spin and its selection rules impose severe restrictions on M1 transitions within IBA-2. The operator (27) can be rewritten as

$$T(\mathrm{M1}) = \sqrt{\frac{3}{4\pi}}\frac{\mu_\mathrm{N}}{c}[\tfrac{1}{2}(g_\mathrm{p} + g_\mathrm{n})\mathbf{J} + \tfrac{1}{2}(g_\mathrm{p} - g_\mathrm{n})(J_\mathrm{p} - J_\mathrm{n})], \tag{28}$$

where the first term is an F scalar and the second an F vector. Clearly only the second term can produce transitions, while both terms generally contribute to diagonal matrix elements, i.e. to static magnetic moments. The Wigner–Eckart theorem applied in F-spin space to Eq. (28) results[29] in the selection rules

$$\Delta F = 0, \pm 1, \quad 0 \not\to 0, \quad F_\mathrm{max} \not\to F_\mathrm{max}. \tag{29}$$

This applies to M1 *transitions*; for static moments there are no F-spin restrictions.

According to Eq. (29) there can be no M1 transitions between states of good $F = F_\mathrm{max}$. This excludes M1 transitions connecting the normal low-lying states as described within IBA-2. Yet such transitions *are* observed, although they are weak, $B(\mathrm{M1}) \approx 10^{-3}$ $(\mu_\mathrm{N}/c)^2$. They are explained, even quantitatively successfully[40,41] as resulting from small admixtures of $F = F_\mathrm{max} - 1$ components in the states dominated by $F = F_\mathrm{max}$.

On the other hand, the mixed-symmetry states with $F = F_\mathrm{max} - 1$ dominating should be strongly connected to the symmetric $F = F_\mathrm{max}$ states. This is indeed the interpretation attached to the strong excitation of the 1^+ scissors states $[B(\mathrm{M1})\!\uparrow \approx 1\ (\mu_\mathrm{N}/c)^2]$ and also to the strong $2^+ \to 2^+$ M1 decays $[B(\mathrm{M1}) \lesssim 0.3\ (\mu_\mathrm{N}/c)^2]$ observed in spherical nuclei[35]. In the latter cases the strong M1s combined with weak E2s (see end of Subsect. 3.3) between the same states are seen in small mixing ratios $\delta(\mathrm{E2/M1})$, which thus serve as signals of mixed symmetry.

3.5. Synthesis of IBA Phenomenology

I conclude this section by relating the excitation energies, E2 rates and M1 rates to the search for a good phenomenological IBA-2 Hamiltonian. The energies and E2s for the normal states are insensitive to the small deviations from pn symmetry present in the wave functions. For a Talmi-type Hamiltonian, Eq. (13), this means that the sum $\chi_\mathrm{p} + \chi_\mathrm{n}$ can be largely deduced from the energies and E2 data. Then the

M1 data reveal the difference $\chi_p - \chi_n$, whence the quadrupole structure parameters χ_p and χ_n can be fixed.

In fact the energy and E2 calculations are substantially the same in IBA-1 as they are in IBA-2. This similarity has been put into systematic form by a scheme of projecting[37] from IBA-2 to IBA-1. The projection carries the IBA-2 states with $F = F_{\max}$ to IBA-1 but it neglects the mixed-symmetry states. The scheme provides an algorithm for transforming any IBA-2 operator into IBA-1. The Majorana operator (15) vanishes in the process.

4. Microscopic Basis of the IBA

Apart from occasional qualitative comments, up to now I have discussed the IBA in general and the proton–neutron version IBA-2 in particular as phenomenological models containing a plethora of parameters. Microscopic considerations I have alluded to, on the one hand, and experience from numerical fitting, on the other hand, can reduce the parameters to a manageable number, witness the Talmi Hamiltonian (13). Below I sketch the microscopic basis of the IBA as it was initially proposed[19] with comments on its later phases[22,24].

4.1. S and D Fermion Pairs

The correlated S and D pairs which give rise to the s and d bosons are created by the operators (k = p, n)

$$S_k^\dagger = \sum_j \alpha'_{kj}\sqrt{\tfrac{1}{2}}[a_{kj}^\dagger a_{kj}^\dagger]_0\,, \quad D_k^\dagger = \sum_{j\le j'} \beta_{kjj'}(1+\delta_{jj'})^{-1/2}[a_{kj}^\dagger a_{kj'}^\dagger]_2 \tag{30}$$

acting on the vacuum $|0\rangle$ of closed shells. The shell-model space encompasses a single major shell for each kind of nucleon, and j suffices to denote the orbitals. The amplitudes α'_{kj} and $\beta_{kjj'}$ are normalized to unity. In the IBA literature the S pair is commonly written not normalized to 1 but with amplitudes $\alpha_{kj}(j+\frac{1}{2})^{1/2}$ in place of α'_{kj} such that

$$\alpha_{kj} = \alpha'_{kj}\sqrt{\frac{\sum_{j'}(2j'+1)}{2j+1}}\,. \tag{31}$$

The fermion Hamiltonian has the structure $H^{\rm F} = H_{\rm p}^{\rm F} + H_{\rm n}^{\rm F} + H_{\rm pn}^{\rm F}$. The amplitudes α'_{kj} and $\beta_{kjj'}$ in the S and D pairs are determined from a two-nucleon shell-model calculation. For protons (neutrons) it is done with $H_{\rm p}^{\rm F}$ ($H_{\rm n}^{\rm F}$) in the space of a full major shell. The resulting lowest 0^+ and 2^+ states, respectively, then define the S and D pairs for each kind of nucleon. Further recourse to the shell model will be only through these S and D pairs.

This is the basic method employed by e.g. Pittel *et al.*[42]. They extract the single-particle energies from experimental data and use the surface delta interaction

(SDI) to define H_k^{F}, which then yields the two-body matrix elements for the shell-model calculation. Table 1 shows their α' and β coefficients for the proton shell $Z = 50$–82, consisting of orbits $1g_{7/2}$, $2d_{5/2}$, $2d_{3/2}$, $3s_{1/2}$ and $1h_{11/2}$. For comparison I have calculated the same coefficients from the empirically determined $Z = 50$–82, $N = 82$ single-particle energies and two-body matrix elements of Wildenthal [43]. The difference between the two sets is seen to be small. The dominance of the lowest single-particle state $1g_{7/2}$ is evident. The pairs are thus not impressively collective.

Table 1. Microscopic structure of S and D pairs for $Z = 50$–82

Amplitude	SDI	Wildenthal
$\alpha'_{7/2}$	0.861	0.891
$\alpha'_{5/2}$	0.378	0.362
$\alpha'_{3/2}$	0.163	0.144
$\alpha'_{1/2}$	0.107	0.107
$\alpha'_{11/2}$	−0.278	−0.207
$\beta_{7/2\ 7/2}$	0.941	0.965
$\beta_{5/2\ 7/2}$	0.148	0.126
$\beta_{3/2\ 7/2}$	0.200	0.161
$\beta_{5/2\ 5/2}$	0.153	0.084
$\beta_{3/2\ 5/2}$	−0.062	−0.050
$\beta_{1/2\ 5/2}$	0.109	0.070
$\beta_{3/2\ 3/2}$	0.047	0.060
$\beta_{1/2\ 3/2}$	0.064	0.057
$\beta_{11/2\ 11/2}$	−0.089	−0.069

When a composite of fermion pairs is strongly bound relative to the energy scale, it behaves as a boson; the ^{4}He atom in low-temperature physics is a case in point. How well can the correlated fermion pairs S and D be viewed as bosons? Rather poorly, because their binding energies $\approx$ 1–2 MeV are of the same order as single-particle excitation energies. Formally we see that the commutator $[S_k, S_k^\dagger]$ is not equal to unity as it would be for bosons, but

$$[S_k, S_k^\dagger] = 1 - 2\sum_j (\alpha'_{kj})^2 \hat{n}_{kj}/(2j+1)\,, \tag{32}$$

where $\hat{n}_{kj} \equiv \sum_m a_{kjm}^\dagger a_{kjm}$ is the number operator for orbital kj. The effect of the "nonbosonic" second term is small whenever the j shells are relatively empty. This is to be expected *a priori* because for sparse occupation the Pauli principle is of minor importance.

4.2. OAI Mapping

The established tehnique of passing from the SD-truncated shell model to the IBA bosons is known as the Otsuka–Arima–Iachello (OAI) mapping[21]. The original work, as well as many subsequent studies, was restricted for mathematical simplicity to a fermion space of a single j shell. In order to have room for many particles, such studies have resorted to artificially large j, say $j = \frac{23}{2}$. The true situation embodied in the shell model, however, requires several j shells, typically a complete major shell as in Table 1.

Fermion states built from the S and D pairs (30) are suitably classified according to their *generalized seniority* $v_{\rm g}$, defined as the number of nucleons not in S pairs[44]. Each kind of nucleon $k =$ p, n is considered separately and I now drop the index k. States (unnormalized) of a total of N pairs

$$(S^\dagger)^N|0\rangle, \quad (S^\dagger)^{N-1}D^\dagger|0\rangle \tag{33}$$

have $v_{\rm g} = 0$ and $v_{\rm g} = 2$ respectively. Continuing to $v_{\rm g} = 4$, I write

$$(S^\dagger)^{N-2}\mathcal{P}[D^\dagger D^\dagger]_J|0\rangle\,, \quad J = 0,\,2,\,4\,, \tag{34}$$

where $\mathcal{P}$ is a projection operator which excludes components with $v_{\rm g} = 0$ and $v_{\rm g} = 2$. Generally such components appear in $D^\dagger D^\dagger$. To appreciate this, consider pairs formed of $j = \frac{3}{2}$ nucleons only. Then the only $D^\dagger D^\dagger$ state is $J = 0$ and it is created by

$$[D^\dagger D^\dagger]_0 \propto [[a^\dagger_{3/2}a^\dagger_{3/2}]_2[a^\dagger_{3/2}a^\dagger_{3/2}]_2]_0 \propto [a^\dagger_{3/2}a^\dagger_{3/2}]_0[a^\dagger_{3/2}a^\dagger_{3/2}]_0 \propto (S^\dagger)^2\,, \tag{35}$$

so that the state has pure $v_{\rm g} = 0$. Note further that the coupling to $J = 0,\,2,\,4$ in Eq. (34) is an imposed restriction since in the general multi-j case $J = 1$ and 3 could also be constructed.

The step from S and D fermion pairs to s and d bosons is now a matter of simple replacement: $S^\dagger \to s^\dagger$, $D^\dagger \to d^\dagger$, with the label $k =$ p, n restored. At the same time the fermion vacuum is replaced by the boson vacuum, both denoted by $|0\rangle$; the pair number N becomes the boson number; and half the generalized seniority becomes the number of d bosons, $n_{\rm d} = \frac{1}{2}v_{\rm g}$. The step definitely embodies an approximation since the S and D fermion pairs do not obey boson commutation relations, recall Eq. (32), while the s and d bosons do.

Having established the passage from fermion states to boson states, how do we find the boson image of a given fermion operator, especially the Hamiltonian $H^{\rm F}$? The OAI procedure consists in the following three steps:

(i) On the basis of general invariance principles (e.g. the Hamiltonian is a scalar) write down a candidate, with undetermined constants, for the boson operator.

(ii) Seek the simplest corresponding pairs of fermion-pair and boson states which yield a nonvanishing matrix element. The normal "zeroth-order" mapping is done with states (33) and their boson counterparts.

(iii) Calculate the fermion matrix element and its boson counterpart, equate the two and compute the constant in the boson operator.

The resultant boson operator is taken to be approximately valid also for other states than those used in fixing it. Numerical applications reveal how good or bad this approximation is. In practice the boson calculation in step (iii) is very easy while the fermion calculation can be rather unwieldy; see e.g. Pittel *et al.*[42] and Lipas *et al.*[45].

As a simple illustration[21,12] of OAI mapping, consider the pairing Hamiltonian for a single j shell of protons or neutrons:

$$H^{\mathrm{F}} = -GS^{\dagger}S\,, \quad S^{\dagger} = \sqrt{\tfrac{1}{2}}\,[a_j^{\dagger}a_j^{\dagger}]_0\,. \tag{36}$$

Step (i) consists in guessing the form of the boson Hamiltonian. Let us restrict the mapping to apply to S pairs, and thence to s bosons, only. With one- and two-body terms we then have, with the parameter names from Eq. (2), the ansatz

$$H^{\mathrm{B}} = \epsilon_{\mathrm{s}} s^{\dagger}s + \tfrac{1}{2}u_0 s^{\dagger}s^{\dagger}ss\,. \tag{37}$$

For step (ii) we have the fermion-pair states $|\mathrm{S}^N\rangle$ and the boson states $|\mathrm{s}^N\rangle$. Then, for step (iii), we have the fermion result[46]

$$\langle \mathrm{S}^N|H^{\mathrm{F}}|\mathrm{S}^N\rangle = -\frac{G}{2j+1}N(2j+3-2N) \tag{38}$$

and the boson result

$$\langle \mathrm{s}^N|H^{\mathrm{B}}|\mathrm{s}^N\rangle = \epsilon_{\mathrm{s}}N + \tfrac{1}{2}u_0(N^2-N)\,. \tag{39}$$

These are not only expectation values but exact eigenvalues. Equating them we find

$$\epsilon_{\mathrm{s}} = -G\,, \quad u_0 = \frac{4G}{2j+1} \tag{40}$$

thus fixing the boson Hamiltonian (37).

Unlike results from OAI mapping in general, Eq. (40) is exact and valid for any boson number $N \leq j + \frac{1}{2}$. It is interesting to observe that the two-body fermion interaction (36) gave rise not only to the one-body boson term $s^{\dagger}s$ in Eq. (37) but also to the boson–boson interaction $s^{\dagger}s^{\dagger}ss$. The latter is seen to be repulsive and it is interpreted as representing a Pauli-principle correction to the negative one-boson energy imaging the attractive fermion pairing interaction. I conclude this example by noting that D pairs/d bosons can be and should be included in the boson mapping of the pairing Hamiltonian (36). The result then contains all the n_{s}- and n_{d}-conserving terms of Eq. (2), i.e. $\hat{v}_2, \hat{v}_0 = 0$, and has been worked out by Otsuka *et al.*[21]. In this context Eq. (2) of course is meant for only protons or only neutrons.

4.3. Results and Refinements of IBA Microscopy

In Subsect. 4.1 I described the determination of the S and D pairs from a given H^{F} by means of a simple two-nucleon shell-model calculation. The resulting pair structure is the same throughout the major shell. As has been emphasized by Talmi (see e.g. Ref. 22), the constant structure is rather accurately realized in singly magic nuclei, such as the long chain of tin isotopes ${}^{102-130}_{50}\mathrm{Sn}_{52-80}$. This is consistent with the pairing description and good generalized seniority. The most striking observed property is the (nearly) constant 2^+ excitation energy. In the simple case of a single j shell it results from Eq. (38) extended to D pairs and thus to seniorities $v = 2, 4, \ldots, 2N$:

$$\langle Nv|H^{\mathrm{F}}|Nv\rangle = -\frac{G}{2j+1}\,(N - \tfrac{1}{2}v)(2j+3-2N-v)\,. \tag{41}$$

This equation gives $E_{\mathrm{ex}}(N, v=2) = G$ as the constant 2^+ excitation energy.

Pittel *et al.*[42] let the structure of the proton (neutron) boson vary with the neutron (proton) number over a major shell. This they accomplish by using four sets of different p (n) single-particle energies and different SDI strengths progressing through the n (p) shell. Apart from the particle–hole switching at midshell, this produces discernible but not dramatic changes in the α and β coefficients defining the pairs. Their values (SDI) shown in Table 1 are for the beginning of the $N = 82$–126 neutron shell. Other attempts at improving the pair description over the shell[47,48] end up in substantial agreement with the constancy predicted by the seniority scheme.

The object of the study by Pittel *et al.*[42] is to map the assumed proton–neutron quadrupole interaction into its boson counterpart present in the Talmi Hamiltonian (13). They treat all the major shells beyond Z, $N = 50$ and calculate the parameters κ, χ_{p} and χ_{n}. In a subsequent effort, Druce *et al.*[49] make similar calculations with specific application to ${}^{194-202}_{80}\mathrm{Hg}_{114-122}$. Not only for the pp and nn but also for the pn fermion interaction they now use the SDI with appropriately fitted strengths. Their boson Hamiltonian is essentially the Talmi Hamiltonian (13). Spectra calculated on IBA-2 with their microscopically determined parameters are in rough agreement with phenomenological IBA-2 fits and with experiment. Yet the calculated and fitted parameter sets display glaring differences. Similar discrepancies appear in Scholten's[47] calculations of ${}^{146-154}_{62}\mathrm{Sm}_{84-92}$. Yoshinaga[50] has extended the OAI procedure to include G pairs, which appears necessary for application to deformed nuclei.

To end with, I shortly consider the "democratic mapping" recently reported by Skouras *et al.*[23] and Bonatsos *et al.*[24]. As one might surmise from the name, the idea is to include not only s and d bosons but all bosons of importance as building blocks of low-lying states. The study is carried out in the sd and fp shells in isospin formalism. The fermion Hamiltonian is defined by empirical single-particle energies and two-body matrix elements of Preedom and Wildenthal[51]. Comparisons are made between exact shell-model calculations and boson calculations with a hierarchy of different bosons, up to 12 out of 28 possible in the sd shell. The conclusion is that with a full slate of bosons the shell model can be very well approximated. A gratifying qualitative result is that the s and d bosons are indeed the most important ones, followed by d′ and g bosons. Yet the results of an sd calculation are inferior to those of the calculation with 12 kinds of boson.

5. Summary

The interacting-boson model/approximation (IBM/IBA) has come of age. In its best established versions IBA-1 and IBA-2 it is a practical tool for the experimentalist for correlating and even predicting data through numerical application. The special cases of the model referred to as dynamic symmetries U(5), SU(3) and O(6) have strong intuitive appeal since they represent the geometric pictures of a spherical vibrator, deformed rotor–vibrator and γ-unstable nucleus, respectively.

Viewed in a wide context of phenomenological collective models, IBA-1 is a one-fluid model like the traditional Bohr–Mottelson model, while IBA-2 is a two-fluid model. Yet the IBA has the special feature of conserved boson number, achieved through the presence of s bosons in addition to the (quadrupole) d bosons. At the level of numerical computations and fitting data, the "algebraic" IBA is not drastically different from the "geometric" model. The practical advantage of the IBA acquired through widespread and user-friendly computer codes, and accumulated routine of application, is now being balanced[52].

The two-fluid nature of IBA-2 entails a richer field of phenomena than those encompassed by IBA-1. The states of mixed symmetry with respect to proton and neutron bosons have been a celebrated case, although their simple interpretation is now challenged by observed fragmentation. Such states occur in deformed nuclei at about 3 MeV (1^+) and in spherical nuclei at about 2 MeV (2^+). In all cases their characteristic property are strong M1 transitions from and to states of full pn symmetry. The F-spin quantum number, which measures the degree of proton–neutron boson symmetry, is a remarkably good quantum number.

Theoretically the most important aspect of the IBA is its derivation from the shell model. S and D pairs of like nucleons are deemed as the most important elementary excitations within the shell-model space and they are mapped into corresponding s and d bosons. While the basic technique (OAI mapping) has been known for a long time, its quantitative results are so far modest. Corrections such as taking fermion G pairs into account via perturbation theory have not helped much. At least for deformed nuclei the G pairs/g bosons seem to be a necessary ingredient on a par with the S and D pairs. A recent development known as democratic mapping has shed light on the possibility of a wider hierarchy of bosons as an alternative to a shell-model calculation.

Although the IBA in its various versions has proved to be a sharp and versatile tool at the phenomenological level, its microscopic justification continues to be a challenge. The recent work by Halse[53] and Nakada *et al.*[54] on ${}^{56,58}_{26}Fe_{30,32}$ may well point the way to a merger of the IBA and the shell model which will justify the former in terms of the latter.

References

1. B. A. Brown, *Nucl. Phys.* **A522** (1991) 221c.
2. F. Iachello and A. Arima, *Phys. Lett.* **53B** (1974) 309.
3. A. Bohr, *Mat. Fys. Medd. Dan. Vid. Selsk.* **26**, No. 14 (1952).
4. A. Bohr and B. R. Mottelson, *Nuclear Structure*, Vol. 2 (Benjamin, New York, 1975).
5. G. Gneuss and W. Greiner, *Nucl. Phys.* **A171** (1971) 449.
6. A. Arima and F. Iachello, *Phys. Rev. Lett.* **35** (1975) 1069.
7. A. Arima and F. Iachello, *Ann. of Phys.* **99** (1976) 253; **111** (1978) 201; **123** (1979) 468.
8. A. Arima and F. Iachello, *Adv. Nucl. Phys.* **13** (1984) 139.
9. P. O. Lipas, in *Collective Phenomena in Atomic Nuclei*, ed. T. Engeland, J. Rekstad and J. S. Vaagen (World Scientific, Singapore, 1984), p. 33.
10. P. O. Lipas, in *International Summer School on Hyperfine Interactions and Physics of Oriented Nuclei 1985*, ed. E Rotter (Charles University, Prague, 1986), p. 175.
11. A. E. L. Dieperink and G. Wenes, *Ann. Rev. Nucl. Part. Sci.* **35** (1985) 77.
12. J. P. Elliott, *Rep. Prog. Phys.* **48** (1985) 171.
13. R. F. Casten and D. D. Warner, *Rev. Mod. Phys.* **60** (1988) 389.
14. F. Iachello and A. Arima, *The Interacting Boson Model* (Cambridge U. P., Cambridge, 1987).
15. D. Bonatsos, *Interacting Boson Models of Nuclear Structure* (Clarendon, Oxford, 1988).
16. P. O. Lipas, in *Nuclear Structure, Reactions and Symmetries*, ed. R. A. Mayer and V. Paar (World Scientific, Singapore, 1986).
17. D. Janssen, R. V. Jolos and F. Dönau, *Nucl. Phys.* **A224** (1974) 93.
18. P. Van Isacker and D. D. Warner, SERC Daresbury Report DL/NUC/P325T (1991).
19. A. Arima, T. Ohtsuka, F. Iachello and I. Talmi, *Phys. Lett.* **66B** (1977) 205; T. Otsuka, A. Arima, F. Iachello and I. Talmi, *Phys. Lett.* **76B** (1978) 139.
20. D. Bohle, A. Richter, W. Steffen, A. E. L. Dieperink, N. Lo Iudice, F. Palumbo and O. Scholten, *Phys. Lett.* **137B** (1984) 27.
21. T. Otsuka, A. Arima and F. Iachello, *Nucl. Phys.* **A309** (1978) 1.
22. F. Iachello and I. Talmi, *Rev. Mod. Phys.* **59** (1987) 339.
23. L. D. Skouras, P. Van Isacker and M. A. Nagarajan, *Nucl. Phys.* **A516** (1990) 255.
24. D. Bonatsos, L. D. Skouras, P. Van Isacker and M. A. Nagarajan, *J. Phys.* **G17** (1991) 1803.
25. J. P. Elliott and A. P. White, *Phys. Lett.* **97B** (1980) 169;
J. P. Elliott and J. A. Evans, *Phys. Lett.* **101B** (1981) 216.
26. F. Iachello and P. Van Isacker, *The Interacting Boson–Fermion Model* (Cambridge U. P., Cambridge, 1991).

27. O. Scholten, *Program-Package PHINT*, Kernfysisch Versneller Instituut Report No. 63 (1979).
28. D. D. Warner and R. F. Casten, *Phys. Rev. Lett.* **48** (1982) 1385.
29. P. O. Lipas, P. von Brentano and A. Gelberg, *Rep. Prog. Phys.* **53** (1990) 1355.
30. R. F. Casten, J. P. Draayer, K. Heyde, P. O. Lipas, T. Otsuka and D. D. Warner, *Algebraic Approaches to Nuclear Structure: Interacting Boson and Fermion Models* (Gordon and Breach, New York, 1992), in press, Ch. 2.
31. G. Puddu, O. Scholten and T. Otsuka, *Nucl. Phys.* **A348** (1980) 109.
32. A. Novoselsky and I. Talmi, *Phys. Lett.* **172B** (1986) 139.
33. T. Otsuka and N. Yoshida, *Program NPBOS*, Japan Atomic Energy Research Institute Report JAERI-M85-094 (1985).
34. A. Richter, in *Contemporary Topics in Nuclear Structure Physics*, ed. R. F. Casten, A. Frank, M. Moshinsky and S. Pittel (World Scientific, Singapore, 1988), p. 127; *Nucl. Phys.* **A507** (1990) 99c.
35. W. D. Hamilton, A. Irbäck and J. P. Elliott, *Phys. Rev. Lett.* **53** (1094) 2469.
36. H. Harter, A. Gelberg and P. von Brentano, *Phys. Lett.* **157B** (1985) 1.
37. W. Frank and P. O. Lipas, *J. Phys.* **G16** (1990) 1653.
38. P. O. Lipas, P. Toivonen and E. Hammarén, *Nucl. Phys.* **A469** (1987) 348.
39. P. O. Lipas and A. Gelberg, *J. Phys.* **G16** (1990) 1669.
40. H. Harter, P. von Brentano, A. Gelberg and T. Otsuka, *Phys. Lett.* **188B** (1987) 295.
41. P. Van Isacker, P. O. Lipas, K. Helimäki, I. Koivistoinen and D. D. Warner, *Nucl. Phys.* **A476** (1988) 301.
42. S. Pittel, P. D. Duval and B. R. Barrett, *Ann. of Phys.* **144** (1982) 168.
43. B. H. Wildenthal, in *Understanding the Variety of Nuclear Excitations*, ed. A. Covello (World Scientific, Singapore, 1991), p. 35.
44. I. Talmi, *Nucl. Phys.* **A172** (1971) 1.
45. P. O. Lipas, M. Koskinen, H. Harter, R. Nojarov and A. Faessler, *Nucl. Phys.* **A509** (1990) 509.
46. H. Lipkin, *Lie Groups for Pedestrians* (North-Holland, Amsterdam, 1965).
47. O. Scholten, *Phys. Rev.* **C28** (1983) 1783.
48. G. Bonsignori, K. Allaart and A. van Egmond, *Prog. Nucl. Part. Phys.* **9** (1983) 431.
49. C. H. Druce, S. Pittel, B. R. Barrett and P. D. Duval, *Phys. Lett.* **157B** (1985) 115; *Ann. of Phys.* **176** (1987) 114.
50. N. Yoshinaga, *Nucl. Phys.* **A503** (1989) 65.
51. B. M. Preedom and B. H. Wildenthal, *Phys. Rev.* **C6** (1972) 1633.
52. D. Troltenier, J. A. Maruhn and P. O. Hess, in *Computational Nuclear Physics 1*, ed. K. Langanke, J. A. Maruhn and S. E. Koonin (Springer, Berlin, 1991), p.105.
53. P. Halse, *Phys. Rev.* **C41** (1990) 2340.
54. H. Nakada, T. Otsuka and T. Sebe, *Phys. Rev. Lett.* **67** (1991) 1086.

COMMENTS AND QUESTIONS ABOUT THE INTERACTING BOSON MODEL

D. J. Rowe, F. McGowan, S. Raman, R. Wyss and V. Zelevinsky.

1 Introduction

The Interacting Boson Model (IBM) has had an enormous influence on nuclear physics. One of its important achievements has been to remove the mystique and psychological barriers that once surrounded the use of group theory and algebraic methods in nuclear physics. Surely noone nowadays doubts that a dynamical system can be very simple when it has an algebraic structure.

The IBM has also provided a systematic classification of a wide variety of data in terms of a small number of parameters. The wide range of successful applications of the model is very impressive. If the model did nothing more, it would have served an important and useful purpose in getting theorists to look more closedly at the systematics of nuclear data. It also challenges us to explain, in physical terms, the reasons for its success.

2 What's in a name?

In his talk, Lipas expressed a preference for calling the model the Interacting Boson Approximation (IBA) rather than the IBM. In doing so its seems he was expressing a commitment to an interpretation of the IBM as a particular approximation to the shell model in which the active space of the shell model is restricted to a subspace in which the valence particles are coupled in either $J = 0$ or $J = 2$ pairs (cf., talk by Otsuka). We suggest that, in phenomenological applications of the model, such a commitment is not essential. A commitment to a particular interpretation need only be made when one wishes to derive the model from a more fundamental model. Moreover, in phenomenological applications, it is not clear to us that the basic building blocks of the model are necessarily even bosons. What is clear is that the so-called IBM is an algebraic model.

An algebraic model has three essential ingredients: a Lie algebra; a set of observables (e.g., a Hamiltonian, quadrupole moment and angular momentum operators) which are expressed as simple polynomials in the Lie algebra; and a represenation or set of representations. For the IBM, the fundamental Lie algebra is u(6). The observables are usually expressed as number conserving boson operators. However, such expressions are only an intermediate step. What is important is that they are elements, or simple polynomials of elements, of u(6). In IBM1, the representations of the model are the fully symmetric representations of the type $[N]$ and, in IBM2, they are of the type $[N_1N_2]$.

Now, it is well known that every finite-dimensional irrep of every classical Lie algebra has a boson realization of the type

$$C_{ij} \sim \sum_n b^\dagger_{ni} b_{nj}, \tag{1}$$

where $[b_{ni}, b^\dagger_{nj}] = \delta_{mn}\delta_{ij}$. These realizations are well known in the mathematical literature as Weil representations. It is also known that the same representations have fermion realizations of the type

$$C_{ij} \sim \sum_n a^\dagger_{ni} a_{nj}, \tag{2}$$

with $\{a_{ni}, a^\dagger_{nj}\} = \delta_{mn}\delta_{ij}$. One can also realize Lie algebras in other ways. Of course, one realization may be more economical than another. For example, a fully symmetric representation of the type $[N]$ can be realized with just one kind of boson (i.e., $C_{ij} = b^\dagger_i b_j$) while a fully antisymmetric representation $[1^N]$ (with $N \leq 6$) of u(6) can be realized with just one kind of fermion (i.e., $C_{ij} = a^\dagger_i a_j$). However, the point we wish to make is that, while a boson interpretation of the IBM may be simple and natural, it is not unique. Maybe there are considerations, experimental or theoretical, which go beyond the simple representations of u(6), e.g., two-nucleon transfer reactions, which would lead one to believe that a bosonic interpretation is essential. If so, we would be very interested to learn about them.

3 Derivations of the IBM

If the basic building blocks of the IBM really are boson, what is their character? Are they fermion pairs or phonons or something else?

The Otsuka-Arima-Iachello (OAI) prescription for deriving an irrep $[N]$ of the IBM is in four steps:

(i) Define $J = 0$ and $J = 2$ fermion pairs of the form

$$S^\dagger = \sum_i \alpha_i (a^\dagger_i a^\dagger_i)_0, \quad D^\dagger_\nu = \sum_{ij} \beta_{ij} (a^\dagger_i a^\dagger_j)_{2\nu} \tag{3}$$

within a given shell model space.

(ii) Truncate the shell model space to the subspace spanned by N nucleon pair states of the form

$$(S^\dagger)^{N-n}(D^\dagger)^n|0\rangle, \quad n = 0, \ldots, N. \tag{4}$$

(iii) Map the truncated shell model space to a boson Fock space such that

$$(S^\dagger)^{N-n}(D^\dagger)^n|0\rangle \to (s^\dagger)^{N-n}(d^\dagger)^n|0\rangle. \tag{5}$$

This defines a boson image of any operator on the truncated shell model space.

(iv) Expand the boson images of physically relevant observables to leading order, or sometimes to second order, in the boson creation and annihilation operators and discard higher order terms.

It is appropriate to enquire if and when the OAI truncation of the shell model space is valid and if the boson expansion of operators can reasonably be restricted to one or two terms. The comparisons, given by Otsuka, with more exactly calculated shell model results suggests that the approximations may be reasonable for a few lowlying states of spherical nuclei. But, we doubt that they would be valid more generally.

A very different interpretation of the IBM goes back to the formulation of a compactified version of the Bohr-Mottelson model by Jannsen, Jolos and Dönau (JJD). It is well known that the Bohr-Mottelson model can be expressed as a non-compact algebraic model involving only $J = 2$ phonon-type d bosons. A $J = 0$ s-boson can then be introduced as a formal device for reducing the Bohr-Mottelson model to a compact algebraic model.

Is it possible that the IBM is a clever device for interpolating between the various limits of the Bohr-Mottelson collective model? One knows, for example, that the rotational model can be realized as the large dimensional limit of the SU(3) model. One knows that U(5) is the symmetry group of the Bohr-Mottelson vibrational model and that a truncated vibrational spectrum is given by the U(5) irreps contained in a U(6) irrep. One has also learned that the Wilets-Jean limit of the collective model is associated with an O(6) symmetry group. Since U(6) is the smallest group that contains all three of these subgroups, one is led to enquire if the IBM is a compact and simple algebraic expression of the essential physics of the Bohr-Mottelson collective model.

Although the OAI and JJD interpretations of the IBM are not orthogonal, they are very different. For OAI, the bosons of the IBM are fermion pairs whereas, for JJD, they are phonons (particle-hole pairs). The OAI scheme adds quadrupole collectivity to the standard $J = 0$ pair coupling models by including $J = 2$ pair coupling whereas JJD start with quadrupole phonons and add $J = 0$ bosons as a formal compactification device. The difference is important because one knows that the effective interaction between nucleons can be expanded in either particle-particle or particle-hole multipoles. However, it turns out that the high multipoles of one expansion correspond to the low multipoles of the other; this is why the popular "pairing plus $Q.Q$" model, which includes the low multipoles from both expansions, was so successful.

The essential difference between the fermion-pair and the truncated Bohr-Mottelson interpretations of the IBM shows up in the way the model should be generalized to gain extra collectivity of states. In the fermion-pair interpretation, the boson number corresponds to twice the extra-core fermion number and is considered to be a conserved quantity. In contrast, it can be made as large as one wishes in the truncated Bohr-Mottelson picture; the larger it is, the less the truncation. It is therefore of considerable interest to know if there is any experimental evidence for the strict conservation of the boson number. We know of none. It is also of considerable interest to know if there is any real justification for the association of the two types of bosons of IBM2 with neutrons and protons, respectively. Is it possible that the extra successes of IBM2, such as they are, can be attributed simply to the extra flexibility one gains from admitting a larger variety of u(6) representations into the model?

4 What is the domain of validity of the IBM?

As Ikuko Hammamoto said in her talk "a model should not explain everything; otherwise it is not a model". We have the impression that many advocates of the IBM tend to lose sight of the model's limitations in their zeal to explain all of nuclear physics. It would be useful if the boundaries of what the IBM can and cannot do were established.

For example, to what extent is it reasonable to expect the IBM to describe M1 transitions. If the model is focussed on pairing and quadrupole collective degrees of freedom with no explicit reference to intrinsic spins, one might conclude that IBM2 contains only the so-called scissors M1 degree of freedom. The failure of the IBM to describe the fragmentation of the scissors vibrational state should therefore be expected.

What about high spin states? How much of the low energy spectrum should the IBM describe – all or only a selected subset of states below some energy? And if the model applies only to a selected subset of low energy states, how is one to make the selection? Is it acceptable to simply omit states that don't fit?

5 Where from here?

We ask the question "has the model reached maturity and, if not, where does it go from here?"

We suggest that the most pressing need is to establish a clear microscopic foundation for the IBM. Such a foundation would define the domain of validity of the model.

A good model in physics is one that has both explanatory and predictive power. In order that the IBM should explain the physical phenomena underlying the states it describes, we must know what it means. This calls for both theoretical and experimental work which probes the physical interpretation of the model; it is not enough to fit data with adjustable parameters.

It is clear that the IBM is very good at extrapolating the properties of some low energy states to new regions. However, we would like to know if the model really has predictive power. For example, if a new island of stable superheavy nuclei were to show up, could the IBM, even in principle, predict what kinds of energy spectra the newly discovered nuclei should have?

Useful experiments might be ones which look for situations in which a given interpretation of the IBM should break down. For example, the $J = 0$ and $J = 2$ pair-coupling scheme of OIA presumably indicates that, for some nuclei, u(6) irreps of the maximally symmetric type $[N]$ should not exist because they would violate the Pauli principle. One might then expect the IBM to break down. The point we make is that one can learn almost as much, perhaps more, about a model from situations where it breaks down as one can from its successes. Ironically, one of the problems the IBM has to contend with is that it appears to be too successful. A model needs to be wrong in some situations if one is to have a hope of explaining its successes when it is right.

There are several ways in which the IBM can and has been extended. A popular way is to add more bosons. In principle one should be able to make a better model in this way. However, in practice, one creates more potential for violating the Pauli principle and one loses much of the simplicity of the model which is its most attractive feature. One can also add giant resonance phonon-type bosons. Both extensions have already been contemplated and pursued.

An extension that we think needs to be considered is to distinguish between bosons corresponding to particles and boson corresponding to holes. One knows that particles and holes are, respectively, associated with positive and negative matter distributions. This suggests, for example, that a $(\lambda, \mu = 0)$ su(3) irrep should correspond to a prolate deformation for particle bosons and to an oblate deformation for hole bosons. This fact is easily accommodated by an appropriate definition of the quadrupole operator when only one type of boson is present. But, when one has both, there would surely be difficulties unless the two types are distinguished.

6 Relationship of the IBM to other models

A lot of work has been done in relating the IBM to the so-called *geometric model*. This is important because one needs to know what a model does better than other models if one is to learn anything from it.

Before pursuing the question, we wish to express a mild distaste for describing the Bohr-Mottelson model as a geometric model to distinguish it from the IBM which is represented as an algebraic model. Certainly the Bohr-Mottelson model has a very attractive physical expression in geometrical terms. However, the fact that it is so amenable to analysis is by virtue of its well-defined algebraic structure. The real distinction between the algebraic structure of the Bohr-Mottelson model and that of the IBM is that the former is non-compact whereas the latter is compact. Both models have natural expressions in group theoretical terms.

It is sometimes forgotten by physicists that a Lie group is, by definition, both a group and an analytic manifold. Thus, algebraic models with Lie group structures bring together the three basic mathematical tools used by theoretical physicists: algebra (group theory), analysis, and geometry.

Comparison between the IBM and Bohr-Mottelson model have frequently been made using coherent states to generate a geometrical surface associated with a given irrep. In effect, the comparisons show, for example, that the geometrical surface associated with an irrep $[N]$ of U(6) is the coset space $U(6)/U(1)\times U(5)$. This is a five-dimensional complex manifold which has much in common with the phase space of the Bohr-Mottelson quadrupole vibrational model. The essential difference is that the former is compact whereas the latter is non-compact. While the two spaces are very different globally, they are much alike locally, provided N is sufficiently large.

The relationship between the mathematical structures of the two models has also been established by showing that the U(6) Lie algebra contracts to a Heisenberg-Weyl Lie algebra in five d-bosons as $N \to \infty$. However, it is important to emphasize that, although the IBM and Bohr-Mottelson model have been shown to have similar geometrical and algebraic structures, the comparisons have very little to say about the physical content of the two models and the ways they relate. They derive simply from the fact that the two models have related group theoretical structures (related by the process of group contraction).

7 What is "The Best Model Ever?"

At the beginning of this workshop, Witek Nazarewicz challenged each of us to establish his or her favourite model as the best model ever. It seems to be agreed among the participants that no single model emerges as the clear winner. This is as it should be because there are many criteria. Moreover, many different models are needed to describe the complexity of phenomena observed in nuclear physics.

We conclude this commentary therefore with a suggestion of three important criteria by which the value of a model can be judged:

(i) how fundamental and how complete it is;

(ii) its physical content; and

(iii) its simplicity.

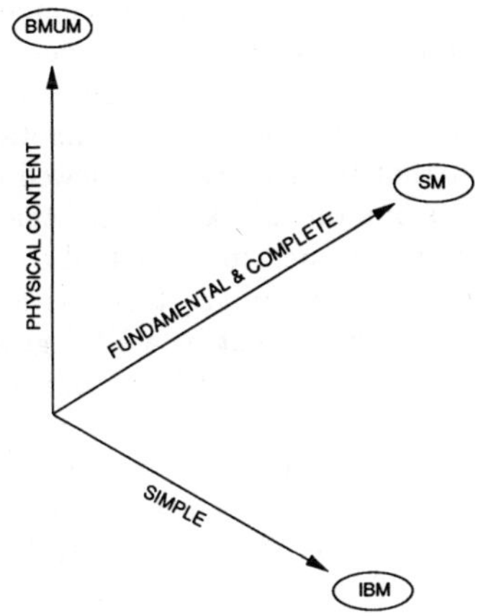

Fig. 1. Three criteria on which the value of a model can be judged.

Thus, a given model can be assigned a place on a three-dimensional plot, as illustrated in Fig. 1. We suggest that according to criterion (i), the shell model would win. According to criterion (ii), it would be hard to beat the Bohr-Mottelson unified model. However, according to criterion (iii), we would give the honours to the IBM and express the hope that its physical content can be more clearly established so that its location on the chart would move further in that dimension.

RESPONSE TO CRITIQUE OF THE IBM BY D. J. ROWE ET AL.

P. O. LIPAS
Department of Physics, University of Jyväskylä
SF-40351 Jyväskylä, Finland

I find the points raised by Rowe *et al.* well taken and have little to add or rebut. Their reviewing comments represent pithy observations of the nature of the model. Nevertheless, I feel the following short comments are called for.

My preference for using the designation IBA rather than IBM does not derive from a commitment to a fundamental rather than phenomenological view of the boson scheme. Although such a distinction between the names was propagated in the early history of the scheme, later practice has rendered the names synonymous. IBA leaves no room for a misunderstanding with commercial connotations.

Rowe *et al.* point out that the IBA can be viewed as a pure algebraic scheme without bosons specified as its building blocks. Indeed the Hamiltonian can be stated as a linear combination of Casimir operators not tied to a boson realization. Other operators and the state vectors can be similarly taken without reference to bosons; the operator for two-nucleon transfer would not be a U(6) generator, but this is the case also in the boson realization. As long as the IBA is considered as a phenomenological model, there is no experimental or theoretical reason forcing the boson interpretation.

Nevertheless, the boson background of the model was assumed as its basic starting point, and disclaiming it would seem an unnatural abstraction. Moreover, I feel the connection to the phonons of the Bohr–Mottelson model (BMM) is an essential assett to the IBA. And finally the microscopy of IBA-2 derives from nucleon pairs, which leads to a compelling boson interpretation.

The question of the role of the s boson vis-à-vis the truncated BMM is brought up with insight in the critique. The question is asked whether the IBM is but a clever device for interpolating between the various limits of the BMM. In qualitative, approximate terms one can say yes; in quantitative terms no, because the BMM, as generalized by the Frankfurt school (Greiner *et al.*), has more latitude than the IBA. The essential difference is that, unlike the IBM, the BMM does not insist on conservation of boson number. And no unequivocal experimental evidence exists in support of boson-number conservation.

Another question raised is whether the extra success of IBA-2 is truly due to the recognition of separate proton and neutron protons. In the standard boson interpretation the admission of U(6) mixed-symmetry representations $[N_1N_2]$ is justified precisely by the presence of two types of boson. Then the inclusion of M1 transitions is a natural extension of the model. Also parameter systematics for isotopic and isotonic chains lend support to the proton/neutron-boson picture.

As for the question about high-spin states, I wish to point out that Iachello and Vretenar [Phys. Rev. **C43** (1991) R945] have recently launched a major extension of the IBM which includes one and two broken pairs. The extension has proved successful for the description not only of high-spin states but also of other observed spectroscopic features not incorporated in standard IBA.

Rowe *et al.* point out the need for distinguishing between particle and hole bosons. In the deformed regime the question relates to prolate and oblate shapes. This extension of the IBA was in fact proposed ten years ago by Dieperink and Bijker [Phys. Lett. **116B** (1982) 77], but it has not been followed up very vigorously. The nucleus ^{104}Ru has served as a test case exhibing triaxiality due to an oblate proton shape and a prolate neutron shape.

The critics ask whether the IBA really has any predictive power. When its parameters are calculated from microscopy, routinely by the OAI procedure, the answer is yes; only the quality of such predictions is generally poor. A weaker class of predictive power results from parameter systematics.

I conclude this "rebuttal" of the critique of my review by applauding the critics' concluding remarks about The Best Model Ever, TBME. The figure illustrating its three criteria is delightfully enlightning.

RECENT ISSUES OF THE INTERACTING BOSON MODEL

Takaharu OTSUKA
Department of Physics, University of Tokyo, Hongo, Bunkyo-ku, Tokyo, 113, Japan

ABSTRACT

Two selected recent developments of the Proton-Neutron Interacting Boson Model are discussed with a review on the microscopic basis of the Interacting Boson Model. The first subject is a derivation of the IBM-2 starting from the G-matrix interaction, and the second is a calculation of several IBM-2 parameters from the microscopic multi-nucleon systems. In the former, ^{56}Fe is considered, while Cd and Xe isotopes are studied in the latter.

1. Introducition

This talk covers some microscopic aspects of the Interacting Boson Model (IBM), being focused on two selected recent works after a brief review on such aspects. This talk consists of the following parts.

1. Introduction
2. Microscopic Basis of the Interacting Boson Model
3. IBM-2 built on G-matrix interaction
4. Microscopic calculation of IBM-2 parameters for the Cd and Xe isotopes
5. Summary

Section 3 is a collaboration with H. Nakada, while sect. 4 is with T. Mizusaki.

2. Microscopic Basis of the Interacting Boson Model

2.1. The Outline

The Interacting Boson Model (IBM) has been introduced, by Arima and Iachello nearly two decades ago, and has presented a unified description of quadrupole collective states [1]. The description is made in terms of a scalar boson of positive parity, called *s*, and a quadrupole boson of positive parity, called *d*. There are two characteristic features in IBM; (i) the total boson number is conserved for a given nucleus, (ii) the boson Hamiltonian consists of single particle energy and two-body interaction terms. By choosing specific values for the parameters in the Hamiltonian, one obtains limiting cases where the solutions can be derived analytically. Those limiting situations are referred to as dynamical symmetries, and are actually comprised of three cases, (i) U(5), (ii) SU(3), and (iii) O(6). The U(5) limit corresponds, in geometrical terms, to anharmonic quadrupole vibrations, the SU(3) to axially symmetric rotation, while the O(6) corresponds to γ-unstable deformation [1].

The underlying microscopic picture of the IBM is given, as already mentioned by Lipas in his review lecture [5], in terms of pairs of nucleons [2-4]. First, collective pairs of valence nucleons of $J^{\pi}=0^+$ (called *S*) and $J^{\pi}=2^+$ (called *D*) are introduced. These pairs are nothing but generalized or extended Cooper pairs. In fact, the S pair in spherical nuclei is exactly the Cooper pair of valence nucleons in the BCS ground state. The D pair absorbs much of the quadrupole strength when the quadrupole (moment) operator is acting on the S pair. The S and D pairs, thus, should be strongly connected through the quadrupole collectivity. I shall discuss more in detail on the structure of the S and D pairs in sect. 2.3. The microscopic basis of the IBM can be given, at least for nuclei without strong static deformation, by the Otsuka-Arima-Iachello (OAI) mapping [4], where these pairs correspond intuitively to the s and d bosons as,

$$\text{S pair} \rightarrow \text{s boson} \quad \text{and} \quad \text{D pair} \rightarrow \text{d boson}. \tag{1}$$

The number of bosons is hence given by the number of pairs. The number of s bosons, N_s, is equal to the number of the S pairs, while the number of d bosons, N_d, is equal to the number of the D pairs. Consequently, the total number of bosons is given by half of the number of valence particles.

The states constructed by the S and D pairs form a subspace truncated from the entire shell-model space. This subspace is called SD (pair) subspace. The states in the SD subspace are mapped onto the IBM states as

$$|S^N; 0^+\rangle \rightarrow |s^N; 0^+), \quad |S^{N-1}D; 2^+\rangle \rightarrow |s^{N-1}d; 2^+), \ldots.. \tag{2}$$

The boson Hamiltonian, H^B, is then determined so that the matrix elements of the Hamiltonian H in the SD subspace can be reproduced approximately by the corresponding sd boson matrix elements as

$$\langle \psi |H| \psi' \rangle \approx (\varphi |H^B| \varphi'), \tag{3}$$

where ψ and ψ' are SD states, and φ and φ' are the corresponding sd boson states obtained by the mapping in eq. (1). This process can be generalized to more general operators, for instance, the quadrupole operator, as will be shown.

2.2. Proton-Neutron IBM or IBM-2

Since the nucleus consists of protons and neutrons, SD pairs of protons, called S_π and D_π , and SD pairs of neutrons, called S_ν and D_ν are introduced based on the above discussion. Consequently, the proton bosons, denoted as s_π and d_π, and neutron bosons denoted as s_ν and d_ν are introduced. The IBM with the proton bosons and the neutron bosons is referred to as the Proton-Neutron IBM or IBM-2 [2-4]. The mapping process in eqs. (2)-(3) can be generalized to the IBM-2, by mapping the proton states (operators) on to proton boson states (operators), and likewise neutron states (operators) on to neutron boson states (operators). The number of proton bosons is denoted as N_π, while the value of N_π is given by half of the number of valence protons, if the proton system is in the first half of the major shell. If it is in the second half of the major shell, pairs of proton holes are mapped on to bosons, and N_π is given by half of the number of proton holes. A similar rule holds for neutrons, and the number of neutron bosons is denoted as

N_ν [4]. Likewise, the suffix π (ν) indicates hereafter quantities referring to protons (neutrons), although this suffix may be omitted for brevity if no confusion occurs.

The original IBM is sometimes called the IBM-1 in order to distinguish it from the IBM-2. Evidently, there is no distinction between proton boson and neutron boson in the IBM-1.

2.3. *Structure of the S and D pairs*

The structure of the S and D pairs is discussed first. The type of nucleon in the following discussions should be kept to either proton or neutron, although not mentioned explicitly.

The S pair creation operator is written as

$$S^\dagger = \Sigma_j \; \alpha_j \frac{1}{\sqrt{2}} [a_j{}^\dagger a_j{}^\dagger]^{(0)} , \tag{4}$$

where $a_j{}^\dagger$ denotes the creation operator of a nucleon on the orbit j, and α_j means the amplitude. Similarly the D pair is written with the β amplitude as,

$$D^\dagger = \Sigma_{i\leq j} \; \beta_{ij} \frac{1}{\sqrt{1+\delta_{ij}}} [a_i{}^\dagger a_j{}^\dagger]^{(2)} . \tag{5}$$

The S and D pairs have been introduced usually in terms of the seniority scheme of shell model [2,6]. I present here a slightly different formulation based on the many-body approach such as BCS or HFB, but the basic physics remains the same. In general, the amplitudes α and β should be determined so that the S and D pairs produce the most optimum description of low-lying quadrupole collective states. Basically the α amplitude of the S pair is determined by the competition between the pairing correlation and the one-body mean field (*i.e.*, single-particle energy). The β amplitude of the D pair can be more sensitive to the dynamical and/or static quadrupole deformation.

In the present method, the BCS (HFB) ground state is obtained first by the BCS (HFB) calculation. This ground state is written generally as

$$|\Psi\rangle = \Pi_{m>0} \; (u_m + v_m \, a_m{}^\dagger \, a_{\bar{m}}{}^\dagger \,) \, |0\rangle \tag{6}$$

where the u's (v's) are the usual u (v) factors, m stands for the magnetic quantum number of single-particle state, and $\bar{m}$ is the time-reversed state of m. Introducing the particle number projector, one projects $|\Psi\rangle$ onto a good particle number 2N as

$$P_N |\Psi\rangle \propto [\, \Sigma_{m>0} \; (v_m \,/\, u_m \,) \; a_m{}^\dagger \; a_{\bar{m}}{}^\dagger \,]^N |0\rangle. \tag{7}$$

This is nothing but N-pair condensate of the Cooper pair. The Cooper pair is denoted as the Λ pair hereafter, and its creation operator is written as

$$\Lambda^\dagger \equiv \mathcal{N} \; \Sigma_{m>0} \; (v_m \,/\, u_m \,) \; a_m{}^\dagger \; a_{\bar{m}}{}^\dagger \tag{8}$$

where $\mathcal{N}$ denotes the normalization constant. Equation (7) can then be rewritten as

$$P_N |\Psi\rangle \propto (\, \Lambda^\dagger \,)^N |0\rangle. \tag{9}$$

We calculate the u and v factors in eq. (6) with the particle number conservation by the method of [7], and hence what is actually determined is $P_N |\Psi\rangle$ rather than $|\Psi\rangle$.

If the nucleus being considered is spherical, u_m and v_m become independent of the magnetic quantum number m, while they are still dependent on the angular momentum j of the single-particle orbit. In this case, the Λ pair is scalar, and carries $J^\pi=0^+$. We then identify the Λ pair as the S pair as,

$$\Lambda \text{ pair} \rightarrow \text{S pair} \tag{10}$$

and thereby

$$| S^N ; 0^+ > \propto P_N |\Psi>. \tag{11}$$

The seniority (v) of this state is v=0 in the generalized seniority scheme.

In order to introduce the D pair in such spherical cases, one can use the broken-pair formalism [9]. The broken pair state is created, in general, by changing one of the Λ pairs in eq. (9) into a pair of nucleons, called *broken pair*, as

$$| N, \rho\sigma > \propto a_\rho{}^\dagger a_\sigma{}^\dagger \, (\Lambda^\dagger)^{N-1} |0>, \tag{12}$$

where ρ and σ stand for single-particle states (or their linear combinations) including magnetic quantum numbers. The ρ-σ pair thus introduced is called a broken pair, because it is produced by destroying one of the condensed Cooper pairs. The states given by eq. (12) are referred to as one-broken-pair states, and are denoted also as (number-projected) TDA states. There are a number of broken pair states due to many combinations of ρ and σ in eq. (12).

In the spherical cases with eq. (10), the Λ pair is nothing but the S pair. The one-broken-pair states can be redefined by a proper linear combination among themselves, so as to have a good angular momentum, as

$$| N, ij; J^\pi > \propto \frac{1}{\sqrt{1+\delta_{ij}}} [a_i{}^\dagger a_j{}^\dagger]^{(J)} \, (S^\dagger)^{N-1} |0>, \tag{13}$$

where J^π stands for the angular momemtum and parity of the broken pair, and i and j refer to the single particle orbits ($i \leq j$). The states in eq. (13) form an orthonormal set of basis states, except that one needs reorthogonalization for $J^\pi=0^+$. In order to determine the structure of the D pair, we calculate matrix elements of the Hamiltonian with respects to such orthonormal bases for $J^\pi=2^+$,

$$< N, ij; 2^+ | \; H \; | N, kl; 2^+ >. \tag{14}$$

The lowest eigensolution of this matrix can be identified as an SD state,

$$| D \, S^{N-1} ; 2^+ > \propto D^\dagger \, (S^\dagger)^{N-1} |0>. \tag{15}$$

Thus, the β amplitudes in eq. (4) are determined [8,9]. In the diagonalization of the matrix in eq. (14), the proton-neutron interaction can be included in the Hamiltonian, and then the proton D pair and the neutron D pair are determined simultaneously. This is the outline how one can determine the structure of S and D pairs in spherical nuclei. Note that the one-broken-pair states in spherical cases are v=2 states in the generalized seniority. I shall present an actual example of spherical calculations with Cd isotopes in sect. 4.1.

If the system is deformed, the Λ pair is no longer scalar, and contains components of various angular momenta. One can then extract the S and D pairs directly from the Λ pair. We decompose the Λ pair into such components as

$$\Lambda^\dagger \;=\; x_0\, S^\dagger \;+\; x_2\, D^\dagger \;+\; x_4\, G^\dagger \;+ \ldots, \tag{16}$$

where the x's are amplitudes, and S, D, G, *etc*. denote the positive-parity pairs of angular momenta 0, 2, 4, *etc*. projected from the Λ pair. The S and D pairs thus fixed can be rewritten in the forms of eqs. (4) and (5), and we use the same notation as in the spherical cases discussed above. In the intermediate situation, the S pair is determined by eq. (16), while the D pair is obtained from eqs. (14)-(15). I shall mention an example of this situation with Xe isotopes in sect. 4.2.

The values of α and β thus calculated vary gradually as functions of the neutron and proton numbers within a major shell. This property has been mentioned in [6,8-10]. In earliest papers of the IBM, as mentioned by Lipas, it was suggested that the α and β amplitudes are determined by solving two-body problems, *i.e.*, N=1 in the above discussion. This is true only in the limit that the changes of α and β are negligible as functions of nucleon number, although it is the case approximately in quite a few situations.

I have presented a basic method to determine the α and β amplitudes, starting from the BCS/HFB ground state. The HFB method has been used in [11] also. There can be several variations in the calculation of α and β, and I show such an example in sect. 3.1.

2.4. OAI Mapping

Once the S and D pairs are fixed, one can construct SD (pair) states. We normally construct them in the order that the number of D pairs increases. The non-orthogonality may occur, and the Schmidt orthogonalization is carried out to have orthogonal basis vectors of the SD subspace. With the orthogonalization, each SD basis state is normalized. The SD basis vectors are then mapped onto the sd boson basis vectors. Some examples of this mapping are shown below,

$$|\, S^N ; 0^+ \rangle \quad \rightarrow \quad |\, s^N ; 0^+) \tag{17a}$$

$$|\, D\, S^{N-1} ; 2^+ \rangle \quad \rightarrow \quad |\, d\, s^{N-1} ; 2^+) \tag{17b}$$

$$|\, D^2\, S^{N-2} ; J^+ \rangle \quad \rightarrow \quad |\, d^2\, s^{N-2} ; J^+), \qquad J = 0, 2, 4, \tag{17c}$$

$$\vdots \qquad\qquad\qquad \vdots$$

I would like to emphasize that the SD pair states and sd boson states in eq. (17) are all normalized. Equation (17) presents the state-to-state mapping which constitutes the first step of the OAI mapping [4].

The next step of the OAI mapping is the mapping of Hamiltonian and other operators. The *boson image* of this mapping is obtained by requiring that SD pair matrix elements of a given operator can be well reproduced by the corresponding sd boson matrix elements [4]. The correspondence here means the one given by eq. (17).

I would like to illustrate this process by an example. The boson image of the proton Hamiltonian can be written as

$$H_\pi^B = E_\pi + \varepsilon_\pi \hat{N}_{d\pi} + \ldots \tag{18}$$

where E_π and ε_π are coefficients to be calculated, and $\hat{N}_{d\pi}$ denotes the number operator of the proton d bosons. Clearly, E_π is a c-number dependent on the total number of proton bosons, N_π, and contributes to the binding energy only. These two coefficients are determined by equations for the proton Hamiltonian H_π;

$$< S^N; 0^+ | H_\pi | S^N; 0^+ > \;=\; (s^N; 0^+ | H_\pi^B | s^N; 0^+) \;=\; E_\pi , \tag{19a}$$

and

$$< D\, S^{N-1}; 2^+ | H_\pi | D\, S^{N-1}; 2^+ > \;=\; (d\, s^{N-1}; 2^+ | H_\pi^B | d\, s^{N-1}; 2^+) \;=\; E_\pi + \varepsilon_\pi . \tag{19b}$$

Other terms of the boson Hamiltonian are determined similarly. The quadrupole operator plays a crucial role in the description of quadrupole collective states. The quadrupole operator of either protons or neutrons is written in the second quantized form as

$$Q^F = \Sigma_{k,m} < k | r^2 Y^{(2)} | m > a_k{}^\dagger a_m , \tag{20}$$

with k and m being single-particle states. The boson image of Q^F is written as

$$Q^B = q \{ d^\dagger s + s^\dagger \tilde{d} + \chi [d^\dagger \tilde{d}]^{(2)} \}, \tag{21}$$

where q and χ are coefficients. The values of q and χ are determined by the equations

$$< D\, S^{N-1}; 2^+ \| Q^F \| S^N; 0^+ > \;=\; (d\, s^{N-1}; 2^+ \| Q^B \| s^N; 0^+) \;=\; q \sqrt{5} \sqrt{N}. \tag{22}$$

and

$$< D\, S^{N-1}; 2^+ \| Q^F \| D\, S^{N-1}; 2^+ > \;=\; (d\, s^{N-1}; 2^+ \| Q^B \| d\, s^{N-1}; 2^+) \;=\; q\, \chi \sqrt{5} . \tag{23}$$

The parameters for the proton system, denoted as q_π and χ_π, and those for the neutron system, denoted as q_ν and χ_ν, are determined separately by considering proton matrix elements and neutron matrix elements, respectively.

2.5. *Renormalization of the boson Hamiltonian (and other operators) due to couplings to states outside the SD pair space*

There are non-vanishing matrix elements connecting SD pair states to other states in general. Such coupling effects should be taken into account by the renormalization of the boson Hamiltonian or extending the boson model, for instance, to the model with s, d and g bosons. Here the g boson means a boson of $J^\pi=4^+$, and corresponds to the $J^\pi=4^+$ pair, called G. I am not going to discuss on the renormalization here because of the limited time, but will show some examples of the renormalization.

2.6. F-spin

The symmetry associated with the proton-neutron degree of fredom in the IBM-2 is represented in terms of F-spin [2,4]. F spin is an isospin-like concept for bosons, and has similarities and differences compared to isospin. In the isospin scheme, each nucleon carries isospin T=1/2, and protons and neutrons are treated as the $T_z=1/2$ and $T_z=-1/2$ states, respectively. Similarly, each boson carries F-spin F=1/2. The proton boson is the $F_z=1/2$ state, while the neutron boson the $F_z=-1/2$ state. This assignment is tabulated in Table 1.

Table 1 F spin quantum numbers

		F	F_z
proton boson	s_π and d_π	1/2	1/2
neutron boson	s_ν and d_ν	1/2	-1/2

The F spin obeys the angular momentum algebra, as the ordinary spin does. Because of the vector addition rule of the angular momentum algebra, a system with N_π proton bosons and N_ν neutron bosons has $F_z=(N_\pi-N_\nu)/2$, and can have F value, $|N_\pi-N_\nu|/2 \le F \le (N_\pi+N_\nu)/2$. The F spin characterizes proton-neutron symmetry in IBM-2 wave functions, and this symmetry property can be seen most clearly in two-boson systems. Table 2 presents F spin classification of two-boson states. First of all, the states with $N_\pi=2$ and $N_\nu=0$ (*i.e.*, $F_z=1$) and the states with $N_\pi=0$ and $N_\nu=2$ (*i.e.*, $F_z=-1$) are IBM-1 states because they contain either proton bosons or neutron bosons. Note that these states are in F=1 multiplets. Most of the states with $N_\pi=N_\nu=1$ in Table 2 are symmetric with respect to the interchange between the proton boson and the neutron boson. Such symmetric states have F=1 and $F_z=0$, and can be transformed by F-spin raising and lowering operators, which are introduced analogously to the angular momentum operators. The states with the maximum F value, $F=(N_\pi+N_\nu)/2$, possess maximal proton-neutron symmetry, and are referred to as totally symmetric states. The totally symmetric states contain proton bosons and neutron bosons coherently, and are equivalent to IBM-1 states of $F_z=\pm(N_\pi+N_\nu)/2$. They therefore can be described by the IBM-1. Lowest collective states are dominated by components consisting of the totally symmetric states, and one does not have to use explicitly proton bosons and neutron bosons in their description. It may be worth noting that the total wavefunctions should be antisymmetric for nucleons, and, by making the isospin part antisymmetric, the rest of the wavefunction becomes symmetric and can gain binding energy. The antisymmetric isospin wavefunction means lower isospin. The situation is reversed in boson systems where the total wavefunction should be symmetric; by keeping the F spin part symmetric, the rest of the wavefunction remains symmetric, coming down in energy. This is the basic reason why lowest states are dominated usually by components of largest F spin in the IBM-2, while the isospin takes minimum value in lowest states. IBM-2 states containing F=0 pair of proton boson and neutron boson are called mixed-symmetry states.

Table 2 F spin classification of two-boson states

	F	F_z
$\mid s_\pi^2 ; 0^+)$	1	1
$\mid s_\pi s_\nu ; 0^+)$	1	0
$\mid s_\nu^2 ; 0^+)$	1	-1
$\mid d_\pi s_\pi ; 2^+)$	1	1
$\frac{1}{\sqrt{2}}\{\mid d_\pi s_\nu ; 2^+) + \mid d_\nu s_\pi ; 2^+)\}$	1	0
$\mid d_\nu s_\nu ; 2^+)$	1	-1
$\frac{1}{\sqrt{2}}\{\mid d_\pi s_\nu ; 2^+) - \mid d_\nu s_\pi ; 2^+)\}$	0	0
$\mid d_\pi^2 ; L^+)$ (L=0, 2, 4)	1	1
$\mid d_\pi d_\nu ; L^+)$ (L=0, 2, 4)	1	0
$\mid d_\nu^2 ; L^+)$ (L=0, 2, 4)	1	-1
$\mid d_\pi d_\nu ; L^+)$ (L=1, 3)	0	0

3. IBM-2 built on G-matrix interaction

In this section, I move on to a concrete calculation where ^{56}Fe is described with the boson Hamiltonian derived from the G-matrix interaction [12]. Although ^{56}Fe is not a best example for the study of quadrupole collective states, it provides us with a meeting point of the collective model and the realistic shell model calculation.

3.1. Large-scale shell-model calculation of ^{56}Fe

Nakada, Sebe and myself have carried out large-scale shell-model calculations for ^{56}Fe and ^{54}Cr in a realistic framework, using a state-of-the-art computer program running on supercomputers [13]. The single particle orbits $1f_{7/2,5/2}$ and $2p_{3/2,1/2}$ are included in this shell model calculation, and the single particle energies are determined from experiment. The two-body interaction matrix elements are taken from Kuo-Brown's G-matrix interaction with 3p-1h correction included [12]. Thus, there is no adjustable parameter in this calculation.

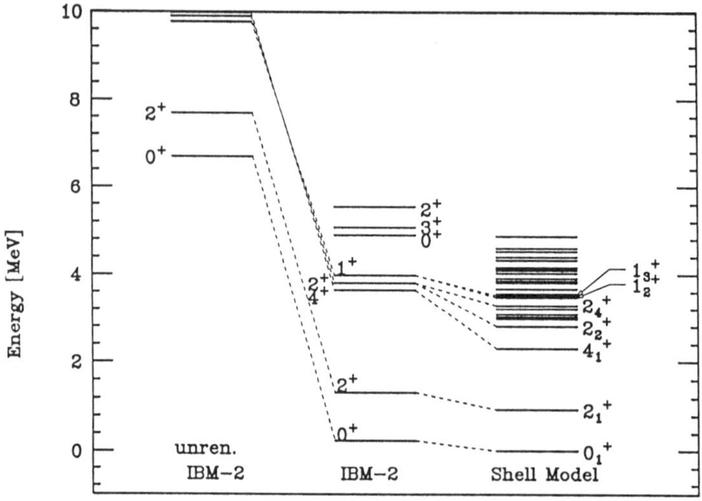

Fig. 1 Unrenormalized (left) and renormalized (middle) IBM-2 spectra compared to the shell-model spectra (right) of ^{56}Fe. All spectra are calculated starting from the Kuo-Brown interaction.

The lowest configuration in the calculation for ^{56}Fe is $(1f_{7/2})^6$ for protons (*i.e.*, $(1f_{7/2})^{-2}$ proton holes) and $(1f_{7/2})^8$ $(2p_{3/2,1/2},1f_{5/2})^2$ for neutrons. Since this configuration consists of a pair of proton holes and a pair of valence neutrons, the corrsponding IBM-2 system is comprised of a proton boson and a neutron boson, *i.e.*, $N_\pi=N_\nu=1$. In other words, the S_π and D_π pairs are defined, respectively, as $J^\pi=0^+$ and 2^+ $(1f_{7/2})^{-2}$ states, while the S_ν and D_ν pairs are collective $J^\pi=0^+$ and 2^+ $(2p_{3/2,1/2},1f_{5/2})^2$ states. In the case of neutrons, because three orbits in the upper pf-shell contribute, we take the most appropriate pairs so that the overlap to the relevant shell-model wave functions becomes maximum. This way of determination of the SD pairs may look different from the prescription in sect. 2.3, but is based on the same physics, *i.e.*, the optimization for describing the low-lying collective states.

In addition to the lowest configuration, the excitation from $1f_{7/2}$ to higher pf-shell orbits is included; the present shell-model calculation includes configurations $(1f_{7/2})^{14-k}$ $(1p_{3/2,1/2},0f_{5/2})^{2+k}$ with k=0,1,2, where k denotes the total number of protons and neutrons shifted from $1f_{7/2}$ [13]. The isospin is conserved as T=2 in shell-model eigenstates discussed in the following. Up to E_X ~ 4 MeV, the discrepancy between experiment and calculation is typically only ~ 0.1 MeV, demonstrating a remarkable agreement between experiment and calculation [13] .

3.2. Derivation of IBM-2 Hamiltonian for ^{56}Fe

We derived the IBM-2 Hamiltonian for ^{56}Fe, by carrying out the mapping in eqs. (1) and (2). Using the SD pairs defined above, we map the nucleon Hamiltonian onto an sd-boson Hamiltonian. In Fig. 1, the energy levels obtained by this IBM-2 Hamiltonian are shown at far left, in comparison to the shell-model energies at far right. Note that both calculations are carried out for the same Kuo-Brown interaction. Although the spacing between the 0_1^+ and 2_1^+ states is almost equal between the two calculations, the binding energy appears to be different, indicating non-negligible effects due to couplings between the SD pairs and the non-SD pairs. We therefore modify the SD states so that such effects can be included. The modification contains two aspects, (i) couplings to pairs of higher angular momenta, and (ii) couplings to states involving higher configurations of $k \geq 1$. We introduce a method where these two aspects are included on an equal footing. For the i-th original SD state $|\varphi_i)$, we construct a series of states $|\varphi_i ; r)$ $(r=1,2, ...)$ as

$$|\varphi_i ; r=1) = \mathbf{N}^{(1)} P^{(1)} \; H |\varphi_i \;),$$

$$|\varphi_i ; r=2) = \mathbf{N}^{(2)} P^{(2)} \; H |\varphi_i \; ; r=1) \; , \; ...... \tag{24}$$

where the $\mathbf{N}$ stand for normalization constants, H denotes the nucleon Hamiltonian, and the projection operator $P^{(k)}$ removes previously determined components such as $|\varphi_i)$, $|\varphi_i; r=1),...,$ $|\varphi_i ; r=k-1)$. The projection operator $P^{(k)}$ has another function as explained below. Note that the zeroth-order renormalized state, $|\varphi_i ; r=0)$, is nothing but the original SD state. The Hamiltonian matrix is then diagonalized in the subspace spanned by the bases $|\varphi_i)$, $|\varphi_i; 1),...,$ $|\varphi_i ; r)$ for a given r. The renormalized SD state of the r-th order is defined as the lowest eigenstate in this diagonalization. Throughout the process of determining $|\varphi_i ; r)$, the $P^{(r)}$ operator is supposed to remove mixing of other renormalized SD states $|\varphi_1; r)$, $|\varphi_2; r)$, ... , $|\varphi_{i-1}; r)$ obtained prior to $|\varphi_i ; r)$. It has been confirmed, in the present case, that the r=1 renormalized bases already produce reasonably good approximation to the exact shell-model calculation, and the r=2 bases give rise to excellent result. Thus, we shall identify, in this work, the r=2 renormalized SD states as the (extended) SD states to be used. The Hamiltonian H causes particle-hole excitation from the orbit $1f_{7/2}$ to the orbits $2p_{3/2,1/2}$ and $1f_{5/2}$. In the shell-model calculation of the energy levels, the k=1 and 2 configurations are included as well as the k=0 configuration. The above renormalized SD bases include such particle-hole excitations also.

The renormalization effects are included by carrying out the mapping in sect. 2.4 in terms of the r=2 renormalized SD bases introduced above. Compared to heavier nuclei, the configuration space is quite limited in ^{56}Fe. The coupling to non-SD pairs, therefore, can be stronger and more state-dependent. Therefore, the coupling should be treated in a more elaborated method than for heavier nuclei. Thus, the renormalization has to be done more carefully than in usual cases, and we adopt a method where specific linear combinations of SD states corresponding F-spin eigenstates in Table 2 are used as the zeroth-order renormalized state $|\varphi_i ; r=0)$. For instance, for $J^\pi = 2^+$, the following zeroth-order basis states are taken,

$$|\varphi_1 ; r=0) \equiv \frac{1}{\sqrt{2}} \{ | D_\pi \, S_\nu \, ; 2^+) + | D_\nu \, S_\pi \, ; 2^+) \}, \tag{25}$$

$$|\varphi_2\,;\,r{=}0) \equiv |\,D_\pi\,D_\nu\,;\,2^+\,)\,, \tag{26}$$

$$|\varphi_3\,;\,r{=}0) \equiv \frac{1}{\sqrt{2}}\{|\,D_\pi\,S_\nu\,;\,2^+\,) - |\,D_\nu\,S_\pi\,;\,2^+\,)\}. \tag{27}$$

By treating renormalized SD basis states, $|\varphi_i\,;\,r{=}2)$ (i=1,3), as the SD states, we determine parameters in the boson Hamiltonian by the prescription shown in sect. 2.4. Figure 1 shows the energy levels calculated by the diagonalizations of (i) the unrenormalized IBM-2 Hamiltonian, (ii) the renormalized IBM-2 Hamiltonian, and (iii) the shell-model Hamiltonian. For the 0_1^+ and 2_1^+ states, the agreement is very good between the renormalized IBM-2 and the shell model up to the binding energy.

3.3. Mixed-symmetry states in ^{56}Fe

The 2_2^+, 1^+ and 3^+ states in the spectrum of the renormalized IBM-2 of Fig. 1 are mixed-symmetry states. The 2_2^+ state corresponds to the 7th state in Table 2, while the 1^+ and 3^+ states to the last state in Table 2. On the other hand, these IBM-2 states do not correspond to any particular individual shell-model states. In fact, they correspond to certain linear combinations of several shell-model states. The mixed-symmetry states (or states comprised mainly of mixed-symmetry components) are not favored by the proton-neutron interaction, and are actually pushed up to higher excitation energy. There are other non-collective states in the region of higher excitation energy, and the mixed-symmetry state is mixed with some of such non-collective states. This results in certain fragmentation. In this fragmentation, the mixing matrix elements are in general small because of the significant difference in structure, and hence the mixing occurs mainly due to small energy differences between the mixed-symmetry state and neighboring non-collective states. Thus, in most cases, the fragmentation is inevitable. On the other hand, the fragmentation should not destroy, in most cases, the identity of the mixed-symmetry states, since the fragmentation is concentrated into a few states near the unperturbed level because of the mixing mechanism mentioned above. This picture leads us to an ansatz that the mixed-symmetry state can be interpreted as a kind of "Doorway State". This ansatz has been verified by looking at the actual distribution of the 2^+ and 1^+ mixed-symmetry renormalized SD bases over shell-model eigenstates [14].

3.4. M1 transitions in ^{56}Fe

The IBM-2 M1 operator is derived also by the mapping of the renormalized SD bases. The nucleon M1 operator is given by

$$M^F = \sqrt{\frac{3}{4\pi}}\{\, g_{s;\pi}\, S_\pi + g_{l;\pi}\, L_\pi + g_{s;\nu}\, S_\nu + g_{l;\nu}\, L_\nu \,\}, \tag{28}$$

where the g's are g-factors, and S and L denote the spin and orbital angular momentum operators. The boson M1 operator is written in the lowest order as

$$M^B = \sqrt{\frac{3}{4\pi}}\{\, g_\pi^B\, J_\pi^B + g_\nu^B\, J_\nu^B \,\}, \tag{29}$$

where the g^B's are coefficients called boson g-factors, and the J^B's stand for the angular momentum operators of either proton bosons or neutron bosons. Based on the method in sect. 2.3, we map the M1 operator separately for protons and neutrons as

$$< D\, S^{N-1}; 2^+ \| M^F \| D\, S^{N-1}; 2^+ > \; = \; (d\, s^{N-1}; 2^+ \| \; M^B \| d\, s^{N-1}; 2^+)\,, \qquad (30)$$

where matrix elements and operators are either for protons or neutrons.

Table 3. Boson g-factors for ^{56}Fe (See eq. (29)).

	unrenormalized	renormalized
g_π	1.26	1.11
g_ν	-0.03	-0.04

The resultant boson g-factors are shown in Table 3. Using these boson g-factors, we calculated the B(M1) value from the 2_2^+ state, which is of mixed-symmetry, to the 2_1^+ state in the renormalized IBM-2 calculation. This B(M1) gives rise to 0.40 (μ_N^2). In the shell model, $B(M1;2_2^+ \rightarrow 2_1^+) + B(M1;2_4^+ \rightarrow 2_1^+)$ yields 0.41 (μ_N^2), close to the IBM-2 value. This summed value is 0.34 ± 0.12 (μ_N^2) experimentally [15]. The comparison among these B(M1) values suggests that the M1 transition is a crucial quantity to search for the mixed-symmetry state [16,17]. The large B(M1) value is a necessary condition for mixed-symmetry states. A more detailed account can be found in [14].

4. Microscopic Calculation of IBM-2 parameters for Cd and Xe isotopes

I shall present results of recent microscopic calculations of IBM-2 parameters for Cd and Xe isotopes. The nucleon Hamiltonian is assumed to consist of the single-particle energy, the monopole, quadrupole and hexadecupole pairing interactions between identical nucleons, and the quadrupole-quadrupole interactions between identical nucleons and between proton and neutron. All parameters such as single-particle energies and interaction strengths are determined by the spectra of single-closed nuclei except the strength of the proton-neutron interaction. The single-particle energies are determined basically by the levels of nuclei with one valence nucleon or hole, with slight variation (up to a few hundreds keV) so as to fit observed levels of one-quasi-particle states of relevant single-closed nuclei, for instance Sn isotopes for neutrons. The strengths of the interactions between identical nucleons are determined so that levels of 2_1^+ and 4_1^+ of single-closed nuclei are reproduced well. The proton-neutron interaction strength thus remains as only adjustable parameter, and its adopted value appears to be similar to what is usually used.

The α and β amplitudes in eqs. (4) and (5) are determined in the method of sect. 2.3. We then calculate the parameters q_τ and χ_τ ($\tau = \pi, \nu$) in eq. (21) by using eqs. (22) and (23). The shell-model effective charges are denoted as e_π and e_ν for protons and neutrons, respectively. By multiplying them to the boson quadrupole operator in eq. (21), one obtains the boson (E2) effective charges as,

$$e_\pi^B \equiv e_\pi q_\pi \quad \text{and} \quad e_\nu^B \equiv e_\nu q_\nu . \tag{31}$$

The boson g-factors defined in eq. (29) are also calculated.

4.1. Calculation for Cd isotopes

The HFB solution is spherical for Cd isotopes. The S pair is determined by eq. (10), while the D pair is obtained by the one-broken pair calculation in eqs. (12)-(15). The configuration space is the major shell between 50 and 82 for neutrons, while $1g_{9/2}$ for protons. Figure 2 shows values of the χ's, e^B's and g^B's. The values of χ_ν appears very different from fitted values previously reported, but in a good agreement with a recent fit by Deleze *et al.* [18]. As a function of the neutron number, χ_ν stays rather constant in the second half of the major shell, but comes down sharply as the neutron number goes down beyond the middle of the shell. For the e^B's, the nucleon effective charges 1.72e [19] and

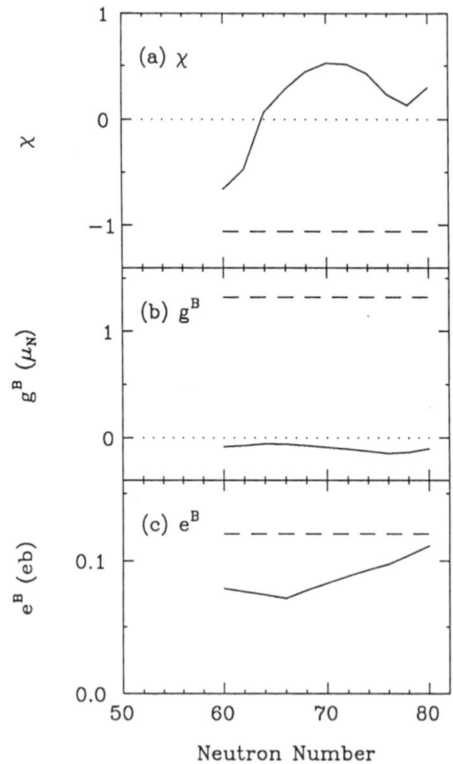

Fig. 2 IBM-2 parameters for Cd isotopes. Solid (dashed) lines are for neutron (proton) bosons.

1.1e are taken for protons and neutrons, respectively. For the g^B's, the spin quenching factor 0.7 is assumed for both protons and neutrons. The large value 1.32 for g_π^B is due to coherent contributions from the orbital and spin terms. Such an effect is quite natural, but has been overlooked in the past. Note that the results in Fig. 2 are tentative.

4.2. Calculation for Xe isotopes

The calculation for Xe isotopes are basically the same as those for Cd isotopes, except that protons are filling the major shell between 50 and 82 instead of $1g_{9/2}$. Because the proton-neutron interaction is more effective in Xe isotopes than in Cd isotopes, the HFB ground state is slightly deformed in Xe isotopes, and the S pair is determined by eq. (16). The most important feature found in Fig. 3 is that χ_v remains almost constant at the value around $-\chi_\pi$, which supports strongly the O(6) structure for a wide range of Xe isotopes. This is rather in contrast to the microscopic calculation of [20] where χ_v changes much faster as a function of the neutron number. A similar rapid and almost linear change is seen in the fitting in [21]. Proton and neutron effective charges are

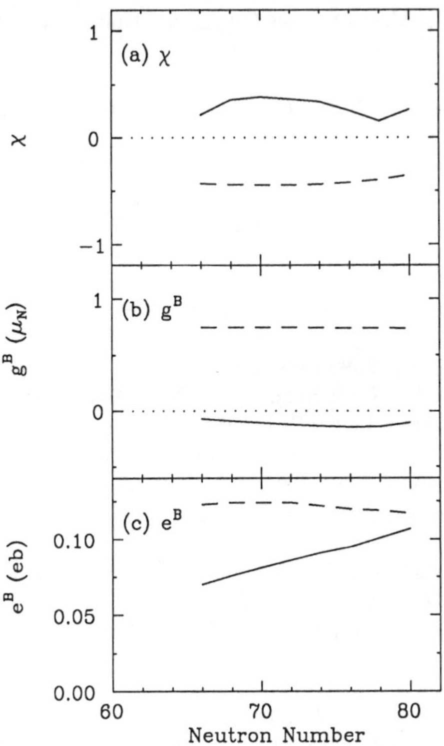

Fig. 3 IBM-2 parameters for Xe isotopes. See the caption of Fig. 2.

assumed to be 1.4e and 1.0e respectively, and spin quenching 0.7 is used. Similarly to Cd calculations, the results in Fig. 3 are tentative.

5. SUMMARY

The IBM-2 has been derived from the Kuo-Brown interaction for the first time, and is shown to be capable of reproducing the shell-model result. Microscopic calculations of IBM-2 parameters are carried out recently for Cd and Xe isotopes based on the HFB+Broken-Pair formalism, resulting in several distinctive features.

ACKNOWLEDGEMENTS

I thank Profs. W. Nazarewitz and J.P. Draayer for the opportunity of presenting this talk. This work has been supported in part by the Grant-in-Aid for General Scientific Research (no. 01540231) by the Ministry of Education, Science, and Culture as well as by Research Center of Nuclear Physics, Osaka University. The numerical calculations have been carried out partly by HITAC S-820/80 at Kanagawa Factory, Hitachi Corp., and partly by VAX 6640 at the Meson Science Laboratory, University of Tokyo.

References

[1] F. Iachello and A. Arima, *The Interacting Boson Model* (Cambridge Univ. Press, New York, 1987), and references therein.

[2] A. Arima, T. Otsuka, F. Iachello and I. Talmi, Phys. Lett. **66B** (1977) 205.

[3] T. Otsuka, A. Arima, F. Iachello and I. Talmi, Phys. Lett. **76B** (1978) 139.

[4] T. Otsuka, A. Arima and F. Iachello, Nucl. Phys. **A309** (1978) 1.

[5] P.O. Lipas, this proceedings.

[6] C.H. Druce *et al.*, Ann. Phys. **176** (1987) 114, and references therein.

[7] J.L. Egido and P. Ring, Nucl. Phys. **A383** (1982) 189.

[8] O. Scholten, Phys. Rev. **C28** (1983) 1783.

[9] K. Allaart *et al.*, Phys. Rep. **169** (1988) 209.

[10] W. Pannert, P. Ring and Y.K. Gambhir, Nucl. Phys. **A443** (1985) 189.

[11] L.C. de Winter *et al.*, Phys. Lett. **179B** (1986) 322.

[12] T.T.S. Kuo and G.E. Brown, Nucl. Phys. **A114** (1968) 241.

[13] H. Nakada, T. Otsuka and T. Sebe, Phys. Rev. Lett. **67** (1991) 1086; T. Otsuka, in Understanding the Variety of Nuclear Excitations, edited by A. Covello (World Scientific, Singapore, 1991), p. 321.

[14] T. Otsuka, in the proceedings of *OLNO-2* published in Hyperfine Int. (1992).

[15] H. Junde *et al.*, Nucl. Data Sheets **51** (1987) 1.

[16] S.A.A. Eid *et al.*, Phys. Lett. **166B** (1986) 267; S.P. Collins. *et al.*, J. Phys. **G 15** (1989) 321.

17] G. Hartung *et al.*, Phys. Lett. **221B** (1989) 109.

[18] M. Deleze *et al.*, Nucl. Phys. (1992) in print.

[19] F.J.D. Serduke, R.D. Lawson and D.H. Gloeckner, Nucl. Phys. **A256** (1976) 45.

[20] A. van Egmond and K. Allaart, Nucl. Phys. **A425** (1984) 275.

[21] G. Puddu, O. Scholten and T. Otsuka, Nucl. Phys. **A348** (1980) 109.

GROUP 11

Schematic Models

AN EXACT SOLUTION TO THE NUCLEAR CRANKING MODEL HAMILTONIAN

W. SATUŁA*
Department of Mathematical Physics, University of Lund,
Box 118, S-221 00 Lund, Sweden

and

Z. SZYMAŃSKI
Institute of Theoretical Physics, University of Warsaw,
ul. Hoża 69, PL-00 681 Warsaw, Poland

ABSTRACT

A method of finding exact solutions to the nuclear cranking-model Hamiltonian by means of group theory is developed. The properties of solutions that are labelled additionally by a new symmetry quantum number are discussed and illustrated in some simple applications.

1. Introduction

One of the most effective methods applied for the study of nuclear properties at high angular momenta is the cranking formalism[1,2]. In its most standard version the cranking method is reduced to the solution of the system described by the Hamiltonian being the sum of the single–particle term and the two–body pairing force. The standard solution to this problem that has been applied in numerous investigations is based on the Hartree–Fock–Bogolyubov (HFB) approximation in the rotating dynamical system. This procedure is known as the cranked–HFB (CHFB) (see e.g. refs.[3–6]). An improved version of this formalism employs the nuclear wave function with sharp numbers of nucleons obtained from the original HFB function by means of the particle number projection[7,8,9].

The approximations described above have been commonly used in the studies of nuclear behaviour at large angular momenta. Consequently, an interesting question arises what could be the relations of these approximate treatment with the rigorously exact solutions to the cranking Hamiltonian. Although it is not certain what is the more basic physical reality: either the rigorous solution or the approximate CHFB solution it seems very useful and interesting to elucidate their possible relations as to analyse the limitations and advantages of both these concepts. The analysis may be especially important at some critical situations in a fast rotating

*On leave of absence from Institute of Theoretical Physics, University of Warsaw, Poland

nucleus where the CHFB method may be insufficient to describe adequately the physical picture.

There have been many attempts to improve the quality of the CHFB approximation. One possible way would be to start with the improved (or even exact if possible) solutions to the nonrotating Hamiltonian. One could employ[9-12] for example the rigorous Richardson[13] solutions[11,12], the broken pair approximation[14,15], or else an improved BCS treatment[16-20]. Another method consists of the search for an exact solution to the full cranking model Hamiltonian in the m-basis for a single j-shell[21,22,23] for the sale of comparison with the CHFB approximation. Possible truncations in the basis that simplify the numerical computations and – on the other hand – remain close to the exact solutions have also been studied (see e.g. refs.[24,25]).

In the present paper we will limit our considerations to the single j-shell. This is a very particular case of the full problem of nuclear structure at high angular momentum. Nevertheless, when it is applied to the intruder subshells in nuclei (such as for example $i_{13/2}$ neutrons or $h_{11/2}$ protons in the rare earth nuclei) it may be able to describe the most essential features in rotating nuclei since these subshells are known to respond most strongly to nuclear rotation in contrast to the normal parity subshells.

This publication, however, is mainly concentrated on the further development of the method of exact solution to the cranking Hamiltonian in a case of the one j-shell. We apply and simplify the method based on the Lie algebras of the group theory which was already employed for the particle–rotor model by *Krumlinde and Szymański* [26] and *Andersson and Krumlinde* [27]. We shall see that the consistent application of the theory of group offers a very simple and elegant method to solve the problem. For example, all the matrix elements turn out to be extremely simple when calculated within the irreducible representations of the resulting symmetry group. Moreover, the associated symmetry provides a new quantum number (in addition to the signature and particle number) that greatly simplifies the diagonalisation of the Hamiltonian as compared to any other method used so far.

The contribution is organized as follows. The group theoretical structure of the cranking Hamiltonian is discussed in chapter 2. In particular, the irreducible representations of the associated group are constructed and the relevant formulae for matrix elements of the Lie algebra generators are given. In chapter 3 some simple examples of numerical calculations applied to nuclear structure are presented. The short summary is given in chapter 4.

2. The Model and its Symmetries

As already mentioned our considerations are limited to a single j-shell which is an intruder among normal–parity states. The starting cranking Hamiltonian (in our notation we adopt $\hbar = 1$):

$$\hat{H}^{\omega} = \hat{H}_{sp} + \hat{H}_{pair} - \omega \hat{j}_x \tag{1}$$

to be solved is a standard one. It contains a single particle part $\hat{H}_{sp}$ with a nuclear deformed axially symmetric field (with respect to, say z–axis), the two–body pairing force $\hat{H}_{pair}$ and the usual cranking term $-\omega\hat{j}_x$ describing the uniform rotation of the intrinsic nuclear system about the axis (say x–axis) perpendicular to the symmetry axis. The single–particle energies (eigenvalues of $\hat{H}_{sp}$) for a single j–shell split in the axially symmetric field into a set of n doubly degenerate levels (with $n = j+1/2$) due to the Kramers degeneracy. The $2n$ states formed in this way constitute the one–particle part of the Hilbert space corresponding to the dynamical system defined by Hamiltonian (1). Using the standard notation for the creation and anihillation operators $c^\dagger_\alpha$ and c_β of the shell–model states one can express the Hamiltonian (1) as certain combinations of the bilinear forms of the type $c^\dagger_\alpha c_\beta$, $c^\dagger_\alpha c^\dagger_{\bar{\alpha}}$ or $c_{\bar{\alpha}} c_\alpha$. Moreover, as has been shown in ref.[26] (see also Appendix), there exist a one to one corresponedence between the above mentioned operators (or rather their particular combinations being the constituents of the cranking Hamiltonian) and a set of operators $A_{k,l}$ $(k, l = 1, 2, \cdots, 2n)$ that fulfil the relations

$$[A_{k,l}, A_{m,n}] = \delta_{l,m} A_{k,n} - \delta_{k,n} A_{m,l}; \quad \text{and} \quad A^\dagger_{k,l} = A_{l,k} \tag{2}$$

characteristic for the generators of the Lie algebra of the unitary group $\boldsymbol{U(2n)}$ in $2n$ dimensions. Hamiltonian (1) can thus be built entirely out of the generators $A_{k,l}$ (cf. ref.[26]):

$$\hat{H}_{sp} = \sum_{i=1}^{n} \varepsilon_i \left(A_{i,i} - A_{i+n,i+n} + 1\right) \tag{3}$$

$$\hat{H}_{pair} = -G \sum_{i,j=1}^{n} (-)^{i+j} A_{i,i+n} A_{j+n,j} \tag{4}$$

$$\begin{gathered}\hat{j}_x = \frac{n}{2}(-)^n \left(A_{1,1} + A_{n+1,n+1} - 1\right) + \\ \frac{1}{2}\sum_{i=2}^{n} \sqrt{(n-i+1)(n+i-1)}\left(A_{i,i-1} + A_{i-1,i} - A_{i+n,i+n-1} - A_{i+n-1,i+n}\right)\end{gathered} \tag{5}$$

The explicit expressions for the generators $A_{k,l}$ in terms of the creation and annihilation operators $c^\dagger_\alpha, c_\beta$ are presented in ref.[26]. A brief compendium of the relevant formulae is also given in the Appendix for the sake of completness. One should also mention here, that in all the applications presented in chapter 3 we are using the standard quadrupole splitting for the single particle energies ε_i of Eq. (3):

$$\varepsilon_i = \kappa \frac{3m_i^2 - j(j+1)}{j(j+1)} \equiv \kappa \frac{3(i-1/2)^2 - (n-1/2)(n+1/2)}{(n-1/2)(n+1/2)} \tag{6}$$

where κ is the energy unit often used in one j–shell calculations and $n = j + 1/2$, $m_i = i - 1/2$ $(i = 1, 2, \ldots n)$ according to our notation (see Appendix). However, the structure of the model allows for applications of any other single–particle energy splitting as the one given by Eq. (6) is often appreciably different from that generated by more realistic axial mean field potentials (e.g. by Woods–Saxon potential).

The underlying unitary symmetry offers the possibility of explicit construction of the basis states according to the method proposed by *Gelfand and Tsetlin*[28] and further developed by *Baird and Biedenharn*[29]. The basis vector – Gelfand–Tsetlin (GT) pattern – could be represented as:

$$\begin{pmatrix} m_{1,2n} & m_{2,2n} & \dots & m_{2n-2,2n} & m_{2n-1,2n} & m_{2n,2n} \\ m_{1,2n-1} & m_{2,2n-1} & \dots & m_{2n-2,2n-1} & m_{2n-1,2n-1} & \\ \dots & \dots & \dots & \dots & & \\ \vdots & \vdots & & & & \\ m_{1,2} & m_{2,2} & & & & \\ m_{1,1} & & & & & \end{pmatrix} \tag{7}$$

where all $m_{k,l}$ elements are integer numbers obeying the following relations:

$$m_{i,k} \geq m_{i,k-1} \geq m_{i+1,k} \quad \text{where} \quad k = 2,3,\dots,2n; \quad i = 1,2,\dots,k-1 \tag{8}$$

The first row in the GT–pattern labels the representation (is fixed) and could be determined by the eigenvalues – highest weights – of the associated Cartan subalgebra. The Cartan subalgebra is formed by the diagonal generators $A_{\alpha,\alpha}$ and, for fermionic systems, one can easily show that the highest weights (representations labels) have the following form $[1]^k\,[0]^{2n-k}$ where $k = 1,\dots 2n$. The symbolic notation used means that the first row of the GT–pattern consists k unities followed by $2n-k$ zeros. Consequently, because of the relations (8), all the $m_{k,l}$ are either unities or zeros. Morover, in each row the unities are placed to the left and they are followed by zeros.

This enable us to use simplified notation for the GT–states:

$$\begin{pmatrix} \nu_{2n} \\ \nu_{2n-1} \\ \vdots \\ \nu_2 \\ \nu_1 \end{pmatrix} \tag{9}$$

where ν_l counts the number of unities in the l-th row of GT–pattern, and

$$k \geq \nu_k \quad \text{and} \quad \begin{cases} \nu_k = \nu_{k-1}; & \text{or} \\ \nu_k = \nu_{k-1} + 1 \end{cases} \tag{10}$$

as a consequence of the relations (8)

The great advantage of the above notation is the simplification of the matrix elements of group generators $A_{i,j}$. Indeed, for diagonal generators $A_{l,l}$ one gets

$$A_{l,l} \begin{pmatrix} \nu_{2n} \\ \nu_{2n-1} \\ \vdots \\ \nu_2 \\ \nu_1 \end{pmatrix} = (\nu_l - \nu_{l-1}) \begin{pmatrix} \nu_{2n} \\ \nu_{2n-1} \\ \vdots \\ \nu_2 \\ \nu_1 \end{pmatrix} \tag{11}$$

while for nondiagonal generators $A_{l,m}$ where $l < m$ (note that $A^\dagger_{m,l} = A_{l,m}$) one has:

$$A_{l,m}\begin{pmatrix} \nu_{2n} \\ \vdots \\ \nu_m \\ \nu_{m-1} \\ \vdots \\ \nu_{l+1} \\ \nu_l \\ \vdots \\ \nu_1 \end{pmatrix} = (-)^{\nu_m-\nu_l}(\nu_m - \nu_{m-1})(\nu_l - \nu_{l-1} - 1)\begin{pmatrix} \nu_{2n} \\ \vdots \\ \nu_m \\ \nu_{m-1}+1 \\ \vdots \\ \nu_{l+1}+1 \\ \nu_l+1 \\ \vdots \\ \nu_1 \end{pmatrix} \tag{12}$$

Note, that all the matrix elements are equal ± 1 or 0.

The dimension of the representation labelled by $p \equiv \nu_{2n}$ is equal $\binom{2n}{p}$. One should stress however, that the states belonging to given irreducible representation labelled by p (*p-irrep*) correspond to different numbers of particles N. This is because the particle number operator

$$\hat{N} = \sum_{\alpha=1}^{n}(A_{\alpha,\alpha} - A_{\alpha+n,\alpha+n} + 1) \tag{13}$$

does not commute with all the group generators. Consequently, N is not specified by representation itself, although the number operator commutes with the Hamiltonian (1). The diagonal form of $\hat{N}$ allows to select easily fixed particle number states within given *p-irrep*. Indeed, using Eq. (9) one obtains

$$N = n - p + 2q, \quad \text{where} \quad q \equiv \nu_n. \tag{14}$$

Thus, according the above formula, the particle number conservation requirement is equivalent to fixing *n-th* row value of GT–state. Simple counting indicates that the dimension of the subset of states (within given *p-irrep*) consisting N particles is equal:

$$\binom{n}{p-q}\binom{n}{q} \tag{15}$$

Finally, it is interesting to observe that states belonging to given *p-irrep* are simultanously the eigenvalues of the signature operator $\hat{R} \equiv \exp(-i\pi\hat{j}_x)$. Since the total Hamiltonian $\hat{H}^\omega$ (cf. Eq. (1)) commutes with $\hat{R}$ its eigenstates are simultaneously the eigenstates of $\hat{R}$ (we assume no degeneracy in the spectrum of $\hat{H}^\omega$).

In addition one can prove that all the states belonging to the basic set of the *p–irrep* are eigenstates of the signature operator $\hat{R}$ with eigenvalues

$$r = i^{p-n} \tag{16}$$

Thus each *p–irrep* is characterised by unique signature (16). As can be seen from Eq. (16) the signature is independent on particle number N. As a consequence of Eq. (16) we have in particular:

- $r = +1(-1)$ – for even number of particles in the case of even (odd) values of $1/2(n-p)$
- $r = -i(+i)$ – for odd number of particles in the case of $p - n = 3(1) \bmod 4$.

The method of diagonalizing $\hat{H}^{\omega}$ (cf. Eq. (1) described by means of the group theoretical formalism offers a substantial simplification in the numerical computations as compared to any standard diagonalisation of $\hat{H}^{\omega}$ (for example in the basis of the wave functions in the m–representation) both because of simple form of matrix elements and straightforward selection rules as well as due to exsistence of additional quantum number p. It is very instructive to see the advantage of our method by considering an example. Let us choose $N = 7$ particles in the $i_{13/2}$ shell ($n = 7$).

p \ N	0	1	2	3	4	5	6	7	8	9	10	11	12	13	14
0								1 $\uparrow$							
1							$\overleftarrow{7}$		$\overleftarrow{7}$						
2						21 $\downarrow$		49 $\downarrow$		21 $\downarrow$					
3					$\overrightarrow{35}$		$\overrightarrow{147}$		$\overrightarrow{147}$		$\overrightarrow{35}$				
4				35 $\uparrow$		245 $\uparrow$		441 $\uparrow$		245 $\uparrow$		35 $\uparrow$			
5			$\overleftarrow{21}$		$\overleftarrow{245}$		$\overleftarrow{735}$		$\overleftarrow{735}$		$\overleftarrow{245}$		$\overleftarrow{21}$		
6		7 $\downarrow$		147 $\downarrow$		735 $\downarrow$		1225 $\downarrow$		735 $\downarrow$		147 $\downarrow$		7 $\downarrow$	
7	$\overrightarrow{1}$		$\overrightarrow{49}$		$\overrightarrow{441}$		$\overrightarrow{1225}$		$\overrightarrow{1225}$		$\overrightarrow{441}$		$\overrightarrow{49}$		$\overrightarrow{1}$
8		7 $\uparrow$		147 $\uparrow$		735 $\uparrow$		1225 $\uparrow$		735 $\uparrow$		147 $\uparrow$		7 $\uparrow$	
9			$\overleftarrow{21}$		$\overleftarrow{245}$		$\overleftarrow{735}$		$\overleftarrow{735}$		$\overleftarrow{245}$		$\overleftarrow{21}$		
10				35 $\downarrow$		245 $\downarrow$		441 $\downarrow$		245 $\downarrow$		35 $\downarrow$			
11					$\overrightarrow{35}$		$\overrightarrow{147}$		$\overrightarrow{147}$		$\overrightarrow{35}$				
12						21 $\uparrow$		49 $\uparrow$		21 $\uparrow$					
13							$\overleftarrow{7}$		$\overleftarrow{7}$						
14								1 $\downarrow$							
Σ	1	14	91	364	1001	2002	3003	3432	3003	2002	1001	364	91	14	1

Table 1 In each box a number specifying the dimension of the irreducible representation of the group $U(14)$ (i.e. for the $i_{13/2}$ shell) is shown for given Cartan weight p and for fixed particle number N. Arrows shown in each box specify the signature r ($\uparrow$, $\downarrow$, $\rightarrow$ and $\leftarrow$ denote $r = i, -i, +1$ and -1 respectively). Bottom row gives sum in each column.

The total number of basic states is $\binom{2n}{N} = 3432$. Thus in order to find eigenvalues of $\hat{H}^{\omega}$ in the m–scheme one would be forced to diagonalize a matrix of this order (or twice the matrices of the order 1716×1716 as a consequence of signature symmetry). On the other hand, Table 1 shows the dimensions of all submatrices to be diagonalized for any N and p in the $i_{13/2}$ shell resulting from our approach (see Eq. (15)). Considering the column corresponding to N=7 in Table 1 one can see that the 3432×3432 matrix becomes block-diagonal (owing to the existence of the symmetry determined by the quantum number p) into submatrices of the order given in the sum:

$$1\uparrow \oplus 49\downarrow \oplus 441\uparrow \oplus 1225\downarrow \oplus 1225\uparrow \oplus 441\downarrow \oplus 49\uparrow \oplus 1\downarrow = 3432 \tag{17}$$

The arrows in the Eq. (17) denote the signature quantum number ($\uparrow$ ($\downarrow$) $\equiv r = +i(-i)$) for the consecutive eight $p - irreps$. The above example gives an idea of the simplifications offered by the present method.

It is worth to mention that the exsistence of the quantum number p is ultimately related to the associated unitary symmetry of the Hamiltonian. Thus, the price we are paying for the simplifications discussed above is the limited variety of Hamiltonians which posseses this symmetry and are simultanously interesting from the point of view of applications to the nuclear structure. In a case of a general two-body Hamiltonian the appropriate symmetry group is a $\boldsymbol{SO}(4\boldsymbol{n})$ group[26]. The detailed structure of this model will be discussed in the forthcoming publication[30].

3. Some Applications to the Nuclear Structure

The model contains three parameters: δ_ε – parameter defining the magnitude of single-particle energy splitting ($\delta_\varepsilon \equiv \varepsilon_{m=j} - \varepsilon_{m=1/2}$ see eq. (6)), G – the strength of the pairing force and ω – the rotational frequency. Taking the dimensionless ratios for example G/δ_ε and $\omega/\delta_\varepsilon$ we are left with two parameters only. In the following we shall nevertheless use the "absolute" values for δ_ε, G, ω and Δ as if there were given – for example – in MeV although the magnitudes of some quantities could result as not very realistic. For example, the magnitude of parameter G may look as unnaturally large which is due to the relatively low number of states contributing to the pairing interaction in our model (see below).

Now for the one-particle states (N=1) the two-body pairing force is inactive and our exact solutions overlap with the well known results of the mean field approximation for a single nucleon in a rotating one-body potential. However, for $N > 1$ the results contain the nontrivial many-body correlations and are thus different from those obtained in the framework of mean-field approximation.

The first problem in adjusting the model to the proper physical situation in the real nucleus is the resonable choice of the pairing force strength G (or else parameter G/δ_ε). In order to estimate the resonable values of G/δ_ε we have plotted in Fig. 1 the values of $\langle \hat{j}_x \rangle$ (expectation value of $\hat{j}_x$ operator for the lowest solution $|\Psi_o\rangle$ of the model) as a function of ω for $G = 0, 0.2, 0.4$ and 0.6 (at δ_ε=6). The limiting value of the derivative

$$\Im = \left(\frac{d\langle j_x\rangle}{d\omega}\right)_{\omega\to 0} \tag{18}$$

gives the moment of inertia $\Im$. For G=0 it is just the rigid moment of inertia $\Im_{rigid}$. For another value of $G \neq 0$ moment of inertia is reduced ($\Im/\Im_{rigid} < 1$). We can see that the realistic ratio of $\Im/\Im_{rigid} \approx 0.43$ occures for $G = 0.4$ (i.e. $G/\delta_\epsilon \approx 0.067$). Thus we accept this value of G as a realistic pairing force strength. The values $G = 0.2, 0.4, 1.0$ and 2.0 will be considered as a weak, moderate (realistic), strong and very strong pairing respectively.

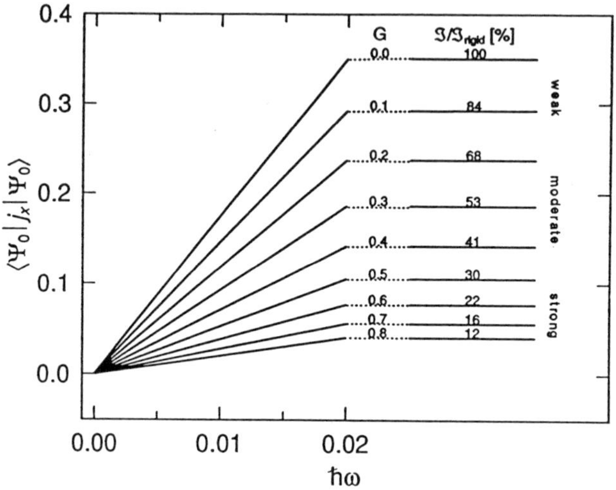

Fig. 1 Angular momentum alignment $\langle j_x\rangle$ plotted versus ω (near $\omega = 0$) for several values of pairing strength G. The figure illustartes the prcedure of adjusting parameter G as to obtain a resonable reduction in the moment of inertia $\Im^{(1)} \equiv \langle j_x\rangle/\omega$ at $\omega \to 0$ due to pairing correlations (see text).

In order to estimate the magnitude of the pairing correlations and the angular momentum alignment we have to introduce some relevant physical quantities. Having the solution for the lowest state in the rotating potential $|\Psi_o\rangle$ we can define the pairing energy E_{pair} as the expectation value $E_{pair} = \langle\Psi_o|\hat{H}_{pair}|\Psi_o\rangle$ Another quantity $\Delta_{exact} = \sqrt{-GE_{pair}}$ is also often used as a meassure of the pair correlation magnitude. In the BCS approximation this quantity overlaps with the gap parameter Δ. As for the angular momentum alignment we shall use $\langle\hat{j}_x\rangle = \langle\Psi_o|j_x|\Psi_o\rangle$ which was already defined above.

Considering the dependence of $\langle\hat{j}_x\rangle$ and Δ_{exact} on ω two possible extreme situations could be expected in principle. In the first scenario (i) one pair of nucleons coupled by the pairing force breaks first. The effect of breaking may be continuous or abrupt (depending on the magnitude of yrast–yrare interaction) but it does not lead to the disappearance of the static pairing correlations i.e. Δ_{exact} does not drop to zero completly. In a system with very many particles the decrease in Δ_{exact} may be even relatively weak. On the other hand the alignment $\langle \hat{j}_x \rangle$ increases up to the value corresponding to the alignment of one pair of particles. This scenario is more of the type of Stephens–Simon alignment[31].

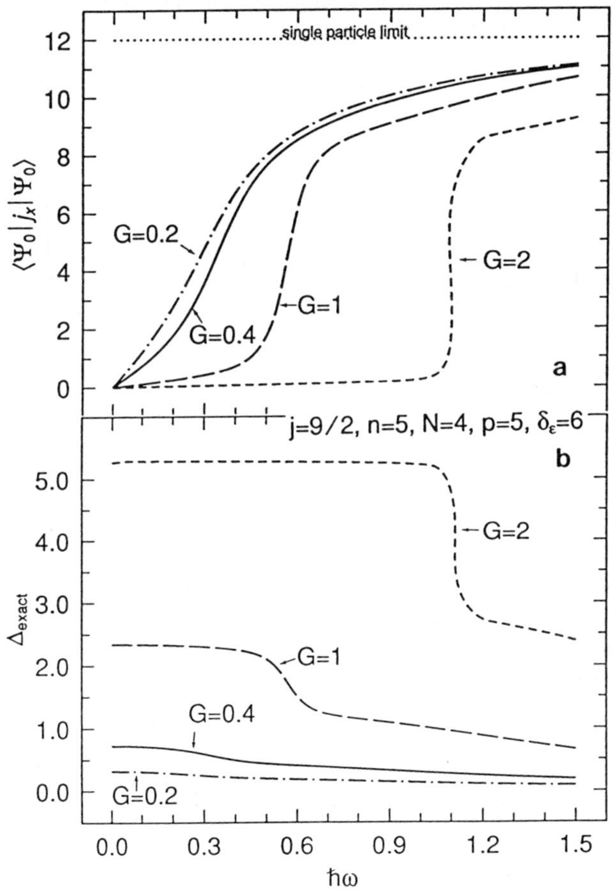

Fig. 2 Dependence of the aligned angular momentum $\langle j_x \rangle$ **(a)** and the parameter Δ_{exact} $(= \sqrt{G\langle H_{pair}\rangle}$ **(b)** plotted versus ω for several values of the pairing strength G. The curves correspond to even values of particle number N.

The alternative scenario (*ii*) corresponds generally to the rapid and massive change in Δ_{exact} with a considerably fast decrease to zero (or to a certain small value corresponding to the pairing vibration). In the same time the alignment increases and reaches its possible maximum value exceeding considerably the one pair alignment. This scenario is usually referred to as the Mottelson–Valatin effect i.e. the pairing phase transition[32].

Fig. 2a illustrates the dependence of the alignment on rotational frequency for various values of the pairing force strength G. Fig. 2b shows the quantity Δ_{exact} as a function of ω for the same set of values of G.

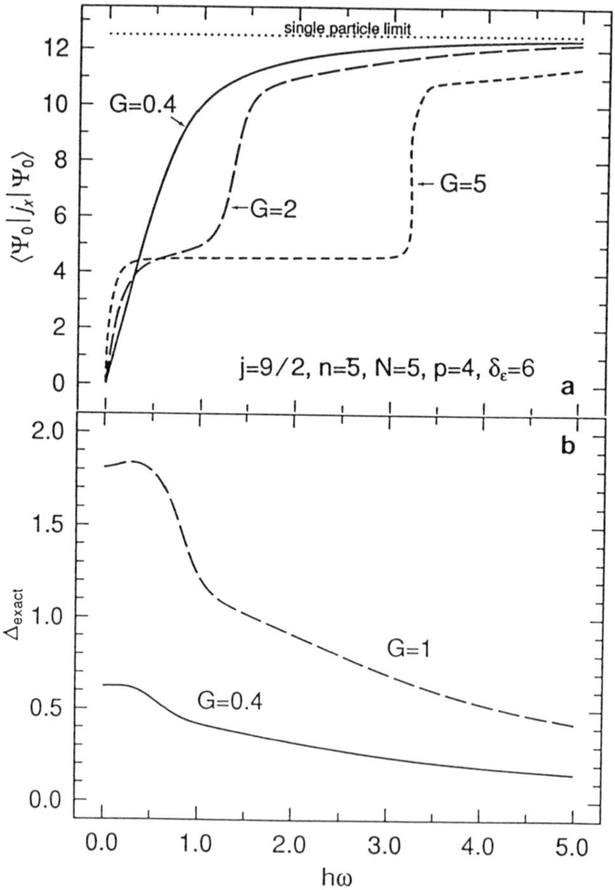

Fig. 3 Same as Fig. 2a,b but for odd particle number

What conclusions can be drown from the exact model concerning the two possible scenarios for the antipairing tendencies due to rotation. For the case either weak or moderate pairing both the quantities $\langle \hat{j}_x \rangle$ and Δ_{exact} vary rather smoothly and no abrupt changes are visible in their dependence on ω. This is probably due to rather dominant competition between the single–particle structure and the rotational coupling with pairing force playing only secondary role. When G increases the abrupt positive jumps in $\langle \hat{j}_x \rangle$ and negative ones in Δ_{exact} become visible in the curves on Fig. 2.

Nevertheless, even in this region the changes do not remind the pairing phase transition. They are rather of the Stephens–Simon type exibiting more the two–quasiparticle change in nuclear structure. For still larger values of ω exceeding the first jump the dependence on ω becomes rather smooth so that no traces of the break of the second, third etc. pairs is visible. We can see that in the region of G considered i Fig. 2 there is no possiblility for the existence of the global disappearence of static pairing correlations in the spirit of the Mottelson–Valatin effect. Such a mechanism is perhaps only possible when G is increased above the value of $G = 2$ i.e. for the very strong pairing.

Fig. 3ab illustrates the behaviour of the alignment $\langle \hat{j}_x \rangle$ and Δ_{exact} for systems with odd particle number $N = 5$. In the mean field approximation the alignment is always plotted for one particle versus ω. However, the mean field approach is not able to describe the alignment for many particle system. The curves on Fig. 3a, on the contrary, give an idea how the alignment of an odd particle number $N > 1$ is affected by the presence of other particles. This behaviour is especially well visible for a case of very strong pairing. For example the curve corresponding to $G = 2$ in Fig. 3a is seen to increase rapidly from $\langle j_x \rangle \approx 4.5$ at $\omega \approx 1.25$ tending gradually to the limiting value of the alignment of 5 particles equal 25/2. This rapid increase in $\langle \hat{j}_x \rangle$ is due to the breaking the paired structure acting among all the 5 (or perhaps 4) particles. Such a influence of the paired particles on the total alignment $\langle j_x \rangle$ is a many–body effect that goes beyond the mean field approximation.

4. Summary

The method for finding an exact solutions to the fermionic systems described by Hamiltonians possesing unitary symmetry has been developed In particular, we limited our considerations to the case of nuclear cranking–model Hamiltonian in the one j–shell space. The consistent applications of the group theory provide a lot of simplifications as compared to any standard method used so far. The resulting matrix elements of the Lie algebra generators turn out to be very simple – there are equal 0 or ± 1 – with a straightforward selections rules. The theory provides also the new symmetry quantum number p (besides those following from the signature and the particle number conservation) being just the Cartan weight which label the irreducible representations of the symmetry group. The existence of p–number greatly simplifies the diagonalisation proces reducing large matrices into block–diagonal forms. Morover, the structure of the model allows to select very easy the fixed particle number subspace within given irreducible representations labelled by quantum number p as well as it provides the solutions all of well defined signture (within given $p - irrep$).

Application of the model to the uniform rotation of the axially symmetric nucleus opens the possibility for testing the validity of the familiar CHFB approximation (or any other approximations) in the whole region of rotational frequency ω. It seems that our model can also be used as a very easy applicable and thus very effective reference model for testing different approximations used in realistic calculations.

The simple applications to the "nuclear pairing phase transition" induced by fast rotation shows that the pairing correlations decrease gradually with increasing rotational frequency. For the optimal value of parameter G (G=0.4 in our example) the decrease is quite smooth and no other effects seem to show up. For strong pairing (G=2. in our example) the Stephens–Simon effect appear. For the abrupt Mottelson–Valatin effect to occur even much stronger values of G would be required. In particular, it is shown that the angular momentum alignment for the odd number of nucleons is moderated by the pairing interaction of all the particles and does not have any more structure of the single particle effect.

This work was partly supported by the Commitee for Scientific Research (KBN) in Poland (contract no 204509101) and by the Swedish Natural Science Research Council (NFR). One of us (Z.S) acknowledges gratefully the hospitality and financial support of the Centre des Recherches Nucleaires in Strasbourg where the part of this work was done. The other (W.S.) strongly acknowledges the NFR postdoc position at the Department of Mathematical Physics of Lund University of Technology.

Appendix

Expressions for the $A_{\alpha,\beta}$ generators of the group $\boldsymbol{U}(2\boldsymbol{n})$ in terms of the original operators $N_{\alpha,\beta}$, $B^{\dagger}_{\alpha,\beta}$ and $B_{\alpha,\beta}$ which in turn depend on the creation and annihilation operators $c^{\dagger}_{\alpha}$ and c_{α} can be found in ref.[26]. Nevertheless, we reproduce these relations here for the sake of completeness. This may be further justified since we have also changed the enumeration of the single–particle levels as compared to ref.[26].

Thus the $N_{\alpha,\beta}$, $B^{\dagger}_{\alpha,\beta}$ and $B_{\alpha,\beta}$ are defined as:

$$N_{\alpha,\beta} = c^{\dagger}_{\alpha} c_{\beta} \tag{19}$$

$$B^{\dagger}_{\alpha,\beta} = c^{\dagger}_{\alpha} c^{\dagger}_{\beta} = (B_{\alpha,\beta})^{\dagger} \tag{20}$$

with $\alpha, \beta = 1, 2, \ldots, 2n$. Here indices α and β run over single–particle states of the j–shell ($n = j + 1/2$). If the single particle levels are split into n–pairs (each twofold degenerate) by the axially symmetric prolate quadrupole field then the lowest pairs contains two degenerate levels with $m = \pm 1/2$. The next lowest pair has $m = \pm 3/2$, and so on. We denote them consecutively by $|1\rangle$ and $|2\rangle = \hat{T}|1\rangle$ ($\hat{T}$ denoting the time–reversal operator) for $m = \pm 1/2$ respectively, $|3\rangle$ and $|4\rangle = \hat{T}|3\rangle$ for $m = \pm 3/2$ and so on. The last pair (highest energy) with $m = \pm j$ is denoted as $|2n-1\rangle$ and $|2n\rangle = \hat{T}|2n-1\rangle$. In this notation the single–particle part $\hat{H}_{sp}$ of the Hamiltonian (cf. Eq. (1)) reads

$$\hat{H}_{sp} = \sum_{i=1}^{n} \varepsilon_i \left(N_{2i-1,2i-1} + N_{2i,2i}\right) \tag{21}$$

The pairing two–body Hamiltonian $\hat{H}_{pair}$ reads

$$\hat{H}_{pair} = -G \sum_{i,j=1}^{n} B^{\dagger}_{2i-1,2i} B_{2j-1,2j} \tag{22}$$

Finally, the j_x–operator for the single particle j–shell has the form

$$\begin{gathered}\hat{j}_x = \frac{n}{2}(-)^n \left(N_{2,1} + N_{1,2}\right) + \\ \frac{1}{2}\sum_{i=2}^{n} \sqrt{(n-i+1)(n+i-1)}\left(N_{2i-1,2i-3} - N_{2i,2i-2} + N_{2i-3,2i-1} - N_{2i-2,2i}\right)\end{gathered} \tag{23}$$

Now, the relations between the $N_{\alpha,\beta}$, $B^{\dagger}_{\alpha,\beta}$ and $B_{\alpha,\beta}$ operators with the $A_{i,k}$ operators are defined by the following transformation

$$\begin{aligned}A_{\alpha,\beta} &= \frac{1}{2}\left(N_{2\alpha-1,2\beta-1} + (-)^{\alpha+\beta} N_{2\alpha,2\beta}\right) \\ &+(-)^{\beta+1}\frac{1}{2}\left(N_{2\alpha-1,2\beta} + (-)^{\alpha+\beta} N_{2\alpha,2\beta-1}\right)\end{aligned} \tag{24}$$

$$\begin{aligned}A_{\beta+n,\alpha+n} &= -\frac{1}{2}\left(N_{2\alpha-1,2\beta-1} + (-)^{\alpha+\beta} N_{2\alpha,2\beta}\right) \\ &+(-)^{\beta+1}\frac{1}{2}\left(N_{2\alpha-1,2\beta} + (-)^{\alpha+\beta} N_{2\alpha,2\beta-1}\right) + \delta_{\alpha,\beta}\end{aligned} \tag{25}$$

$$\begin{aligned}A_{\alpha,\beta+n} &= \frac{1}{2}\left(B^{\dagger}_{2\alpha-1,2\beta-1} + (-)^{\alpha+\beta+1} B^{\dagger}_{2\alpha,2\beta}\right) \\ &+(-)^{\beta}\frac{1}{2}\left(B^{\dagger}_{2\alpha-1,2\beta} + (-)^{\alpha+\beta+1} B_{2\alpha,2\beta-1}\right)\end{aligned} \tag{26}$$

$$A_{\beta+n,\alpha} = \left(A_{\alpha+n,\beta}\right)^{\dagger} \tag{27}$$

with $\alpha, \beta = 1, 2, \ldots, n$

Applying the above transformation (24)–(27) to expressions (21)–(23) one obtains straightforwardly formulae (3)–(5) that express cranking Hamiltonian in terms of generators $A_{\alpha,\beta}$.

References

1. D.R. Inglis, *Phys. Rev.* **96** (1954) 1059; *Phys. Rev.* **97** (1955) 701.

2. A. Bohr and B.R. Mottelson, *Mat. Phys. Medd. Dan. Vid. Selsk.* **30** (1955) no 1.
3. P. Ring, R. Beck and H.J. Mang, *Z. Phys.* **231** (1969) 10.
4. H.R. Dalafi, B. Banarjee, H.J. Mang and P. Ring, *Phys. Lett.* **44B** (1973).
5. A.L. Goodman, *Nucl. Phys.* **A230** (1974) 466; *Nucl. Phys.* **A265** (1976) 113.
6. I. Hamamoto, *Nucl. Phys.* **A271** (1976) 15.
7. A. Faessler, K.R. Sanhya–Devi, F. Grümmer, K.W. Schmid and R.R. Hilton, *Nucl. Phys.* **A256** (1976) 106.
8. R. Bengtsson and H.–B. Håkansson, *Nucl. Phys.* **A357** (1981) 61.
9. J.L. Egido and P. Ring, *Nucl. Phys.* **A383** (1982) 189; *Nucl. Phys.* **A388** (1982) 19.
10. M. Hasegawa and S. Tazaki, *Phys. Rev.* **C35** (1987) 1508.
11. M. Hasegawa and S. Tazaki, *Phys. Lett.* **226B** (1989) 1.
12. S. Tazaki, K. Muramatsu, R. Hayakawa and M. Hasegawa, *Phys. Rev.* **C43** (1991) 596.
13. R.W. Richardson and N. Sherman, *Nucl. Phys.* **52** (1964) 27; R.W. Richardson, *J. Math. Phys.* **6** (1965) 1034.
14. B. Lorazo, *Nucl. Phys.* **A153** (1970) 225.
15. K. Allaart and E. Boeker, *Nucl. Phys.* **A168** (1971) 630.
16. J.Y. Zeng and T.S. Cheng, *Nucl. Phys.* **A405** (1983) 1; J.Y. Zeng, T.S. Cheng, L. Cheng and C.S. Wu, *Nucl. Phys.* **A411** (1983) 4; *Nucl. Phys.* **A414** (1984) 253; *Nucl. Phys.* **A421** (1984) 125.
17. C.S. Wu and J.Y. Zeng, *Phys. Rev.* **C39** (1989) 666.
18. C.S. Wu and J.Y. Zeng, *Phys. Rev.* **C40** (1989) 998.
19. C.S. Wu and J.Y. Zeng, *Phys. Rev.* **C41** (1990) 1822.
20. J.Y. Zeng, C.S. Wu, L. Cheng and C.Z. Lin, *Phys. Rev.* **C41** (1990) 2911.
21. K.F. Pál, N. Rowley and M.A. Nagarajan, *Nucl. Phys.* **A470** (1987) 285.
22. N. Rowley, K.F. Pál and M.A. Nagarajan, *Nucl. Phys.* **A493** (1989) 13.
23. K.F. Pál, M.A. Nagarajan and N. Rowley, *Nucl. Phys.* **A500** (1989) 221.
24. J.A. Sheikh, M.A. Nagarajan, N. Rowley and K.F. Pál, *Phys. Lett.* **223B** (1989) 1.
25. J.A. Sheikh, N. Rowley and M.A. Nagarajan, *Phys. Rev.* **C42** (1990) 787.
26. J. Krumlinde, Z. Szymański, *Nucl. Phys.* **A221** (1974) 93.
27. C.G. Andersson and J. Krumlinde, *Nucl. Phys.* **A334** (1980) 486.
28. I.M. Gelfand and M.L. Zetlin, *Dokl. Akad. Nauk SSSR* **71** (1950) 825.
29. G.E. Baird and L.C. Biedenharn, *J. Math. Phys.* **4** (1963) 1449.
30. W. Satuła, Z. Szymański, W. Nazarewicz, O. Burglin and J. Dudek, to be published.

31. F.S. Stephens and R.S. Simon, *Nucl. Phys.* **A183** (1972) 257.
32. B.R. Mottelson and J.G. Valatin, *Phys. Rev. Lett.* **5** (1960) 511.

TOY MODELS WITH IDENTICAL BANDS

STEVEN A. MOSZKOWSKI
Department of Physics
UCLA, Los Angeles, CA 90024

ABSTRACT

A toy model with quasispin or generalised seniority (known to be valid for semi-magic nuclei) is shown to have identical bands. In particular we discuss nucleons in a 2D degenerate s, d shell, interacting via surface delta interactions. We comment on the effect of single-particle splitting.

1. Introduction

The microscopic origin of identical bands is still a puzzle. Such bands were discovered, rather unexpectedly, in superdeformed nuclei, where in some cases the moments of inertia change by only 0.1% in going from nucleus to nucleus [1,2]. However, the phenomenon also seems to occur in less deformed nuclei, though not with the same degree of precision. Some of the same physics may apply even in spherical nuclei. For example, the even-A and even-Z isotopes of Sn (Z = 50) have nearly identical excitation [3] energies for the two lowest states, 2^+ and 4^+. This near equality is a consequence of the approximate validity of quasispin symmetry, known also as generalized seniority [4]. Quasispin symmetry is satisfied exactly if two conditions are met:
1. No splitting between single-neutron levels.
2. Surface Delta Interaction between neutrons.

A Quadrupole-Quadrupole interaction between neutrons violates quasispin symmetry, while the SDI commutes with quasispin [5].

It was suggested by M. Mukerjee [6] that quasispin symmetry is, in fact, at the root of identical bands, even in deformed nuclei. We will now describe a model with nucleons in a two dimensional potential, in which exactly identical bands occur. We then give an interpretation of this result in terms of the interacting boson model. In the present paper, we consider mainly the effect of adding a pair of nucleons.

2. Two Dimensional Oscillator Shell Model with SDI

Consider nucleons in an oscillator shell, but in two dimensions. There is only a single component of angular momentum. Let us suppose that the single particle potential is a pure harmonic oscillator. Thus the single particle levels are degenerate. Now turn on a surface delta residual interaction. It is well known that in the two particle spectrum only a single state of each L_z value is shifted in energy, while all other states remain at the same energy. Thus the SDI acts like a pairing interaction, but not only in $L = 0$ states, but also $L = 2$, (i.e. $L_z = + 2$, or -2), etc. So, let us consider two particular examples, namely nucleons in the oscillator p shell, and then nucleons in the s,d shell. We will obtain identical bands for both cases. However, for the p shell, the result is rather trivial, since it is independent of details of the nuclear force.

3. Two and Four Nucleons in p Shell

For nucleons in the p shell, the SDI reduces to an ordinary delta interaction, since all radial wavefunctions in the shell are identical. (Provided we have no spin-orbit coupling.) For two identical nucleons with attractive delta interaction, there will be an $L = 0$ ground state at an energy of 2G (measured downward), where G is the coupling strength. The $L = 2$ state (actually two states, since L_z can have either sign) occurs at G. The remaining L = 0 state is not shifted.

Thus the spectrum is as follows:
E/G = 2 (0), 1 (2), 0 (0')
Here values of L are indicated in parentheses.
Note that the $L = 2$ state occurs at an excitation energy of 1 (in units of G.) Next we consider configurations made of 2 neutrons and 2 protons, but subject to the same delta interaction. A straightforward shell model calculation gives the following energies for the Yrast states:
E/G = 10 (0), 9 (2), 6 (4)

So the spacing of the first excited state does not change in going from 2 to 4 particles. The Yrast states form a pure rotational band. The moment of inertia is independent of both L and N, the number of nucleons. On the other hand, it is known that in superdeformed nuclei, the moment of inertia changes appreciably with angular momentum. (Generally, it decreases in the $A = 150$ region, but increases in the $A = 190$ region.) So the p shell model does not teach us much interesting physics. Furthermore, we get pure rotational bands for any interaction, provided only there is no spin-orbit coupling.

4. Nucleons in p Shell with Spin-Orbit Interaction

If we add a one body spin orbit coupling ϵ, then the excitation energies are no longer independent of the number of particles. For simplicity, we suppose a pure pairing interaction, acting only if two particles couple to zero orbital angular momentum. Let us consider the fraction by which the excitation energy of the lowest 2^+ state changes as we go from n to n+2 particles. First of all, for $n = 2$, this fractional change is no more than 15 percent (decrease), regardless of the relative strength of spin-orbit and pairing interaction. Also, in the weak coupling limit, we find that

$$\frac{(E_2 - E_0)_{n+2}}{(E_2 - E_0)_n} = 1 - \frac{8G}{n(n+2)\epsilon} \tag{1}$$

Thus for large number of particles, the spacing again becomes identical [6].

5. Two and Four Nucleons in sd Shell

Next we consider two, and then four nucleons in the oscillator s,d shell. Again we suppose there is no spin-orbit coupling. Further, we make an implied assumption regarding the shape of the single particle potential by requiring that the s and d states are degenerate. An arbitrary interaction between nucleons, including pure delta or quadrupole-quadrupole, will split this degeneracy. However, an SDI will not. That is the interaction we will consider here. For two identical nucleons, the spectrum is similar to that for the p shell, in the sense that for each L, only a single state is shifted. Of course, there are more states than for the p shell. Thus the spectrum is as follows:

E/G = 3 (0), 2 (2), 1 (4), 0 (0',2',0")

The spacing between the states happens to be equal. This is an artifact of the two dimensional model, and we do not attach any great significance to it. What is more significant is what happens when we go to four particles, two neutrons and two protons. We list here the energies of only a subset of states. First of all, as can be seen from the spacings, the Yrast states do NOT form an identical band.

L	0	2	4	6	8
E/G	14.30	13.61	12	9	6

Thus the energy of the first excited state is almost 1/3 smaller than for the two particle case. However, in the four particle spectrum there is also another band of states at the following energies: which has the same spacing as for two particles.

E/G = 6 (0), 5 (2), 4 (4)

(Note added 4/9: The four particle band just mentioned has spatial symmetry described by [22], while the Yrast band has maximum spatial symmetry [4]. This follows purely from the charge independence of the interaction (and absence of spin-orbit coupling), independent of details. The spacing in this excited band is the same as the symmetric band [2] of the two particle system. Quasispin symmetry requires that the spacings for [2] and [22] are the same. Presumably the same phenomenon should occur also for more general toy models, e.g. nucleons in a 2 dimensional p,f shell, or in a 3 dimensional s,d shell.)

6. Interpretation in Terms of Interacting Boson Model

Let us look at the results of the last section, i.e. nucleons in two dimensional s,d shell, from the standpoint of the interacting boson model. We consider two identical nucleons as a single boson, and thus the four particle configurations is equivalent to two bosons. First consider again the two particle = 1 boson spectrum. This is given by:

$e_i/G = 3(s), 2(d), 1(g), 0(s', s", d)$

Thus a single boson of each L has its energy shifted. All other bosons remain at zero energy. Note that, unlike in usual boson model calculations, we have not truncated the model space.

Now, for an SDI between nucleons, the interactions between bosons have a very simple form, at least if at least one s boson is involved. First, we just mention the results for 4 like nucleons = 2 like bosons. For an SDI, there is simply no interaction between these bosons [7]. Physically, this has its origin in that two spin-up (or spin-down) nucleons cannot be at the same point, and thus cannot interact. Thus, the excitation energy of the L = 2 state will be just the energy difference between s and d bosons, independent of the number of bosons present. This is just another way of describing the quasisymmetry property of excitation spectra for semi-closed shell even Z even A nuclei, which we have already mentioned.

More to the point of our discussion, we now turn to the spectrum of two neutrons and two protons, i.e. a neutron boson plus a proton boson. First of all, all diagonal matrix elements between bosons are equal to 4G. That is since the matrix elements of the SDI are all equal = G, and we have two neutrons interacting with two protons. Of more interest here are the off diagonal matrix elements. Now, as Druce and the author showed some years ago, for an SDI, the boson-boson interaction matrix elements can be expressed in terms of the structure of the bosons themselves [7]. Thus:

$$V(ss,ii) = V(si,is) = 4G * (e_i/e_s) \tag{2}$$

where i refers to any boson. In particular, we have:

$$V(ss,s's') = V(ss',s's) = 0 \tag{3}$$

Since the s' bosons are not shifted by the SDI, they also do not interact with s bosons (except for the diagonal matrix element.) It is thus easy to see that if we start with a band containing a single s or d or g boson, then adding an s' boson does not change the spacing. In particular the energies for states involving s' bosons are given as below:

E/G = 7 (ss'), 6 (ds'), 5 (gs')

As can be seen, we again get an identical band. However, it must be pointed out that the simple boson model as outlined here gives absolute energies which differ by one unit from the exact shell model results. So, to summarize this discussion, the s' boson is uncoupled from other bosons, and thus we can get identical bands, which, of course, are not necessarily rotational!

7. Classical Interpretation of Identical Bands in Deformed Nuclei

In order to understand better the physical origin of identical bands, it is useful to consider a viewpoint developed by M. Schmeing [8] for deformed nuclei. The usual approach, say for calculating the moment of inertia, is to use the mean field of the deformed intrinsic states. However, as Schmeing noted, if the nucleus is rotating rapidly, then the nucleons may rather feel the time average of the potential, i.e. we should look at the nucleus in the lab, rather than the body system. In order to understand this better, consider a classical analogue of a rotating deformed nucleus. Imagine you are at rest inside a hollow football (American), which is rotating about an axis through its center but perpendicular to its symmetry axis. Obviously, if you are located near the tip of the football, the wall will bump into you and push you along with the rotation. But if you are near the center, then, absent friction, you won't feel a thing! Thus, classically, the space separates into two regions, which differ in how a particle located there responds to rotation.

The classical situation has its counterpart in quantum mechanics. Thus a nucleon in an orbit such as (402)5/2 is rather decoupled from the rotation. That is since it tends to be located more in the nuclear interior. There is, however, a further point that must be made here. When we add a nucleon (or two) to a nucleus, what stops this nucleon from polarizing the other nucleons slightly, so that the moment inertia changes, by say, 1that the identical band phenomenon does contain some interesting physics, going beyond the usual self consistency condition between density and single particle potential. The extra condition is that the effective interactions are assumed to be density dependent in such a way that they act mainly only at the surface. If this condition is satisfied, then a nucleon added in an interior orbit does not interact with the other nucleons, and thus no polarization results. Of course, if a nucleon is added into an orbit which can get to the surface, (i.e. most orbits) then, of course, it will polarize the rest of the nucleus, leading to a change in the moment of inertia.

8. Conclusions

To summarize, the present author and also M. Mukerjee [6] have considered several toy models which lead to identical bands. The main physical ingredient here is the requirement that the interaction commutes with quasispin. The SDI has this property, at least for the case of degenerate orbits. It remains to generalize our results to more realistic nuclei where the orbits are not degenerate. In addition, it will be interesting to relate the boson description to one involving motion of individual nucleons in a deformed nucleus.

9. References

1. T. Byrski et al., Phys. Rev. Lett. **64** 1650(1990)
2. R.V.F. Janssens and T.L. Khoo, Ann. rev. Nucl. Part. Sci. **41** 321(1991)
3. I. Talmi, Nucl. Phys. **A172** 1(1971)
4. A. Kerman, Ann. Phys. **12** 300(1961)
5. R. Arvieu and S.A. Moszkowski, Phys. Rev. **145** 830(1966)
6. M. Mukerjee, talk presented at this workshop.
7. C.H. Druce and S.A. Moszkowski, Phys. Rev. C **33** 330(1986)
8. N. Schmeing, talk presented at Strasbourg conference, (March 1991)

PAIRING AND IDENTICAL BANDS IN A TOY MODEL

MADHUSREE MUKERJEE
Kellogg Radiation Laboratory, Caltech, 106-38
Pasadena, CA 91125, USA

ABSTRACT

A p-shell with L = 0, 1 and 2 pairing is coupled to a rotor in 2D. Some pairing phases are found to exist at extremely high spin. Identical-energy bands occur for phases which force pairs into unaligned orbitals.

This paper suggests that BCS pairing is critical to the existence of identical bands[1,2] in nuclei. A nuclear valence shell is simulated by a 2-level model with large degeneracy (a la Valatin[3]); and coupled to a deformed rotating core. All possible pairing interactions are allowed within the shell, and the pairing gaps and spectra are studied for different phases. Identical energy bands exist in phases (such as singlet) which force additional pairs to occupy time-reversal-invariant orbitals. They can be understood as a decoupling of the rotor from the particle number degree of freedom in these particular phases.

Further, some pairing phases are found to exist at extremely high spin. Once alignment occurs, pairing gaps may exist but identical bands do not.

1. The Model

The model consists of identical particles in 2 D with orbital angular momentum $l = 1$ and spin $s = 1/2$ with pairing interactions in all possible channels. A degeneracy Ω is assigned to take care of many-particle effects; so each operator has an extra quantum number ν, where $\nu = 1, 2, \ldots, \Omega$. The single-particle operators create the following states:

Operator	l_z	s_z	j_z	$l_z s_z$
$a^\dagger_{\uparrow\nu} \equiv a^\dagger_{3/2\nu}$	1	$\frac{1}{2}$	$\frac{3}{2}$	$\frac{1}{2}$
$a^\dagger_{\downarrow\nu} \equiv a^\dagger_{1/2\nu}$	1	$-\frac{1}{2}$	$\frac{1}{2}$	$-\frac{1}{2}$
$\bar{a}^\dagger_{\uparrow\nu} \equiv a^\dagger_{-1/2\nu}$	-1	$\frac{1}{2}$	$-\frac{1}{2}$	$-\frac{1}{2}$
$\bar{a}^\dagger_{\downarrow\nu} \equiv a^\dagger_{-3/2\nu}$	-1	$-\frac{1}{2}$	$-\frac{3}{2}$	$\frac{1}{2}$

(1)

In what follows the index ν will often be omitted. This model can also be looked at as a $j = 3/2$ shell; in this form it has a venerable history[3,4,5,6].

There are 6 types of pairs (each operator has a suppressed ν index):

$$\begin{aligned} s^\dagger &= (a^\dagger_\uparrow \bar{a}^\dagger_\downarrow - a^\dagger_\downarrow \bar{a}^\dagger_\uparrow)/\sqrt{2} \\ t^\dagger &= a^\dagger_\uparrow \bar{a}^\dagger_\uparrow, \qquad t^\dagger_0 = (a^\dagger_\uparrow \bar{a}^\dagger_\downarrow + a^\dagger_\downarrow \bar{a}^\dagger_\uparrow)/\sqrt{2}, \quad \bar{t}^\dagger = a^\dagger_\downarrow \bar{a}^\dagger_\downarrow \\ d^\dagger &= a^\dagger_\uparrow a^\dagger_\downarrow, \qquad \bar{d}^\dagger = \bar{a}^\dagger_\downarrow \bar{a}^\dagger_\uparrow. \end{aligned} \tag{2}$$

The shell is further coupled to a (2 D) rotor, with I the total (conserved) angular momentum of the combined system, and J_z is the (2-D) angular momentum of the shell: $J_z = \sum_\nu [\frac{3}{2} a_\uparrow^\dagger a_\uparrow + \frac{1}{2} a_\downarrow^\dagger a_\downarrow - \frac{1}{2}\bar{a}_\uparrow^\dagger \bar{a}_\uparrow - \frac{3}{2}\bar{a}_\downarrow^\dagger \bar{a}_\downarrow]$. The hamiltonian is:

$$H = H_{\rm sp} + H_{\rm pair} + H_{\rm rot} \tag{3}$$

where

$$\begin{aligned} H_{sp} &= -2\epsilon \sum_\nu l_z s_z a_{l_z s_z}^\dagger a_{l_z s_z} = \epsilon \sum_\nu (-a_\uparrow^\dagger a_\uparrow + a_\downarrow^\dagger a_\downarrow + \bar{a}_\uparrow^\dagger a_\uparrow - \bar{a}_\downarrow^\dagger a_\downarrow) \\ H_{\rm pair} &= -V_S \sum_{\nu,\nu'} s_\nu^\dagger s_{\nu'} - V_T \sum_{\nu,\nu'} \vec{t}_\nu^\dagger . \vec{t}_{\nu'} - V_D \sum_{\nu,\nu'} d_\nu^\dagger d_{\nu'} - V_D \sum_{\nu,\nu'} \bar{d}_\nu^\dagger \bar{d}_{\nu'} \\ H_{\rm rot} &= \frac{1}{2}{\rm a}(I - J_z)^2. \end{aligned} \tag{4}$$

There is a pairing, a spin-orbit splitting, and a shell-rotor coupling. The Hamiltonian is linearised as follows:

$$\begin{aligned} &H_{\rm lin} = H_{\rm sp} + H_{\rm pair,lin} - \omega J_z - \mu N, \\ &H_{\rm pair,lin} = -\textstyle\sum_\nu [\Delta_S(s_\nu^\dagger + s_\nu) + \Delta_T(t_\nu^\dagger + t_\nu) + \Delta_0(t_{0\nu}^\dagger + t_{0\nu}) + \Delta_{\bar{T}}(\bar{t}_\nu^\dagger + \bar{t}_\nu) \\ &+\Delta_D(d_\nu^\dagger + d_\nu) + \Delta_{\bar{D}}(\bar{d}_\nu^\dagger + \bar{d}_\nu)] \end{aligned} \tag{5}$$

where a chemical potential has been thrown in to control particle number. There are a total of 8 (coupled) self-consistency equations, six for the gaps,

$$\begin{aligned} &\Delta_S = V_S \Omega \langle s \rangle \\ &\Delta_T = V_T \Omega \langle t \rangle \qquad \Delta_0 = V_T \Omega \langle t_0 \rangle \qquad \Delta_{\bar{T}} = V_T \Omega \langle \bar{t} \rangle \\ &\Delta_D = V_D \Omega \langle d \rangle \qquad \Delta_{\bar{D}} = V_D \Omega \langle \bar{d} \rangle \end{aligned} \tag{6}$$

and two for the particle number and angular velocity:

$$\begin{aligned} N &= \Omega \langle a_\uparrow^\dagger a_\uparrow + a_\downarrow^\dagger a_\downarrow + \bar{a}_\uparrow^\dagger \bar{a}_\uparrow + \bar{a}_\downarrow^\dagger \bar{a}_\downarrow \rangle \\ \omega &= {\rm a}(I - \langle J_z \rangle). \end{aligned}$$

Finally, the total energy of the system is given by

$$E = \epsilon(-n_\uparrow + n_\downarrow + \bar{n}_\uparrow - \bar{n}_\downarrow) - \frac{\Delta_S^2}{V_S} - \frac{\Delta_T^2 + \Delta_0^2 + \Delta_{\bar{T}}^2}{V_T} - \frac{\Delta_D^2 + \Delta_{\bar{D}}^2}{V_D} + \frac{\rm a}{2}(I - \langle J_z \rangle)^2. \tag{7}$$

where $n_\uparrow = \langle a_\uparrow^\dagger a_\uparrow \rangle$, etc. are the occupation numbers of the four single-particle states.

2. Results

The results are illustrated in Fig. 1 and 2. We choose $V_S = V_T = V_D = .1$ MeV to show the competitiveness of the different phases under rotation. $\Omega = 10$ gives a total allowed particle number of 40, roughly the capacity of a heavy nuclear shell. The rotor strength ${\rm a} = .02$ MeV gives a singlet gap breakup at $I = 16$ a la Mottelson and Valatin[7], for all N. $\epsilon = .1$ MeV is chosen strong enough to compete with pairing. For half-filled shell, the gap energy $= \sqrt{2}\Delta_S$ turns out to be $\simeq 1$ MeV.

Fig. 1 shows the system energy E in the pure singlet phase (a) for $N = 10, 12, 20$ and 22 as a function of I. Note that while these different particle numbers have different total E in the ground state, they have identical spectra.

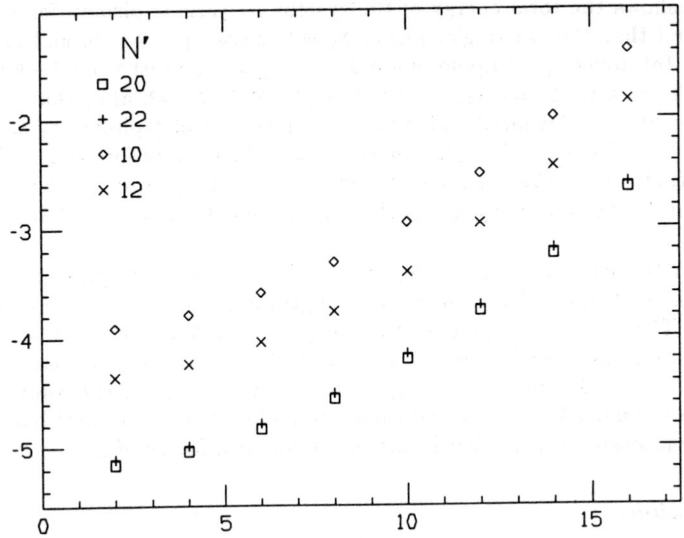

Figure 1. Total Energy E vs. angular momentum I for pure Δ_S or Δ_0 phase for $N = 20, 22, 10, 11 (\Omega = 10, V_S = V_T = .1, \epsilon = .1, \mathrm{a} = .02)$.

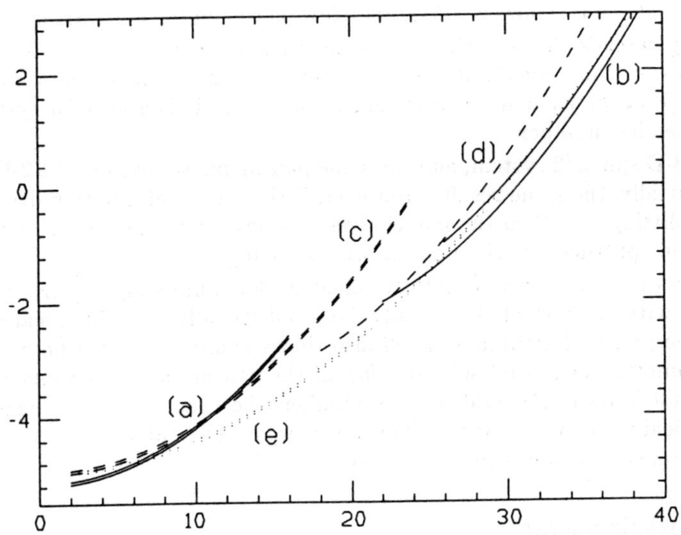

Figure 2. E vs. I for (a) Δ_S going to (b) mixed $\Delta_S = -\Delta_0$ phase, (c) $\Delta_T = -\Delta_{\bar{T}}$ phase going to (d) $\Delta_{\bar{T}}$ phase (dashed line) (e)$\Delta_D = -\Delta_{\bar{D}}$ phase. In each case, the lower curve is for $N = 20$, the upper for $N = 22$. $(\Omega = 10, V_s = V_T = 2V_D = .1, \epsilon = .1, \mathrm{a} = .02)$. Only $N = 20$ is shown for phase (d).

Fig. 2 shows the total energy of the system for various phases, for $N = 20$ and 22. We find that the (a) singlet phase $\Delta_S \neq 0$ exists upto $I = 16$ and is yrast till $I = 6$. The (e) mixed quadrupole phase $\Delta_D \simeq -\Delta_{\tilde{D}} \neq 0$ is yrast from $I = 8$ till $I = 24$. The (b) phase, with $\Delta_S = -\Delta_0$, is yrast from $I = 26$ to $I = 42$; after that the aligned (no-gap) solution is favoured. There is a (c) mixed triplet phase, $\Delta_T = -\Delta_{\tilde{T}}$, and also a (d) partially aligned triplet phase, $\Delta_T \neq 0$ shown only for N=20. For $I > 42$, in the partially aligned regime, one further gets (f) a $\Delta_{\tilde{D}}$ pairing gap for $N = 22$ (not shown). At $I = 78$, we finally get perfect alignement for $N = 22$ and this gap, too, dies.

The phases listed above are independent of interaction strengths in other (inactive) channels. The relative interaction strengths determine which phase is yrast at a given I. The (a), (b), (c) phases are largely unaligned and show identical bands; the (e) phase shows identical bands till $I = 40$. As I is further increased, there is a transition to partially aligned pairing phases (d) and (f). Here the rotor is directly coupled to the particle number and there are no longer any identical bands. These results are insensitive to the level splitting to about a factor of ± 2.

3. Discussions

In this model, pairing persists to over 4 times the spin at which pure singlet pairing breaks up. In the partially aligned phases, it can exist at even higher spin. There is a succession of transitions between different phases, till we reach perfect alignment. The model however does not allow pairing in the perfectly aligned phase.

Pairing spreads the particles around over all the levels to maximise scattering interactions, so their contributions to $\langle J_z \rangle$ tend to cancel. Hence *the rotor becomes decoupled from the particle number*, and its spectrum is identical for systems with different particle number.

For a 3-D spin 3/2 system, and the same pairing phases used in the 2-D problem we get, trivially, the same results. However, if there is a deformation-driven single-particle splitting (ϵ'), then the model gives identical bands for $\epsilon = \epsilon'$; that is, if the deformation splitting cancels the spin-orbit splitting.

The model is in principle exactly solvable. For pure singlet pairing the rotor commutes with the rest of the hamiltonian, and trivially gives identical rotational bands. The pairing hamiltonian itself also shows identical spectra because of quasispin symmetry. The exact solutions for all the pairing phases discussed above is, however, not known. For odd particle number, the exact solutions suggest bands with identical moments of inertia. This opens up a rich field of study, since this toy model contains a built-in supersymmetry, SU(6/4).

4. Acknowledgements

I would like to thank S. A. Moszkowski, S. Schramm and S. E. Koonin for helpful discussions. This work is supported by NSF Grant No. PHY90-13248.

5. References

1. T. Byrski *et al.*, *Phys. Rev. Lett.* **64** (1990) 1650.
2. R. V. F. Janssens and T. L. Khoo, *Ann. Rev. Nucl. Part. Sci.* **41** (1991) 321.
3. J. G. Valatin, *Lectures in Theor. Physics*, Vol. IV, (University of Colorado, Boulder, CO, 1961), p. 1.
4. J. Krumlinde and Z. Szymanski, *Ann of Phys.* **79** (1973) 201.
5. S. Y. Chu *et al.*, *Phys. Rev.* **C12** (1975) 1017.
6. R. Piepenbring *et al.*, *Nucl. Phys.* **A349** (1980) 77; *Nucl. Phys.* **A378** (1982) 77; and *Z. Phys.* **A318** (1984) 49.
7. B. R. Mottelson and J. G. Valatin, *Phys. Rev. Lett.* **5** (1960) 511.

ATTENDEES LIST

INTERNATIONAL WORKSHOP ON NUCLEAR STRUCTURE MODELS
MARCH 16-25, 1992

Y AKOVALI
OAK RIDGE NATIONAL LABORATORY
PHYSICS DIVISION
BLDG 6000 MS 6-371
OAK RIDGE TN 37831

CHAIRUL BAHRI
LOUISIANA STATE UNIVERSITY
DEPT OF PHYSICS & ASTRONOMY
BATON ROUGE LA 70803-4001

CYRUS BAKTASH
OAK RIDGE NATIONAL LABORATORY
PHYSICS DIVISION
BLDG 6000 MS 6-371
OAK RIDGE TN 37831

BRUCE BARRETT
UNIVERSITY OF ARIZONA
DEPARTMENT OF PHYSICS
BUILDING 81
TUCSON AZ 85721

B RAGNAR BENGTSSON
UNISOR/ORAU
OAK RIDGE NATIONAL LABORATORY
BLDG 6008 MS 6-374
OAK RIDGE TN 37831-6374

DAVE STADELMAN
BICRON
12345 KINSMAN ROAD
NEWBURY OH 44065

J F BERGER
COMMISSARIAT A L'ENERGIE ATOMIQUE
SERVICE PTN
BP 12-91680 BRUYERES-LE-CHATEL
FRANCE

C R BINGHAM
UNIVERSITY OF TENNESSEE
401 PHYSICS BUILDING
KNOXVILLE TN 37996-1200

GEORGE BLANAR
RESEARCH SYSTEMS DIVISION
LECROY CORPORATION
700 CHESTNUT RIDGE ROAD
CHESTNUT RIDGE NY 10977-6499

MATTHIAS BRACK
INST FUR THEORETISCHE PHYSIK
UNIVERSITAT REGENSBURG 31
D-8400 REGENSBURG
GERMANY

M CARPENTER
ARGONNE NATIONAL LABORATORY
PHYSICS DIVISION
9700 SOUTH CASS AVENUE
ARGONNE IL 60439

H KEN CARTER
UNISOR/ORAU
OAK RIDGE NATIONAL LABORATORY
BLDG 6008 MS 6-374
OAK RIDGE TN 37831-6374

OCTAVIO CASTAÑOS
INSTITUTO DE CIENCIAS NUCLEARES
UNIVERSIDAD NACIONAL AUTONOMA
CIRCUITO EXTERIOR CU A POSTAL 70-543
MEXICO DF CP 04510

DAVE CULLEN
OAK RIDGE NATIONAL LABORATORY
PHYSICS DIVISION
BLDG 6000 MS 6-371
OAK RIDGE TN 37831

DAVID MARCHESSAULT
DIGITAL EQUIPMENT CORPORATION
412 EXECUTIVE TOWER DRIVE
KNOXVILLE TENNESSEE 37923

JERRY DRAAYER
LOUISIANA STATE UNIVERSITY
DEPT OF PHYS & ASTRONOMY
BATON ROUGE LOUISIANA 70803

JUTTA ESCHER
LOUISIANA STATE UNIVERSITY
DEPT OF PHYS & ASTRONOMY
BATON ROUGE LOUISIANA 70803

AMAND FAESSLER
UNIVERSITY OF TÜBINGEN
MORGENSTELLE 14
D-27400 TÜBINGEN
GERMANY

D H FENG
DREXEL UNIVERSITY
DEPT OF PHYS & ATMOS SCIENCES
PHILADELPHIA PA 19104

JERRY D GARRETT
OAK RIDGE NATIONAL LABORATORY
PHYSICS DIVISION
BLDG 6000 MS 6-371
OAK RIDGE TN 37831

ALAN GOODMAN
TULANE UNIVERSITY
PHYSICS DEPARTMENT
NEW ORLEANS LA 70118

MICHAEL W GUIDRY
UNIVERSITY OF TENNESSEE
401 DEPARTMENT OF PHYSICS
KNOXVILLE TN 37996-1200

EDITH HALBERT
OAK RIDGE NATIONAL LABORATORY
PHYSICS DIVISION
BLDG 6008 MS 6-374
OAK RIDGE TN 37831

IKUKO HAMAMOTO
LUND INSTITUTE OF TECHNOLOGY
DEPT OF MATHEMATICAL PHYSICS
PO BOX 118 LUND S-22100
SWEDEN

P H HEENEN
PHYSIQUE NUCLEAIRE THEORIQUE
CP22G
UNIVERSITE LIBRE DE BRUXELLES
BELGIUM

DAN HOREN
OAK RIDGE NATIONAL LABORATORY
PHYSICS DIVISION
BLDG 6000 MS 6-368
OAK RIDGE TN 37831-6368

MARTIN JARRIO
GEORGIA INSTITUTE OF TECHNOLOGY
SCHOOL OF PHYSICS
ATLANTA GA 30332

NOAH R JOHNSON
OAK RIDGE NATIONAL LABORATORY
PHYSICS DIVISION
BLDG 6000 MS 6-371
OAK RIDGE TN 37831

MICHAEL KIRSON
DEPARTMENT OF NUCLEAR PHYSICS
WEIZMANN INSTITUTE OF SCIENCE
REHOVOT
ISRAEL

JAN KORMICKI
UNISOR/VANDERBILT UNIVERSITY
DEPT OF PHYS & ASTRONOMY
STATION 1807-B
NASHVILLE TN 37235

K KUMAR
TENNESSEE TECH UNIVERSITY
DEPARTMENT OF PHYSICS
COOKEVILLE TN 38505

P LIPAS
DEPARTMENT OF PHYSICS
UNIVERSITY OF JYVÄSKYLÄ
SEMINAARINKATU 15
SF-40100 JYVÄSKYLÄ 10 FINLAND

VICTOR MADSEN
JIHIR/ORNL
PHYSICS DIVISION
BLDG 6008 MS 6-374
OAK RIDGE TN 37831

E MARSHALEK
UNIVERSITY OF NOTRE DAME
DEPARTMENT OF PHYSICS
NOTRE DAME IN 46556

JOACHIM MARUHN
UNIVERSITY OF FRANKFURT
PHYSICS DEPARTMENT
ROBERT-MAYER-STRASSE 8-10
6000 FRANKFURT 1 GERMANY

FRANCIS MCGOWAN
OAK RIDGE NATIONAL LABORATORY
PHYSICS DIVISION
BLDG 6000 MS 6-371
OAK RIDGE TN 37831

STEVEN A MOSZKOWSKI
UNIV CAL-LA
DEPARTMENT OF PHYSICS
405 HILGARD AVENUE
LOS ANGELES CA 90024

M MUKERJEE
CALIFORNIA INST OF TECHNOLOGY
KELLOGG RADIATION LAB 106-38
PASADENA CA 91125

WITEK NAZAREWICZ
JIHIR/ORNL
BLDG 6008 MS 6-374
OAK RIDGE TN 37831-6374

V OBERACKER
VANDERBILT UNIVERSITY
DEPT OF PHYSICS & ASTRONOMY
STATION 1807-B
NASHVILLE TN 37235

E OSNES
DEPARTMENT OF PHYSICS
UNIVERSITY OF OSLO
PO BOX 1048 BLINDERN
N-0316 OSLO NORWAY

TAKAHARU OTSUKA
DEPARTMENT OF PHYSICS
UNIVERSITY OF TOKYO
HONGO BUNKYO-KU
113 TOKYO JAPAN

RODNEY B PIERCEY
DEPT OF PHYS & ASTRON
MISSISSIPPI STATE UNIV
DRAWER 5167
MISSISSIPPI STATE MS 39762

RICK SEYMOUR
OXFORD INSTRUMENTS INCORPORATED
ANALYTICAL SYSTEMS DIVISION
601 OAK RIDGE TURNPIKE
OAK RIDGE TN 37830

PHILIPPE QUENTIN
CSNSM
BAT 104-108
91405 ORSAY CAMPUS
FRANCE

S RAMAN
OAK RIDGE NATIONAL LABORATORY
PHYSICS DIVISION
BLDG 6010 MS 6-354
OAK RIDGE TN 37831

L L RIEDINGER
ASSOC VICE CHANCELLOR OF RES
UNIVERSITY OF TENNESSEE
404 ANDY HOLT TOWER
KNOXVILLE TN 37996

RUSSELL L ROBINSON
OAK RIDGE NATIONAL LABORATORY
PHYSICS DIVISION
BLDG 6000 MS 6-368
OAK RIDGE TN 37831-6368

LUIS ROBLEDO
DEPT FISICA TEORICA C-XI
FACULTAD DE CIENCIAS
UNIVERSIDAD AUTONOMA DE MADRID
E-28049 MADRID SPAIN

STANISLAW ROHOZINSKI
INST OF THEORETICAL PHYSICS
WARSAW UNIVERSITY
UL HOZA 69
00 681 WARSZAWA
POLAND

DAVID J ROWE
UNIVERSITY OF TORONTO
DEPARTMENT OF PHYSICS
TORONTO ONTARIO
CANADA M5S 1A7

F SAKATA
INSTITUTE FOR NUCLEAR STUDY
UNIVERSITY OF TOKYO
TANASHI TOKYO 188
JAPAN

WOJTEK SATULA
DEPT OF MATHEMATICAL PHYSICS
UNIVERSITY OF LUND
PO BOX 118
S-22100 LUND SWEDEN

K H SCHMID
INST FUR THEORETISCHE PHYSIK
UNIVERSITAT TUBINGEN
AUF DER MORGENSTELLE 14
D-7400 TUBINGEN GERMANY

PAUL HRICH
SOLON TECHNOLOGIES INC
5100 OLD CHESTNUT COURT
WOODSTOCK GA 30188

M R STRAYER
OAK RIDGE NATIONAL LABORATORY
PHYSICS DIVISION
BLDG 6003 MS 6-373
OAK RIDGE TN 37831

A SUSTICH
ARKANSAS STATE UNIVERSITY
DEPT OF COMP SCI MATH & PHYS
PO BOX 70
STATE UNIVERSITY AR 72467-0070

SUDHA R SWAMINATHAN
DEPARTMENT OF PHYSICS
UNIVERSITY OF MICHIGAN
500 E UNIVERSITY
ANN ARBOR MICHIGAN 48109-1090

ZDZISLAW SZYMANSKI
UNIVERSITY OF WARSAW
INST FOR THEORETICAL PHYSICS
UL HOZA 69
WARSAW PL-00-681 POLAND

S UMAR
VANDERBILT UNIVERSITY
DEPT OF PHYSICS AND ASTRONOMY
STATION 1807-B
NASHVILLE TN 37235

MICHAEL VALLIERES
DREXEL UNIVERSITY
DEPT OF PHYS & ATMOS SCIENCE
PHILADELPHIA PA 19104

N R WALET
UNIVERSITY OF PENNSYLVANIA
DEPT OF PHYSICS
DAVID RITTENHOUSE LABORATORY
PHILADELPHIA PA 19104-6396

JOHN WOOD
GEORGIA INSTITUTE OF TECHNOLOGY
SCHOOL OF PHYSICS
ATLANTA GA 30332

C Y WONG
OAK RIDGE NATIONAL LABORATORY
PHYSICS DIVISION
BLDG 6003 MS 6-373
OAK RIDGE TN 37831

RAMON WYSS
JIHIR/ORNL
PHYSICS DIVISION
BLDG 6008 MS 6374
OAK RIDGE TN 37831

V ZELEVINSKI
NIELS BOHR INSTITUTE
BLEGDAMSVEJ 17
COPENHAGEN 0
DENMARK

JING YE ZHANG
401 PHYSICS DEPARTMENT
UNIVERSITY OF TENNESSEE
KNOXVILLE TENNESSEE 37996-1200